MINERALOGY OF URANIUM AND THORIUM

ROBERT J. LAUF

Schiffer Publishing Ltd
4880 Lower Valley Road • Atglen, PA 19310

Other Schiffer Books by the Author:
Collector's Guide to the Amphibole Group, 978-0-7643-4870-9
Introduction to Radioactive Minerals, 978-0-7643-2912-8
Collector's Guide to the Zeolite Group, 978-0-7643-4675-0
Collector's Guide to the Feldspar Group, 978-0-7643-4329-2

Published by Schiffer Publishing, Ltd.
4880 Lower Valley Road
Atglen, PA 19310
Phone: (610) 593–1777; Fax: (610) 593–2002
E–mail: Info@schifferbooks.com
Web: www.schifferbooks.com

Library of Congress Control Number: 2015957261

Designed by Molly Shields
Type set in DIN Black/Times New Roman
ISBN: 978-0-7643-5113-6
Printed in China

DEDICATION

This book is dedicated to the memory of Professor Clifford Frondel (1907–2002), whose exhaustive work did so much to place our understanding of radioactive minerals on a firm footing.

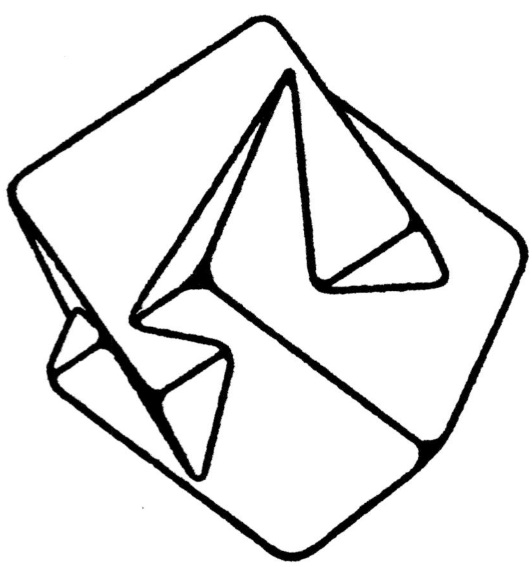

CONTENTS

FOREWORD

The 1938 discovery that a small amount of mass could be converted to a great deal of energy, by the splitting of unstable (fissile) nuclei interacting with a neutron, catapulted the elements thorium and uranium to the status of great international strategic importance. The low-abundance 235 isotope of uranium is fissile and is a fuel for reactors to produce electricity, for production of U-based nuclear weapons, and for reactor-based creation of strategic elements including plutonium. Plutonium-238 is also used in thermoelectric generators that power many satellites and most of space exploration, including rovers on Mars. Thorium is a fertile element, meaning that it can be converted in a reactor to fissile isotopes that are then used as a fuel. Although a few uranium and thorium minerals had been described prior to the beginning of the atomic age, and industrial applications included use as coloring agents, society's interest in these unusual elements greatly accelerated mineralogical studies in the 1950s, and with modern scientific equipment, mineralogical studies of uranium and thorium have again flourished over the past two decades. New species are now being reported at the rate of several each year, with most of these found by reexamination of classic localities. Uranium and thorium mineralogy is thus a rapidly expanding class of minerals that exhibit tremendous chemical and structural complexity, and in the case of uranium, considerable beauty.

The mineralogy of uranyl, the hexavalent form of uranium that exists in near-surface environments, is particularly diverse and fascinating. Uranyl combines with a broad range of oxyanions to form diverse structures and compositions, and there is a disproportionately large number of uranyl species relative to the low crustal abundance of uranium. This is due in part to its unique crystal chemical behavior, and in part to geochemical processes that have led to rich uranium deposits in many geographic areas of the world. Understanding uranium mineralogy is essential for understanding the genesis of uranium deposits, as well as the fate of uranium in contaminated environments such as Cold War weapons sites and uranium mining operations.

Robert Lauf presents here a highly authoritative examination of the mineralogy of uranium and thorium, with detailed coverage of the chemical composition, crystal structure, and localities of each species. Also present is a historical account of the discovery and subsequent work of many of the species, with the type locality identified. This book is mandatory reading for any researcher interested in uranium or thorium mineralogy, as it collects a vast quantity of scientific information in a well-organized and highly accurate form. It is by far the most authoritative effort of this nature for uranium and thorium minerals, and I am thankful that Dr. Lauf has been able to dedicate a huge effort to the advancement of the science of uranium and thorium mineralogy.

The treatment of historical details makes for highly interesting reading, and you'll find here an abundance of information about the distribution of these strategic elements worldwide. Significant uranium deposits are listed according to country, geochemical type, size, and status. Lists of radioactive minerals are given according to the country of their first discovery. Whether it is the structural diversity, chemical complexity, physical beauty as captured in Dr. Lauf's photographs, or locality species diversity that fascinates you, this volume will satisfy.

—Prof. Peter C. Burns, University of Notre Dame

PREFACE

This book is intended to serve as a systematic guide to uranium and thorium minerals and their occurrences for the serious student or mineral collector seeking to better understand the specimens in his or her collection. It will also be of use to the exploration geologist who wants an up-to-date reference covering the mineralogy of radioactive raw materials. It is *not intended* as a guidebook for directing amateurs to specimen collection sites.

The book assumes some basic knowledge of chemistry, crystallography, and geology, as well as familiarity with the commonly used terms in those fields. It also assumes that the reader has one or more of the standard general reference volumes on mineralogy, such as one of the editions of *Dana's Mineralogy*, the *Handbook of Mineralogy*, or the *Encyclopedia of Minerals*. For this reason, the decision was made not to repeat standard quantitative data such as optical properties and density. No attempt has been made to repeat exhaustive locality information for each mineral, as excellent online databases exist for this purpose. The decision was also made not to simply list the minerals alphabetically, as in the encyclopedias, but rather to treat them in a taxonomic framework so that their relationships and differences can be better understood. While no single book can do justice to the vast literature in this field, I have attempted to summarize the key historical information and modern interpretations while suggesting appropriate references for further reading.

The general treatment begins with a brief historical discussion of the discovery of radioactive elements and the development of uranium and thorium ore deposits. The tools and methods used to measure the radioactivity of these minerals are described and demonstrated.

Next, the geochemical conditions that produce significant deposits of primary and secondary minerals are discussed, so that when important localities are described, the reader can have a fuller understanding of their geological setting and history. Major countries are then arranged geographically, and include both those with commercially important ore deposits and those with "specimen" localities of greatest interest to collectors.

Entries for individual minerals and groups emphasize their chemical and structural relationships, supported by the liberal use of clear, easily understood crystal structure drawings. Important localities and associated minerals are given, along with some historical context, to aid the collector in studying and understanding individual specimens of the type he or she is likely to encounter. For this reason, most photos show the entire specimen, or at least enough of the matrix to place the featured mineral or crystals into some geological context.

The minerals are arranged into four somewhat arbitrary groupings, with the expectation that this treatment will be generally meaningful to most mineral collectors:

1. Primary minerals (i.e., those with tetravalent uranium or thorium)
2. Secondary minerals (generally those with hexavalent uranium)
3. Other minerals containing essential uranium or thorium
4. Minerals that may contain uranium or thorium as significant impurities.

The book concludes with an extensive bibliography for further reading or personal research on particular subtopics and a recommended "core library" of important and reasonably accessible reference works. Whenever possible, I have consulted the original descriptions for each mineral (most of which are now available in several online archives), and these are included in the bibliography along with citations for their corresponding English-language abstracts where appropriate.

Research in this field is ongoing, and new minerals continue to be described in the literature; the information herein is believed to be current as of late 2014.

A WORD ABOUT SAFE HANDLING

Many common minerals contain toxic elements (chromium, arsenic, mercury, lead, etc.), so collectors should always exercise good "chemical" hygiene when handling *any* mineral specimen. This means keeping the materials away from food and drink and washing hands after handling. I do not recommend using gloves for the simple reason that the cost of using a new pair of disposable gloves every time specimens are handled will add up, whereas reusable gloves will become contaminated and thereby defeat their purpose. Washing is simple and effective.

Inhalation or ingestion is a greater concern because the mucous membranes lining the lungs and intestines are much more sensitive to damage from alpha radiation than is the skin. To avoid inhalation hazards, be especially careful to avoid creating airborne dust and debris when cleaning or trimming a sample, and thoroughly clean the work area afterwards. Avoid displaying specimens on an open shelf where they will need frequent dusting.

It is my belief that in general, the mere possession of a reasonable quantity of uranium or thorium minerals in a properly stored collection does not in itself present a significant health hazard. This belief is based on the likely dose of radiation that one would receive from this source compared to the "background" radiation we receive from all other sources. However, common sense dictates that these materials be treated with respect. Large amounts of radioactive minerals (especially thorium minerals) should not be stored close to where people will spend a lot of time (under your bed or next to your desk, for example!) and all radioactive minerals should be carefully labeled and *kept away from children*. Lead sheets can be used to shield the cabinet, but very highly radioactive minerals containing thorium will emit high-energy gamma, which is more difficult to shield. A good Geiger counter is a useful investment for the uranium collector.

It is also conceivable (but unlikely) that a large amount of these minerals stored for long periods in an enclosed space could create a buildup of radon gas. Inexpensive radon test kits are available, and any collector who has a concern about this issue should test his working or storage environment to make sure that radon is not accumulating to troublesome levels.

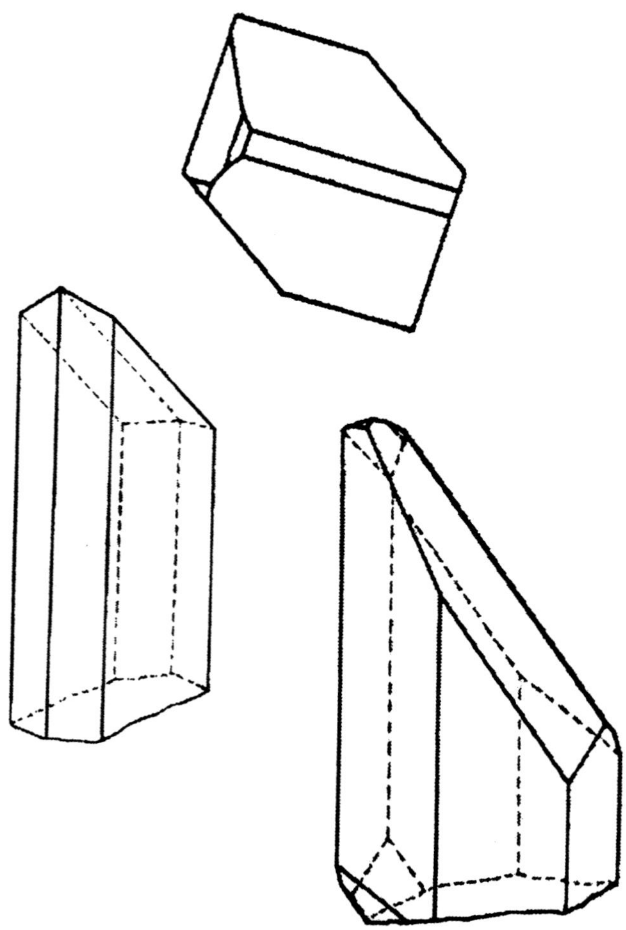

ACKNOWLEDGMENTS

The literature of radioactive minerals is vast and deep; it covers a lot of territory (and languages) and many years of scientific studies as well as the careful observations of enlightened collectors. Assembling a useful and representative compendium of information from such varied sources has required a great deal of help from many friends and colleagues.

First, Prof. Peter Burns, University of Notre Dame, provided extensive advice, encouragement, and technical data throughout the course of this project, without which it would have been difficult to fully realize the book in its present form.

Second, I am grateful to the following colleagues who graciously provided technical information, photographs, and numerous helpful discussions:

Christian Biagioni, University of Pisa; Joel Brugger, University of Adelaide; Robert Finch, Argonne National Laboratory; Teresa Fortney, Oak Ridge Public Library; Dermot Henry, Museum Victoria; Tony Kampf, Natural History Museum of Los Angeles County; Helen Kerbey, The Mineralogical Society of Great Britain and Ireland; Kurt Mereiter, Vienna University of Technology; Stefano Merlino, University of Pisa; Kevin Murphy, The Mineralogical Society of Great Britain and Ireland; Arvid Pasto, materials scientist, Sparks, Nevada; Jim A. Paschis, Ore Microscopist, Boulder, Colorado; Paula Piilonen, Candian Museum of Nature; Jakob Plašil, University of Bern; William "Skip" Simmons, University of New Orleans; Pavel Škacha, mineralogist, Czech Republic; J. Alex Speer, Mineralogical Society of America; and Reinhard Wegner, Universidade Federal da Paraiba.

Third, the following mineral dealers have supplied crucial materials for my reference collection along with valuable anecdotal information on localities and history: John Betts; Dudley Blauwet, *Mountain Minerals*; Sharon Cisneros, *Mineralogical Research*; Richard Dale; Jordi Fabre; the late Gilbert Gautier; Eric Greene, *Treasure Mountain Mining*; Leonard Himes and Michael Jacob, *Minerals America*; Shelley and Dan Lambert, *Lambert Minerals*; Rob Lavinsky, *Arkenstone*; Tony Nikischer, *Excalibur Mineral Company*; Jaye Smith, *The Rocksmiths*; and Sergey Vasiliev, *Systematic Mineralogy*.

Finally, I thank Prof. Carl Francis, Harvard University, whose memorial essay on Clifford Frondel inspired me to undertake this project, and whose encouragement motivated me to finish it.

PHOTO CREDITS

Maps were prepared by Richard Maxey using Adobe Illustrator® and original vector files from the Map Resources® Countries collection. Unless otherwise noted in the captions, all photos were taken by the author using specimens from his personal research collection.

Thanks are due to the following sources for permission to reprint copyrighted material: the Mineralogical Society of America for photographs appearing in *American Mineralogist*; the Mineralogical Society of Great Britain and Ireland for photographs appearing in *Mineralogical Magazine*; and the Czech Geological Society for photographs appearing in *Journal of Geosciences*.

Specific sources are acknowledged in the individual figure captions.

CRYSTAL STRUCTURE MODELS

Crystal structure drawings were made with CrystalMaker® 2.0 software, using data files from the American Mineralogist Crystal Structure Database. I thank Tom Watkins and Andrew Payzant, Oak Ridge National Laboratory, for their help in learning to use this system to full advantage to produce an essential feature of the book.

Through a special arrangement with CrystalMaker, CrystalViewer software has been made available for free download to purchasers of this book, along with a folder containing cmdf files of all the crystal structures shown herein. Visit our website at www.schifferbooks.com/crystalviewer/index.php and follow the instructions to download the files and view the structures in three dimensions for a richer appreciation of their details.

CrystalMaker® SOFTWARE

ABBREVIATIONS AND DEFINITIONS

Å: Ångstrom units (10^{-10} m)
a,b,c: unit cell edges in the *x,y,z* directions
apfu: atoms per formula unit
cm: centimeters
cm^{-1}: inverse centimeters (for wavelength of light in terms of wavenumber)
cpm: counts per minute
DTA: differential thermal analysis
EDX: energy-dispersive X-ray (analysis)
ESR: electron spin resonance
FOV: field of view
HR: high resolution
Ga: billion years
IR: infrared
km: kilometers
LWUV: long wave ultraviolet
Ma: million years
mm: millimeters
nm: nanometers (10^{-9} m)
P: pressure
ppm: parts per million
RE, REE: rare-earth, rare-earth elements
SEM: scanning electron microscopy
SWUV: short wave ultraviolet
T: temperature
t: tons (generally metric tons, 1,000 kg)
TEM: transmission electron microscopy
TGA: thermogravimetric analysis
TL: type locality
UV: ultraviolet
XRD: X-ray diffraction
Z: number of formula units per unit cell
µm: micrometers (10^{-6} m)
~: approximately
□: vacancy (in chemical formulas)
φ: unspecified coordinating ligand (O^{2-}, OH^-, H_2O, etc.)

INTRODUCTION

DISCOVERY AND EXPLOITATION OF URANIUM AND THORIUM

The earliest known localities for uranium minerals are the vein deposits at Joachimsthal, Bohemia (now Jachymov, Czech Republic) and elsewhere in the Erzgebirge district. Those mines had been worked for silver and cobalt for hundreds of years, and by the early part of the eighteenth century, pitchblende was recognized by the miners as a distinct substance, more than fifty years before the German chemist and mineralogist M. H. Klaproth showed that it contained a new chemical element. Klaproth produced "uranium" powder in 1789 by reducing pitchblende with charcoal.

This product was assumed to be the element but was actually UO_2. About fifty years later, the French chemist E. Peligot correctly identified Klaproth's material as an oxide and went on to produce elemental uranium by reducing UCl_4 with potassium. The French chemist Henri Becquerel discovered that uranium was radioactive in 1896, and two years later, P. Curie, M. Curie, and G. Bemont isolated the element radium from pitchblende mined at Jachymov. Later work on radioactive decay by Ernest Rutherford, Frederick Soddy, Bertram Boltwood, and others pioneered modern techniques of isotopic dating. Their research also provided key evidence to refute Lord Kelvin's calculation that suggested the Earth was no more than 20-30 million years old.

In 1828 the Swedish chemist J. J. Berzelius first isolated the element thorium from a silicate mineral, thorite ($ThSiO_4$), found in a pegmatite on an island in Langesund Fjord, southern Norway (Frondel 1958). M. Curie and C. G. Schmidt, working independently, discovered the radioactivity of thorium in 1889.

Modern readers naturally associate uranium and thorium with nuclear energy for power generation, weapons, and medical treatments. However, compounds of uranium have been used for over 2,000 years in more traditional industries including ceramics, catalysis, and steelmaking. Colored glass dating from around 79 AD contains uranium oxide, and its use as a pigment in glass and ceramics continued well into the twentieth century.

At present, the main non-nuclear use of uranium is for anti-tank weapons and other ballistic penetrators, whose performance takes advantage of the extremely high density of the material and its pyrophoric behavior on impact. These weapons make use of *depleted uranium*, which is essentially a very slightly radioactive waste product that remains after most of the fissile isotopes have been removed for nuclear fuel. Because of environmental and health concerns, depleted uranium is gradually being replaced by tungsten for ballistic applications.

Non-nuclear applications of thorium are fairly limited: the high visible emissivity of ThO_2 when heated led to its widespread use as a mantle in gas lanterns, and some nickel-based superalloys are strengthened by fine ThO_2 precipitates. Thorium is also added to tungsten electron guns and welding electrodes to decrease the electron work function, making it easier to establish and maintain a stable electron beam or arc. Over the longer term, the potential value of thorium

Figure 1. "Vaseline glass," shown here in natural light, gets its distinctive yellow-green color from small amounts of uranium oxide.

Figure 2. In ultraviolet light, vaseline glass displays intense green fluorescence.

Figure 3. The bright orange glaze on this Fiestaware saucer contains uranium as a colorant.

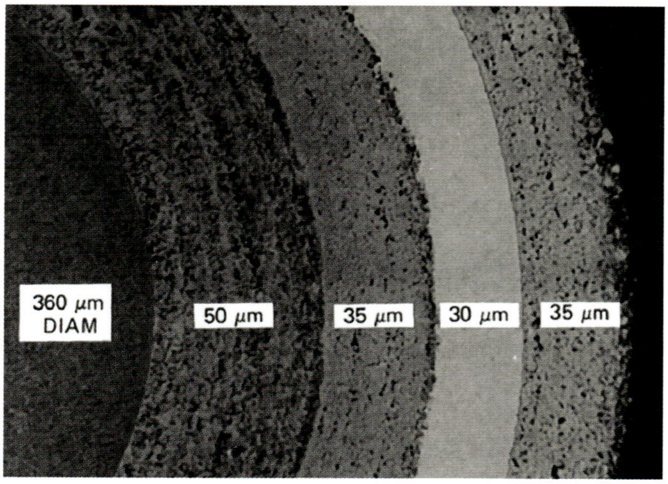

Figure 4. Cross section of a fuel particle for the High-Temperature Gas-Cooled Reactor, showing the fuel kernel surrounded by layers of SiC and pyrolytic carbon, designed to contain the fission products. The entire coated particle is less than 1 mm diameter. Photo courtesy of Oak Ridge National Laboratory

depends on the adoption of "thermal breeder reactor" designs such as the High-Temperature Gas-Cooled Reactor (HTGR) and others. No thorium isotopes are fissile, but a thermal breeder reactor exploits the overall reaction: $^{232}Th + ^{1}n \rightarrow ^{233}U$ in which ^{232}Th absorbs a thermal neutron to become the unstable isotope ^{233}Th, which decays to an isotope of protactinium and then to ^{233}U, a fissile isotope that can be separated and fed back into a reactor to generate power. Thermal breeder reactors have been intensively researched for over forty years, and the overall neutronics and fuel chemistry of these reactors are well understood (see, for example, Lauf et al., 1984). Prototype reactors based on thorium fuel cycles have been built in several countries, including: *AVR* in Julich, Germany; *Peach Bottom* and *Fort St. Vrain* in the US; *Dragon* project at Winfrith, UK; and *Kimini* near Kalpakkam, India.

WORLDWIDE ECONOMIC RESERVES

Uranium

The International Atomic Energy Agency (IAEA) continuously monitors the status of the uranium industry worldwide and issues public reports on the status of uranium resources. IAEA documents of particular interest include an excellent map of the global distribution of uranium deposits (Finch et al. 1995). This map and the accompanying guidebook provide information on 582 uranium deposits with an emphasis on those containing at least 500 metric tons of U and an average grade of 0.03% U or greater. An update was published in 2009, and the IAEA now maintains a current database on its web site. A biennial joint report of the Organization for Economic Cooperation and Development (OECD) Nuclear Energy Agency and the IAEA, also known as the "Red Book," provides a statistical profile of the uranium industry in the areas of exploration, resource estimates, production, and reactor-related requirements.

As with all mineral resources, a particular uranium deposit might or might not be economical to develop or operate depending on the current price of uranium. For this reason, uranium reserves are generally tabulated as Reasonably Assured Resources (RAR) and Estimated Additional Resources (EAR) based on several price points: < \$40/kgU; < \$80/kgU; < \$130/kgU; and < \$260/kgU. Mines may be mothballed and later reopened as prices fluctuate, and a deposit that has been completely mined out may later yield more uranium if higher prices justify reprocessing the mine tailings to recover some of the uranium that was not extracted in the first processing. Finally, more uranium may be recovered from tailings or wastewater during environmental reclamation at a mine that has ceased production.

The following statistical "snapshot" of worldwide uranium mining activities and reserves, representing data that are believed to be current as of 2011, is based on data from the recent Red Book (OECD 2012). Note that tabular data may not sum precisely because of independent rounding.

Table 1. Reasonably Assured Resources among countries with major resources (tU)[a]

Country	<\$40/kgU	<\$80/kgU	<\$130/kgU	<\$260/kgU
Australia	0	961,500	1,158,000	1,180,100
Canada	237,900	292,500	319,700	421,900
Kazakhstan	17,400	244,900	319,900	402,400
United States	0	39,100	207,400	472,100
Namibia	0	5,900	234,900	362,600
Niger	5,500	5,500	339,000	340,600
Russian Federation	0	11,800	172,900	218,300
South Africa	0	96,400	144,600	192,900
Brazil	137,900	155,700	155,700	155,700
Ukraine	2,800	44,600	86,800	143,300
China	45,800	88,500	109,500	109,500
Uzbekistan	46,600	46,600	64,300	64,300

[a]Tonnes uranium; data from OECD (2012)

Uranium deposits may be exploited by several methods, depending on the size, type, and placement of the deposit. These methods include underground mining, open-pit mining, *in-situ* leaching (ISL), heap leaching or in-place leaching, and extraction as a co-product or by-product of mining for another metal.

Table 2. Reasonably Assured Resources by production method (tU)a				
Method	**<$40/kgU**	**<$80/kgU**	**<$130/ kgU**	**<$260/ kgU**
Open-pit mining	28,600	99,600	868,700	1,286,000
Underground mining	313,800	428,900	875,800	1,272,000
In situ leaching	80,400	384,000	527,000	537,700
Co-product/ by-product	71,100	1,102,300	1,150,700	1,198,900
Unspecified mining method	0	0	33,100	84,100
Total	493,900	2,014,800	3,455,500	4,378,700
[a]Data from OECD (2012)				

In 2010, production of uranium in North America represented about 21% of the worldwide total. Canada was the world's second-largest producer, at around 9,775 tU, and was the dominant North American producer. The MacArthur River mine, Saskatchewan, is the world's largest high-grade uranium mine. The ore from that mine is crushed and treated underground to make a high-grade ore slurry, which is pumped to the surface and taken in specially designed trucks to the Key Lake mill for processing.

Production in the United States was 1,630 tU, most of which was produced by ISL. Six small underground mines closed in 2009, leaving four mines in production feeding one operational mill, White Mesa, Utah. In 2011, there were five ISL facilities in production and five partially permitted and licensed. Two mills were on standby status and a third mill was permitted and licensed but not yet operating.

Brazil was the only country in South America producing in 2009–2010; production was 347 tU in 2009 and 148 tU in 2010. In Argentina, efforts were underway to restart production at the Sierra Pintada mine of the San Rafael complex, but environmental and regulatory issues remained to be satisfied.

Production in the European Union was 358 tU in 2010, of which the Czech Republic represented 254 tU. Finland is about to start production at the Talvivaara Ni-Zn-Cu-Co mine, where bio-heap leaching is used to extract metals from black schist hosted ore. Although the overall U grade is extremely low, the U is efficiently leached to form a solution containing 15-25 mgU/liter. The plan is to build a solvent extraction circuit to recover by-product uranium with annual production of ~350 tU.

In the Russian Federation, 2011 production was expected to be ~3,350 tU, but will likely increase in the future as more projects are developed, particularly in the Elkon district.

Production in Central Asia increased significantly as Kazakhstan became the world's largest producer, with 17,803 tU in 2010 and ~20,000 tU projected for 2011. Output also increased in Uzbekistan, to 2,874 tU in 2010.

Australia suffered a decrease from 8,422 tU in 2008 to 5,918 tU in 2010, attributed mainly to operational problems. Production is expected to grow when the Four Mile ISL mine comes on stream, and planned expansion of Olympic Dam could boost total production to over 16,000 tU in the future.

In Africa, production was ongoing in Namibia, Niger, and South Africa, and in 2009 production began in Malawi at the Kayelakera mine. Overall output from Africa was ~9,600 tU in 2011 but is expected to increase substantially as development activities continue in Namibia and in Niger, particularly at the large sandstone deposit at Imouraren.

Thorium

Thorium reserves are fairly abundant worldwide, but are not studied as intensively as those of uranium. Recent estimates of total thorium resources, which include both economic reserves and additional resources, place the figure at about 7 million tonnes. Reported levels of ore reserves, Table 3, should be considered very conservative because of limited reporting from some countries (particularly China and the former Soviet Bloc) and because weak market demand for thorium has historically limited exploration activities (OECD 2012).

Table 3. Identified thorium resources (total resources in tonnes Th *in situ*)	
Country	**Resources (t Th)**
CIS[a]	1,500,000
India	846,000
Turkey	744,000–880,000
Brazil	606,000–1,300,000
Australia	521,000
United States	434,000
Egypt	380,000
Norway	320,000
Venezuela	300,000
Canada	172,000
South Africa	148,000
China[b]	>100,000
Russian Federation[b,c]	>100,000
Greenland	86,000–93,000
Finland	60,000
Russian Federation[d]	55,000
Kazakhstan[b]	>50,000
Sweden	50,000
[a]Excluding Russian Federation, European part	
[b]Estimated	
[c]Asian part	
[d]European part	

CHARACTERIZING RADIOACTIVE MINERALS

Students of mineralogy (and most mineral collectors as well) are familiar with the standard tools used to characterize the physical properties of minerals. These properties include hardness, tenacity, color, refractive index, birefringence, dispersion, density, chemical composition, and crystal structure. In the case of uranium and thorium minerals, natural radioactivity provides another property to be studied.

There are three predominant naturally occurring isotopes of uranium and thorium, ^{238}U, ^{235}U, and ^{232}Th. Each has a characteristic decay chain, with the end products ^{206}Pb, ^{207}Pb, and ^{208}Pb respectively. All three isotopes decay by emitting an alpha particle, which is a ^{4}He nucleus. Subsequent isotopes in each decay chain may emit alpha particles, beta particles (electrons), or gamma rays. Alpha and beta particles have a very limited range because they are easily scattered by air or other matter, whereas gamma rays are quite penetrating. One can see from the differences in atomic mass that the three decay chains represent total emission of 8, 7, and 6 alpha particles, respectively. Note that natural materials also contain minute amounts of the isotope ^{234}U, which is an indirect decay product of ^{238}U. Because the half-life of ^{234}U (245,500 years) is shorter by a factor of 18,000 than that of ^{238}U (4.468 billion years), at secular equilibrium the concentration of ^{234}U present at any time only represents about 55 ppm of the total uranium present.

RADIATION DETECTORS

Many types of radiation detectors exist, the two most common being scintillation detectors and Geiger counters. The scintillation detector uses a photomultiplier tube to detect minute flashes of light emitted by a phosphor when it is struck by a high-energy particle. The Geiger counter detects ionization events in a gas by measuring pulses of current between two high-voltage electrodes. Activity may be expressed simply as counts/minute (cpm), typically cpm *above background*, or it may be converted to other units that are more suitable for particular needs, e.g., controlling worker exposure, ore grading, etc.

Figure 5. The Radian RKSB-104 is a small, battery-operated Geiger counter that measures both gamma and surface beta radiation. Here it is being used to measure the activity of a sample of dendritic uraninite in pegmatite.

Figure 6. General view of the autoradiography setup when using instant sheet film. Note that at no time is the film pulled out of the dark slide (as it would be when the film is used in a microscope or other optical system).

AUTORADIOGRAPHY

The ability of radioactive materials to darken a photographic film that is completely shielded from light has been known for well over a century. Traditionally, a single photographic film was wrapped in dark paper and the radioactive sample was placed on top of the paper for some period of time, after which the negative was developed in the darkroom. Working at Oak Ridge National Laboratory in the late 1970s, Hobart Keating demonstrated that the technique can be accomplished quickly and easily using single sheets of Polaroid® "instant" film, of the type that until recently was used in scanning electron microscopes and other laboratory equipment. All of the autoradiographs in this book were prepared using instant film; readers should note, however, that this type of film is no longer being made, and when existing stocks are sold, it will become unavailable for good.

The procedure is carried out as shown in Figure 6. A single sheet of Polaroid® Type 57 4 × 5" Black and White Sheet Film is placed on a flat surface. The side that normally faces the camera lens is facing upward. The specimen is centered on the sheet and the time is noted. The setup is left undisturbed for four to twenty-four hours. Then, the film sheet is slid into the camera back, the lever is flipped to "process" and the sheet is pulled back out to start the development. After about twenty-five seconds, the film is pulled apart to reveal the autoradiograph. To get the best results, it is usually necessary to make several exposures,

Figure 7. A slab of black dendritic uraninite altering to orange gummite in pegmatite, from the Ruggles mine, New Hampshire. The activity of this sample is ~16,000 cpm measured at a distance of about 1 cm.

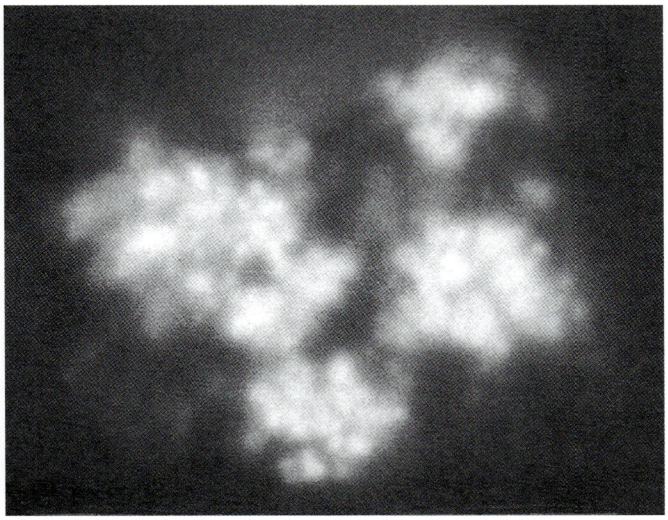

Figure 8. Autoradiograph of the sample in the previous figure after an eight-hour exposure.

Figure 9. A hand-sized sample of liebigite crystals on matrix from the Schwartzwalder mine, Colorado. The activity of this sample is ~700 cpm.

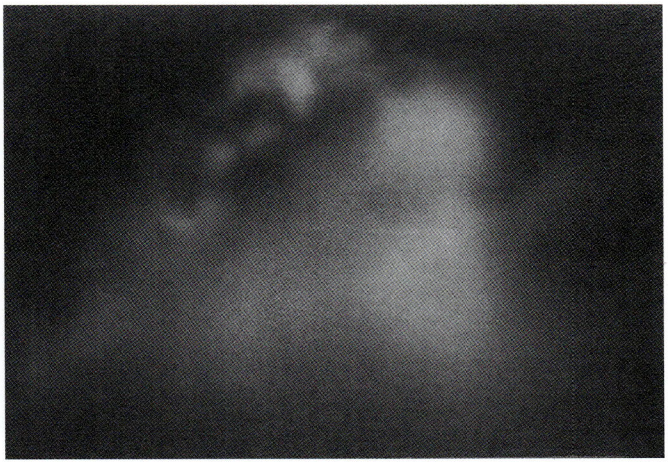

Figure 10. Autoradiograph of the sample in the previous figure after a twenty-four-hour exposure. Comparing the uraninite and liebigite samples shows clearly that the exposure correlates with the sample's activity.

doubling the time in increments such as four, eight, and sixteen hours, and then selecting the best image.

Exposure time will vary based on the sample's overall level of radioactivity. In general, a sample containing a significant amount of uraninite will be "hotter" than a sample comprising only secondary uranium minerals. Thorium minerals often have significant gamma activity; because the film is less sensitive to gamma, one can enhance the response somewhat by placing a metal backing plate (e.g., aluminum or copper) under the sheet of film. The metal sheet serves as a down-converter, reacting to the gamma by yielding X-rays and photoelectrons, many of which will be emitted back in the direction of the film, and are more efficiently detected by this type of film (H. Keating *pers. comm*). Exposure time will also be different if a film of a different speed is used. Type 57, for example, is rated ISO 3000 speed.

Figure 11. A massive piece of nearly pure uraninite, partially altering to red-orange wölsendorfite and other secondary oxides. The activity of this sample is ~13,000 cpm. RJL354

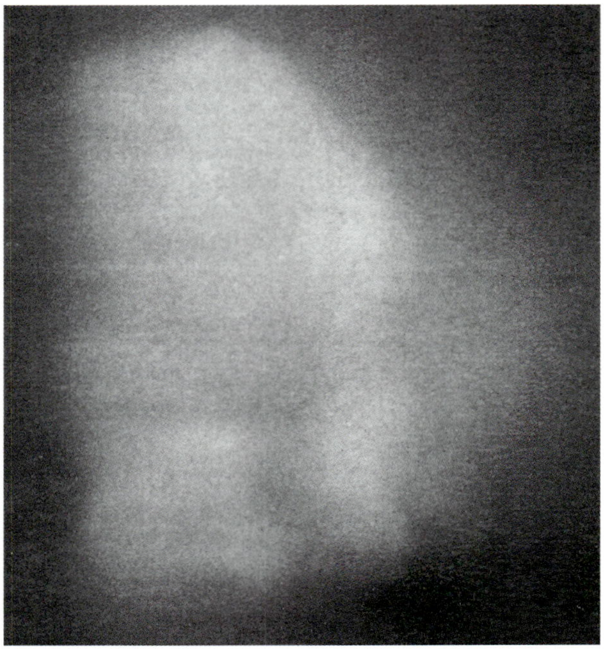

Figure 12. Autoradiograph of the sample in the previous figure after a six-hour exposure. This photo illustrates the limited penetration range of much of the alpha and beta radiation: notice that the rightmost quarter of the specimen was held about 5 mm away from the film by the step on the fracture surface, and this small standoff was enough to greatly reduce the intensity of exposure under that area.

Figure 13. Brown to black prismatic thorite crystals in matrix from Pacoima Canyon, California. The activity of this sample is ~4300 cpm. RJL566

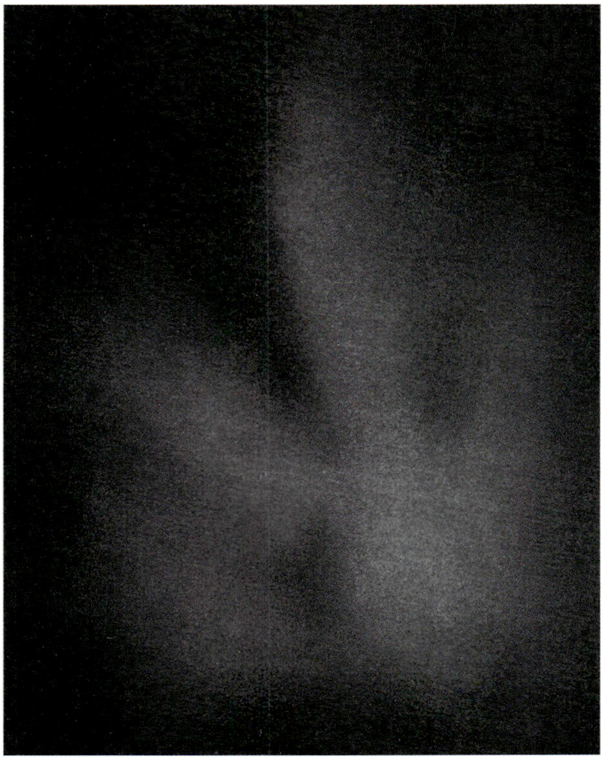

Figure 14. Autoradiograph of the sample in the previous figure after an eight-hour exposure with an aluminum plate under the film to enhance sensitivity to γ-rays.

ORE MICROSCOPY

Because of its economic importance, uranium ore is frequently studied using the standard techniques of reflected light microscopy. The general procedure may be summarized as follows (Paschis 2009):

A representative sample is selected using a hand lens or stereo microscope. The sample is sectioned with a diamond saw and checked with a γ-radiation detector to make sure that uranium mineralization is still present in the specimen. The specimen is mounted in a clear resin matrix, e.g., epoxy, typically forming a cylinder ~3 cm in diameter, with the sawn surface of the sample on one of the flat ends.

The face is ground on 200- 325-, and 1200-grit diamond laps, then on glass plates using 12- 9-, and 3-µm alumina, then on rotating laps charged with 6- 3-, and 1- µm diamond. Final buffing is done on cotton using a slurry of 0.05-µm alumina.

Microscopic examination is done using a reflected-light polarizing microscope. Images done with plane-polarized light can show the mineral's color, pleochroism, cleavage, relative reflectivity, and relative hardness (as indicated by polishing relief). Images taken with crossed polarized light can show optical properties such as color, anisotropy, internal reflectance, or degree of ore and gangue oxidation.

Figure 15. A prepared mount, showing samples mounted in clear medium and polished to reveal the microstructure for examination. Photo courtesy of Jim A. Paschis, ore microscopist

Figure 16. Digital ore micrograph in reflected light of a sample from the Beaverlodge district, Saskatchewan, shows the cracked radial structure of the uraninite var. pitchblende (light gray) with hematized calcite (darker gray). FOV is ~2 mm wide. Photo courtesy of Jim A. Paschis

FORMATION AND CHARACTERISTICS OF RADIOACTIVE DEPOSITS

GEOCHEMISTRY OF URANIUM AND THORIUM

Uranium and thorium are widely distributed throughout the earth's crust. Uranium, with an average crustal abundance of 2.4 ppm, is more plentiful than such familiar elements as silver, gold, arsenic, mercury, molybdenum, antimony, and tin. Thorium, at 12 ppm, is roughly as plentiful as lead and boron.

Relative to their cosmic abundance, the crustal abundances of uranium and thorium represent a huge concentration factor and it is interesting to consider how this concentration came about and how large-scale processes acted (and continue to act) to redistribute uranium and thorium and influence their mineralogy, as discussed in detail by Hazen, Ewing, and Sverjensky (2009). Uranium and thorium abundances in Type CI carbonaceous chondrites (0.0074 and 0.02 ppm, respectively) may be taken as an estimate of their original concentrations in the solar nebula. As material accretion continues, eventually asteroids become large enough to undergo differentiation into a metallic core and a basalt-like crust. The high stability of U and Th oxides causes these "lithophile" elements to segregate into the primitive silicate crust of the asteroid at concentrations represented by eucrite meteorites (0.07 to 0.15 and 0.3 to 0.8 ppm, respectively). These concentrations are roughly comparable to those seen in mid-ocean ridge basalts on earth. U and Th became further concentrated by magmatic hydrothermal activity associated with tectonic movements of the earth's crust during perhaps the first billion years; their concentrations in granite are roughly 10 and 30 ppm, respectively.

In primary rocks and magmas, uranium is normally present in the quadrivalent state, U^{4+}. During early crystallization of the magma, the low concentration of U prevents it from crystallizing as distinct mineral phases. At the same time, the large size and charge of the U^{4+} ion effectively keep it from becoming incorporated in the usual rock-forming silicate minerals as they crystallize. Thus, uranium tends to become concentrated in the hydrothermal solutions and pegmatites that remain as the cooling magma becomes consolidated. Small amounts of uranium may be taken into solid solution in various accessory minerals, particularly zircon, monazite, and thorite. The bulk of the uranium will ultimately crystallize as uraninite, UO_2, although in some hydrothermal vein deposits the primary uranium mineral is coffinite, $U(SiO_4)$. Within the pegmatites, uranium may be present as solid solutions in thorite, thorianite, monazite, zircon, brannerite, and various niobate-tantalate minerals. In these deposits, uranium is normally associated with thorium, zirconium, and rare-earth elements (REE). By contrast, in hydrothermal deposits of magmatic origin, uranium has become separated from these elements and is more typically associated with cobalt, nickel, bismuth, and arsenic. In these settings uraninite is again the most abundant primary uranium compound, but it typically forms as fine-grained, massive pitchblende.

The geochemical behavior of uranium has important similarities and differences compared to that of thorium, cerium, and zirconium, and it is instructive to examine these in detail (Frondel 1958). The similarities arise from their general chemical behavior, particularly their strong affinity for oxygen. The dissimilarities relate to the small size of the Zr^{4+} ion compared to U^{4+}, Th^{4+}, and Ce^{4+} and, more importantly, to the fact that only uranium has a higher valence state, U^{6+}. In their quadrivalent states, all of these elements show the same general behavior in magmas and tend to occur together, either as separate accessory minerals in the same pegmatite or as substitutional solid solution species in the same mineral.

In contrast to thorium, cerium, and zirconium, only uranium has a stable hexavalent state, usually forming the uranyl ion, $(UO_2)^{2+}$; importantly, this is the stable oxidation state under normal atmospheric conditions. This ion is a linear $[O\text{-}U\text{-}O]^{2+}$ complex that behaves as a distinct unit in crystal structures. Its large "dumbbell" shape prevents it from substituting for other divalent ions, giving rise to a whole collection of unique secondary minerals in which the uranyl ion is a crucial part of the structure. The unique structural features and crystal chemistry of uranyl compounds are the subject of extensive research involving both natural uranium minerals and synthetic compounds (Burns, Miller, and Ewing 1996; Burns 2004). Furthermore, the uranyl ion tends to be fairly soluble, so once primary uraninite is weathered or otherwise oxidized, the uranyl species can migrate over large distances before precipitating as carbonates, phosphates, sulfates, silicates, or vanadates. Because Th^{4+} has no higher valence state that would correspond to $(UO_2)^{2+}$, the primary mineral thorianite, ThO_2, is quite stable in an oxidizing environment and there are no corresponding secondary thorium minerals.

When primary uraninite crystallizes, even in an oxygen-poor environment, it always contains some oxygen in excess of the stoichiometric ratio, that is, the formula would be expressed as UO_{2+x}, where x ranges from about 0.15 to perhaps as much as 0.6. Once uraninite is exposed to the air, or to oxygenated groundwater, further oxidation proceeds steadily. Oxidation kinetics is made faster by the breakdown of the uraninite crystal structure from internal radiation damage to the crystal structure (*metamictization*) as well as by the process of *auto-oxidation*. The concept of auto-oxidation is based on the fact that the products of radioactive decay, primarily lead, do not have as high an affinity for oxygen as uranium does; thus, as more lead is produced within the uraninite, the uranium ions react to what appears to be a more oxidizing condition by moving from U^{4+} to U^{6+}. If sulfur is available, the radiogenic lead typically forms galena, often seen as microscopic inclusions within the uraninite.

The mobility of uranium in the environment is highly dependent on its oxidation state, with U^{6+} being considerably more mobile than U^{4+}. Based on extensive thermodynamic calculations, Langmuir (1978) has shown that U^{4+} can only reach significant concentrations in ground water as the fluoride complex. This complex is only stable in reduced, acidic waters (pH below about 3) containing sufficient fluoride. In the natural environment, ground waters satisfying all of these conditions are very rare and consequently U^{4+} may be generally regarded as substantially immobile.

When uraninite becomes oxidized and water is present, a number of soluble uranyl complexes can be formed depending on factors such as the pH and the other anions that are present in the water. Some soluble uranyl species are listed in the following table. It must be kept in mind that in an aqueous solution there will usually be several competing ionic species present at different concentrations (defined by the equilibrium constant of the particular reactions), so Table 4 presents the dominant species in a particular pH range.

Table 4. The dominant uranyl species in aqueous solutions

Solution acidity	**pH < 4**	**pH = 6**	**pH > 7**
Main species	$(UO_2)^{2+}$	$[(UO_2)OH]^+$	$[(UO_2)_3(OH)_5]^+$
		$[(UO_2(HPO_4)_2]^{2-}$	$[(UO_2)(HPO_4)_2]^{2-}$
		$(UO_2)CO_3$	$[(UO_2)(CO_3)_2]^{2-}$
		$[(UO_2)SiO(OH)_3]^+$	

The solubilities of uranyl minerals follow a fairly regular sequence in order of decreasing solubilities: carbonates > sulfates > phosphates and arsenates > silicates > vanadates. It is not surprising, then, that the most commonly found uranyl minerals are the vanadates carnotite and tyuyamunite, followed by autunite-group phosphates and arsenates, and silicates, particularly uranophane. The large number of potential combinations of the uranyl ion with other metals and each of the major anions creates a fascinating diversity of known secondary minerals, and new ones continue to be described today.

In uranium ore deposits such as "roll fronts," the presence of organic matter such as wood, plant detritus, and hydrocarbons are important reductants to convert U^{6+} to U^{4+}. The electrochemical processes that give rise to particular uranium

Figure 17. Uraninite cubes to about 1 cm across, on massive uraninite, in the process of alteration to curite and other secondary uranium minerals. The specimen is from Shinkolobwe, Congo. RJL1700

Figure 18. Rear of the specimen shown in the previous photo, showing the formation of "gummite" rims on the black uraninite. Note also the velvety yellow crust of uranophane. RJL1700

deposits can in some cases be facilitated by the action of anaerobic bacteria. These microorganisms have evolved a respiration cycle that relies not on air but rather on exploiting other chemical reactions in their environment. For instance, the organism might split molecular hydrogen, short-chain fatty acids, or other substrates (using a hydrogenase enzyme) into two H^+ ions and two electrons. Those electrons travel through some metabolic pathway and are then released to an electron acceptor, which is typically a reducible metal ion of some kind. It has been demonstrated that processes of this type can drive the reduction of uranyl species to form uraninite in the presence of suitable organic material to act as the hydrogen donor. The overall reaction may be simplified as follows: $H_2 + (UO_2)^{2+}$ (uranyl ions) $\rightarrow 2H^+ + UO_2$ (uraninite).

CLASSIFICATION OF RADIOACTIVE MINERAL DEPOSITS

The inherent complexity of the natural world invariably frustrates human attempts to create simple, consistent classification systems, and uranium deposits are no exception. Many schemes have been proposed (e.g., Heinrich 1958) and each has its own merits and shortcomings. Purely descriptive terms based on the local geology are less likely to provoke controversy; as McMillan (1978) points out, ". . . genetic ideas have a tendency to change while the observed wallrock association does not!" On the other hand, the main purpose of studying ore deposits is to understand their genesis so that exploration for similar deposits might be focused on geological settings that have the greatest chance of being productive. Thus it is most common to classify deposits based on both the geologic setting and the probable mechanisms of formation, with the understanding that in many cases the geologic evidence might be subject to more than one equally plausible genetic interpretation.

For the mineral collector, as opposed to the exploration geologist, a general understanding of the geology will aid the interpretation of a mineral specimen: What are the associated minerals? What is the matrix? Is the specimen typical of the locality? Does the specimen nicely illustrate some unique features of that locality? How much geologic "history" is preserved and displayed by the sample at hand? The following simplified discussion of uranium deposits is based largely on the classification used by the IAEA and major mining companies (Dahlkamp 1993; OECD 2008; Kyser and Cuney 2008; Fayek 2013).

1. Unconformity-related deposits

Unconformity-type deposits were first recognized as a distinctive class in the early 1970s. They occur close to unconformities between metasediments of Archean-Paleoproterozoic age and overlying Paleo-Mesoproterozoic sandstone. The U deposits can lie in the sandstone or in the basement rocks, but within a few hundred meters of the unconformity. They are hosted near faults and are normally associated with brecciation. They are interpreted to have formed from brines at 150–250°C. Known occurrences are restricted to Proterozoic rocks, and most of them formed shortly after the assembly of large continents. Unconformity-related deposits make up ~12% of total U resources and 33% of current production, and they include some of the largest and richest deposits (Kyser and Cuney 2008).

Deposits of this type are represented by major uranium mining districts in northern Saskatchewan, Canada, and in Northern Territory, Australia. At one time these two districts represented about 16% of the world's uranium reserves (Nash et al. 1981). The two deposits have important similarities and differences, and the question of how they were formed continues to be debated. Both deposits lie at or near an unconformity where Lower Proterozoic metasediments are overlain by Mesoproterozoic quartz sandstone. The mineralization seems to be locally controlled by faulting, fracturing, or collapse breccias.

Mineralogy of the northern Saskatchewan deposits is fairly complex. For example, the mineralization at Rabbit Lake is postulated to have occurred in three stages (Hoeve and Sibbald, 1978) as follows. Stage 1: veins of pitchblende and coffinite with adularia, chlorite, quartz, hematite, calcite, quartz, and sulfides; chloritization and hematization of the wall rock indicates oxidizing ore fluids. Stage 2: formation of veins of euhedral quartz with little uranium under somewhat reducing conditions. Stage 3: impregnation of sooty pitchblende and coffinite along fractures, also under reducing conditions.

Mineralogy of the Northern Territory deposit is somewhat simpler and wall rock alterations are more uniform, suggesting that mineralization took place under uniformly reducing conditions (Edwards and Atkinson 1986). The primary ore mineral is uraninite, with lesser amounts of coffinite, brannerite, and *thucolite*. The gangue mineral is predominantly chlorite, with hematite plentiful at some sites. At the Ranger 1 mine, economically significant secondary minerals (mainly saleéite, along with sklodowskite, gummite, and metatorbernite) occur in a thick oxidation zone.

2. Sandstone deposits

Sandstone uranium deposits are rarely the source of good specimens for the collector, but are of great importance because they represent the second-largest share of economic ore reserves and production at present. Their economic value lies partly in their large size and partly in the low cost of exploiting them. In some cases the uranium is recovered by *in-situ leaching*, in which wells are drilled to the permeable host layer. Solutions are pumped in some of the wells to create soluble uranyl species, which are then pumped out of adjacent wells.

These deposits occur in medium- to coarse-grained arkosic sandstone, often lying between layers of impermeable mudstone.

Most occurrences are younger than 440 Ma, about the time that organisms invaded the continents. They were likely formed at low temperatures from slightly modified basinal brines. Sandstone deposits represent ~28% of global resources and 32% of current production (Kyser and Cuney 2008).

Sandstone-type deposits include those often referred to as "Colorado Plateau type." Examples occur throughout the Colorado Plateau itself as well as in Wyoming, South Dakota, New Mexico, Texas, and California; similar occurrences include: Beverley, Australia; Akouta, Niger; and Mokum, Kazakhstan. Within this category, however, individual deposits show numerous variations in their mineralogy and probable origin:

Tabular/peneconcordant deposits lie roughly parallel to the bedding of the host sandstone. The main ore minerals, uraninite and coffinite, tend to be closely associated with organic material, which might be carbonized or coalified wood, plant debris, or humic material. The soluble uranyl species were presumably captured by adsorption onto the organic matter and then chemically reduced by the carbonaceous material and other reducing species such as H_2S, which could have been biogenically derived.

Roll front deposits, sometimes called geochemical cell deposits, usually have a crescent-shaped cross section that can extend laterally for some distance. In this type of deposit, concentrations of uraninite and coffinite lie at or near oxidation-reduction interfaces where oxygenated, uranium-rich groundwater came into contact with a reducing environment containing carbonaceous material and pyrite. The changing redox conditions (and pH) cause the precipitation of reduced uranium species and sometimes vanadium and molybdenum minerals in the interface zone (De Voto 1978). These deposits were probably formed below about 50°C and it is likely that communities of microorganisms played a role in reducing the soluble U^{6+} to form the insoluble uranium minerals (Suzuki and Banfield 1999).

Tectonic/lithologic deposits occur in sandstone adjacent to permeable fault zones. Most are of limited size, and all have variable ore grades.

The mineralogy of sandstone-type deposits is complex and depends, among other things, on the presence or absence of vanadium. As Heinrich (1958) points out, vanadium provides a convenient internal measure of the geochemical conditions because it has such a wide variety of minerals, each of which is only stable over a narrow range of oxidation potential. In nonvanadiferous ores, numerous secondary uranium minerals may be found, including oxides, sulfates, carbonates, members of the autunite group, and various uranyl silicates. In vanadiferous ores the uranium minerals include carnotite, tyuyamunite, uranophane, autunite, zippeite, schroeckingerite, andersonite, becquerelite, and schoepite. If the ore is incompletely oxidized, other vanadium minerals may include paramontroseite, corvusite, melanovanadite, duttonite, simplotite, sherwoodite, and doloresite. If the ore is fully oxidized, the vanadium minerals tend to include navajoite, hewettite, pascoite, rossite, metarossite, hummerite, steigerite, fervanite, volborthite, calciovolborthite, and pintadoite.

3. Vein deposits

Vein deposits occur as veins in metamorphic or other rock types; this category includes deposits that don't easily fit into the other categories, and in many cases their origins are complex. As a result, the mineral assemblages can vary greatly from one subtype to the next. Vein deposits represent ~7% of global resources and 10% of global production (Kyser and Cuney 2008). Notable examples are found at Shinkolobwe, Democratic Republic of Congo, and at Jachymov, Czech Republic. Rich, Holland, and Petersen (1977) provide detailed descriptions of a large number of hydrothermal vein deposits throughout the world.

In hypothermal deposits, formed at 300-500°C, uranium and thorium tend to be closely associated in the minerals davidite, brannerite, and thorian uraninite. Commonly associated minerals include ilmenite, hematite, pyrite, magnetite, and molybdenite. Other minerals that are sometimes seen are huebnerite, cobaltite, cassiterite, and arsenopyrite. Some examples of hypothermal deposits include the following:

Davidite veins are exemplified by the deposit at Radium Hill, South Australia. The deposit is postulated to have developed in several stages: 1. replacement of micaceous shear rock by quartz-biotite-hematite-ilmenite mineralization; 2. intrusion of REE-rich pegmatites containing allanite and xenotime; 3. movement along the shears, causing brecciation and biotization; 4. intrusion of amphibolites along faults; and, 5. introduction of quartz stringers containing davidite. The davidite also developed intergrowths with the existing ilmenite-hematite mineralizations. Davidite vein deposits also occur at several sites in the Mavuzi district, Mozambique, where davidite crystals exceeding a foot in diameter were found.

Brannerite deposits, presumed to be of hypothermal origin, occur at the California mine, Chaffee Co., Colorado; at the West Walker River, near Coleville, California; and in the molybdenite deposit at Chateau-Lambert, Vosges, France. Note that the California mine, Colorado, contains considerable molybdenite and beryl and may be interpreted as a pegmatite deposit.

Mesothermal deposits, which formed at intermediate temperatures and depths (200-300°C), include several very productive mining districts such as Great Bear Lake, Northwest Territories; Shinkolobwe, Congo; and Goldfields, Saskatchewan, where large tonnages of high-grade ore have been recovered. There are three main types of radioactive mesothermal deposits: Ni-Co-Ag type; pitchblende-pyrite type; and monazite or thorite veins. The Ni-Co-Ag type deposits are further subdivided by some authors into numerous subtypes based on details of the mineral assemblage.

In general, the mesothermal vein deposits usually contain a gangue of carbonate minerals and/or coarse quartz. Hematite is very common and often abundant enough to color the gangue minerals and impregnate the wall rock. Pyrite and chlorite are common; barite and fluorite, along with arsenopyrite and other sulfosalts, are rarer. Many of these deposits contain significant amounts of gold, silver, or platinum-group metals. The usual base metals are bismuth, copper, lead, cobalt, and nickel; zinc and molybdenum are sometimes found as well. The predominant copper mineral is usually chalcopyrite; bismuth is typically present as the native metal or bismuthinite. The cobalt and nickel minerals vary greatly from one deposit to another and may include niccolite, cobaltite, gersdorffite, glaucodot, skutterudite, smaltite, chloanthite, safflorite, rammelsbergite, maucherite, polydymite, vaesite, cattierite, and siegenite. At Shinkolobwe, selenium is abundant in the Co-Ni minerals, particularly as Se-rich digenite, whereas at Goldfields, selenides of copper and lead are found.

The normal radioactive mineral in mesothermal deposits is uraninite (var. pitchblende) that is usually low in thorium but may contain REE. From measurements of the UO_2 unit cell parameters, the estimated U^{4+}/U^{6+} ratio indicates formation at intermediate temperatures. In most cases the uraninite is botryoidal or spherulitic; a notable exception is at Shinkolobwe, where it is granular to euhedral in small cubic crystals.

Ni-Co-Ag uranium deposits include: the Radium vein at the Caribou mine, near Nederland, Colorado; the Black Hawk silver district, New Mexico; the Coeur d'Alene district, Idaho; Great Bear Lake area, Northwest Territories; Cornwall, England; the Erzgebirge district, Czech Republic; and the Shinkolobwe mine, Congo.

Pitchblende-pyrite type deposits include the Ralston Buttes district, Jefferson Co., Colorado, which produced 20,000,000 pounds U_3O_8; the principal mine was the Schwartzwalder. Other examples include: the Central City district, Gilpin Co., Colorado; Theano Point, Montreal River district, Ontario; the Goldfields region, Saskatchewan; Placer de Guadalupe, Chihuahua, Mexico; Margnac, France; the Darwin-Katherine area, Northern Territory, Australia; and Serra de Jacobina, Bahia, Brazil.

Thorium veins are the least common type of radioactive mesothermal deposit. In the Lemhi Pass district of Idaho and Montana, the metamorphic country rock (argillite, quartzite, and mica schist) is cut by fissures with veins containing copper minerals. These veins contain bornite, chalcopyrite, chalcocite, pyrite, quartz, hematite, gold, silver minerals, and thorite. Associated with these are quartz-hematite veins containing barite, goethite, and thorite. Thorogummite is sometimes present in the iron oxides and small amounts of monazite and allanite may be present in some of the veins.

A *monazite deposit* at Steenkampskraal, South Africa, contains dark veins of corroded and sericitized feldspar with a fine-grained mixture of quartz, pyrite, hematite, monazite, apatite, and leucoxene. In some places the vein material is as much as 75% monazite and the ore contains as much as 5% ThO_2. The mineralization is interpreted as having occurred in three general stages (Heinrich 1958): 1. Feldspar, monazite, apatite, magnetite, and possibly ilmenite; 2. Quartz, pyrite, and possibly chalcopyrite; and 3. Sericite.

The *radioactive zirconium deposits* of Poços de Caldas, Brazil, contain an unusual assemblage of baddeleyite/zircon mixtures ("caldasite") containing up to several percent uranium, as well as magnetite ore containing veins of basnaesite and thorogummite. The presence of baddeleyite, along with zeolite pockets, suggests that the mineralization took place under conditions that fall within the mesothermal range.

Uraniferous vein deposits formed by low-temperature (100-200°C) hydrothermal processes may be quartzose, fluoritic, or both, and are relatively sulfide-poor. The usual uranium species is botryoidal to fine-grained pitchblende. Individual sites may be completely silicic, completely fluoritic, or virtually any intermediate composition. In a few cases, for example in the Thomas Range, Utah, a single fluorite pipe grades downward into a chalcedony replacement mass. The gangue minerals may include clays, alunite, iron and manganese oxides, and small amounts of calcite or dolomite. Pyrite and marcasite are often present, along with small amounts of galena, chalcopyrite, and sphalerite, but overall the veins contain fairly small amounts of metallic sulfides.

Epithermal thorite veins contain thorite and thorogummite and relatively little uranium. The gangue minerals include feldspar, smoky quartz, siderite, fluorite, barite, and hematite.

Siliceous pitchblende-sulfide veins occur at the Gray Eagle mine and other lead-silver deposits in the Boulder Batholith, Montana. The gangue material is primarily quartz and the ore is mainly pitchblende (replacing pyrite). In several mines in the Boulder district secondary minerals are found in the oxidized parts of the deposits; these include uranophane, uranophane-β, meta-autunite, metatorbernite, metazeunerite, and phosphuranylite. Important deposits in Portugal and France are also examples of this type.

Massive fluorite deposits in the Thomas Range, Utah, don't have commercially significant uranium concentrations, but at several of the sites minor amounts of secondary minerals such as carnotite are found in the upper parts.

Fluorite-sulfide veins in the Jamestown district, Colorado, were long mined for fluorite. Localized concentrations of both uraninite (pitchblende) and uranothorite occur in purple fluorite breccia pipes. Secondary deposits of torbernite are also found in the area. Veins of fluorite, pitchblende, pyrite, and ilsemannite are found at the Miracle Mine and the Kergon deposit in Kern Co., California; autunite, torbernite, and other uranyl minerals are locally abundant. At the Rexspar deposit, British Columbia, Canada, the fine-grained uraninite contains thorium and REE and is associated with fluorite, celestite, mica, and pyrite. The hydrothermal concentration of radioactive species, REE, fluorine, and strontium is interpreted to be of vulcanogenic origin (McMillan 1978).

Fluorite-quartz-sulfide veins in the Marysvale district, Utah, contain chalcedony or coarse quartz along with fluorite, adularia, and hematite. Pitchblende, pyrite, marcasite, and jordisite are present, as well as mercury minerals of the metacinnabar-tiemannite series. In the oxidized zone, numerous secondary minerals are found, including autunite, torbernite, metatorbernite, phosphuranylite, uranophane, uranophane-β, schroekingerite, johannite, uranopilite, zippeite, tyuyamunite, and rauvite. Fluorite vein deposits in the Wölsendorff region, Germany, are also examples of this type.

Thorite veins and breccia pipes in the Wet Mountains, Custer Co., Colorado, contain quartz, feldspar, barite, and iron oxides, with smaller amounts of fluorite, siderite, and various sulfides. There are finely disseminated veinlets and pods of thorite and thorogummite, and xenotime can be locally abundant. Many REE are present but uranium concentrations are quite low.

Veins oxidized in place typically begin as a vein or other hydrothermal deposit containing pitchblende, often with pyrite or other metallic sulfides. Oxidation of the sulfides introduces sulfuric acid, H_2SO_4, into meteoric or ground water, which readily attacks the uraninite. The uranium is preferentially leached as soluble $(UO_2)^{2+}$. Radiogenic lead and radium tend to form very insoluble sulfates and are therefore not mobilized, but remain as minute radioactive fracture fillings in the corroded uraninite. The most common secondary mineral species that form tend to be uranyl phosphates, sulfates, and silicates. Often, Ca^{2+} and Cu^{2+} are present as well, so autunite, torbernite, and uranophane are particularly widespread examples. Other commonly seen species include uranocircite, saleéite, parsonsite, kasolite, zippeite, uranopilite, and schroekingerite. Vanadates such as carnotite are fairly rare unless the primary ore minerals contain vanadium. If the primary vein material contains significant arsenic, then arsenate minerals such as zeunerite and uranospinite may form.

The secondary uranyl oxides such as schoepite, masuyite, and becquerelite tend to form directly on the uraninite as alteration layers or pseudomorphs. Where there is a lot of ground water flow, these minerals tend not to survive in place for very long periods but instead are ultimately leached away.

The thickness or depth of the oxidized zone can vary considerably; for example, at Shinkolobwe, unoxidized ore begins at a depth of about 50 m, whereas in the Gunnar deposit, Goldfields, Saskatchewan, secondary minerals (particularly uranophane) are found at a depth of about 300 m. In some cases the lower limit of the oxidation zone is fairly well defined by the water table, but there are many exceptions to this generality. When oxidation extends for a great distance below the present water table it may be attributed to several possibilities: 1. a seasonally fluctuating water table; 2. large variations in permeability created by localized fractures; 3. a water table that was previously much lower but has since been raised by climatic changes (Heinrich 1958).

Important examples of oxidized vein deposits include: Spokane, Washington; Marysvale, Utah; Goldfields, Saskatchewan, Canada; Shinkolobwe, Congo; and many localities in Portugal and France.

4. Metasomatic deposits

Metasomatic deposits occur in deformed rock (typically granites or metasediments) that have been altered by Na or Ca metasomatism and hematization (with albitization and formation of sodic amphibole or pyroxene). The deposits tend to be small and of low to medium grade (Dahlkamp, 1993), and represent ~9% of global resources. Examples include: Krivoj Rog, Ukraine; Ross Adams, Alaska; Lagoa Real, Brazil; and Valhalla, Australia. These deposits are especially common. Within this broad category are many subtypes, including the following:

Deposits of thorianite-uraninite in marble, as masses of lime-rich amphibolite ("metapyroxenite") near intrusive contacts. In the marble a skarn mineral assemblage is typically seen, and the recrystallized calcite is often salmon-pink. The chief radioactive species is uranoan thorianite or thorian uraninite; minor amounts of thorite, monazite, allanite, melanocerite, and other REE species may occur. In the lime-rich amphibolites the usual radioactive species is uraninite; pyrochlore and uranothorite are sometimes seen. Examples are found in the Bancroft region, Ontario; Easton, Pennsylvania; and the Ft. Dauphin area, Madagascar.

Contact allanite deposits occur in contact zones between intrusives and various sedimentary deposits. In some of these occurrences allanite and other REE minerals are abundant but only weakly radioactive, whereas in others the allanite is responsible for considerable radioactivity. Deposits of this type include the Whalen mine in west-central Alaska; Grand Calumet Island, Quebec; the Kyshtymsk district, Ural Mountains, Russia; and at Bastnaes and at the Ostanmossa mine, Sweden. At the Mary Kathleen mine, Queensland, Australia, the allanite itself is not radioactive, but disseminated grains of uraninite are present and in the oxidized zone uranophane, uranophane-β, and *gummite* occur.

Metasomatic monazite deposits found in northern Lemhi Co., Idaho, consist of monazite-bearing carbonate layers in a country rock of mica schists and gneisses. The carbonate lenses consist of calcite, stained in places by hematite and containing monazite along with actinolite, barite, siderite, pyrite, magnetite, ilmenorutile, garnet, zoisite, and apatite. The honey-colored monazite forms fine to coarse, generally euhedral, crystals and irregular masses in concentrated zones that can reach several meters thick and 100 m long. The overall mineral assemblage suggests a metasomatic origin (Heinrich, 1958) but the deposits also show similarities to some carbonatite vein deposits in Montana.

Oxidized contact deposits contain various secondary minerals such as autunite, metatorbernite, and metazeunerite. In some localities (e.g., near Austin, Nevada; Deer Lodge, Montana; and at Brooks Mountain, Seward Peninsula, Alaska) the original contacting bodies and the resulting "primary" contact zone mineralogy are fairly well understood. A notable exception is the important deposit at Mt. Spokane, Washington; there, the primary phases are not well understood and details of the genesis of this deposit remain incompletely known.

5. Breccia-complex deposits

Breccia-complex deposits are represented mainly by the Olympic Dam, Australia, iron oxide-copper-gold (IOCG) deposit, which is the world's largest with ~2,200,000 tU. It lies in a hematite-rich granite breccia complex. The central core is barren hematite-quartz breccia, flanked to the west and east by zones of hematite-rich breccia and granitic breccia ~1 km wide and running ~5 km in a NW-SE direction. The Cu-U mineralization is hosted by the hematite-rich breccia. Granitic breccia, which extends up to 3 km farther, holds Fe, Cu, U, Au, Ag, REE, and F. Only Cu, U, Au, and Ag are actively recovered, and the ore grade is ~0.04% U. The complex was formed by hydraulic fracturing, tectonic faulting, chemical corrosion, and gravity collapse in the near-surface eruptive environment of a crater complex when boiling or explosive contact of water or brines with magma drove eruptive episodes (Kyser and Cuney 2008). Gangue minerals are predominantly quartz and fluorite. The main U minerals are pitchblende, subordinate coffinite, and locally minor brannerite. The Cu occurs mainly as sulfides in assemblages such as bornite-chalcocite-chalcopyrite-pyrite (Dahlkamp 1993).

Olympic Dam, along with several smaller deposits, represents ~15% of global U reserves and ~10% of global production. Because the U ore grade is so low, it would be uneconomic to exploit were it not for the large values of Cu, Au, and Ag that are co-produced.

6. Intrusive deposits

Intrusive deposits comprise disseminated primary U minerals such as uraninite, U-rich thorianite, and uranothorite in rocks of intrusive magmatic or anatectic origin. Sub-classes are based on the host rock petrology, with the following examples: alaskite, Rössing, Namibia; quartz monzonite, Bingham, Utah; carbonatite, Phalaborwa, South Africa; peralkaline syenite, Kvanefjeld, Greenland; and pegmatite, Madawaska/Bancroft, Ontario (Dahlkamp 1993). Intrusive deposits are generally of low grade but may be large (e.g., Rössing) and they represent ~5% of global reserves and 3% of production.

In granitic and intermediate rocks, most of the radioactivity lies in the accessory minerals zircon, titanite, apatite, allanite, xenotime, and monazite. Minor radioactive accessory minerals include uranothorite, thorianite, euxenite, thorite, pyrochlore, chevkinite, fluorite, bastnaesite, and davidite. The radioactivity usually varies significantly across a single pluton, with concentration of uranium and thorium often seen at the outer margins of the pluton and/or in the last material to crystallize, as expected based on the large size and charge of the U^{4+} and Th^{4+} ions.

Uranium and thorium are fairly common minor constituents in pegmatites. As with the plutonic deposits described previously, pegmatites are rarely of economic value as uranium ore but are the main source for many of the primary uranium and thorium minerals and the U- and Th-containing complex oxides sought by collectors. Pegmatites typically occur in "swarms" or well-defined groups associated with a larger pluton that represents the source of the original magma. Many radioactive species are found in pegmatites; some are primary minerals formed directly from magmatic or hydrothermal processes, whereas others are secondary minerals formed *in situ* through the action of meteoric or ground waters. The primary radioactive minerals often found in pegmatites include uraninite and to a much lesser extent thorianite. The complex mixed oxides (pyrochlore and samarskite groups, the aeschynite series, davidite, chevkinite, and others) are often found, and in fact pegmatites are the main source of many of these minerals. Phosphates such as monazite and, more rarely, xenotime, as well as silicates such as thorite, allanite, gadolinite, and zircon are fairly widespread primary pegmatite minerals. Well-known localities for primary minerals include the region around Bancroft, Ontario; Charlebois Lake and Foster Lake areas, Saskatchewan; the Ruggles pegmatite, Grafton, New Hampshire; and near Idaho Springs, Colorado. In Northern Karelia, Russia, there are at least five separate pegmatites that contain uraninite (Heinrich 1958).

Most secondary uranium minerals found in pegmatites can be attributed to the action of air and water on the primary minerals. The most commonly seen species include autunite, torbernite, and uranophane, along with a mixture of alteration products called gummite; other hydrous uranyl oxides, carbonates, arsenates, and silicates are sometimes found in pegmatites. Noteworthy examples of this type of deposit are the localities in New England; near Spruce Pine, North Carolina; the region between Sabugal and Guarda, Portugal; and Salamanca, Spain.

Carbonatites are magmatic carbonate-silicate rocks that contain mainly calcite or dolomite (and sometimes other carbonate minerals) with lesser amounts of silicates, oxides, and phosphates, and which are thought to have been deposited from a carbonate-rich fluid. They are almost always associated with nepheline syenite ring complexes, which can range up to several km across. Carbonatites often form the lower part of the necks of some explosive-type volcanoes, but can also form other structures such as dikes, cone sheets, and breccia zones. Often these structures are surrounded by a contact metasomatic aureole in which the original country rocks have been altered to syenites by the action of alkalic solutions, a process called *fenitization*. A flow structure can sometimes be seen in the parallel orientation of biotite, apatite, rutile, or amphibole crystals.

Carbonatites have relatively high concentrations of phosphorus, chlorine, fluorine, titanium, zirconium, niobium, cerium-group REE, sulfur, barium, and strontium, along with appreciable thorium and (to a lesser degree) uranium. They are mineralogically complex; typical radioactive species include pyrochlore, monazite, thorite, bastnaesite, thorogummite, xenotime, and betafite. Notable localities include: Alno Island, Sweden; the Fen area, near Telemark, Norway; Tororo, Uganda; Oka, Quebec; thorium deposits in Gunnison Co., Colorado; and at the Sulfide Queen mine in San Bernardino Co., California.

7. Volcanic associated deposits

Volcanic associated deposits occur with felsic extrusive rocks, in continental extensional environments, or in caldera complexes. They are related to faults and shears within the volcanic units and the U is typically associated with Mo, F, Th, and REE (Plant et al. 1999). They form from U-enriched felsic magmas that are later enriched by hydrothermal fluids. Examples range in age from Proterozoic to Tertiary and include: the Streltsovsk caldera, Russia; the McDermitt caldera, Nevada; Nopal I, Peña Blanca district, Mexico; and Xiangshan, China. They have grades of 0.04 to 4% U and represent ~3% of global production (Kyser and Cuney 2008).

The host rock lithology varies from one deposit to another, and may include rhyolite, ignimbrite, dacite flows, rhyolitic breccia, granodiorite, etc. Hydrothermal alteration may be intense and can include devitrification, argillitization, zeolitization, and silicification. At Nopal I, for example, the host rocks have been completely altered. The U minerals occur as disseminations or coatings in a pipe-shaped deposit filled with silicified tectonic breccia ~20 × 40 m wide and >100 m deep (Dahlkamp 1993).

8. Quartz-pebble conglomerate deposits

Quartz-pebble conglomerate deposits represent very ancient alluvial formations, or *paleoplacers*. At the time these deposits were formed, roughly 2-3 Ga ago, there was much less free oxygen in the atmosphere, which allowed the U^{4+} species (uraninite or brannerite) to survive long enough for the placer to become buried under additional sedimentary layers, which then protected them from oxidation as the atmosphere evolved to its present state. The typical mineral assemblage is quartz, gold, pyrite, uraninite, brannerite, zircon, chromite, monazite, leucoxene, and platinum-group minerals (PGM) such as osmium-iridium alloys, isoferroplatinum, and sperrylite. Examples of quartz-pebble conglomerate deposits include: Witwatersrand, South Africa; Elliot Lake, Canada; Jacobina, Brazil; and Tarkwa, Ghana.

Quartz-pebble conglomerates represent ~6% of global U reserves but only ~2% of production. In some cases the gold and PGM values determine the economics of mining. For instance, at Elliot Lake, which was mined solely for U, the ore grade is ~0.14% U, whereas at Witwatersrand the grade is only ~0.008% and the U is recovered as a by-product of gold mining (Kyser and Cuney 2008).

9. Surficial uranium deposits

Surficial deposits represent Tertiary to Recent concentrations of U in sediments or soil. The largest contain much calc tufa or other cementitious phases and would be properly called calcretes. The U is present largely as carnotite. The deposits can form from low-temperature ground water in arid environments or under cool, wet conditions in peat bogs and limestone caverns. They amount to ~2% of global reserves but nearly 5% of global production. Ore grades are variable, with 0.2% U at Yeelirrie, Australia, and 0.06% at Langer Heinrich, Namibia (Kyser and Cuney 2008).

10. Collapse breccia pipe deposits

Collapse breccia pipes occur only in the Arizona Strip and consist of relatively small, circular collapse structures cutting through strata that include limestone, shale, and permeable sandstone. The pipes can be 10–300 m across and up to 1,000 m deep, and are filled with breccia fragments and matrix. The U mineralization (primarily pitchblende, coffinite, and their alteration products) forms matrix impregnations that are somewhat similar to those of sandstone deposits. Associated minerals can include sulfides, arsenides, and locally oxides, carbonates, and sulfates of Fe, Cu, Ni, Co, Mo, Pb, Zn, As, and trace amounts of Ag, Au, Hg, Sb, and V. Gangue minerals include calcite, dolomite, siderite, barite, collophane, chalcedony, quartz, gypsum, and anhydrite. The genesis of these deposits is not completely understood.

Although the individual deposits are small, 2 to 2,500 tU, some are attractive because of their high grade, 0.3-1%. Known reserves in the Arizona Strip district are ~15,000 tU. In addition, Au, Ag, and/or Cu have been recovered from some pipes (Dahlkamp 1993).

11. Phosphorite deposits

Phosphorite deposits include syngenetic U in phosphate-rich sediments, oolitic marine phosphate beds, and U-rich apatite in land-pebble phosphatic sandstone. In most of these, there are no uranium minerals *per se*, but rather cryptocrystalline fluor-carbonate apatite that contains some U. These deposits tend to be fairly low grade on average (<20 to 300 ppm U) and contain elements such as Se and As that would complicate mining and extraction, so they aren't generally exploited, but because of their large size they represent a significant potential resource. Some U is produced as a by-product of phosphate production. The largest deposits are in Morocco; worldwide resources are estimated at 22,000,000 tU (Dahlkamp 1993; Kyser and Cuney 2008).

12. Black shale and seawater

These sources are noted simply because their overall size is very large, but there are presently no plausible means of extracting them economically. For example, the U contained in seawater represents a staggering *4 billion tU*, but the feasibility of extraction would require a price greater than $300/kgU. Similarly, the Chattanooga Shale in the southeastern US, and the Ronneburg black shale, Germany, together contain perhaps 4.2 million tU but at very low grades (<400 ppm), making them unlikely candidates for exploitation.

Uraniferous coals, lignites, and black shales are found in large areas of the western US as well as in Europe and Madagascar. These deposits can be extensive but rarely contain recognizable uranium minerals.

13. Placer deposits

The relative ease with which uraninite is oxidized virtually precludes even highly crystalline material from surviving long enough to be transported intact to a modern placer deposit. Thorium minerals are stable with respect to oxidation, but they are frequently metamict and this tends to reduce their resistance to weathering and alteration. As a result, most radioactive placer deposits of commercial interest contain mixtures of various minerals such as monazite, zircon, thorite, euxenite, samarskite, and xenotime. Individual rounded grains of these minerals make up the bulk of the "heavy sand" concentrates, whose overall composition reflects both the source rock and the intensity of weathering and transport.

A placer deposit of some interest to collectors lies in the interior of Sri Lanka. In the Peak district, small, unabraded black cubes of thorianite are recovered as a by-product of placer gem mining for sapphire, spinel, etc. This occurrence has been known since the early 1900s and yielded over nine tons of thorianite (Heinrich 1956).

PROVINCES AND EPOCHS

There is ample evidence that certain petrographic provinces have their own characteristic radioactivity, just as others might have notable enrichments in other chemical elements. For example, acid igneous rocks from the western margin of the Canadian Shield (e.g., Great Slave Lake and Great Bear Lake areas) are about five times as radioactive as acid rocks from the interior parts of the shield in Ontario and Quebec. Various authors have proposed the concept of a "uranium metallogenic province" as a broad but indefinitely bounded area in which uranium-rich rocks are relatively abundant; the uranium deposits within the province might be of more than one type and age (Klepper and Wyant 1956; Dahlkamp 1993). This concept suggests that the non-uniform distribution of radioactive species arose early in the development of the earth's crust and persists as large-scale features to the present time. Dahlkamp (1993) further defines a *uranium district* (economic) as "areas within uranium provinces containing clusters of deposits, perhaps with the restriction to deposits of the same type."

Similarly, the ages of many uranium deposits seem to be concentrated in three major groups: Laramide and Tertiary; Hercynian; and Precambrian, although there were likely several separate epochs of uranium mineralization during the Precambrian.

Table 5. Major uranium provinces of the world[a]

NORTH AMERICA

Canada
Athabasca Basin - Uranium City region, Saskatchewan
Elliot Lake - Quirke Lake - Blind River, Ontario
Bancroft, Ontario
Great Bear Lake, Northwest Territories
Baker Lake - Thelon Basin, Northwest Territories
Makkovik - Seal Lake, Newfoundland
Kelowna - Beaverdell, British Columbia
United States
Colorado Plateau
Wyoming Basins - Black Hills
South Texas Coastal Plains
Colorado Rocky Mountains/Front Range, Colorado
Northern Rocky Mountains/Mount Spokane region, Washington - Idaho
Appalachian - Piedmont, North Carolina - Virginia - Pennsylvania
Alaska Panhandle/Bokan Mountain, Alaska
Mexico
Sierra de Peña Blanca, Chihuahua
La Sierrita - Burgos Basin, Nuevo Leon

SOUTH AMERICA

Argentina
Sub-Andean Zone
Brazil
Atlantic Shield/Brazilian Orogenic Belt
Paraná Basin
Bolivia - Peru
Andean Volcanic Belt

EUROPE AND RUSSIA

Poland - Germany - Czech Republic - France - Spain - Portugal[b]
Bohemian Massif, Poland - Czech - Germany
Black Forest - Voges, Germany - France
Massif Central - Armorican Massif, France
Cornwall, England
Iberian Meseta, Spain - Portugal
Bulgaria
Western and eastern Balkan metallogenic zone
Russia
Caucasus region
Streltsov, Transbaikal
Sweden
Svecofennian Belt (Arjeplog - Arvidsjaur - Sorsele)
Hotagen - Olden Window
Greenland
Ketilidian Belt, south Greenland
Ukraine
Krivorozhsky/Krivoj Rog region

CENTRAL ASIA

Kazakhstan
Pribalkash
Kokchetav
Pakistan
Eastern Baluchistan (Baghal Chur)
Uzbekistan
Kyzylkum
Karamazar

EAST ASIA

China
Zhungel - Tianshan, northwest China
Yinshan - Liaohe, northeast China
Qilian - Qinling, central China
West Yunnan, southern China
Jiangnan Old Land, southest China
India
Singhbhum Thrust Belt
Japan
Central and western Honshu (Tono-Ningyo toge)

AUSTRALIA

Pine Creek Geosyncline, Northern Territory
Westmoreland - Pandanus Creek, Northern Territory - Queensland
Mount Isa Geosyncline, Queensland
Georgetown Inlier, Queensland
Lachlan Fold Belt, Queensland
Arunta Block - Ngalia Basin - Amadeus Basin, Northern Territory
Lake Frome Embayment, South Australia
Stuart Shelf, South Australia
Yilgarn Block, Western Australia

AFRICA

Algeria - Niger
Hoggar - Air Massif - Agades Basin
Congo - Zambia
Katanga Synclinorium
Gabon
Massif de Chaillu - Franceville Basin
Namibia
Damara Belt
South Africa
Witwatersrand Basin - Kaapvaal Craton

[a]Data from Dahlkamp (1993)
[b]These provinces are part of the Moldanubian and Saxo-Thuringian metallogenic zone of the Hercynian orogenic belt and adjoining continental sedimentary basins.

IMPORTANT RADIOACTIVE MINERAL LOCALITIES

Although many, if not most, countries have at least some U occurrences, this chapter will focus on countries that have significant U resources and/or have produced specimens that mineral collectors are likely to encounter. For each of these countries, tabular data are given, where practical, on the size, location, and status of large uranium resources that are known to exist, and, where appropriate, a map showing the location of major mines or districts. The data are mainly from Finch et al. (1995) and IAEA (2009), but have been updated wherever possible with information from recent editions of the OECD Red Book or other sources.

Locality discussions emphasize occurrences that are of particular interest to the collector because of the mineral specimens they have produced. Type localities (TL) are noted for any radioactive minerals first described from the country. These are presented in chronological order of their descriptions, partly for historical interest and partly to convey a sense of the general levels of research activity over time.

It must be emphasized that the information herein is absolutely not intended to be used as a guide for field collectors seeking specific places to collect and directions to the sites! The reasons should be obvious: First, active mines normally don't grant access to mineral collectors. Second, many notable mines have been depleted, closed, and often completely reclaimed to the point that not even waste dumps exist on the surface today. Third, most localities are either on private property or on government-protected lands or parks. And perhaps most importantly, not only are abandoned mines inherently unsafe, but for sites in the developing world the most dangerous part is simply traveling in the host country.

Note that all references in this book to "Congo" refer to the Democratic Republic of Congo (formerly "Belgian Congo" or "Zaire").

NORTH AMERICA

CANADA

Table 6. Selected uranium deposits in Canada[c]

Deposit	Type and Method[a]	Status[b]	Size (tU)
Saskatchewan			
Beaverlodge	Vein, U	DEP	20,000–50,000
Cigar Lake	Unconformity, U	DOR	>100,000
Cluff Lake	Unconformity, U and OP	DEP	20,000–50,000
Collins Bay	Unconformity, OP	DEP	20,000–50,000
Dawn Lake	Unconformity, OP	DOR	5,000–20,000
Eagle Point	Unconformity, U	PRO	20,000–50,000
Gunnar	Vein, OP	DEP	5,000–20,000
Key Lake	Unconformity, OP	PRO	>50,000
Lorado	Vein, U	DOR	500–1,500
McArthur River	Unconformity, U	PRO	>100,000
McClean Lake	Unconformity, U	PRO	0,000–25,000
Midwest	Unconformity, U	DOR	5,000–20,000
Rabbit Lake	Unconformity, OP	DEP	5,000–20,000
West Bear	Unconformity	DOR	500–1,500
Ontario			
Agnew Lake	Quartz-pebble conglomerate, U	DEP	1,500–5,000
Bicroft	Intrusive pegmatite, U	DEP	1,500–5,000
Denison	Quartz-pebble conglomerate, U	DOR	>50,000
Madawaska	Intrusive, U	DEP	5,000–20,000
Stanleigh	Quartz-pebble conglomerate, U	DOR	20,000–50,000
Northwest Territories			
Eldorado	Vein, U	DEP	1,500–5,000
Kiggavik	Unconformity, OP	DOR	5,000–20,000
Labrador and Newfoundland			
Kitts	Volcanic, U and OP	DOR	1,500–5,000
Michelin	Volcanic, OP	DOR	5,000–20,000
British Columbia			
Blizzard	Sandstone, OP	DOR	1,500–5,000

[a] U = underground mining; OP = open-pit mining; ISL = *in situ* leaching; HL = heap leaching; IPL = in-place leaching; CP = extraction as a co-product or by-product of mining for another metal; NS = not specified.

[b] DEP = depleted; DOR = dormant; PRO = producing

[c] There are ~120 known U deposits in Canada, of which ~40 were in the exploration or development phase as of 2009. A selection of well-defined deposits is presented here.

Data primarily from Finch et al. (1995) and IAEA (2009).

Uranium mining in Canada began with the discovery in 1930 of the Port Radium deposit on the eastern shore of Great Bear Lake, Northwest Territories. The deposit is primarily mesothermal veins of the Ag-Co-Ni type with Bi and Cu, containing veins and stringers of pitchblende. The Eldorado mine was worked from 1933 to 1940 and reopened in 1942 to supply uranium to British and American defense programs. The nearby Contact Lake mine was a small producer of native silver and pitchblende, and many similar deposits have been reported from the district (Heinrich 1958).

Figure 19. Important uranium localities and districts in Canada:
1 Blizzard
2 Athabasca Basin: Beaverlodge, Gunnar, Lorado, Maurice Bay
3 Athabasca Basin: Cigar Lake, Eagle Point
4 Athabasca Basin: Collins Bay
5 Athabasca Basin: Dawn Lake, Key Lake, McArthur Lake, McClean Lake, Midwest, Rabbit Lake, West Bear
6 Great Bear Lake: Eldorado, Port Radium
7 Kiggavik
8 Michelin
9 Kitts
10 Blind River - Elliot Lake district: Agnew Lake, Denison, Stanleigh
11 Bicroft and Madawaska

The Blind River-Elliot Lake district, Ontario, was discovered in 1949 and produced uranium for more than forty years. The deposit is a quartz pebble conglomerate and although the ore was fairly low grade, the favorable location and low processing costs made the deposit economical to exploit (Heinrich 1958). The last production facility at Elliot Lake closed in 1996.

At present, all active uranium production is from unconformity-related deposits in the Athabasca basin, northern Saskatchewan. The Athabasca basin represents the most significant high-grade, low-cost U resource currently mined anywhere in the world, and is estimated to contain ~375,000 tU. These deposits formed through the circulation of saline diagenetic brines in the middle Proterozoic Athabasca sedimentary basin, and the redox interaction of these oxidized brines with reduced basement fluids. The circulation within the sandstone and basement of reactive, oxidized, highly saline, Ca-Na-Mg brines, generated at the base of the thick,

Figure 20. Hand sample of quartz-pebble conglomerate from the Elliot Lake district, Ontario. The pebbles range up to ~2 cm; darkening of the matrix is caused by minute inclusions of brannerite, sulfides, and other minerals. RJL4528

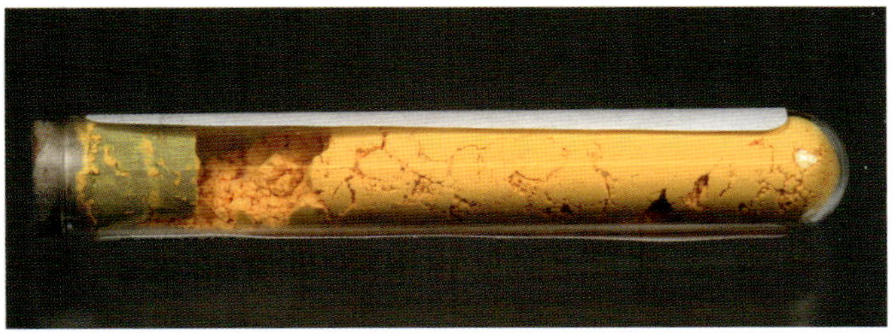

Figure 21. A sample of yellow cake, powdered sodium diuranate, the product of uranium ore processing by the Eldorado Mining & Refining Co., Ltd., Beaverlodge Operations, Uranium City, Saskatchewan. This historical specimen was obtained by E. Wm. Heinrich around 1950. RJL4279

Athabasca sandstone basin, which was relatively free of organic matter and K-feldspar-free, caused the scavenging of uranium even from refractory minerals, such as monazite. The nature of the uranium precipitation mechanism isn't well understood, but probably involves a reduction process. The uranium ore occurs close to the unconformity that separates the Athabasca Group sediments and the underlying Archean-Paleoproterozoic crystalline basement rocks, but some occurs as much as 450 m into the basement, (e.g., in the Eagle Point deposit). The ore zones are spatially controlled by faults that crosscut this unconformity and are rooted in graphite-rich metamorphic rocks (Gaboreau et al. 2007).

Within the Athabasca district, the McArthur River mine is the world's largest uranium mine. Canada is presently the world's second-largest uranium producer. Annual output in 2011, around 8,600 tU, represents about 18% of the global total.

Mines in the Athabasca district sometimes contain weathered vein material in which various secondary minerals occur, e.g., uranophane, metazeunerite, cuprosklodowskite, kasolite, etc. (Hogarth 1951).

For collectors, the unique deposits of the Haliburton-Bancroft area, Ontario, are of particular interest. Heinrich (1958, pp. 217-21) notes that they comprise lenticular fissure fillings, which various authors have called syenite pegmatite, calcite-fluorite vein-dikes, calcite-fluorite pegmatites, and calcite-fluorite-apatite veins. These deposits are hosted in various Grenville metasedimentary rocks, banded gneisses, syenitized gneisses, and syenite, which are cut by syenitic or granitic pegmatites. The deposits range from ~30 cm to 4 m thick. The main radioactive mineral is uraninite containing high levels of REE and Th (11-13% ThO_2), which typically forms large cubic crystals in the central granular calcite-fluorite-apatite rock. Associated minerals include biotite, titanite, scapolite, magnetite, tourmaline, hematite, and various sulfides, along with radioactive pyrochlore, allanite, and zircon.

Figure 22. Ore sample from the Rabbit Lake mine, Athabasca basin, Saskatchewan, showing orange crusts of masuyite, with kasolite and uranophane (yellow) and black uraninite. This sample has been in the author's collection for over thirty years. RJL583

Figure 23. Crude black uraninite crystal in calcite with dark grains of fluorite and yellow-green apatite, from the Cardiff Uranium mine, Wilberforce, Ontario, Canada. RJL579

Figure 24. Felted masses of pale yellow uranophane-β, from the Faraday mine, Bancroft, Ontario. RJL218

The Richardson deposit, near Wilberforce, was discovered in 1922 and explored off and on through 1949. It has been traced for nearly 1 km and shows the following zoning (Heinrich 1958):

1. A thin, fine-grained border zone of equigranular red feldspar with accessory magnetite, uraninite, biotite, hornblende, titanite, etc.

2. A wall zone of red oligoclase-antiperthite crystals up to ~10 cm, with minor uraninite, magnetite, zircon, tourmaline, and fluorite.

3. A somewhat discontinuous intermediate hornblende zone, with hornblende crystals occasionally reaching 2 m long.

4. A core of banded, granular calcite, fluorite, and apatite.

Within the intrusives and the metamorphosed Grenville limestones are found several dozen radioactive species, of which fine crystals of betafite from the Silver Crater mine are perhaps best known to collectors. The Madawaska mine (formerly Faraday mine) has produced world-class uranophane and uranophane-β specimens. Excellent simple or modified cubic uraninite crystals are found at the Cardiff mine. Robinson and Sabina (1955) reported a systematic study of uraninite and thorianite occurrences at a large number of localities in the Grenville province, including the Cardiff mine. Other species noted from the district include: euxenite, fergusonite, thorianite, thorite, uranpyrochlore, and zircon var. *cyrtolite*, a radioactive variety characterized by curved faces and pale tan color (Kennedy 1979).

Figure 25. Complex betafite crystals forming a circular group ~2 cm wide, on corroded hornblende from the Halo uranium mines, Hogan property, Haliburton Co., Ontario. RJL2151

Figure 26. A large group of radioactive zircon crystals (cyrtolite) from the Davis mine, Bancroft, Ontario. Note the slightly curved faces that are typical of this variety. RJL2660

MEXICO

Table 7. Known uranium deposits in Mexico

Deposit	Type and Method[a]	Status[b]	Size (tU)
Baja California			
Santo Domingo	Phosphorite	DOR	>50,000
Chihuahua			
Peña Blanca–El Nopal	Volcanic, OP	DOR	1,500–5,000
Durango			
Coneto	Volcanic	DOR	500–1,000
Nuevo Leon			
El Chapote-Diana	Sandstone	DOR	1000–2,500
La Coma	Sandstone	DOR	500–1,000
La Sierrita	Sandstone, OP and U	DOR	2,500–5,000
Oaxaca			
Santa Caterina Tayata	Volcanic	DOR	<500
Sonora			
Los Amoles	Volcanic, OP and U	DOR	1000–2,500

[a] U = underground mining; OP = open-pit mining; ISL = *in situ* leaching; HL = heap leaching; IPL = in-place leaching; CP = extraction as a co-product or by-product of mining for another metal; NS = not specified.

[b] DEP = depleted; DOR = dormant; PRO = producing

Data primarily from Finch et al. (1995) and IAEA (2009).

The Sierra Peña Blanca district, which lies 50 km north of Chihuahua City, is an example of uranium deposits related to volcanic rocks. Three isolated deposits have been studied extensively and each has its own characteristic genesis: Nopal I is a hydrothermal type; Las Margaritas is a mixed supergene-exhalative type; and Puerto 3 is a supergene type (Godell 1981; Aniel and Leroy 1985; George-Aniel, Leroy, and Poty 1991). Many secondary uranium species are found in the Peña Blanca deposits, particularly uranophane, uranophane-β, and weeksite. Las Margaritas is the type locale for margaritasite, the Cs analogue of carnotite (Wenrich et al. 1982).

Similar U-rich volcanics occur in the Chihuahua-Sacramento valley. Analysis of samples from the San Marcos area found uranophane, tyuyamunite, metatyuyamunite, and uraninite (Reyes-Cortes et al. 2007).

Figure 28. An ore sample from Las Margaritas, showing yellow secondary minerals, primarily uranophane, on rhyolitic matrix. Sample is 7 cm wide.

Figure 27. Important uranium localities in Mexico:
1 Santo Domingo, Baja California
2 Peña Blanca-El Nopal, Chihuahua
3 Coneto, Durango
4 El Chapote-Diana, Nuevo Leon
5 La Sierrita, Nuevo Leon
6 Santa Catarina Tayata, Oaxaca
7 Moctezuma, Sonora
8 Santa Eulalia, Chihuahua

Table 8. Radioactive minerals first described from Mexico

Species	Type Locale	Year
Moctezumite	Moctezuma mine, Sonora	1965
Cliffordite	San Miguel mine, Sonora	1969
Schmitterite	Moctezuma mine, Sonora	1971
Margaritasite	Margaritas mine, Peña Blanca, Chihuahua	1982

There are several localities of interest to mineral collectors:

The Moctezuma mine (locally called "La Bambolla"), near Moctezuma, Sonora, is a gold-tellurium deposit that was mined for gold in the 1930s and '40s. A large number of new minerals have been described from this mine and the nearby San Miguel mine, including three uranyl tellurites: cliffordite, schmitterite, and moctezumite.

Elsewhere in Sonora, excellent specimens of torbernite are found at La Luz. Sklodowskite, haiweeite, and nováčekite have been reported from several locales in the Santa Eulalia district, Chihuahua.

Figure 29. Cliffordite (yellow earthy blebs) with mackayite (olive green), on matrix from Moctezuma, Sonora, Mexico. Mackayite is $Fe^{3+}Te_2O_5(OH)$. RJL338

Figure 31. Torbernite crystals covering a surface ~3 × 4 cm, from La Luz, Sonora, Mexico. RJL2860

Figure 30. Crudely crystallized moctezumite forming a surface coating on matrix from the TL at the Moctezuma (Bambolla) mine, Sonora, Mexico, showing the range of color from bright orange to brownish. RJL2891

UNITED STATES

There are about 1,000 documented radioactive mineral occurrences in the United States, and many have been thoroughly studied (Heinrich 1958). Presenting tabular data here for such a large number of mines is impractical; however, the reader can refer to Finch et al. (1995) and IAEA (2009) for tabular summaries of about 180 of the commercial deposits.

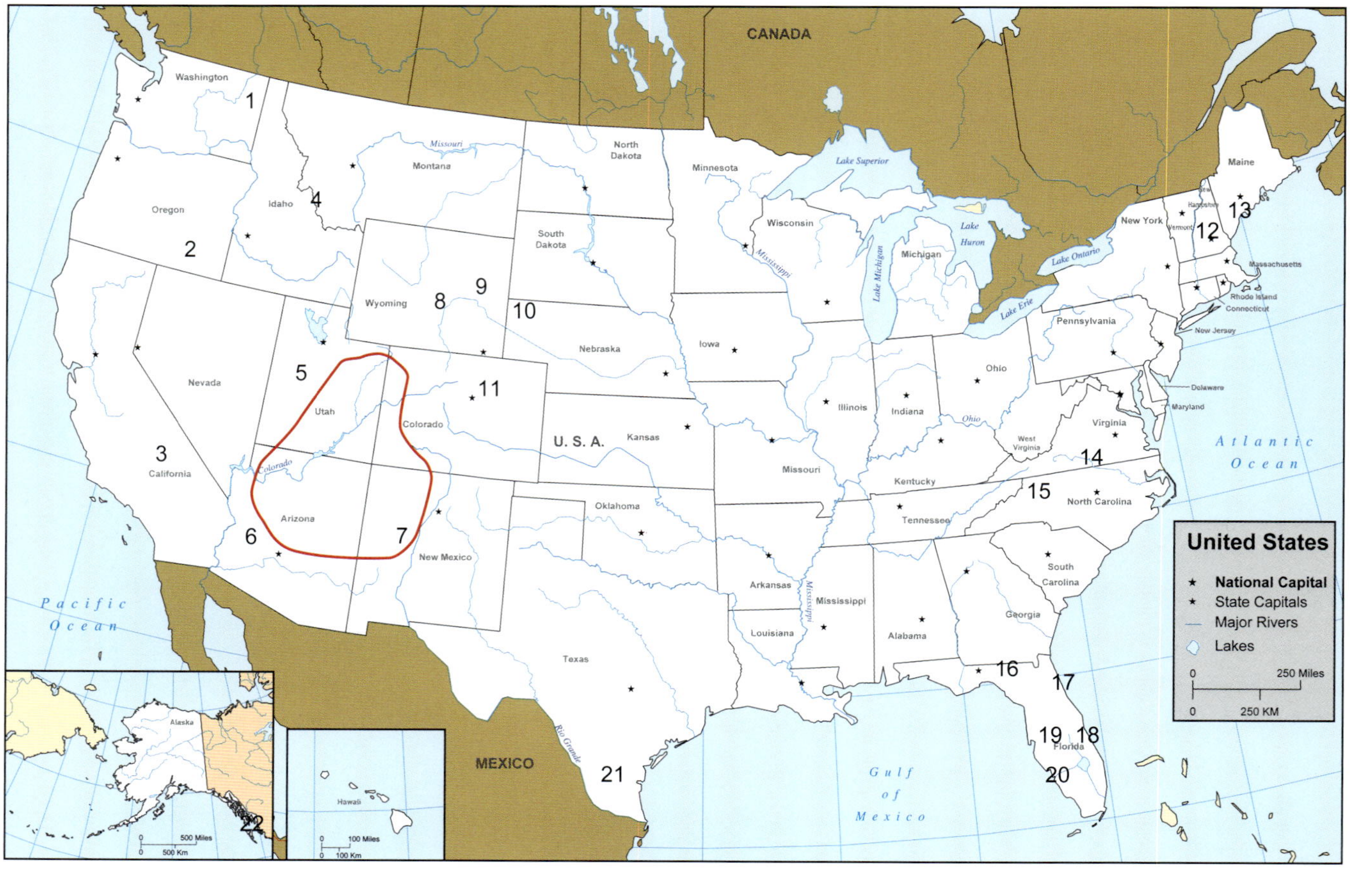

Figure 32. Important uranium localities and districts in the United States (the Colorado Plateau is outlined in red):
1 Spokane Co., Washington: Dahl, Daybreak, Huffman, Midnite, Sherwood
2 White King mine, Oregon
3 Haiwee Reservoir, California
4 Lemhi Pass, Idaho-Montana
5 Thomas Range, Utah
6 Anderson, Arizona
7 Grants - Ambrosia Lake, New Mexico
8 Lucky Mc mine, Wyoming
9 Bear Creek, Wyoming
10 Crow Butte, Nebraska
11 Schwartzwalder mine, Colorado
12 Ruggles mine, New Hampshire
13 Topsham, Maine: Russell Brothers quarry, Swamp quarries, Yedlin locality
14 Swanson, Virginia
15 Spruce Pine, North Carolina
16 North Florida district (phosphorites)
17 Northeast Florida district
18 East Florida district
19 Central Florida district
20 South Florida district
21 Gulf Coast basin: Alta Mesa, Alta Verde, Benevides, Bruni, Burns Ranch, Butler-Weddington, Chapman, Clay West, Holiday-El Mesquite, Kingsville Dome, Lamprecht-Felder, Mt. Lucas, O'Hearn, Rhode Ranch, Rosita
22 Bokan Mountain, Alaska

Table 9. Characteristics of uranium deposits in the United States[a]

Type	Number	Average size (tU)
Sandstone	152	2,500–5000
Vein	5	2,000–3,000
Collapse breccia	11	1,000–2,500
Volcanic	3	5,000–10,000
Phosphorite	5	100,000
Intrusive	1	500–1,000
Metasomatite	1	10,000–25,000
[a]Data from IAEA (2009)		

One can see from Table 9 that in terms of sheer number of deposits and their size, sandstone deposits are by far the largest resource and host the only operating mines in the US at this time. Individual phosphorite deposits are large but fairly low grade, making them generally uneconomic. Estimated sandstone resources in the US (OECD 2012) range from 39,000 tU (recoverable at <$80/kgU) to ~400,000 tU (recoverable at <$260/kgU).

The Colorado Plateau contains well over a thousand documented uranium occurrences, including small prospects, unnamed sites, and mines in all stages from active to depleted. Most but not all of the resources are sandstone deposits (the huge Schwartzwalder mine being a notable exception). The localities may be organized into various districts, with the understanding that a district is not precisely defined and there is occasionally some overlap, as can be seen from the following lists:

Figure 33. Detail of the distribution of uranium districts and mines of the Colorado Plateau and surrounding areas:
1 Paradox Basin, Colorado-Utah
2 Repete mine, Utah
3 White Canyon district, Utah
4 San Rafael Swell, Utah
5 Thomas Range, Utah
6 Uravan district, Montrose Co., Colorado
7 Grants - Ambrosia Lake district, New Mexico
8 San Juan Basin
9 Sun valley mine, Coconino Co., Arizona
10 Hillside mine, Yavapai Co.., Arizona
11 Schwartzwalder mine
12 Pitch mine
13 Wamsutter, Sweetwater Co., Wyoming
14 Laramide Basin, Wyoming - Colorado

The Paradox Basin extends from western Colorado into eastern Utah. Some important deposits on the Colorado side include: Carpenter Flats, Calamity - Outlaw - Flaptop, John Brown, King Solomon, Legin - Charles T - Deremo, Long and Horse Mesas, Long Park, Lower and Radium groups, Monogram Mesa, Sharp Canyon, and Slick Rock. Utah localities include: Abbe, Happy Jack, Hershey, Hideout, Shinarump, and Vanura.

The Uravan district, Montrose Co., Colorado, extends north and east of the Paradox district and includes hundreds of documented occurrences; it is also the type locality for carnotite and metatyuyamunite.

The White Canyon district, Utah, is subdivided into Red Canyon and White Canyon. Red Canyon localities include: Chocolate Drop group, Blue Lizard, Bullseye, Copperhead open pit, George H. group, Giveaway - Simplot, Joe Bishop, Posey, Radium King - Allen group, Windy Day, and Yankee Girl. White Canyon localities include: Badger, Deer Flat area, Jomac, Happy Jack, Hopefull, Lone Butte, Notch, White Canyon No. 1, Widowmaker, and Yellow Jack No. 10.

The San Rafael Swell, Emery Co., Utah, includes: Temple Mountain, Allied uranium mines, Block Mountain mine, Green Vein group, Hertz, Moroni Hunt, School Section No. 36, Snow claim, and Wickiup.

The Grants - Ambrosia Lake district, New Mexico, includes: Ann Lee, the F 33 mine, Flat Top, Haystack, Hanosh, Homestake, Marquez, Mt. Taylor, San Fidel, and Sections 15, 22, 23, 25, 28, 30, 33, and 35.

The San Juan Basin, New Mexico, includes: Bernabe - Montano, Borrego Pass, Church Rock, Cliffside (Section 36), Crownpoint, Dalton Pass, Faith, Jackpile, Lee, Mariano Lake, Marquez Canyon, Nose Rock, Pinenut, Poison Canyon, Rio Puerco, Ruby, San Antonio Valley, and West Largo.

The Laramide Basin covers a very large area of southern Wyoming and extends into far northern Colorado, and includes: Bear Creek, Bison Basin, Bill Smith, Cedar Hills, Christenson Ranch, Highland, Irigary, Leuenberger, Moore Ranch, Nine Mile Lake, North Butte, Peterson Ranch, Petrotomics/Dave, Reno Creek, Ruth, Ruby Ranch, South Morton Ranch, and Shirley Basin in Wyoming, and Rob Rollo in Colorado.

About three dozen radioactive minerals were first described from localities in the US, including at least six in the last decade, Table 10.

Several districts may be considered classic localities based on their historical importance, quality of the specimen material, and familiarity to the collecting community.

In New England, swarms of radioactive pegmatites have been known since the late 1800s. Two distinctive types of specimens are especially noteworthy: At the Swamp mine, Topsham, Maine, uraninite forms sharp, complex crystals, typically up to a centimeter across. At the Ruggles pegmatite, Grafton Center, New Hampshire, uraninite forms dendritic aggregates in white feldspar, and colorful "gummite" crusts are formed where these rocks are exposed to surface weathering.

Table 10. Radioactive minerals first described from the United States

Species	Type Locale	Year
Carnotite	Montrose Co., Colorado	1899
Phosphuranylite	Flat Rock pegmatite, North Carolina	1879
Uvanite	Temple Rock, Emery Co., Utah	1914
Rauvite	Temple Rock, Emery Co., Utah	1922
Clarkeite	Spruce Pine district, North Carolina	1931
Andersonite	Hillside mine, Yavapai Co., Arizona	1951
Bayleyite	Hillside mine, Yavapai Co., Arizona	1951
Swartzite	Hillside mine, Yavapai Co., Arizona	1951
Umohoite	Freedom No. 2 mine, Piute Co., Utah	1953
Coffinite	La Sal No. 2 mine, Mesa Co., Colorado	1954
Rabbittite	Lucky Strike No. 2 mine, Emery Co., Utah	1955
Abernathyite	Fuemrol No. 2 mine, Emery Co., Utah	1956
Boltwoodite	Delta mine, Emery Co., Utah	1956
Metatyuyamunite	Jo Dandy mine, Montrose Co., Colorado	1956
Heinrichite	White King mine, Lakeview Co., Oregon	1958
Metaheinrichite	White King mine, Lakeview Co., Oregon	1958
Weeksite	Autunite No. 8 claim, Juab Co., Utah	1960
Coconinoite	Sun Valley mine, Coconino Co., Arizona	1966
Zellerite	Lucky Mc mine, Fremont Co., Wyoming	1966
Metazellerite	Lucky Mc mine, Fremont Co., Wyoming	1966
Colbaltzippeite	Happy Jack mine, Emery Co., Utah	1976
Magnesiozippeite	Lucky Strike No. 2 mine, Emery Co., Utah	1976
Natrozippeite	Happy Jack mine, Emery Co., Utah	1976
Zinczippeite	Hillside mine, Yavapai Co., Arizona	1976
Haiweeite	Haiwee Reservoir, Inyo Co., California	1981
Metahaiweeite	Haiwee Reservoir, Inyo Co., California	1981
Haynesite	Repete mine, San Juan Co., Utah	1991
Blatonite	Jomac mine, San Juan Co., Utah	1998
Oswaldpeetersite	Jomac mine, San Juan Co., Utah	2001
Larisaite	Repete mine, San Juan Co., Utah	2004
Holfertite	Starvation Canyon, Thomas Range, Utah	2006
Markcooperite	Aga mine, Otto Mountain, California	2010
Meisserite	Blue Lizard mine, Utah	2013
Belakovskiite	Blue Lizard mine, Utah	2014
Bluelizardite	Blue Lizard mine, Utah	2014
Plašilite	Blue Lizard mine, Utah	2014
Bobcookite	Blue Lizard mine, Utah	2015
Fermiite	Blue Lizard mine, Utah	2015
Oppenheimerite	Blue Lizard mine, Utah	2015
Wetherillite	Blue Lizard mine, Utah	2015

Figure 34. A complex uraninite crystal, ~1 cm tall, with attached quartz and feldspar, from Swamp mine, Maine. RJL2281

Figure 35. A representative sample of dendritic uraninite with orange-yellow gummite from the Ruggles mine, New Hampshire. RJL3130

Figure 36. Bright green, scaly torbernite on granitic rock from Spruce Pine, North Carolina. RJL2862

The radioactive pegmatite at Spruce Pine, North Carolina, continues to yield excellent specimens of nicely crystallized torbernite as well as other colorful secondary minerals such as uranophane, meta-autunite, and fourmarierite.

In the western US, the Colorado Plateau region hosts commercially important uranium ore deposits and is also a rich source of uranium minerals for the collector. Species first described from the Colorado Plateau are: abernathyite, andersonite, bayleyite, blatonite, boltwoodite, carnotite, haynesite, holfertite, larisaite, metatyuyamunite, oswald-peetersite, rabbittite, swartzite, umohoite, and weeksite.

Commercially significant deposits of autunite and uranophane were discovered in the Mt. Spokane area, Washington, in the 1950s. Rich mineralized veins extend along a contact between granite and argillites, and in places autunite is found in the granite as much as 6 m from the contact. The Daybreak mine, in particular, is noteworthy for the quality and extremely large size of its autunite crystals. In the best specimens, dark green tabular crystals form beautiful fanlike aggregates or "books" several cm across, and are arguably the world's finest examples of the species.

Figure 37. Typical sample of sandstone-type ore, in which the sand grains are cemented together by yellow carnotite, from the Grants district, New Mexico. RJL69

Figure 38. Tabular autunite forming radiating "books" from Spokane, Washington, shown in natural light. RJL 659

Figure 39. The sample in the previous photo shown in LWUV light. RJL659

SOUTH AMERICA

ARGENTINA

Table 11. Known uranium deposits in Argentina

Deposit	Type and Method[a]	Status[b]	Size (tU)
Cerro Solo, Chubut	Sandstone, OP	DOR	1,500–5,000
Don Otto, Salta	Sandstone, OP	DEP	500–1,500
Dr. Baulles, Mendoza	Sandstone, OP	DOR	10,000–25,000
Huemul, Mendoza	Sandstone, U	DEP	500–1,500
La Estela, San Luis	Vein, OP	DEP	500–1,500
Rodolfo, Cordoba	Surficial, OP	DOR	1,000–2,500
Schlagintweit, Cordoba	Vein, OP	DOR	1,000–2,500

[a] U = underground mining; OP = open-pit mining; ISL = *in situ* leaching; HL = heap leaching; IPL = in-place leaching; CP = extraction as a co-product or by-product of mining for another metal; NS = not specified.

[b] DEP = depleted; DOR = dormant; PRO = producing

Data primarily from Finch et al. (1995) and IAEA (2009).

Figure 40. Important uranium localities in Argentina:
1 Don Otto
2 Rodolfo
3 Schlagintweit
4 Dr. Baulles
5 Huemul
6 Cerro Solo

Uranium exploration in Argentina began in 1951-52 and led to the discovery of the sandstone deposits at Huemul, Don Otto, and Los Berthos. Airborne surveys in the late 1950s and early 1960s identified the Los Adobes sandstone deposit in Patagonia. Vein deposits at Schlagintweit and La Estela were discovered and mined in the 1960s.

In the 1970s, follow-up exploration near the known deposits in Patagonia identified new deposits at Cerro Condor and Cerro Solo. Uranium production has been virtually dormant since around 1999, but exploration activities have picked up since around 2007.

An early description of the sandstone deposit at Huemul was given by Gordillo et al. (1966): "The uraniferous ore-bodies are located in conglomerates, arkosic conglomerates, and arkosic sandstone beds of the lower portion of the Diamantiano. The sandstones are generally enclosed between beds of siltstone and claystone; they are always associated with an asphaltic material originating, it is thought, from the oil in the adjacent small oil field of Pampa Amarilla . . . Uraninite is the principal uranium mineral; it occurs with copper and iron sulfides such as chalcopyrite, chalcocite, covellite, bornite, idaite, pyrite, marcasite, and small amount of galena and sphalerite . . . No primary vanadium minerals are known yet. It is thought that the probable source of this element is the asphaltic material. In the alteration zone above the water table, many hydrated yellow minerals like tyuyamunite, sengierite, carnotite, cuprosklodowskite, autunite, bayleyite, andersonite, etc., are found in close association with copper minerals such as malachite, azurite and chalcantite."

Interesting mineral specimens have been produced from time to time. Perhaps the most notable find is the very rare species uranosilite, which occurs with weeksite on coffinite ore at Huemul, Mendoza.

Most of the sandstone deposits have mineral assemblages that one would expect in such settings. For example, minerals reported at Cerro Solo include autunite, carnotite, tyuyamunite, and uranophane.

Figure 41. A sample of coffinite ore, from Mendoza, Argentina, covered with minute yellow blebs of uranosilite and darker yellow weeksite. RJL3808

BRAZIL

Table 12. Known uranium deposits in Brazil

Deposit	Type and Method[a]	Status[b]	Size (tU)
Amorinopolis	Sandstone	DOR	1,500–5,000
Espinharas, Paraiba	Metasomatic, U and OP	DOR	5,000–20,000
Figuera, Parana	Sandstone, OP	DOR	5,000–20,000
Itataia, Ceara	Metasomatic, OP	DOR	>50,000
Lagoa Real, Bahia	Metasomatic, U and OP	PRO	>50,000
Poços de Caldas, Minas Gerais	Volcanic, OP	DOR	20,000–50,000

[a] U = underground mining; OP = open-pit mining; ISL = *in situ* leaching; HL = heap leaching; IPL = in-place leaching; CP = extraction as a co-product or by-product of mining for another metal; NS = not specified.

[b] DEP = depleted; DOR = dormant; PRO = producing

Data primarily from Finch et al. (1995).

Uranium prospecting, which began in 1952, led to the discovery of significant deposits at Poços de Caldas, Minas Gerais, and at Jacobina, Bahia. The Jacobina deposit is a quartz-pebble conglomerate similar to those in Witwatersrand, South Africa, and Blind River, Canada (Bateman 1958). Further exploration in the 1970s led to the discovery and evaluation of eight deposits, of which Osamu Utsumi and Lagoa Real were developed into mining projects. Although Brazil holds significant uranium resources, annual mine output is fairly low, partly because the economics of some large phosphorite deposits will depend on the markets for the phosphate minerals from which uranium will be extracted as a by-product.

Figure 42. Important uranium localities in Brazil:
1 Itataia, Ceara
2 Lagoa Real, Bahia
3 Poços de Caldas, Minas Gerais
4 Figuera, Parana
5 Jacobina
6 Teofilo Otoni; Malacachaeta; Corrego do Urucum; São Jose da Safira; Jaguaraçu

Table 13. Radioactive minerals first described from Brazil

Species	Type Locale	Year
Coutinhoite	Córrego do Urucum, Minas Gerais	2004
Carlosbarbosaite	Jaguaraçu pegmatite, Minas Gerais	2012

Several noteworthy localities in Brazil have produced fine specimens for collectors:

The Corrego do Urucum pegmatite, Galileia, Minas Gerais is the type locale for coutinhoite. Phosphuranylite, saleeite, uraninite, uranocircite, meta-uranocircite, weeksite, and wölsendorfite are also found there.

The uranium deposits at Teofilo Otoni, Minas Gerais, have produced superb specimens of haiweeite, along with autunite, parsonsite, and uranocircite. Elsewhere in Minas Gerais, well-crystallized autunite is found at Malacacheta and at São Jose da Safira.

Figure 43. Meta-autunite crystals to ~6 mm on pegmatite, from São Jose da Safira, Minas Gerais. RJL2565

Figure 44. Autunite crystals forming a stacked aggregate ~3 cm wide, from the São Pedro mine, Minas Gerais. RJL1559

The granitic pegmatite at Perus, São Paulo, has yielded numerous radioactive species, including: autunite, meta-autunite, bassetite, chernikovite, haiweeite, phosphuranylite, phurcalite, torbernite, metatorbernite, uranophane, uranophane-β, and weeksite.

Fine crystals of nováčekite are found at the Brumado mine, Bahia, along with sklodowskite, thorutite, zeunerite, and metazeunerite.

EUROPE

CZECH REPUBLIC

Table 14. Known uranium deposits in the Czech Republic

Deposit	Type and Method[a]	Status[b]	Size (tU)
Brevniste, Bohemia	Sandstone, U	DEP	10,000–25,000
Brzikov, Western Moravia	Vein, U	DOR	1,500–5,000
Dylen	Vein, U	DEP	2,500–5,000
Hamr, Bohemia	Sandstone, U	DOR	25,000–50,000
Horni Slavkov	Vein, U	DEP	2,500–5,000
Hvezdov, Northern Bohemia	Sandstone, U and ISL	DOR	10,000–25,000
Jachymov	Vein, U	DEP	10,000–25,000
Jasenice	Vein, U and OP	DEP	2,500–5,000
Javornik-Zalesi	Vein, U	DEP	2,500–5,000
Licomerice-Brezinka	Vein, U	DEP	2,500–5,000
Mimon, Northern Bohemia	Sandstone	DEP	10,000–25,000
Okrouhla Radoun, Moravia	Vein, U	DEP	2,500–5,000
Olsi, Southern Moravia	Vein, U	DEP	2,500–5,000
Osecna-Kotel, Bohemia	Sandstone, U and ISL	DOR	25,000–50,000
Pribram, Central Bohemia	Vein, U	DEP	25,000–50,000
Rožna, Southern Moravia	Vein, U	PRO	10,000–25,000
Straz, Northern Bohemia	Sandstone, ISL	PRO	25,000–50,000
Vitkov II	Vein, U	DEP	10,000–25,000
Vnitrosudetska Panev	Coal, U	DEP	2,500–5,000
Zadni Chodov	Vein, U	DEP	10,000–25,000

[a] U = underground mining; OP = open-pit mining; ISL = *in situ* leaching; HL = heap leaching; IPL = in-place leaching; CP = extraction as a co-product or by-product of mining for another metal; NS = not specified.

[b] DEP = depleted; DOR = dormant; PRO = producing

Data primarily from Finch et al. (1995) and IAEA (2009).

The Erzgebirge ("Ore Mountains") forms a natural boundary between Saxony in Germany and Bohemia in the Czech Republic. Mining in this region began with the discovery of silver at Freiberg in 1170 and has continued to the present. In the 1700s mining was directed mainly to cobalt for ceramic colorants. In the 1800s, important products were cobalt and bismuth, with silver as a valuable by-product. Uranium ore was first mined at Jachymov strictly as a source of radium. Under Soviet control in the period following World War II, peak production of uranium from the Erzgebirge represented perhaps 3-4% of worldwide production (Heinrich 1958). The location of Jachymov is also placed on the map of Germany to show its relationship to the rest of the Erzgebirge.

Ore deposits in the district are generally classified as veins associated with granite, and include: pyrometasomatic iron-copper deposits; tin-tungsten veins; pyritic lead veins; silver-cobalt-bismuth-uranium mineral veins; and veins of iron and manganese oxides. In these veins, all of the uraninite is pitchblende. Weathering of these ores has produced a fascinating array of secondary minerals (Ondruš et al. 2003a)

Figure 45. Important uranium localities in the Czech Republic:
1 Jachymov
2 Vitkov II and Zadni Chodov
3 Pribram
4 Brevniste
5 Northern Bohemia: Hamr, Straz, Mimon, and Osecna-Kotel
6 Vnitrosudetska Panev
7 Brzikov
8 Rožna

Figure 46. Olive-green pseudojohannite forming a small rosette < 0.25 mm wide on matrix; original authors studied material from the TL at Jachymov, Czech Republic. RJL3066

Table 15. Radioactive minerals first described from the Czech Republic

Species	Type Locale	Year
Uraninite	Jachymov	1727
Torbernite	Jachymov	1793
Johannite	Jachymov	1830
Zippeite	Jachymov	1845
Voglite	Elias mine, Jachymov	1853
Uranopilite	Jachymov	1854
Schroeckingerite	Jachymov	1873
Uranophane-β	Jachymov	1935
Meta-uranopilite	Jachymov	1935
Nickel-zippeite	Jachymov	1976
Albrechtschraufite	Jachymov	1984
Jachymovite	Jachymov	1996
Čejkaite	Geschieber vein, Jachymov	2003
Pseudojohannite	Rovnost shaft, Jachymov	2003
Phosphowalpurgite	Smrkovec, western Bohemia	2004
Šreinite	Horni Halže, near Meděněc	2007
Agricolaite	Giftkiesstollen adit, Jachymov	2011
Sejkoraite-(Y)	Rovnost shaft, Jachymov	2011
Behounekite	Geschieber vein, Jachymov	2011
Adolfpateraite	Svornost mine, Jachymov	2012
Linekite	Geschieber vein, Jachymov	2013
Štěpite	Geschieber vein, Jachymov	2013
Vysokýite	Geschieber vein, Jachymov	2013
Mathesiusite	Svornost mine, Jachymov	2014

and further research might well yield additional new species from this prolific region.

The Czech Republic, and especially the Jachymov district, is the TL for several dozen minerals, among which are many radioactive species (Table 15) listed in their order of discovery from 1727 to the present.

Several useful publications have been devoted to the mineralogy of the Jachymov district and are highly recommended for the interested student of the subject. These include: an issue of *Lapis*, Vol. 24, No. 3, 2002, (in German); a special issue of the *Journal of the Czech Geological Society*, Vol. 48, No. 3-4, 2003, (in English); and a two-volume set *Uranium Minerals of the Czech Republic and Their Occurrences*, 2007 (in Czech) by Pauliš, Kopecký, and Černý.

Other radioactive species reported from Jachymov include: andersonite, arsenuranospathite, autunite, becquerelite, boltwoodite, coffinite, compreignacite, cuprosklodowskite, curienite, dewindtite, kahlerite, kamotoite-(Y), kasolite, liebigite, magnesiozippeite, marecottite, masuyite, meta-autunite, metakirchheimerite, metalodevite, metanovacekite, metaschoepite, metatorbernite, metatyuyamunite, meta-uranocircite, meta-uranospinite, metazellerite, metazeunerite, novacekite, parsonsite, phosphuranylite, rabbittite, rabejacite, richetite, schoepite, sklodowskite, soddyite, sodium-zippeite, torbernite, troegerite, tyuyamunite, urancalcarite, uranocircite, uranophane, uranospathite, uranosphaerite, uranospinite, vandendriesscheite, vochtenite, walpurgite, weeksite, widenmannite, wolsendorffite, yingjiangite, zellerite, zeunerite, and znucalite. An amorphous oxidation product of pitchblende found in the Elias mine was locally called *eliasite*; it is considered identical to *gummite* and is not a valid mineral, but rather an undifferentiated mixture of kasolite and other phases (Hlousek and Tvrdy 2002).

ENGLAND

There are no commercially important uranium deposits anywhere on the British Isles; all uranium minerals were extracted as a by-product of mining for other metals.

The mines of Cornwall are of great historical interest as the source of many "classic" mineral specimens preserved in museums and private collections. Tin and copper smelting were well established in this district during the Bronze Age, and into modern times these metals continued to account for the bulk of commercial production, along with significant amounts of lead, silver, tungsten, and arsenic (Embrey and Symes, 1987). Uranium ranks among many other industrial metals that were recovered in lesser amounts; for the collector, several Cornish mines have produced fine specimens of torbernite and bassettite among others.

Table 16. Radioactive minerals first described from England

Species	Type Locale	Year
Bassetite	Wheal Basset	1915
Uranospathite	Wheal Basset	1915
Tristramite	Wheal Trewavas	1983
Vochtenite	Wheal Basset	1989

The broad geological setting may be described as sedimentary deposits that were heavily deformed by tectonic activity and later intruded by granites approximately 300 million years ago. The important ore deposits consist of fracture-filling or "lode" mineralizations driven by hydrothermal processes that distributed various elements in several distinct episodes that likely occurred after most of the granite had solidified. Localized fractures and fissures provided pathways for the fluids to move through the surrounding rock, so the structure of the fracturing influenced the distribution of the lodes. As a general rule, tin and copper deposits tend to lie in E-W lodes, whereas lead, zinc, and iron came later in roughly N-S lodes. The temperatures at which various deposits were formed may be rationalized to some extent by comparing the minerals in a lode with its distance from the nearby intrusive bodies. Closest to the intrusion, the upper range of the hypothermal suite (500-250°C) typically includes arsenopyrite, wolframite, cassiterite, molybdenite, and hematite. As temperatures decrease, scheelite, stannite, chalcopyrite, and sphalerite become more prevalent. The uraniferous lodes typically formed under mesothermal (350-150°C) conditions as pitchblende/uraninite along with nickeline, smaltite, cobaltite, bismuthinite, and argentite. This model is, however, an oversimplification that fails to account for a number of local variables. In particular, it is difficult to apply a largely steady-state hydrothermal model to what in many cases must have been overlapping, episodic mineralization events.

A large number of Cornish mines have yielded uranium minerals (Embrey and Symes 1987) while they were operating and their waste dumps continue to yield specimens to this day. Wheal Edward has produced pitchblende along with zeunerite and other secondary minerals. Commercial quantities of pitchblende were mined at Wheal Trenwith in the early 1900s.

Figure 47. A nice example of torbernite from the Gunnislake mine, Cornwall, with label dated 1889. RJL1929

The Tincroft mine, which dates from the early 1700s and was closed in 1913, has yielded torbernite. Torbernite has been found at Wheal Butler, which was an important copper producer until it was abandoned in the market slump of the late 1870s. The Tolcarne mine, which was worked for tin in the early 1600s, produced some good specimens of torbernite. Small amounts of torbernite and autunite had been known from the area of South Terras; around 1880 a rich vein was found and this mine became a leading producer of uranium for the next 30-40 years. Nearby, St. Austell Consols, a tin mine, also produced commercial quantities of uranium ore. Wheal Trewavas is the type locale for tristramite, a U-containing member of the rhabdophane group (Atkin, Basham, and Bowles 1983). The Old Gunnislake mine was started in the 1700s and has produced the best of the Cornish torbernite specimens.

Wheal Basset is one of the "classic" localities of Cornwall; it is the type locale for both bassetite and vochtenite and has also produced specimens of torbernite and uranospathite.

At present, there is no active metal mining in Cornwall, but the hundreds of abandoned mine sites are still of great interest. Many are picturesque hiking destinations as well as important sites for industrial archaeology. The British government has been proactive in identifying and protecting remaining sites that are considered to be of special importance, and in these places uncontrolled collecting or other disturbance is now limited by law. At the same time, many dumps are being consumed by land reclamation as well as by reprocessing the waste to recover the metal values that remain. Nonetheless, the ore and waste dumps remain a valuable source of materials for hobby collectors and for ongoing scientific study. It is interesting to note that vochtenite was described (Zwaan et al., 1989) seventy years after underground workings at Wheal Basset ended.

Figure 48. Another carefully curated old-timer: torbernite crystals to ~5 mm on matrix from Wheal Basset, dated 1851. RJL4286

FRANCE

Table 17. Selected uranium deposits in France[c]

Deposit	Type and Method[a]	Status[b]	Size (tU)
Bellezane	Vein, U and OP	DEP	2,500–5,000
Bertholene	Unconformity, U and OP	DEP	500–1,500
Cerilly deposits	Sandstone	DEP	1,500–5,000
Coutras	Sandstone	DOR	10,000–25,000
Escarpier	Vein, U	DEP	2,500–5,000
Fanay	Vein, U and OP	DEP	2,500–5,000
Grury	Vein, U and OP	DEP	500–1,500
Hyvernevesse	Vein, U and OP	DEP	500–1,500
La Besse	Vein, U and OP	DEP	500–1,500
La Chapelle-Largeau-Bel Air	Vein, U	DEP	500–1,500
La Commanderie	Vein, U and OP	DEP	1,500–5,000
Le Bernardan-Mailhac Sur Benai	Vein, U and OP	DEP	5,000–20,000
Le Cellier	Vein, OP	DEP	1,500–5,000
Le Chardon	Vein, U	DEP	2,500–5,000
Le Fraisse Les Gorges	Vein, U and OP	DEP	1,000–2,500
Les Bois Noirs	Vein, U and OP	DEP	5,000–10,000
Les Brugeauds	Vein, U and OP	DEP	1,500–5,000
Les Loges	Vein, OP	DEP	500–1,000
Les Pierres-Plantees	Vein, U and OP	DEP	1,000–2,500
Lignol	Vein, U and OP	DEP	500–1,500
Lodeve, Mas d'Alary	Sandstone, OP	DEP	1,000–2,500
Lodeve, Mas Lavayre	Sandstone, U	DEP	10,000–25,000
Margnac	Vein, U and OP	DEP	5,000–10,000
Montulat	Vein, OP	DOR	1,500–2,500
Pen Arran	Vein, U	DEP	500–1,000
Piegut-Cote Moreau	Vein, U and OP	DEP	<500
St. Pierre du Cantal	Sandstone, OP	DEP	1,000–2,500
Teufelsloch	Black shale	DOR	1,000–2,500
Treville and Ouest	Sandstone	DOR	1,500–5,000

[a] U = underground mining; OP = open-pit mining; ISL = *in situ* leaching; HL = heap leaching; IPL = in-place leaching; CP = extraction as a co-product or by-product of mining for another metal; NS = not specified.

[b] DEP = depleted; DOR = dormant; PRO = producing

[c] Some smaller deposits have been omitted.

Data primarily from Finch et al. (1995) and IAEA (2009).

Figure 49. Important uranium localities and districts in France:
1 Pen Arran
2 Le Bernardan-Mailhac sur Benai, Fanay, Bellezane, and Margnac
3 La Besse
4 Margabal
5 Lodeve
6 Les Bois Noirs
7 Autun - Grury
8 Le Val d'Ajol, Val St. Lambert, Kruth, and Ronchamps
9 Cap Garonne

Figure 50. Sample of parsonsite ore from Grury, Saone et Loire. Although the individual crystals are minute and very pale, the crystal-lined fissures are so numerous that the parsonsite makes up a significant fraction of the ore at Grury. RJL2647

Figure 51. Chartreuse, tabular autunite crystals to ~5 mm, from the TL at Autun, France. RJL3051

Uranium exploration began in 1946 and activities continued at a high level until peaking at 3,420 tU in 1987, after which uranium production began to decline. About thirty significant deposits were developed, most of which were epithermal veins associated with granite. The last operating mines in France were Lodeve (closed in 1997) and Le Bernardan (closed in 2001).

Because of the large number of localities and the long history of uranium mining there, France has been a rich source of superb specimens of many radioactive species, including a number of type minerals, Table 18.

Table 18. Radioactive minerals first described from France

Species	Type Locale	Year
Autunite	Autun, Saone-et-Loire	1824
Meta-autunite	Autun, Saone-et-Loire	1824[a]
Compregnacite	Margnac, Haute-Vienne	1964
Metalodevite	Rivieral deposit, Lodeve	1972
Rameauite	Margnac, Haute-Vienne	1972
Deloryite	Cap Garonne, Var	1992
Fontanite	Rabejac deposit, Lodeve	1992
Rabejacite	Rabejac deposit, Lodeve	1993
Deliensite	Mas d'Alary, Lodeve	1997
Leydetite	Mas d'Alary, Lodeve	2013

[a]Autun, France, is considered the TL for meta-autunite, although at the time autunite was named, the modern understanding of dehydration and nomenclature of "meta-" forms had not been developed.

The Margabal mine, Aveyron, is notable for the occurrence of spectacular torbernite specimens. The site had been worked in the late 1950s and abandoned. Exploration by collectors in 1997 yielded several thousand specimens. In some cases, plates and stalactitic groups of truly stunning dimensions have been recovered (Schwab 1998).

Boisson and Leydet (1998) provide a comprehensive report on radioactive mineral localities in France, illustrated with dozens of superb photomicrographs to help distinguish the minerals and better appreciate their complex associations. Two of the more prolific localities are:

Margnac: agrinierite, becquerelite, billietite, compregnacite, curite, ianthinite, phosphuranylite, rameauite, sabugalite, schoepite, and wölsendorfite.

Rabejac: arsenouranospathite, becquerelite, billietite, calcurmolite, coffinite, cuprosklodowskite, fontanite (TL), heinrichite, iriginite, kahlerite, liebigite, masuyite, nováčekite, rabejacite (TL), rutherfordine, saléeite, schoepite, seelite (co-TL), sklodowskite, natrouranospinite, trögerite, tyuyamunite, umohoite, urancalcarite, uraninite, uranopilite, uranospinite, vandenbrandeite, vandendriesscheite, zeunerite, and zippeite.

In a detailed study of the Margnac and Fanay deposits, Leroy (1978) concluded that emplacement of the granite occurred about 350-360 Ma ago, followed by intrusion of lamprophyres at 285 Ma, corresponding to brittle fracture tectonic activity. Pitchblende and pyrite mineralization immediately followed emplacement of the lamprophyres and micaceous episyenitization of the granite at a minimum age of 275 Ma, and a temperature of about 345°C. The fluid then became more water-rich during precipitation of microcrystalline quartz and hematization of pyrite; quartz deposition and transformation of pitchblende to coffinite took place as temperatures fell from 330 to 140°C. In the final stage, fluorite, barite, and calcite were deposited beginning at ~135°C and continuing to a lower temperature. Secondary minerals, including gummite, are of fairly recent age, starting around 30 Ma and "are still continuing to some extent."

Figure 52. Torbernite aggregates on smoky quartz from the Margabal mine, Aveyron. Sample is ~6 cm wide. RJL3987

Figure 53. A rich example of the late-1990s torbernite find at Margabal: this group, ~5 cm wide, shows the characteristic habit of very thin, tabular crystals forming slightly divergent stacks, or "books." RJL2144

GERMANY

Table 19. Known uranium deposits in Germany

Deposit	Type and Method[a]	Status[b]	Size (tU)
Baden-Württemberg			
Menzenschwand	Vein, U	DOR	2,500–5,000
Muellenbach	Sandstone, U	DOR	2,500–5,000
Bavaria			
Grossschloppen	Vein, U	DOR	1,000–2,500
Waeldel	Vein, U	DOR	1,000–2,500
Saxony			
Annaberg	Vein, U	DEP	500–1,000
Antonsthal	Vein, U	DEP	500–1,000
Freital	Coal, U	DEP	2,500–5,000
Johanngeorgenstadt	Vein, U	DEP	5,000–10,000
Köngistein	Sandstone, U and ISL	DEP	10,000–25,000
Ronneburg ore field	Vein, U and OP	DEP	>100,000
Schlema	Vein, U	DEP	5,000–20,000
Schneckenstein	Vein, U	DEP	500–1,000
Tellerhauser	Vein, U	DEP	1,000–2,500
Zobes	Vein, U	DEP	5,000–10,000
Thuringia			
Culmitzsch	Sandstone, OP	DEP	10,000–25,000

[a] U = underground mining; OP = open-pit mining; ISL = *in situ* leaching; HL = heap leaching; IPL = in-place leaching; CP = extraction as a co-product or by-product of mining for another metal; NS = not specified.

[b] DEP = depleted; DOR = dormant; PRO = producing

Data primarily from Finch et al. (1995) and IAEA (2009).

Table 20. Radioactive minerals first described from Germany

Species	Type Locale	Year
Trögerite	Weisser Hirsch mine, Schneeberg	1873
Uranosphaerite	Weisser Hirsch mine, Schneeberg	1873
Uranospinite	Weisser Hirsch mine, Schneeberg	1873
Walpurgite	Weisser Hirsch mine, Schneeberg	1873
Zeunerite	Weisser Hirsch mine, Schneeberg	1873
Metazeunerite	Weisser Hirsch mine, Schneeberg	1877
Metauranocircite	Streuberg mine, Vogtland, Saxony	1877
Hugelite	Michael mine, Black Forest	1913
Nováčekite	Schneeberg, Saxony	1951
Metanováčekite	Schneeberg, Saxony	1951
Wölsendorfite	Wölsendorf, Bavaria	1957
Metakahlerite	Sophia Mine, Wittichen, central Black Forest	1958
Metakirchheimerite	Sophia Mine, Wittichen, central Black Forest	1958
Arsenuranospathite	Krunkelbach deposit, Menzenschwand	1959
Bergenite	Streuberg mine, Vogtland, Saxony	1959
Widenmannite	Michael mine, Black Forest	1961
Hallimondite	Michael mine, Black Forest	1965
Joliotite	Krunkelbach deposit, Menzenschwand	1976
Phurcalite	Streuberg mine, Vogtland, Saxony	1978
Asselbornite	Schneeberg, Saxony	1983
Uranosilite	Krunkelbach deposit, Menzenschwand	1983
Uranotungstite	Krunkelbach deposit, Menzenschwand	1985
Orthowalpurgite	Schmiedestollen, Wittichen, Saxony	1995
Chadwickite	Sophia Mine, Wittichen, central Black Forest	1998
Arsenovanmeersscheite	Krunkelbach deposit, Menzenschwand	2007
Nielsbohrite	Krunkelbach deposit, Menzenschwand	2009
Heisenbergite	Krunkelbach deposit, Menzenschwand	2012

In the years prior to reunification in 1991, East Germany was a major uranium producer, with cumulative output estimated at over 200,000 tU. All mines have now ceased operation, although reclamation work continues to yield small quantities of uranium. For example, decommissioning of the Königstein mine in Saxony in 1999 yielded around 30 tU.

For collectors, Germany has been a rich source of fine specimens, as well as numerous type minerals, Table 20.

Figure 54. Important uranium localities and districts in Germany:
1 Muellenbach
2 Wittichen
3 Menzenschwand
4 Wölsendorf, Bavaria
5 Grossschlopen-Waeldel
6 Zobes-Bergen district
7 Culmitzsch
8 Johanngeorgenstadt-Schlema-Schneeberg area
9 Ronneburg ore field
10 Freital
11 Jachymov, Czech Republic

The German part of the Erzgebirge region includes many well-known districts in Saxony, including Johanngeorgenstadt, Annaberg, Schneeberg, Freiberg, and Marienberg. As in Jachymov, these mines have been worked mainly for Pb, Ag, Co, Ni, and Bi, beginning with the discovery of silver at Freiberg in 1170. Uranium minerals, primarily pitchblende, have been mined at Schneeberg, Annaberg, and Johanngeorgenstadt. The most economic veins contain Ag-Co-Bi-U and are normally richest in the surrounding metamorphic rocks, becoming barren as they extend into the granite. However, at the Weisser Hirsch mine, Schneeberg, pitchblende is also found in veins in the granite. In a typical deposit, several veins 15-60 cm thick may lie in parallel and extend from ~100 m to over 3 km. The uppermost levels contain abundant Ag ores, with Co-Ni-B ores becoming more prevalent at intermediate levels, and pitchblende predominant at the lower levels (Heinrich 1958, pp. 282-88).

Mines in Saxony, especially the Weisser Hirsch and Streuberg mines, have produced many new species: Weisser Hirsch is the TL for trögerite, uranosphaerite, uranospinite, walpurgite, zeunerite, and metazeunerite. Streuberg is the TL for bergenite, meta-uranocircite I and II, phurcalite, and uranocircite; autunite, billietite, meta-autunite, phosphuranylite, and renardite are also found there.

In the Black Forest, Baden-Württemberg, torbernite, autunite, and zeunerite were known since the late 1700s. Notable

Figure 55. A classic example of botryoidal pitchblende from the Vater Abraham mine, Marienberg, Saxony. RJL3727

Figure 56. Tabular autunite crystals thickly covering matrix from Helemann-Stoller, near Johanngeorgenstadt, Germany. RJL973

Figure 57. A cabinet-sized sample from Bergen, Saxony, with a bright yellow mass of bergenite ~15 mm wide in a quartz-lined cavity. RJL2248

mines include: the Rappenloch mine, Eisenbach; Michael mine, Weiler; the St. Anton and Sophia mines, Wittichen; the Unverhofft Gluck mine, Reinerzau; and the Krunkelbach Valley deposit, Menzenschwand. Most of these are Ag-Co-Ni-Bi-U vein deposits, containing a gangue of barite-fluorite-quartz. At Wittichen, for example, early fluorite and quartz were followed by barite, native silver, and bismuth, then by Ni-Co minerals, and finally by fluorite, carbonate, and Ag ore. The veins are hosted by granite, and pitchblende veinlets extend into the wallrock granite (Heinrich 1958, pp. 288-89).

The Krunkelbach Valley deposit is the TL for arsenovanmeersscheite, heisenbergite, joliotite, nielsbohrite, uranosilite, and uranotungstite; several dozen uranium minerals have been reported there. An extensive study of the Menzenschwand deposit, with hundreds of photos, was recently published by Markl and Wolfsried (2011). Also in the Black Forest, Wittichen is the TL for chadwickite, metakahlerite, metakirchheimerite, and orthowalpurgite.

In the Wölsendorf region, Bavaria, fluorite veins in granite have been mined since 1850. Notable deposits include the Edna, Eberhard, Roland, Johannes, and Marie mines. The sequence of mineralization is quite complex, beginning with carbonates, followed by quartz/chalcedony, then fluorite, barite, and quartz, the galena and chalcopyrite, then carbonates, and finally barite, quartz, fluorite, hematite, pyrite, and pitchblende. The pitchblende occurs as nodules and veinlets in fluorite, and disseminations in the surrounding granite. In some areas, the veins are oxidized to form complex assemblages of secondary Pb, Cu, and U minerals (Heinrich 1958, pp. 352-53).

Figure 58. Large and very rich specimen of uranocircite from Streuberg, Saxony, ~ 5 × 7 cm. RJL3112

Figure 59. Thumbnail-sized specimen of bright chartreuse meta-uranocircite platelets on pegmatite, from Bergen, Saxony. RJL229

Figure 60. Tabular amber walpurgite crystals, ~200 μm long, from the Schmiedestollen mine, Germany. RJL835

PORTUGAL

Uranium development began in 1912 with the discovery of the Urgeiriça U-Ra deposit. Radium was mined until 1944 and uranium production began in 1951. Urgeiriça, along with about 100 nearby deposits, formed an important mining district running in an arc from Coimbra to Viseu to Guarda. Another cluster of deposits lies about 50 km south, near Portalegre. All of the Portugese deposits are granitic and were mined by either underground or open-pit methods. Several mines were listed as producing in the mid-1990s (Finch et al. 1995); however, more recently all mines appear to be dormant and reclamation activities are ongoing.

The vein deposits lie mainly in Hercynian granite porphyry that intrudes older schists and is, in turn, intruded by diabase dikes. Orebodies are often near cross-fractures that follow ancestral faults. The veins are epithermal and believed to be related to Alpine orogeny.

In the Viseu region, the veins fill breccia zones and consist of microcrystalline quartz, reddish jasper, microbotryoidal pitchblende, pyrite, galena, sphalerite, and other sulfides. Disseminated pitchblende also occurs in the altered granite wallrock.

In the Guarda region, the veins comprise white, fine-grained quartz, granular pitchblende, pyrite, and chalcopyrite. Vein breccias are cemented by cherty quartz and pyrite.

In the Beira region, the deposits consist of vuggy fissure fillings and brecciated veins that follow tension fractures.

Figure 61. Important uranium localities in Portugal:
1 Coimbra
2 Viseu
3 Guarda
4 Portalegre
5 Nisa
6 Sabugal

Figure 62. Vein of opaque green metatorbernite crystals in matrix, from Forte Velho, near Guarda. RJL247

Figure 63. Stacked autunite "books" forming a solid crust ~ 5 cm wide, from Sabugal, Portugal. RJL660

Table 21. Known Uranium Deposits in Portugal

Deposit	Type and Method[a]	Status[b]	Size (tU)
Bica	Vein, U	DEP	
Castelejo	Vein, OP	DEP	
Chavelhos	Vein	DOR	
Cunha Baixa	Vein, U and OP	DEP	
Freixiosa	Vein, U	DOR	
Mestras	Vein	DOR	
Nisa	Vein, OP	DOR	2,500–5,000
Pedreiros	Vein	DEP	
Pinhal do Souto	Vein, U	DOR	
Quinta do Bispo	Vein, OP	DEP	
Urgeiriça	Vein, OP	DEP	1,000–2,500

[a] U = underground mining; OP = open-pit mining; ISL = *in situ* leaching; HL = heap leaching; IPL = in-place leaching; CP = extraction as a co-product or by-product of mining for another metal; NS = not specified.

[b] DEP = depleted; DOR = dormant; PRO = producing

Data primarily from Finch et al. (1995) and IAEA (2009).

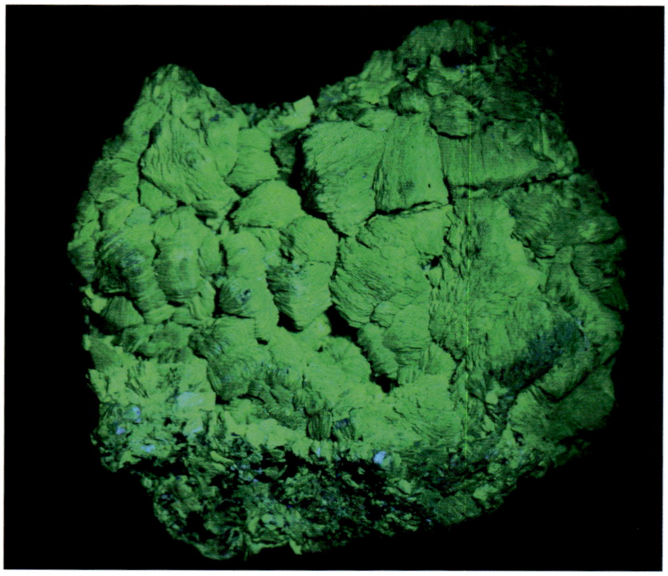

Figure 64. The specimen in the previous photo, viewed in LWUV. RJL660

They contain coarse white and black quartz with microbotryoidal pitchblende in veinlets, geodes, and as powdery coatings.

The veins have been oxidized to considerable depth (up to 95 m in some areas). Secondary minerals including autunite, meta-autunite, sabugalite, saléeite, phosphuranylite, torbernite, uranophane, and various sulfates and hydrous oxides can be found (Heinrich 1958, pp. 335-56).

For the collector, some classic material has come from Portugal, including fine autunites and metatorbernites. Sabugal, Portugal, is the TL for sabugalite, another member of the autunite group.

SPAIN

Figure 65. Important uranium localities in Spain:
1 Andujar, Jaen
2 El Pedregal-Intermedia, Badajoz
3 Mazarete, Guadalajara
4 Mina Fe, Salamanca

Table 22. Known uranium deposits in Spain

Deposit	Type and Method[a]	Status[b]	Size (tU)
Andujar, Jaen	Vein, OP	DOR	
El Pedregal, Badajoz	Vein, OP	DOR	1,000–2,500
Intermedia, Badajoz	Vein, OP	DOR	
Mazarete, Guadalajara	Sandstone, OP	DOR	2,500–5,000
Mina Fe, Salamanca	Vein, OP	DEP	5,000–10,000
Zarcina	Vein	DOR	500–1,000

[a] U = underground mining; OP = open-pit mining; ISL = *in situ* leaching; HL = heap leaching; IPL = in-place leaching; CP = extraction as a co-product or by-product of mining for another metal; NS = not specified.

[b] DEP = depleted; DOR = dormant; PRO = producing

Data primarily from Finch et al. (1995) and IAEA (2009).

Figure 66. Yellow microcrystalline uranospathite forming thin coatings on matrix from Pedro Alvaro, Salamanca. RJL715

Radioactive veins associated with granite extend southeastward from Portugal into Spain. The earliest uranium discoveries were in Salamanca in the late 1950s, followed by additional deposits farther south in Badajoz. All historical uranium mines in Spain are either dormant or depleted, and many are in various stages of remediation or reclamation. However, exploration activities have continued and suggest that the resources in Spain could amount to ~30,000 tU (OECD 2012).

Although the supply of high-quality specimen material from Spain is generally less than that of Portugal, locales in Badajoz have produced specimens of uraninite (var. pitchblende) and colorful secondary minerals including autunite, sabugalite, saléeite, torbernite, metatorbernite, and uranospathite.

Figure 67. Pale yellow sabugalite crystals on granitic matrix from Oliva de la Frontera, Badajoz. RJL2648

SWITZERLAND

At present, there are no commercially active uranium mines in Switzerland. However, the deposit at La Creusaz, near Les Marecottes village in the Western Alps, is of some interest to collectors as the type locale for marecottite; it is scientifically important as an example of the active formation of secondary minerals as a result of the interactions between primary minerals and acid mine drainage. The La Creusaz prospect was discovered in 1973 and explored by drilling and by digging shallow trenches and tunnels. The primary mineralization contained uraninite and pyrite, with minor amounts of Bi, Pb, Se, and other species. The type specimen of marecottite was described from a specimen that had formed on stockpiled ore in an exploration tunnel where it had been left since 1981 and subject to acid mine drainage waters (Brugger et al. 2003). Other recently formed secondary sulfates found in this material include members of the zippeite group, jachymovite, and several potentially new minerals.

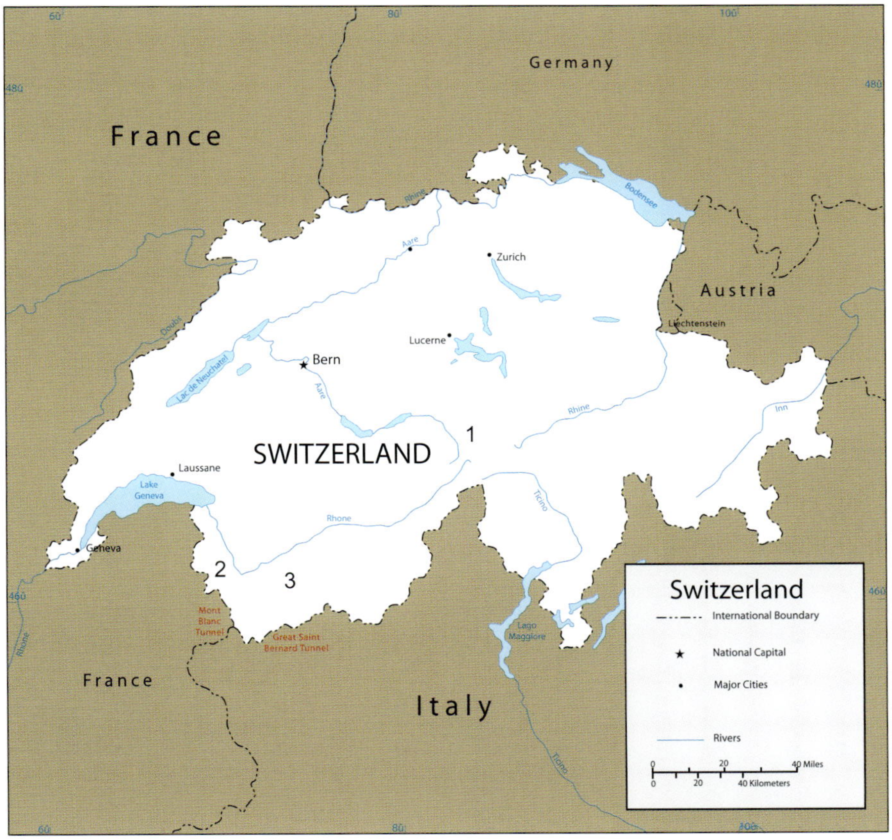

Figure 68. Improtant uranium localities in Switzerland:
1 Grimsel area, Bern
2 La Creusaz, Les Marecottes
3 Cleuson, Valais

Table 23. Radioactive minerals first described from Switzerland

Species	Type Locale	Year
Grimselite	Grimsel area, Bern	1972
Marecottite	La Creusaz, Les Marecottes	2003
Cleusonite	Cleuson, Valais	2005
Françoisite-(Ce)	La Creusaz, Les Marecottes[a]	2010

[a]Co-type locale with Radium Ridge, South Australia

Figure 69. A small ore sample, ~6 mm wide, with dark orange, powdery marecottite and other secondary minerals, from the TL at La Creusaz, Les Marecottes, Valais. RJL3058

CENTRAL ASIA

RUSSIA

Russia and several other republics of the former Soviet Union host significant uranium ore deposits as well as interesting mineral assemblages. Many radioactive species were first described from Kazakhstan, Kyrgyzstan, Tajikistan, and Uzbekistan, as well as from Russia; however, locality information was often missing or inaccurate during the Soviet period. Recent academic work in Russia (Pekov 1998; Pekov 2007) has provided welcome details to fill the gaps in our knowledge of all these localities.

With cumulative production of about 150,000 tU (through 2011) and current annual production of ~3,500 tU, the Russian Federation ranks as the sixth largest uranium producer in the world and accounts for ~6% of global production. Intensive exploration, beginning in 1944, identified over 100 uranium deposits within fourteen districts of the Russian Federation. The most important deposits lie within four major districts: the Streltsovsk district, which includes 19 volcanic deposits, some of which are currently being mined; the Trans-Ural and Vitim districts, where sandstone deposits are being developed for production by ISL; and the Elkon district, where a number of metasomatite deposits are in the planning stage.

Although nearly 20 radioactive minerals have been described from Russia, Table 25, few are known to form good crystals, so the vast majority of them are of interest primarily to the rare species collector. Most can't be conclusively identified without sophisticated microanalytical tools.

Figure 70. Important uranium localities and districts in Russia:
1 Beshtau (Pyatigorsk)
2 Shargadykskoye and Stepnoye (Kalmyk)
3 Padma
4 Mt. Karnasurt, Lovozero massif
5 Chelyabinsk district: Pit 50, Pits 75-76, Sanarskoye
6 Primorskoye
7 Ust-Uyukskoye
8 Buriat district: Radionovskoye, Tetrakhskoye, Vitlansskoye
9 Chita region: Dal'nee, Gornoye, Oktyabr'skoye Mo-U deposit, Olovskoye, Shirondukuevskoye, Stepnoye (Chita), Streltsovskaya, Tulukuevskoye, Vesenneye, Yubileinoye
10 Aleksandrovskii Golets Mo-U occurrence

Table 24. Selected uranium deposits in Russia

Deposit	Type and Method[a]	Status[b]	Size (tU)
Antei	Volcanic	PRO	25,000–50,000
Badyelskoye	Lignite, U	DOR	5,000–10,000
Beryozovoye	Vein, U and ISL	DOR	1,500–5,000
Beshtau	Vein, U	DEP	500–1,000
Briketno	Lignite, U and ISL	DOR	5,000–10,000
Bykogorskoye	Vein, U	DEP	1,000–2,500
Dalmatovskoye	Sandstone, ISL	PRO	10,000–25,000
Dal'nee	Volcanic, U	PRO	1,500–5,000
Dobrovol'noye	Sandstone, ISL	DOR	5,000–10,000
Gornoye	Vein, U and ISL	DOR	5,000–20,000
Kedrovoye	Sandstone, U	DOR	2,500–5,000
Khiagdinskoye	Sandstone, ISL	PRO	10,000–25,000
Labyshkoye	Vein, U	DOR	1,000–2,500
Lastochka	Volcanic, U	DOR	2,500–5,000
Luchistoye	Volcanic	PRO	10,000–25,000
Novogodnoye	Volcanic, U	PRO	2,500–5,000
Primorskoye	Sandstone, U and ISL	DOR	5,000–10,000
Sanarskoye	Surficial, OP	DEP	2,500–5,000
Shirondukuevskoye	Volcanic, U	DOR	5,000–10,000
Solonechnoye	Volcanic, U	DOR	1,500–2,500
Stepnoye (Kalmyk)	Phosphorite, U	DOR	10,000–25,000
Stepnoye (Chita)	Sandstone, ISL	DOR	10,000–25,000
Streltsovskoye	Volcanic, U	PRO	50,000–100,000
Tulukuevskoye	Vein, U	DEP	25,000–50,000
Ust-Uyukskoye	Sandstone, U and ISL	DOR	10,000–25,000
Vesenneye	Volcanic, U	DOR	500–1500
Vitlaunskoye	Surficial	DOR	2,500–5,000
Yubileinoye	Volcanic, U	PRO	5,000–10,000
Zherlovoye	Vein	PRO	2,500–5,000

[a] U = underground mining; OP = open-pit mining; ISL = *in situ* leaching; HL = heap leaching; IPL = in-place leaching; CP = extraction as a co-product or by-product of mining for another metal; NS = not specified.

[b] DEP = depleted; DOR = dormant; PRO = producing

Data primarily from Finch et al. (1995) and IAEA (2009).

Table 25. Radioactive minerals first described from Russia

Species	Type Locale	Year[a]
Aeschynite-(Ce)	Pits 75–76, Ilmeny Mts., S. Urals	1825
Samarskite-(Y)	Blyumovskoye Pit (Pit 50), Ilmeny Mts., S. Urals	1840
Karnasurtite-(Ce)	Mt. Karnasurt, Lovozero massif, Kola Peninsula	1947
Lermontovite	Beshtau uranium deposit, near Pyatigorsk, N. Caucasus	1948
Iriginite	Aleksandrovskii Golets Mo-U deposit	1951
Moluranite	Aleksandrovskii Golets Mo-U deposit	1951
Vanuranylite	Ust-Yuk V-Se-U deposit, Tuva, Siberia	1955
Yttropyrochlore-(Y)	Alakurtti granite pegmatite, N. Karelia	1957
Niobo-aeschynite-(Ce)	Cherry Mountains, Southern Urals	1960
Yttrobetafite-(Y)	Alakurtti granite pegmatite, N. Karelia	1961
Plumbobetafite	Burpala massif, Mama River basin, Siberia	1964
Pseudo-autunite	Vuoriyarvi massif, N. Karelia	1964
Aeschynite-(Nd)	Tatarskii massif, Krasnoyarsk, Siberia	1966
Umbozerite	Mt. Karnasurt, Lovozero massif, Kola Peninsula	1971
Bauranoite	Oktyabrskoye Mo-U deposit, near Krasnokamensk	1973
Calciouranoite	Oktyabrskoye Mo-U deposit, near Krasnokamensk	1973
Metacalciouranoite	Oktyabrskoye Mo-U deposit, near Krasnokamensk	1973
Tuliokite	Kirovskii apatite mine, Khibiny massif, Kola Peninsula	1990
Diversilite-(Ce)	Yukspor Mt., Khibiny massif, Kola Peninsula	2003
Beshtauite	Beshtau uranium deposit, near Pyatigorsk, N. Caucasus	2012

[a]Some dates may differ from those given in Pekov (1998, 2007) in cases where the mineral might have been noted earlier, partially described, given a different name, etc.

Figure 71. Dark needles of umbozerite in alkaline rock from Mt. Panqurayv, Lovozero massif, Kola Peninsula. RJL1495

The rare Th mineral umbozerite, $Na_3Sr_4Th[Si(O,OH)_3]_8$, forms dark, needle-like crystals in alkaline rocks of the Lovozero massif, Kola Peninsula. It has been reported from several localities there, including on Mt. Karnasurt and at the Umbozero mine, Mt. Alluaiv.

KAZAKHSTAN

Figure 72. Some important uranium localities in Kazakhstan:
1 Vostok
2 Zazernoye
3 Bektau-Ata
4 Mynkaduk
5 Kanzhugan-Moynkum
6 Kyzylsai
7 Bota-Burum
8 Koldyat
9 Irkol
10 Karamurun

Table 26. Selected uranium deposits in Kazakhstan[c]

Deposit	Type and Method[a]	Status[b]	Size (tU)
Assarchik	Sandstone, ISL	DOR	10,000–25,000
Balkashinskoe	Vein, U	DEP	2,500–5,000
Bota Burum	Vein, U	DEP	5,000–10,000
Daba	Volcanic	DOR	5,000–10,000
Djusandalynskoe	Vein	DOR	5,000–10,000
Djydelinskoe	Volcanic, U and OP	DEP	1,500–5,000
Grachevskoye	Vein, U	DEP	10,000–25,000
Irkol	Sandstone	DOR	20,000–50,000
Ishimskoye	Vein, U	DEP	5,000–10,000
Kanzhugan	Sandstone, ISL	PRO	50,000–100,000
Karamuran[d]	Sandstone, ISL	PRO	50,000–100,000
Koldyat	Coal	DOR	25,000–50,000
Kyzylsai	Vein, U	DEP	2,500–5,000
Kyzyltu	Sandstone, ISL	DOR	10,000–25,000
Manybaiskoye	Vein, U	DEP	10,000–25,000
Moynkum	Sandstone, ISL	PRO	50,000–100,000
Mynkaduk	Sandstone, ISL	PRO	>100,000
Semizbai	Sandstone, ISL	DOR	10,000–25,000
Suluchenkoye	Sandstone, ISL	DOR	25,000–50,000
Uvanas	Sandstone, ISL	PRO	10,000–25,000
Vostok	Vein, U	PRO	10,000–25,000
Zazernoye	Vein, U	DOR	10,000–25,000

[a] U = underground mining; OP = open-pit mining; ISL = *in situ* leaching; HL = heap leaching; IPL = in-place leaching; CP = extraction as a co-product or by-product of mining for another metal; NS = not specified.

[b] DEP = depleted; DOR = dormant; PRO = producing

[c] There are currently about 70 deposits in various stages of development; larger and further-developed resources are emphasized here.

[d] Includes Karamurun North and Karamurun South

Data primarily from Finch et al. (1995) and IAEA (2009).

Figure 73. Blocky davidite-(La) crystals, with a pale surface alteration in places, from Bektau-Ata, Karagandy province, Kazakhstan. Sample is ~5 cm wide. RJL2801

Table 27. Radioactive minerals first described from Kazakhstan

Species	Type Locale	Year
Natrouranospinite	Bota-Burum deposit, Chu-Ili Mts., SW Balkhash	1957
Mourite	Kyzylsai Mo-U deposit, Chu-Ili Mts., SW Balkhash	1962
Saryarkite-(Y)	Akkuduk rare-metal deposit, near Mointy	1962
Sedovite	Kyzylsai Mo-U deposit, Chu-Ili Mts., SW Balkhash	1965
Natroboltwoodite	Kyzylsai ore field, SW Balkhash region	1975
Chistyakovite	Bota-Burum deposit, Chu-Ili Mts., SW Balkhash	2006
Uramarsite	Bota-Burum deposit, Chu-Ili Mts., SW Balkhash	2007

Uranium has been mined in Kazakhstan since 1957, when open-pit mining began at the Kurdai deposit. For the next twenty years, about fifteen separate deposits, primarily vein-type, were mined by either open-pit or underground methods and yielded about 5,000 tU in total. Mining of extensive sandstone-type deposits by ISL began in 1978 and in 2009 production exceeded 14,000 tU/year, making Kazakhstan the world's leading U producer (OECD 2012).

Radioactive species first described from Kazakhstan are given in Table 27, with TL information generally from Pekov (1998).

KYRGYZSTAN

Table 28. Known uranium deposits in Kyrgyzstan

Deposit	Type and Method[a]	Status[b]	Size (tU)
Mailisai, Osh	Limestone-bitumen, U	DEP	500–1,000
Mailisu, Osh	Limestone-bitumen, U	DEP	1,000–2,500
Shakaptar	Vein	DEP	500–1,000
Tura-Kavak	Coal	DEP	2,500–5,000

[a] U = underground mining; OP = open-pit mining; ISL = *in situ* leaching; HL = heap leaching; IPL = in-place leaching; CP = extraction as a co-product or by-product of mining for another metal; NS = not specified.

[b] DEP = depleted; DOR = dormant; PRO = producing

Data primarily from Finch et al. (1995).

Table 29. Radioactive minerals first described from Kyrgyzstan

Species	Type Locale	Year
Tyuyamunite	Tyuya-Muyun Cu-V-U deposit, Ferghana Valley	1912
Uramphite	Tura-Kavak uranium-coal deposit	1957

Two small underground mines at Mailisai and Mailisu, Osh region, produced uranium from a limestone-bitumen deposit; both are now depleted. Deposits at Aramsu, Sogul, and Sumsar are in the exploration phase.

Radioactive species first described from Kyrgyzstan are listed in Table 29.

The Tyuya-Muyun deposit played a very important role in the radium industry in Russia in the early 1900s. Chervinsky (1925) describes the locality as follows: "Tyuya-Muyun is a prominent rounded hill traversed by the gorge of the Aravan river and is a northern spur of the Alai Mountains in Russian Turkestan. In the aboriginal language the name means camel's hump. It is situated 62 kilometres southeast of the railway station Fedchenko and about the same distance south of the town of Andijan in Fergana . . . The ore forms metasomatic gash-veins and fills Karst caverns in limestone, which around the ore is dolomitized. The limestone beds, of Carboniferous age, are much dislocated and inverted. They are bordered by carbonaceous Silurian slates with a graptolitic fauna. To the north and to the south of the mine these pass into siliceous jasper-like slates with interbedded limestones. The whole series of sediments is penetrated by intrusive sills of diabase and dikes of keratophyre, together with volcanic tuffs and breccias . . . The mine was probably worked in the bronze age, since a stone hammer (of diorite) has been found in the old workings; more recently it was worked for copper by the Chinese. Only the upper oxidized portion ('iron hat') has been exposed, and the underground workings extend to a depth of only 75 metres. The ore has been taken mainly from two caverns ('bug holes'), of which the upper, measuring 16 × 8 × 8 metres, is known as the 'yellow cavern' and the lower as the 'green cavern.'"

Heinrich (1958) notes that by 1935, five uraniferous and thirty barren, barite veins were reported. Ore bodies range in thickness and length from a few cm to 1.5 m. The ore is fine-grained and massive, banded, brecciated, or vuggy. Thin layers and lenses of tyuyamunite are associated with volborthite, calcite, malachite, chrysocolla, barite, goethite, and quartz. Some of the barite is U-free but contains some radium, i.e., $(Ba,Ra)SO_4$. Ra-rich calcite is also reported.

Figure 74. Important uranium localities in Kyrgyzstan:
1 Mailisai, Osh
2 Shakaptar, Osh
3 Tyuya-Muyun, Osh
4 Tura-Kavak, Northern Tien Shan Region

TAJIKISTAN

Table 30. Known uranium deposits in Tajikistan			
Deposit	**Type and Method**[a]	**Status**[b]	**Size (tU)**
Adrasman	Vein, U	DEP	500–1,000
Taboshary	Vein, U	DEP	1,000–2,500

[a] U = underground mining; OP = open-pit mining; ISL = *in situ* leaching; HL = heap leaching; IPL = in-place leaching; CP = extraction as a co-product or by-product of mining for another metal; NS = not specified.

[b] DEP = depleted; DOR = dormant; PRO = producing

Data primarily from Finch et al. (1995) and IAEA (2009).

Table 31. Radioactive minerals first described from Tajikistan		
Species	**Type Locale**	**Year**
Przhevalskite	Dzerkamar deposit, Karamazar Mts.	1946
Chernikovite	Karakat deposit, Karamazar Mts.	1955
Metanatroautunite	Kuruk uranium deposit, near Khodzhent	1957
Calcioursilite	Oktyabrskoye deposit, near Khodzhent	1958
Magnioursilite	Oktyabrskoye deposit, near Khodzhent	1958
Bismutopyrochlore	Mika pegmatite vein, Tau Mt., Eastern Pamirs	1999
Arapovite	Darai-Pioz Galcier, Alai Range	2003

A small vein-type deposit at Taboshary was mined by underground methods and is now depleted, as is a smaller deposit at Adrasman.

Radioactive species first described from Tajikistan are given in Table 31.

Figure 75. Important uranium localities and districts in Tajikistan: 1 Karamazar Mountains: Adrasman, Dzherkamar, Karakat 2 Khodzhent area: Kuruk, Oktyabrskoye, Taboshary 3 Dara-i-Pioz glacier

UZBEKISTAN

Uranium prospecting began in 1952 in the Central Kyzylkum province, and in 1953 the Uchkuduk deposit was discovered. As the Uchkuduk type mineralizations and their pattern of occurrence were identified, more focused prospecting discovered a large number of new deposits, including Sugraly, Bukinai, Ketmenchi, Beshkak, and Lyavlyakan. Mining efforts were initiated in 1964 at Uchkuduk by both open-pit and underground methods. Industrial-scale *in-situ* leaching began as early as 1965. Identified conventional resources are ~140,000 tU, and Uzbekistan accounted for about 5% of global U production in 2010 (OECD 2012).

Several radioactive species have been described from localities in Uzbekistan, Table 33.

Figure 76. Important uranium localities and districts in Uzbekistan:
1 Kendyktyube, Sugraly
2 Central Kyzylkum: Alendy, Aktau, Aulbeck, Bakhaly, Bukinai, Varadzhan
3 Bukhara: Beshkak, Djantuar, Kanimekh, Koscheka, Rudnoye, Uchkuduk
4 Severny Kanimekh, Severny Mayazak, Terekuduk
5 Agron, Shark, Tutly
6 Tashkent: Alatanga, Kattasai, Chauli, Mailikatan
7 Charkasar

Table 32. Known uranium deposits in Uzbekistan

Deposit	Type and Method[a]	Status[b]	Size (tU)
Agron	Sandstone	PRO	5,000–10,000
Aktau	Sandstone	DOR	5,000–10,000
Alatanga	Vein, U	DEP	2,500–5,000
Alendy	Sandstone, ISL	PRO	10,000–25,000
Altyntau	Black shale	DOR	500–1,000
Aulbek	Sandstone	PRO	10,000–25,000
Bakhaly	Sandstone, ISL	DOR	1,000–2,500
Beshkak	Sandstone, ISL	PRO	5,000–10,000
Bukinai	Sandstone, ISL	PRO	25,000–50,000
Charkasar	Vein, U	DEP	1,000–2,500
Chauli	Volcanic, U	DEP	2,500–5,000
Djantuar	Black shale, U and OP	DOR	10,000–25,000
Kanimekh	Sandstone, ISL	PRO	10,000–25,000
Kattasai	Vein, U	DEP	1,000–2,500
Kendyktyube	Sandstone, ISL	DOR	10,000–25,000
Ketmenchi	Sandstone, ISL	PRO	10,000–25,000
Koscheka	Black shale, U and OP	DOR	1,000–2,500
Lyavlyakan	Sandstone, ISL	PRO	5,000–10,000
Mailikatan	Vein, U	DEP	1,000–2,500
Rudnoye	Black shale, U and OP	DOR	2,500–5,000
Severny Kanimekh	Sandstone, ISL	PRO	10,000–25,000
Severny Mayzak	Sandstone	DOR	2,500–5,000
Shark	Sandstone, ISL	PRO	1,000–2,500
Sugraly	Sandstone, ISL	PRO	25,000–50,000
Terekuduk	Sandstone, ISL	DOR	500–1,000
Tutly	Sandstone, ISL	PRO	2,500–5,000
Uchkuduk	Sandstone, OP and ISL	PRO	25,000–50,000
Varadzhan	Sandstone, ISL	DOR	500–1,000
Yuzhny Bukinai	Sandstone, ISL	PRO	10,000–25,000

[a] U = underground mining; OP = open-pit mining; ISL = *in situ* leaching; HL = heap leaching; IPL = in-place leaching; CP = extraction as a co-product or by-product of mining for another metal; NS = not specified.

[b] DEP = depleted; DOR = dormant; PRO = producing

Data primarily from Finch et al. (1995) and IAEA (2009).

Table 33. Radioactive minerals first described from Uzbekistan

Species	Type Locale	Year
Arsenuranylite	Cherkasar deposit, near Pap	1954
Strelkinite	Kendyktas Mts., Central Kyzylkum region	1965
Vyacheslavite	Dzantuar and Rudnoye deposits, Central Kyzylkum	1984

Figure 77. A small but historical specimen: holotype material from the Cherkasar deposit, Uzbekistan, originally studied by Larisa Belova, showing powdery orange-yellow arsenuranylite masses scattered on matrix with yellow schoepite. The sample is ~1 cm across. As is often the case, Soviet-era descriptions often omitted exact localities, leading to confusion in the mineral trade. This sample, for instance, was identified by the vendor in 1994 as having come from Rudnyi, Kazakhstan. RJL1674

AUSTRALIA

Figure 78. Important uranium localities and districts in Australia:
1 Kintyre
2 Yeelirrie
3 Mulga Rock
4 Rum Jungle
6 South Alligator River
6 Koongarra
7 Ranger, Jabiluka, Nabarlek
8 Westmoreland
9 Valhalla
10 Mary Kathleen
11 Ben Lomond
12 Olympic Dam
13 Crockers Well, Mt. Painter, Beverley
14 Radium Hill
15 Lake Boga granite

Uranium was first produced in Australia in 1906 at Radium Hill, which was named by Mawson (1906) in his description of the new mineral davidite from that locality. Exploration picked up around 1944 and led to the development of the Rum Jungle, South Alligator River Valley, Radium Hill, and Mary Kathleen deposits. Further exploration in the late 1960s led to the discovery of major deposits at Lake Frome and the Ngalia basin (Langford 1978). In the 1970s major discoveries included Ranger, Jabiluka, and Nabarlek in Northern Territory,

Table 34. Known Uranium Deposits in Australia

Deposit	Type and Method[a]	Status[b]	Size (tU)
Northern Territory			
Angela	Sandstone	DOR	5,000–20,000
Bigrlyi	Sandstone, OP	DOR	1,500–5,000
Hades Flat	Unconformity	DOR	500–1,500
Jabiluka	Unconformity, U	DOR	>50,000
Koongarra	Unconformity, OP	DOR	5,000–20,000
Nabarlek	Unconformity, OP	DEP	5,000–20,000
Napperby	Surficial duricrust, OP	DOR	1,500–5,000
Ranger	Unconformity, OP	PRO	>50,000
Rum Jungle	Unconformity, U and OP	DEP	1,500–5,000
Walbiri	Sandstone	DOR	500–1,500
South Australia			
Beverley	Sandstone, ISL	DOR	5,000–20,000
Crockers Well	Vein	DOR	1,500–5,000
East Kalkaroo	Sandstone	DOR	500–1,500
Goulds Dam	Sandstone	DOR	500–1,500
Honeymoon	Sandstone	DOR	500–1,500
Malbooma	Coal	DOR	500–1,500
Mt. Painter	Vein, OP	DOR	5,000–20,000
Olympic Dam	Breccia complex, U	PRO	>50,000
Paralana-Pepegoona	Sandstone	DOR	500,–1,500
Radium Hill	Vein, U	DEP	500–1,500
Queensland			
Ben Lomond	Volcanic, U and OP	DOR	1,500–5,000
Mary Kathleen	Metasomatic, OP	DEP	5,000–20,000
Maureen	Metasomatic, U and OP	DOR	1,500–5,000
Twoges	Volcanic	DOR	500–1,500
Valhalla	Volcanic	DOR	5,000–20,000
Westmoreland	Sandstone	DOR	5,000–20,000
Western Australia			
Kintyre	Unconformity, OP	PRO	20,000–50,000
Lake Maitland	Surficial duricrust	DOR	1,500–5,000
Lake Raeside	Surficial duricrust	DOR	500–1,500
Lake Way	Surficial duricrust, ISL	DOR	1,500–5,000
Manyingee	Sandstone, ISL	DOR	1,500–5,000
Mulga Rock	Sandstone	DOR	5,000–20,000
Turee Creek	Unconformity	DOR	500–1,500
Yeelirrie	Surficial duricrust, OP	DOR	20,000–50,000

[a] U = underground mining; OP = open-pit mining; ISL = *in situ* leaching; HL = heap leaching; IPL = in-place leaching; CP = extraction as a co-product or by-product of mining for another metal; NS = not specified.

[b] DEP = depleted; DOR = dormant; PRO = producing

Data primarily from Finch et al. (1995).

Table 35. Radioactive minerals first described from Australia

Species	Type Locale	Year
Davidite-(La)	Radium Hill, South Australia	1906
Ulrichite	Lake Boga granite, Victoria	1988
Spriggite	Mt. Painter No. 2, South Australia	2003
Lakebogaite	Lake Boga granite, Victoria	2008
Françoisite-(Ce)	Mt. Painter No. 2, South Australia[a]	2010
Paulscherrerite	Mt. Painter No. 2, South Australia	2011

[a]Co-type locale with La Creusaz, Switzerland

Yeelirrie in Western Australia, and Olympic Dam in South Australia. Olympic Dam has been operating since 1988; it is the world's largest known uranium deposit and is Australia's largest underground mine. Uranium is produced along with gold as a co-product of copper mining. Beverley, the second-largest uranium resource in Australia, is a sandstone type deposit; there uranium is extracted by *in-situ* leaching. The Ranger open-pit mine, operating since 1981, is an unconformity-related deposit in which the main ore mineral is pitchblende, but it is best known to collectors for superb saléeite crystals (Henry et al. 1999).

The Mt. Painter deposit, Radium Ridge, Northern Flinders Ranges, South Australia, is of great historical interest, and although it was never a large commercial success, it has produced a great variety of secondary uranium minerals. Documented species include billietite, boltwoodite, francoisite-(Nd), schoepite, metaschoepite, soddyite, metatorbernite, uranophane-β, and weeksite. The No. 2 Workings at Radium Ridge is the type locality for spriggite (Brugger, Ansermet, and Pring 2003), françoisite-(Ce) (Meisser et al. 2010), and paulscherrerite (Brugger et al. 2011). Note that Radium Ridge and Radium Hill are different places!

Radium Hill, South Australia, ~100 km southwest of Broken Hill, New South Wales, was worked for radium intermittently from 1906 to 1915, producing 350 mg of $RaBr_2$ from 95 t of concentrate. Mining for uranium began in the mid-1950s. The mineralization follows fractures in Archean metasediments and igneous rocks. Gneisses and quartzites

Figure 79. Bright yellow, tabular saléeite crystals, to ~6 mm, on matrix from the Ranger mine, Northern Territory. RJL2863

were first intruded by a sill of metamorphosed amphibolite, then by a transgressive amphibolite, and last by several types of pegmatites, one of which (a Na-rich aplitic type) might be related to ore deposition. A number of individual lodes exist in the area. A lode typically begins as a sericitic shear in banded gneiss. Next, clusters of Fe and Ti minerals appear, with or without uranium, e.g., ilmenite, rutile, and hematite. Davidite is concentrated centrally in the thicker parts. Secondary carnotite forms along fractures and grain boundaries. The age of the davidite, estimated by Pb isotope dating, is 1,730 Ma (Heinrich 1958, pp. 253-54). Radium Hill is the type locality for davidite-(La).

The South Alligator Valley uranium field, Northern Territory, comprises over twenty identified deposits, many of which were actively mined for uraninite during the 1950s and '60s. In some mines, important gold values were also

Figure 80. Massive davidite-(La), ~4 cm wide, from the TL at Radium Hill, South Australia. The davidite is coated with a thin layer of carnotite. RJL2890

encountered. More recent interest in uranium during the 1980s led to the discovery of significant gold, platinum, and palladium mineralization at Coronation Hill. The rare species threadgoldite is widespread in the district (Henry et al. 2005). Other secondary minerals noted there include autunite, meta-autunite, becquerelite, curite, dewindtite, dumontite, kasolite, masuyite, phosphuranylite, rutherfordine, saléeite, soddyite, torbernite, metatorbernite, uranophane, uranopilite, and vandendriesscheite (Threadgold 1960).

The Lake Boga Granite, Victoria, is the type locale for ulrichite and lakebogaite (Birch, Mumme, and Segnit 1988; Mills et al. 2008); several other secondary species are found there, including torbernite, saléeite, and metanatroautunite ("sodium autunite").

Figure 81. Green ulrichite crystals forming a small rosette in a cavity, from the TL in the Lake Boga granite, Victoria, Australia. RJL2890 macro

AFRICA

CONGO, DEMOCRATIC REPUBLIC OF THE

Table 36. Known uranium deposits in Congo

Deposit	Type and Method[a]	Status[b]	Size (tU)
Kalongwe	Vein	DOR	<500
Kambove West	Vein	DEP	1,000–2,500
Kamoto	Vein	DOR	1,500–2,500
Kasompi East	Vein	DOR	500–1,000
Musonoi	Vein	DEP	2,500–5,000
Shinkolobwe-Kasolo	Vein	DEP	25,000–50,000
Swambo	Vein	DOR	500–1,000

[a] U = underground mining; OP = open-pit mining; ISL = *in situ* leaching; HL = heap leaching; IPL = in-place leaching; CP = extraction as a co-product or by-product of mining for another metal; NS = not specified.

[b] DEP = depleted; DOR = dormant; PRO = producing

Data primarily from Finch et al. (1995) and IAEA (2009).

The sheer quantity and quality of specimens has made the Democratic Republic of the Congo the premier locality for students of uranium mineralogy. This country should not be confused with the neighboring Republic of the Congo (formerly French Congo, or Congo-Brazzaville). The country of interest here is the former Belgian Congo, which was later named Zaire, and later still re-named Democratic Republic of the Congo (DRC). For simplicity, it will be referred to as "Congo." To further complicate matters, the name of the copper-rich Katanga province in the southern part of the country was changed to Shaba province but now is once again generally called Katanga. (Both names are derived from native words for copper.) Collectors may encounter various combinations of these names on labels, particularly with older specimens.

The rich copper deposits of Katanga were known and exploited by natives since ancient times. Within this area are several significant uranium occurrences, each with its own distinctive geology and suite of secondary minerals.

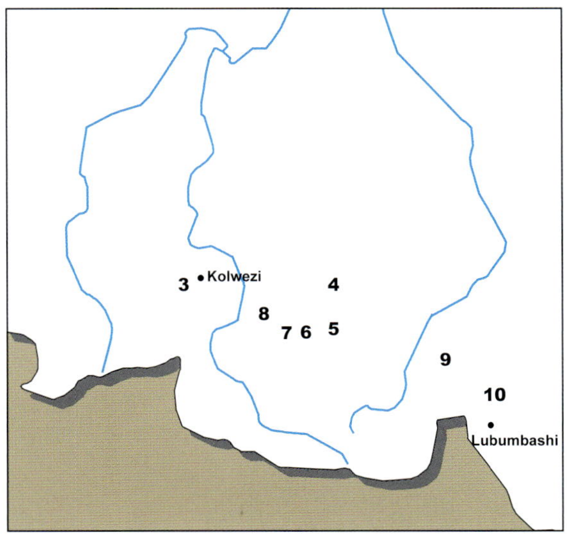

Figure 82. Map of the Democratic Republic of Congo showing the overall locations of:
1 Kobokobo pegmatite, Kivu province
2 The uranium-copper district, Katanga province
The inset detail shows the notable localities in Katanga:
3 Kolwezi district, including Musonoi, Kamoto, and Kalongwe
4 Kambove
5 Shinkolobwe-Kasolo
6 Swambo
7 Mindigi
8 Kasompi
9 Luishia
10 Luiswishi

The first documented find of uranium was at the Luiswishi copper mine in 1913. However, the amount of uranium ore was not economically important. The major deposits at Shinkolobwe were discovered in 1915 and were mined for radium from about 1921 through 1936. Mining was resumed in 1944 and continued until 1959. Long after the mine was closed, scientists have continued to describe new minerals from archived ore samples (Deliens et al. 1981).

Uranium was found at the Kalongwe copper mine associated with chalcocite and chalcopyrite orebodies in 1930. Although uranium has not been actively mined at this locale, it has produced specimens of numerous secondary uranium minerals.

The deposit at Swambo was discovered in 1956, somewhat by accident, as it was not part of a copper mine or prospect. Swambo is the type locality for swamboite and has also yielded fine soddyite crystals.

The Kamoto East mine is noteworthy because of a unique combination of uranium, copper, and rare earth elements that yielded several new U-REE species (Deliens et al. 1990).

The Musonoi uranium-copper-cobalt deposit contains an interesting suite of secondary uranyl selenates in the surface alteration zone. The selenium is apparently derived from weathering of copper sulfide ores that contain some Se and Te in substitution for sulfur.

The uraniferous pegmatite at Kobokobo, Kivu province, which was mined for beryl and columbite, produced at least a half-dozen new aluminum uranyl phosphates and several containing thorium as well.

Over fifty uranium minerals were first described from localities in Congo, Table 37.

Figure 83. A thumbnail-sized sample from Kamoto East, showing a flattened spray of bright yellow laths of kamotoite-(Y) with other secondary U species on uraninite. RJL2699

Figure 84. Pearly, olive-green marthozite forming scaly masses on Se-rich digenite, from Musonoi. RJL1145

Table 37. Radioactive minerals first described from Congo

Species	Type Locale	Year
Curite	Shinkolobwe, Katanga	1921
Kasolite	Shinkolobwe, Katanga	1921
Becquerelite	Shinkolobwe, Katanga	1922
Dewindtite	Shinkolobwe, Katanga	1922
Soddyite	Shinkolobwe, Katanga	1922
Parsonsite	Shinkolobwe, Katanga	1923
Schoepite	Shinkolobwe, Katanga	1923
Dumontite	Shinkolobwe, Katanga	1924
Fourmarierite	Shinkolobwe, Katanga	1924
Sklodowskite	Shinkolobwe, Katanga	1924
Ianthinite	Shinkolobwe, Katanga	1926
Saleeite	Shinkolobwe, Katanga	1932
Vandenbrandeite	Kalongwe, Katanga	1932
Cuprosklodowskite	Shinkolobwe, Katanga	1933
Sharpite	Shinkolobwe, Katanga	1938
Billietite	Shinkolobwe, Katanga	1947
Masuyite	Shinkolobwe, Katanga	1947
Richetite	Shinkolobwe, Katanga	1947
Studtite	Shinkolobwe, Katanga	1947
Vandendriesscheite	Shinkolobwe, Katanga	1947
Sengierite	Luiswishi, Katanga	1949
Wyartite	Shinkolobwe, Katanga	1959
Demesmaekerite	Musonoi, Katanga	1965
Guilleminite	Musonoi, Katanga	1965
Marthozite	Musonoi, Katanga	1969
Roubaultite	Shinkolobwe, Katanga	1970
Derriksite	Musonoi, Katanga	1971
Eylettersite	Kobokobo, Kivu	1971
Phuralumite	Kobokobo, Kivu	1979
Ranunculite	Kobokobo, Kivu	1979
Threadgoldite	Kobokobo, Kivu	1979
Upalite	Kobokobo, Kivu	1979
Mundite	Kobokobo, Kivu	1981
Swamboite	Swambo, Katanga	1981
Bijvoetite-(Y)	Shinkolobwe, Katanga	1982
Lepersonnite-(Gd)	Shinkolobwe, Katanga	1982
Metavanmeersscheite	Kobokobo, Kivu	1982
Triangulite	Kobokobo, Kivu	1982
Vanmeersscheite	Kobokobo, Kivu	1982
Metastudtite	Shinkolobwe, Katanga	1983
Oursinite	Shinkolobwe, Katanga	1983
Sayrite	Shinkolobwe, Katanga	1983
Kamitugaite	Kobokobo, Kivu	1984
Urancalcarite	Shinkolobwe, Katanga	1984
Moreauite	Kobokobo, Kivu	1985
Kamotoite-(Y)	Kamoto East, Katanga	1986
Protasite	Shinkolobwe, Katanga	1986
Althupite	Kobokobo, Kivu	1987
Francoisite-(Nd)	Kamoto East, Katanga	1988
Shabaite-(Nd)	Kamoto East, Katanga	1989
Astrocyanite-(Ce)	Kamoto East, Katanga	1990
Piretite	Shinkolobwe, Katanga	1996

Figure 85. A classic association from the Musonoi mine: dark green, tabular torbernite and apple green, acicular cuprosklodowskite. Specimen is ~3 cm wide. RJL4526

GABON

Table 38. Known uranium deposits in Gabon

Deposit	Type and Method[a]	Status[b]	Size (tU)
Bangombé, Haut-Ogoue	Sandstone	DOR	<500
Boyindzi, Haut-Ogoue	Sandstone, U	DEP	2,500–5,000
Mikoulongou, Haut-Ogoue	Sandstone, OP	DEP	1,000–2,500
Mounana, Haut-Ogoue	Sandstone, U and OP	DEP	5,000–20,000
Okelobondo, Haut-Ogoue	Sandstone, U	DEP	5,000–10,000
Oklo, Haut-Ogoue	Sandstone, U and OP	DEP	10,000–25,000

[a] U = underground mining; OP = open-pit mining; ISL = *in situ* leaching; HL = heap leaching; IPL = in-place leaching; CP = extraction as a co-product or by-product of mining for another metal; NS = not specified.

[b] DEP = depleted; DOR = dormant; PRO = producing

Data primarily from Finch et al. (1995) and IAEA (2009).

The uranium-vanadium deposits of Mounana, in the province of Haut-Ogoue, were discovered in 1956 by geologists of the French Atomic Energy Commission. In all there are six U deposits in the Franceville basin, of which four (Boyindzi, Mounana, Oklo, and Okelobondo) were economically important, with total U resources estimated at 40,000 tU (Gauthier-Lafaye and Weber 1989). Open pit mining began in 1961 and annual production peaked at around 1,250 tons of uranium in the late 1970s. The last areas mined were an underground working at Okelobondo and an open pit at Mikouloungou, which closed in March 1999. Total uranium production was about 26,000 tons. The Franceville basin is especially noteworthy for two reasons. First, the oxidized zone at Mounana has produced an extraordinary variety of uranyl vanadates and many new vanadium minerals. Second, very rich orebodies at Oklo, Okelobondo, and Bangombé formed natural nuclear reactors about 2 Ga ago, a phenomenon that has not been documented anywhere else in the world.

The Mounana mine is well known to collectors as a classic source of secondary uranium minerals. It is the type locality for mounanaite, $PbFe_2^{+3}(VO_4)_2(OH)_2$, and superb specimens of orange francevillite on velvety brown mounanaite are among the unique and sought-after mineral assemblages found only at this mine. The Mounana deposit forms a fault wedge in coarse sandstones of Middle Cambrian age. The ore body was about 150 m deep and up to 40 m wide. The mine was worked as an open pit to a depth of 70 m, and then by underground methods. The ore deposit formed two distinct zones (Cesbron and Bariand 1975):

1. the deep or "black ore" zone where U and V are in their lower oxidation states and form one mineral assemblage; and

2. the oxidized zone, about 40 m thick, where the mineral suite was predominantly aluminum, lead, and barium uranyl vanadates.

Within the black ore zone there is a uranium-rich area where the main ore minerals are uraninite and some coffinite. Together with galena, sphalerite, iron and copper sulfides, and

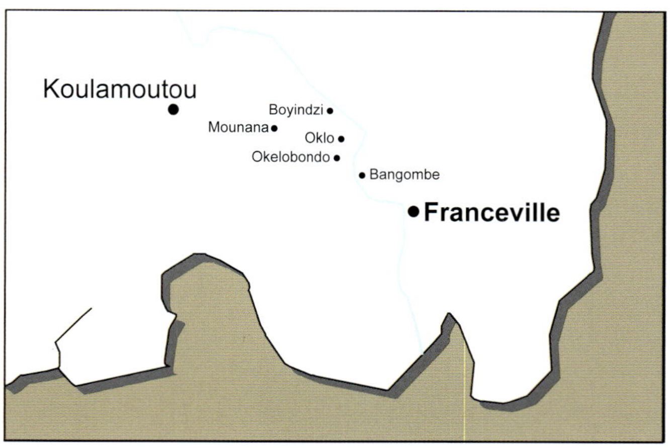

Figure 86. Map of Gabon with the area around Franceville indicated. The inset detail shows the locations of major occurrences relative to the towns of Franceville and Koulamoutou.

asphaltic material, these minerals form the cementing material in the sandstone. Accessory minerals in this zone included zircon, tourmaline, rutile, and barite. Also within the black ore zone was a vanadium-rich pillar containing a number of rare vanadium oxides and hydroxides, including karelianite, montroseite, roscoelite, nolanite, duttonite, lenoblite, and bariandite.

The oxide zone contained an extraordinary suite of uranyl vanadates, typically filling cracks up to several cm thick, in the friable sandstone. From this zone seven new minerals were described in the period from 1957 to 1971, including four radioactive species, Table 39. The following species have been recovered from this deposit: carnotite, francevillite, curienite, vanuralite, metavanuralite, chervetite, mottramite, mounanaite, schubnelite, uranocircite, torbernite, kasolite, uranopilite, and johannite. The oxide zone was completely mined out in the mid-1970s.

Table 39. Radioactive minerals first described from Gabon

Species	Type Locale	Year
Francevillite	Mounana	1957
Vanuralite	Mounana	1963
Curienite	Mounana	1968
Metavanuralite	Mounana	1970

Although the Oklo mine is much less important as a source of minerals for the collector, it is a most unique geological suite that deserves particular mention. The Oklo Natural Fission Reactor phenomenon was first brought to light in 1972, when routine assaying at a French enrichment plant showed that the processed ore was slightly depleted in ^{235}U. Further isotopic analysis and the detection of fission products confirmed that spontaneous nuclear fission had occurred at Oklo, prompting widespread scientific interest in the site. It is the only locality where isotopic concentrations differ greatly from the ratios expected from "primordial" concentrations and natural decay. Understanding this site requires a combination of classic geochemistry and nuclear physics, and from neutronic calculations it became clear why natural reactors have not been found in other places where equally rich U orebodies occur. The conditions at Oklo and Bangombé that were necessary to support self-sustaining nuclear fission are summarized as follows (Naudet, 1991):

1. Total U concentrations of at least 10% in a layer 2 meters thick
2. Uranium ore seams at least 0.5 meter thick
3. Water (as a neutron moderator) to U ratio of around 6%
4. The presence of neutron reflectors (e.g., quartz in the sandstone)
5. Low concentrations of neutron absorbers (B, Li, Mn, V, and heavy rare earths).

The role of hydrothermal activity is described by Gauthier-Lafaye, Weber, and Ohmoto (1989): "The ore which hosts the fission reactors at Oklo is characterized by a gangue of clay minerals, whereas the normal ore of Oklo consists of sandstones. These clay minerals are the result of hydrothermal alteration of sandstone; hydrothermal circulation was induced by the heat produced by fission reactions. The reactor ores are further characterized by very high uranium contents (20–60%) and high uranium accumulation. By means of silica depletion and by uranium enrichment of the normal ore (without uranium depletion), hydrothermal convective

Figure 87. A superb example of francevillite on mounanaite, ~5 cm wide, from the type locality at Mounana, Gabon. RJL1353

circulation, induced by the fission reactions, created very high grade ore next to the reactors. Hydrothermal processes are thus critical to the propagation of the fission reactions."

The mineral assemblage at Oklo includes massive uraninite as well as small grains and veinlets dispersed in sandstone and bitumen. Coffinite is the second most abundant uraniferous mineral at the reactor sites and has generally formed as a replacement product of uraninite. Uranyl minerals that have been identified at these sites include uranophane, torbernite, fourmarierite, rutherfordine, and wölsendorfite. Uranyl minerals identified in sandstone beneath the Bangombé reactor include francoisite-(Nd), uranopilite or zippeite, and schoepite, along with other partially characterized compositions. Perhaps most interesting is the occurrence here of minerals clearly derived from fission product elements. Specifically, sulfides and arsenides of Ru, Rh, and Pd have been found as tiny grains within uraninite (Hollinger 1992). Several conferences and hundreds of research papers have been devoted to this topic and were recently reviewed by Janeczek (1999). Actual mining at Oklo ended in late 1997.

NAMIBIA

Namibia is currently the fourth largest uranium producer, with ~8% of global output. Production capacity at the Rössing mine is about 4,000 tU/year, with actual production running around 2,500 tU/year. Several other significant deposits exist, the most

Table 40. Known Uranium Deposits in Namibia

Deposit	Type and Method[a]	Status[b]	Size (tU)
Auris	Intrusive	DOR	500–1,000
Engo Valley	Sandstone	DOR	5,000–20,000
Goanikontes	Intrusive	DOR	5,000–10,000
Hakskeen	Surficial, OP	DOR	500–1,000
Klein Trekkopje	Surficial, OP	DOR	25,000–50,000
Langer Heinrich	Surficial, OP	PRO	25,000–50,000
Mile 72	Surficial, OP	DOR	<500
Rössing	Intrusive, U and OP	PRO	>100,000
Tumas	Surficial, OP	DOR	2,500–5,000
Valencia	Intrusive	DOR	10,000–25,000

[a] U = underground mining; OP = open-pit mining; ISL = *in situ* leaching; HL = heap leaching; IPL = in-place leaching; CP = extraction as a co-product or by-product of mining for another metal; NS = not specified.

[b] DEP = depleted; DOR = dormant; PRO = producing

Data primarily from Finch et al. (1995) and IAEA (2009).

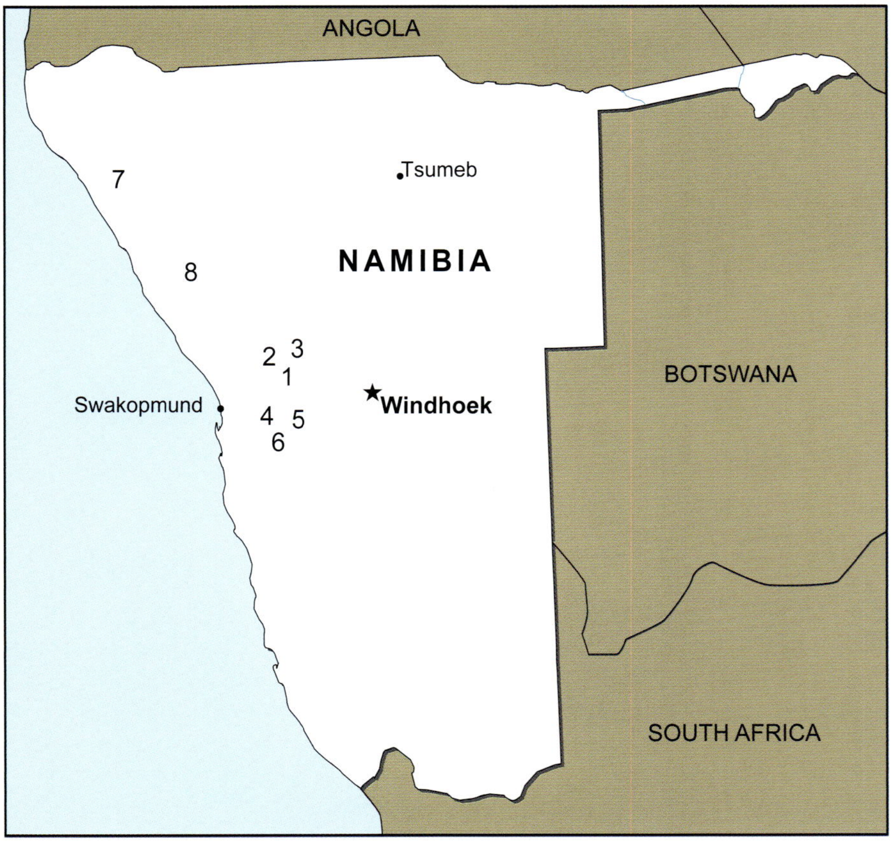

Figure 88. Important uranium deposits in Namibia:
1 Rössing
2 Mile 72
3 Klein Trekkopje
4 Goanikontes
5 Langer Heinrich
6 Tumas
7 Engo Valley
8 Auris

economically favorable being the low-grade surficial duricrust deposit at Langer Heinrich, which is mined as an open pit with a current capacity around 2,000 tU/year (OECD 2012).

The Rössing deposit is located in the Namib Desert about 70 km northeast of Swakopmund. The presence of radioactive minerals was known as early as 1928, based on simple autoradiography tests of rocks from the area. Detailed exploration programs were conducted off and on through 1973, and mine development began in 1974.

The main orebody at Rössing is a uraniferous alaskite (granite containing only quartz and alkali feldspar), whose texture ranges from aplitic (sugary) to granitic to pegmatitic, with the pegmatitic texture generally predominating. All of the primary minerals and most of the secondary minerals occur within the alaskite. The primary minerals (uraninite with minor betafite) represent about 60% of the uranium present. The remaining 40% is contained in the secondary minerals, which include "gummite," uranophane, uranophane-β, torbernite and metatorbernite, carnotite, metahaiweeite, and thorogummite (Berning et al. 1976).

The Goanikontes area is an important source of fine materials for the collector. This area, now known as the Etango deposit, comprises the Anomaly A, Oshiveli, and Onkelo uranium-enriched zones. These three prospects form a contiguous zone over ~6 km. The alaskite ore is similar to that found at Rössing, and occurs from the surface to depths

Figure 90. Pearly green metazeunerite crystals on matrix, from Erongo. RJL2701

Figure 89. Straw-yellow boltwoodite forming dense subparallel masses in calcite-lined cavity, from Goanikontes, Namibia. RJL2590

Figure 91. Uraniferous opal on tabular feldspar crystals to ~3 cm long, from the Erongo Mountains, Namibia. RJL4139

Figure 92. Specimen in the previous photo viewed in LWUV light.

Figure 93. Specimen in the previous photo viewed in SWUV light. Note the more intense yellow-green fluorescence in SWUV as compared to LWUV.

up to 400 m (OECD 2012). This locality has produced colorful boltwoodite specimens, with radiating acicular crystals ranging from opaque straw yellow to translucent orange forming spherical aggregates on dark calcite. The area has also yielded a small number of brilliant glassy yellow weeksite specimens. Although still "micro" crystals, they are spectacular examples of the species.

A notable uranium occurrence is found in the Erongo granite, a district better known for spectacular schorl and beryl crystals. The occurrence lies on the farms Anibib 136 and Omandumba 137. Tabular, emerald green zeunerite crystals are found, up to 1 cm across but typically smaller; these quickly dehydrate to metazeunerite after removal from the pocket. Other U minerals have been found in this general area, including metanováčekite, walpurgite, uranophane, and fluorescent hyaline opal (Cairncross and Bahmann 2006).

NIGER

Uranium has been produced from sandstone-type deposits in Niger since 1970. The country currently produces ~4,000 tU/year, or ~8% of global output, which is roughly equivalent to that of Namibia. All of the U resources in Niger are sandstone-type deposits with average grades of 0.07 to 0.39% U, and total resources as of 2011 are estimated at ~445,000 tU. Active exploration is ongoing; in 2011 uranium exploration was being conducted by foreign companies on 160 concessions. Uranium represents about one-third of Niger's export revenue (OECD 2012).

Because the country's uranium resources are all sandstone deposits, little or no material is produced that would be of interest to the mineral collecting community.

Table 41. Known uranium deposits in Niger

Deposit	Type and Method[a]	Status[b]	Size (tU)
Abakorum	Sandstone	DOR	10,000–25,000
Afasto Est	Sandstone	DOR	5,000–20,000
Afasto Ouest	Sandstone	DOR	20,000–50,000
Akola	Sandstone	PRO	25,000–50,000
Akouta Nord	Sandstone, U	PRO	25,000–50,000
Akouta Sud	Sandstone	DOR	10,000–25,000
Arcadie	Sandstone	DOR	5,000–10,000
Argus	Sandstone	DOR	1,000–2,500
Ariege	Sandstone	PRO	10,000–25,000
Arlette	Sandstone, OP	DEP	10,000–25,000
Nord Somair	Sandstone	DOR	10,000–25,000
Takriza	Sandstone	DEP	5,000–10,000
Tamou	Sandstone	PRO	10,000–25,000
Taossa	Sandstone	DOR	2,500–5,000
Taza Nord	Sandstone	DEP	2,500–5,000
Taza Sud	Sandstone	DEP	5,000–10,000
Teguidda	Sandstone	DOR	10,000–25,000

[a] U = underground mining; OP = open-pit mining; ISL = *in situ* leaching; HL = heap leaching; IPL = in-place leaching; CP = extraction as a co-product or by-product of mining for another metal; NS = not specified.

[b] DEP = depleted; DOR = dormant; PRO = producing

Data primarily from Finch et al. (1995) and IAEA (2009).

SOUTH AFRICA

South Africa began producing uranium in 1952 and has been an important source in the years since then. In the late 1970s and early 1980s, it was the second or third largest producer in the world, but by 2009 had dropped to eleventh place. In the early 1980s, peak production was over 6000 tU/year, representing ~14% of global output.

Although seven of the thirteen recognized deposit types have been found, the vast majority of South Africa's uranium resources are in quartz-pebble conglomerate and associated tailings, with some additional resources in sandstone and coal-hosted deposits. Historically the majority of uranium was produced as a by-product of gold (or in some cases copper) mining. Only two pure uranium producers have operated there, viz., the Beisa mine in the early 1980s, and the Dominion Reefs uranium mine, near Klerksdorp, which produced in the late 2000s (OECD 2012).

It was known as early as 1915 that gold ores in the Witwatersrand contained radioactive minerals. Uraninite was specifically identified among the ore concentrates by R. A. Cooper in 1923, an observation that was of no economic interest at the time. But when interest in uranium suddenly increased in 1944, a major exploration and sampling effort began, which focused on the Blyvooruitzicht, Vogelstruisbult, Western Reefs, and East Daggafontein gold mines. The samples were taken from areas that were much higher grade (in U content) than run-of-the-mine ore, and even then didn't appear to be economic based on their uranium values alone. The low grade of these ores is offset by the value of the gold being extracted, making co-production of Au and U feasible.

The Witwatersrand Supergroup contains both Au- and U-bearing conglomerates; those of the Central Rand Group are the most important and represent ~87% of South Africa's lower-cost uranium resources. These sediments were deposited

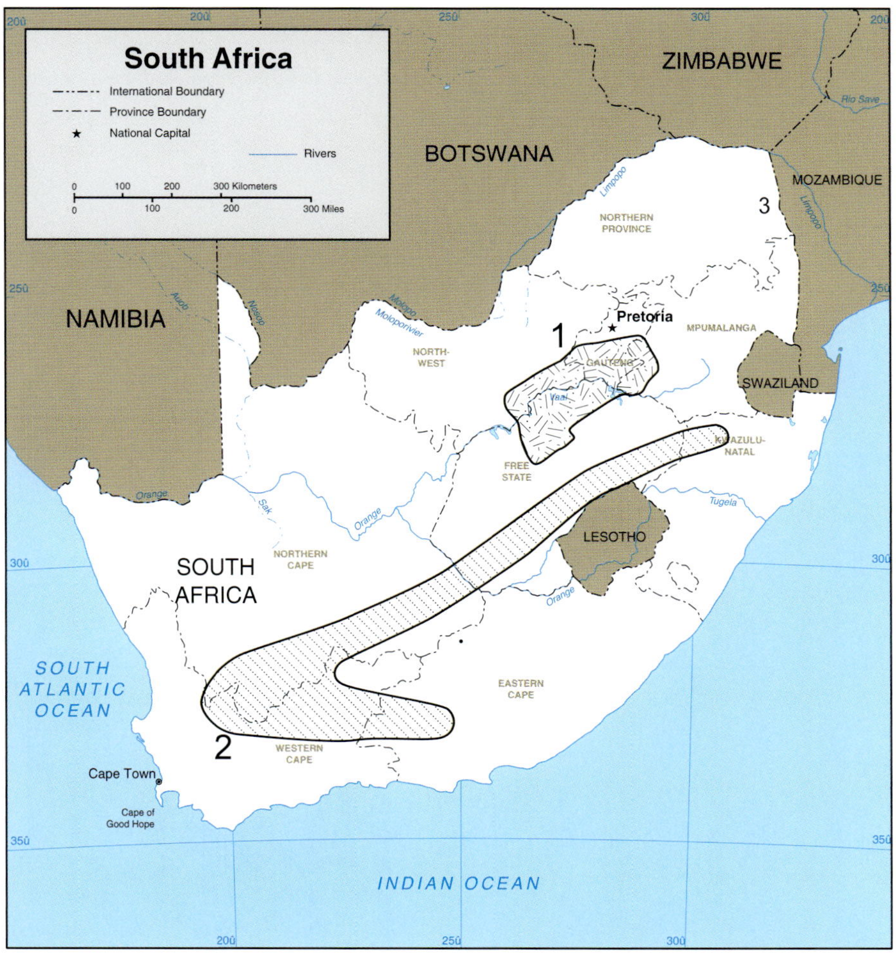

Figure 94. Important uranium deposits in South Africa; stippling indicates the approximate extent of the Witwatersrand and Karoo provinces:
1 Witwatersrand province (includes all quartz-pebble conglomerates)
2 Karoo Sandstone province
3 Phalaborwa

Table 42. Known uranium deposits in South Africa

Deposit	Type and Method[a]	Status[b]	Size (tU)
Blyvooruitzicht	Quartz-pebble conglomerate, CP	DOR	
Buffelsfontein	Quartz-pebble conglomerate, CP	PRO	5,000–10,000
Chemwes-Stilfontein	Quartz-pebble conglomerate, CP	DOR	
Driefontein	Quartz-pebble conglomerate, CP	DOR	
Free State Geduld	Quartz-pebble conglomerate, CP	DOR	5,000–10,000
Harmony	Quartz-pebble conglomerate, CP	DOR	
Hartebeestfontein	Quartz-pebble conglomerate, CP	PRO	5,000–10,000
Henkries	Quartz-pebble conglomerate, CP	DOR	1,000–2,500
Loraine	Quartz-pebble conglomerate, CP	DOR	
Phalaborwa	Intrusive, U and OP	DOR	2,500–5,000
President Brand	Quartz-pebble conglomerate, CP	PRO	2,500–5,000
President Steyn	Quartz-pebble conglomerate, CP	PRO	5,000–10,000
Randfontein	Quartz-pebble conglomerate, CP	DOR	50,000–100,000
St. Helena	Quartz-pebble conglomerate, CP	DOR	10,000–25,000
Vaal Reefs	Quartz-pebble conglomerate, CP	PRO	50,000–100,000
Vaal River area	Quartz-pebble conglomerate, CP	PRO	50,000–100,000
West Rand	Quartz-pebble conglomerate, CP	DOR	
Western Areas	Quartz-pebble conglomerate, CP	DOR	
Western Deep Levels	Quartz-pebble conglomerate, CP	DOR	10,000–25,000
Western Holdings	Quartz-pebble conglomerate, CP		

[a] U = underground mining; OP = open-pit mining; ISL = *in situ* leaching; HL = heap leaching; IPL = in-place leaching; CP = extraction as a co-product or by-product of mining for another metal; NS = not specified.

[b] DEP = depleted; DOR = dormant; PRO = producing

Data primarily from Finch et al. (1995) and IAEA 2009).

Figure 96. Ore microscope view in polished section from the Carbon Leader Reef, Central Rand district, South Africa. Minute particles of gold are visible; larger bright masses are galena. The small cube (lower center) is uraninite containing bright specks of galena formed from radiogenic lead as mentioned earlier in the discussion of auto-oxidation. Photo courtesy of Jim A. Paschis

in a fluvial fan environment. Their formation involved two essential conditions:

1. the absence of vegetation, allowing uninhibited denudation of the parent rocks; and,

2. a setting that created a sedimentological environment where recurrent reworking and winnowing processes efficiently concentrated the heavy minerals.

The major U-bearing reefs (the Basal, Beisa, Vaal, and Carbon Leader Reefs) are small-pebble (8–15 mm) conglomerates. They are gray to dark gray, with a coarse fraction of waterworn pebbles in a finer-grained matrix of quartz, silicates, and heavy minerals. The U-bearing minerals include uraninite, coffinite, brannerite-type species, and U-rich zircon. The uraninite was a primary phase of detrital origin, whereas brannerite and coffinite formed after the sediments were deposited, by reaction of the uraninite with circulating solutions. In general, coffinite and zircon are rare in these rocks, and most of the uranium is contained in the uraninite and brannerite (Ford 1993).

Figure 95. A sample of quartz-pebble conglomerate from the Carbon Leader Reef, Witwatersrand Supergroup, collected at the working face over 2 km underground at Driefontein by Dr. Tommy Phelps in December 1998. Specimen is ~6 cm wide. RJL4535

SYSTEMATIC MINERALOGY

PRIMARY MINERALS

OXIDES

Thorianite-Uraninite Series

The thorianite-uraninite system forms a complete series of solid solutions from ThO_2 to UO_2, based on studies of synthetic materials. In nature, however, compositions with Th:U ratios close to 1:1 are relatively rare (the "uranothorianites" of Madagascar are close to a 1:1 composition).

Structure

Uraninite and thorianite are isometric, with the fluorite structure; in uraninite, each U^{4+} is surrounded by eight O^{2-} ions at the corners of a cube and each O^{2-} ion is surrounded by four U^{4+} ions at the corners of a regular tetrahedron. The fluorite structure arises because the ionic radii of U^{4+} (0.97 Å) and O^{2-} (1.32 Å) closely match those of Ca^{2+} (0.99 Å) and F^- (1.33 Å), respectively.

The unit cell dimension *a* has been reported for a large number of natural and synthetic uraninites. Frondel (1958), reviewing the extensive literature up to that time, presented data for over a dozen synthetic uraninites, with *a* ranging from 5.455 to 5.4708 Å, and over 200 natural uraninites with *a* ranging from 5.368 to 5.555 Å. He cautioned that the results for synthetic compositions are generally "of only moderate accuracy because of failure to state or control the temperature at which the measurements were made, lack of evidence that the composition actually was stoichiometric UO_2 and otherwise pure, and inadequate refinement of the experimental data to remove systematic error." For the natural uraninite and pitchblende, "These results mostly are of very low accuracy and were made for the most part on unanalyzed material. The observed wide variation in *a* is due principally to the varying extent of oxidation of U^{4+} to U^{6+}. . . Additional factors are the content of radiogenic lead, the content of Th, rare earths, or other elements in substitution for U, the amount of O or OH present interstitially in valence compensation for the U^{6+}, and the amount of structural damage by internal alpha-particle bombardment."

Berman (1955) noted a trend in unit cell size versus mode of formation, where samples of pitchblende from hydrothermal veins have a smaller cell size than uraninites from pegmatites. She interpreted this as follows: "The relatively small cell size of pitchblende is due to the following factors: (a) Relatively high oxidation, due to the fine-grain or fine fibrous permeable nature of the material and the resulting large surface area. The oxidation involves the conversion of U^{IV} to U^{VI} with valence compensation effected by the entrance of oxygen into the vacant eight-fold position of the fluorite-type structure; the U^{VI} ion being relatively smaller than the U^{IV} ion, greater oxidation tends to decrease the unit cell size. (b) Relative freedom from larger thorium and rare earth ions, (U^{IV} = 1.05 Å; Th^{IV} = 1.10 Å; Ce^{IV} = 1.02 Å), which are ordinarily present in relatively large amounts in uraninite from pegmatite. Pitchblende and relatively highly oxidized uraninite in general give diffuse and faint diffraction effects. This is due both to line broadening from small particle size, and also . . . to variation in degree of oxidation within the sample, with accompanying variation in cell size. The cell size also should vary with the amount of radiogenic lead present (Pb^{IV} = 0.84 Å; U^{IV} = 1.05 Å)."

The ionic radius of Th^{4+} is almost 5% greater than that of U^{4+} and consequently the lattice parameter of thorianite is greater than that of uraninite. In data summarized by Frondel (1958), reported *a* values for synthetic ThO_2 range from 5.5961 to 5.586 Å, and in natural samples *a* ranges from 5.58 to 5.505 Å.

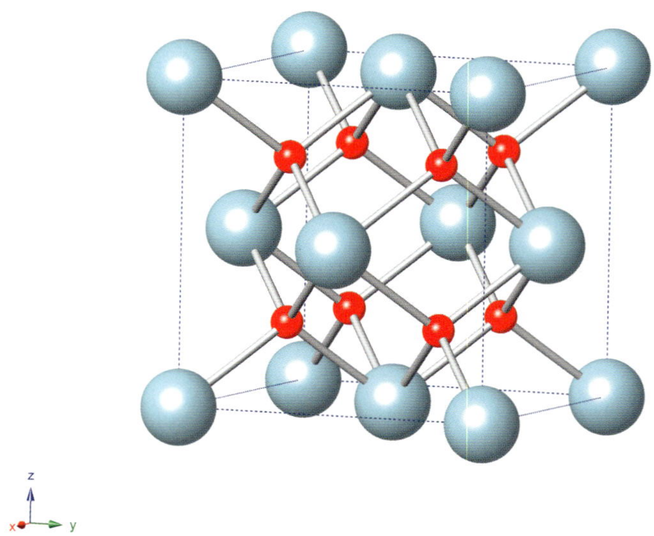

Figure 97. Ball and stick model of uraninite, showing the coordination of the U atoms (blue-gray) and oxygen (red).

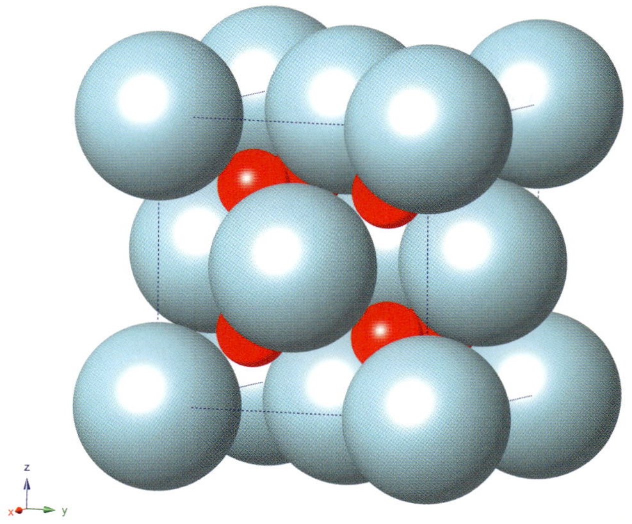

Figure 98. Space-filling model of uraninite, providing a physically realistic view of the relative sizes of the ions.

A continuous series exists between the two end members, uraninite (U>Th) and thorianite (Th>U). They are treated separately here, following the 50% convention for binary systems.

Thorianite ThO_2

Thorianite occurs chiefly in pegmatites and granites, or in sediments and placers derived from these deposits. It was described by Dunstan (1904) from the Balangoda district, Sri Lanka, where sharp, unaltered crystals are recovered from stream gravels as a by-product of gem mining. Thorianite has been found *in situ* in the source rock of these alluvial deposits, a disintegrated pegmatite near Bambarabotuwa. Dunstan noted that thorite, $ThSiO_4$, also occurs in the gem gravels. Some interesting historical context is provided by his comment that, "It is obvious that apart from the scientific interest attaching to the determination of their composition,

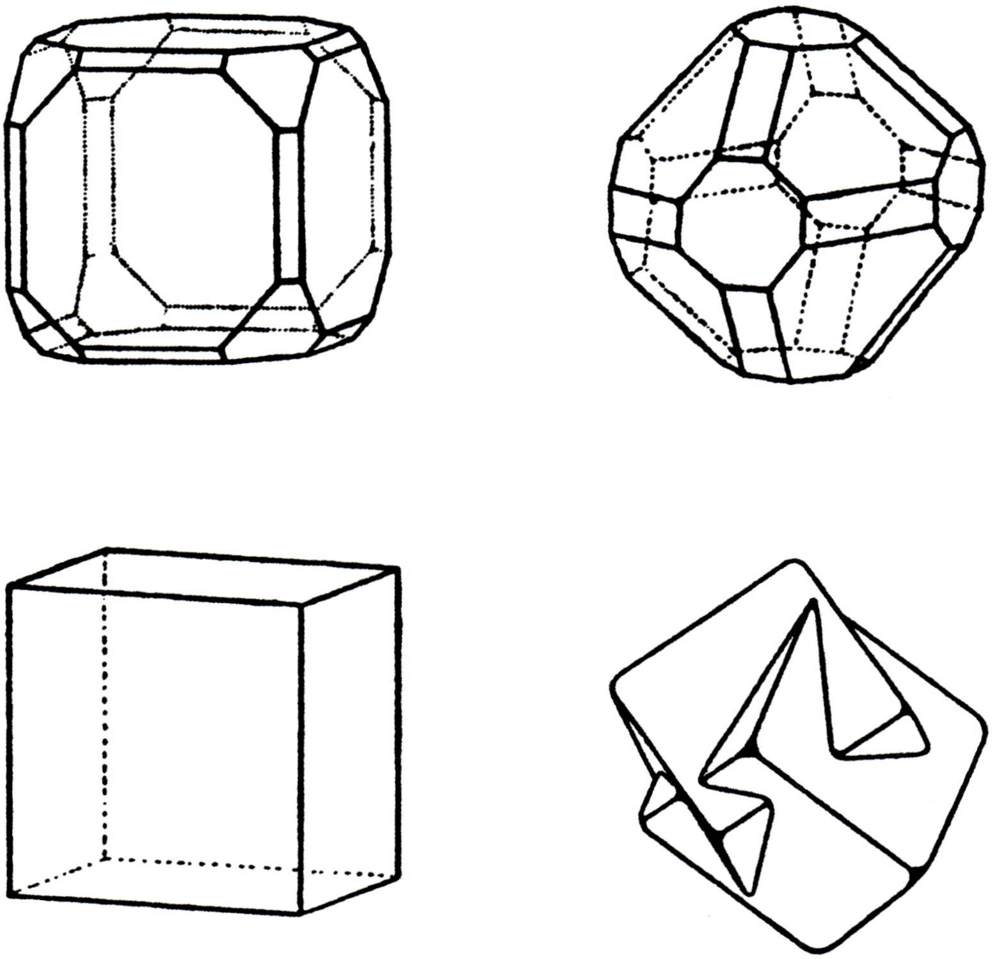

Figure 99. Drawings of some natural crystals, modified from Goldschmidt (1922, 1923). Top: uraninite from Arendal, Norway (left) and Middletown, Connecticut (right). Bottom: two examples of thorianite from Sri Lanka.

the discovery in Sri Lanka of two minerals rich in thoria, now so largely employed for the manufacture of incandescent gas mantles, may be of considerable commercial importance."

Thorianite usually forms simple black to brownish black cubic crystals and penetration twins. A poor cleavage may be seen on {001}; fracture is uneven to conchoidal (particularly in metamict samples). Thorianite typically contains 5–10% UO_2 and up to 8% REE oxides, as well as up to 5% radiogenic lead. Because Th^{4+} has no higher valence (corresponding to U^{6+}), thorianite is chemically stable in the environment, although the crystals are usually metamict to some degree and can undergo further alteration as any substitutional U^{4+} is oxidized to U^{6+}.

Thorianite is commonly found in placer deposits, typically accompanied by thorite, zircon, and ilmenite. In addition to the TL, alluvial occurrences include: Betroka and Andalobe, Madagascar; Travancore, India; in black sands of a gold placer in the Boshogosh River, Siberia, Russia; in the Missouri River near Helena, Montana; in the Scott River, Siskiyou Co., California; and in black sands in the Nixon Fork and Wiseman districts, Alaska.

Small amounts occur in granite in the Congo; with uraninite in pegmatite on Charlebois Lake east of Lake Athabasca, Saskatchewan; in syenitic pegmatite near Bhaluchuan, Odisha, India (Singh et al. 2014); and with baddeleyite as an accessory in carbonatite in Phalaborwa, eastern Transvaal, South Africa (Hiemstra 1955).

Metasomatic occurrences include: in serpentine at the contact of limestone and pegmatite at Easton, Pennsylvania; and in basic skarns and pyroxene skarns at Betroka and elsewhere in Madagascar.

U-rich varieties (U:Th ranging from 0.94 down to 0.43) occur at many localities in Quebec and Ontario, in pegmatites, calcite-fluorite-apatite veins transitional to pegmatites, and metasomatized zones in crystalline limestones (Robinson and Sabina 1955).

Figure 100. Typical black, cubic thorianite crystals: a single crystal ~1 cm tall from Fort Dauphine, Madagascar (left), and a cluster of 3 mm crystals from the Balangoda district, Sri Lanka (right). RJL632, RJL331.

Nearly pure end-member thorianite (a=5.603Å) occurs as small, light orange cubic crystals with pyrochlore as an accessory in calcite-diopside-hornblende and calcite-garnet-vesuvianite rocks in the Central massif of the Turja Peninsula, Russia. The low U content of this material contrasts with that of U-rich thorianite typically found in carbonatites. Bulakh and Mazalov (1974) conclude, "Evidently, such a change from essentially pure thorianite in the phlogopite-bearing, calcite-garnet-vesuvianite and other early metasomatites to the uraniferous varieties in the later carbonatites is in complete agreement with the general geochemical characteristic of thorium and uranium—a gradual decrease in the ratio Th/U with time in natural evolutionary series of rocks and minerals. This tendency of thorium and uranium has been clearly recognized in the formation of the alkalic massifs of the Turja peninsula."

Arnott (1950) reported unit cell dimensions of natural thorianites from Sri Lanka and from Easton, Pennsylvania.

Uraninite UO_2

Uraninite was first noted as a distinct substance in the silver-cobalt mines of Jachymov, Czech Republic, in the early 1700s. It was given a variety of names that referred to its dark, pitch-like appearance, e.g., Schwartz Beck-Erz (Erz = ore), Beck-Blände, Swart Blende, and Pechblende. The element uranium was discovered in this ore by Klaproth, who then named the mineral Uranerz. Haidinger (1845, p. 549) applied the name Uranin in his Handbook. In modern usage, uraninite is the correct name for natural UO_2 in all its forms.

Pitchblende refers to a compact, microcrystalline variety, which often shows concentric banding and a botryoidal or rounded surface when it has developed as a surface crust. The crystallite size is very small, typically 10^{-2} to 10^{-6} mm. The particles or fibers may be randomly oriented; alternatively, there may be a preferred orientation in which fibrous crystallites, elongated along [111], lie perpendicular to the surface of the crusts or globules. The fine grain size may be attributed to rapid crystallization at fairly low temperatures, or to the crystallization of gel masses, but it could be a consequence of direct crystallization in a highly oxidized state, i.e., UO_{2+x}, where the strain introduced into the structure by interstitial oxygen prevents the formation of macroscopic crystals. In summary, the relationship of pitchblende to uraninite is analogous to that of chalcedony to quartz (Frondel 1958; Brooker and Nuffield 1952).

Uraninite can form in several distinct geological settings: 1. In pegmatites, typically as well-formed crystals or dendritic growths associated with monazite, columbite-tantalite, and zircon; 2. In high temperature hydrothermal veins, typically as massive pitchblende associated with cassiterite, arsenopyrite, and Co-Ni-Bi-As minerals; 3. In moderate-temperature hydrothermal veins with galena and other sulfides; and 4. In sandstone deposits with V, Cu, and other metal ores.

In pegmatites, the uraninite crystals usually contain appreciable levels of thorium (up to ~10%), rare earths (up to ~15%), and other impurities, because the size and valence of these ions prevent them from entering into the main rock-forming silicates as they crystallize, so they are scavenged by the uraninite. Pitchblende, on the other hand, is usually fairly pure because it has crystallized from hydrothermal fluids that don't contain these elements. Both varieties of uraninite may contain a significant amount of radiogenic lead (up to ~20%) depending on the geologic age of the deposit (Frondel 1958). A few noteworthy pegmatite locales are of interest to the collector: Lustrous black crystals are found at Topsham, Maine; and Spruce Pine, North Carolina. Dendritic uraninite in pegmatite, usually associated with colorful alteration products ("gummite") is well known from the Ruggles pegmatite, Grafton, New Hampshire.

Pitchblende is found in vein deposits at Jachymov, Czech Republic (TL for uraninite) and in tin-bearing veins in Cornwall, England. Pitchblende is also a frequent component of sandstone or "Colorado Plateau type" deposits.

Uraninite is also found in quartz-pebble conglomerates that represent detrital deposits (paleoplacers) derived from the weathering of Archean-age granite. All of these deposits are very old (>2.2 Ga); they became rare after the Paleoproterozoic because the rising level of oxygen in the atmosphere would have prevented uraninite from surviving long periods of erosion and transport in surface waters (Kyser and Cuney 2008). Deposits of this type are low grade but tend to be very large. Examples occur in the Blind River - Elliot Lake region, Canada; Witwatersrand, South Africa; and Sierra de Jacobina, Brazil (Dahlkamp 1993).

Figure 101. Simple cubic uraninite crystal in calcite with dark grains of fluorite and a yellow-green apatite crystal from the Cardiff uranium mine, Wilberforce, Ontario, Canada. RJL578

Figure 102. Rough uraninite cube ~2 cm tall, from the Richardson mine, Wilberforce, Ontario. RJL3979

Figure 103. Complex uraninite crystal ~1 cm across, in granular calcite with dark brown phlogopite and other phases, from the Payne property, Quebec. RJL349

Figure 104. Dendritic uraninite in pegmatite with orange gummite from the Ruggles mine, Grafton Center, New Hampshire. This slab is ~15 cm wide and extremely radioactive at about 94,000 cpm at contact! RJL3130

Figure 105. Thumbnail-sized sample of massive black uraninite var. pitchblende from Moab, Utah. RJL267

SILICATES

The primary uranium and thorium silicates include coffinite, huttonite, and thorite. Of these, coffinite and thorite are tetragonal members of the zircon group. Huttonite is the monoclinic dimorph of thorite and is structurally related to the phosphate minerals monazite and cheralite.

Thorogummite was long regarded as a partially hydrous member of the group with the proposed formula $Th(SiO_4)_{1-x}(OH)_4$. In a critical review of hydrous, thorite-like minerals, Frondel (1953) concluded that, "The ill-defined thorite-like minerals thorogummite, nicolayite, hydrothorite, maitlandite, mackintoshite, and hyblite are found to be minor chemical variants of a single phase, for which the name thorogummite has priority." Thorogummite was described from the Barringer Hill pegmatite, Blufton, Texas, by Hidden and Mackintosh (1889). Thorogummite, like coffinite, is generally very fine grained and metamict. It occasionally forms pseudomorphs after tetragonal thorite crystals, suggesting that the mineral is an alteration or weathering product. In 2014 the IMA accepted the proposal to discredit thorogummite, and stated, "This name has been used to describe heterogeneous mixtures of secondary, non-crystalline minerals produced by the alteration, hydration, or metamictization of thorite."

Thorite $(Th,U)SiO_4$

Thorite was described by Berzelius in 1829 based on material found by M. T. Esmark in 1828 on Løvøy Island, Langesund Fjord, Norway; it was in this material that Berzelius discovered the element thorium [see Gahn et al. (1817) for a description of early analytical work involving isolation of thorium oxide from rare earth minerals and distinguishing it from zirconia].

The formula for thorite, expressed as the theoretically pure end member, would be $ThSiO_4$, by analogy to zircon. However, thorite nearly always contains significant uranium, and the accepted formula is $(Th,U)SiO_4$. Thorite occasionally forms large, usually brown, crystals that are rarely as sharp as those of zircon because of alteration and metamictization.

Figure 108. Rough tan crystal of thorite var. uranothorite from Bancroft, Ontario, Canada. The crude, rounded lozenge shape is typical of samples from this locality. RJL157

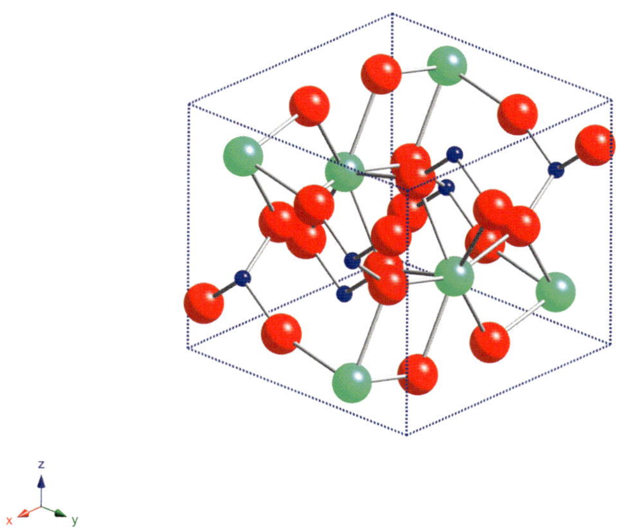

Figure 106. Crystal structure of thorite, shown as a ball and stick model in which large green atoms represent thorium, large red atoms represent oxygen, and small dark blue atoms represent silicon.

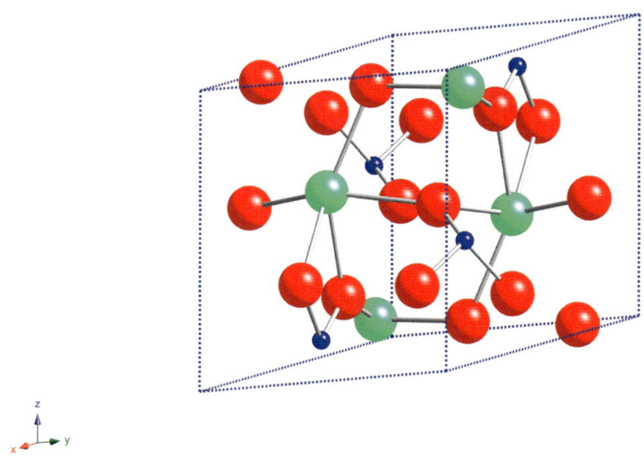

Figure 107. Crystal structure of huttonite, the monoclinic dimorph of thorite, shown as a ball and stick model with atoms represented in the same colors for comparison to the thorite structure shown in figure 106.

Figure 109. Group of elongated gray-brown crystals of thorite var. uranothorite in pegmatite from Pacoima Canyon, San Gabriel Mountains, Los Angeles Co., California. The resemblance to zircon can be seen in the somewhat pyramidal terminations and generally prismatic habit of the crystals. RJL566

Figure 110. Crude, brick-red crystals of thorite var. orangeite in pegmatite, associated with purplish zircon, from Aroplany, Tulear, Madagascar. RJL3126

Major substitutional impurities are U (up to ~25%), REE (up to ~8%), radiogenic Pb for Th, and PO_4 (up to ~13%) for SiO_4.

Good crystals are found at several locales in the Bancroft district, Ontario, Canada; and at Arendel, Norway. The variety *orangeite*, found at Aroplany, Madagascar, was named for its distinctive orange-yellow color. Thorite is a constituent of black sand concentrates from gold placers in Biggs and Merced Counties, California (George 1951).

A Nb-Ti-rich thorite from Mont Saint-Hilaire, Quebec, presented the first example of a zircon-group mineral, with elevated Nb (0.2 apfu) and Ti (0.19 apfu), and evidence of the $(SiO_4)^{4-} \leftrightarrow (OH)_4^{4-}$ "hydrogarnet" substitution. It is slightly metamict; structural refinement showed that it is tetragonal with a=7.058 and c=6.226 Å. It is isostructural with other members of the zircon group, with a unit cell ~4% smaller than that of end-member thorite because of the substitution of high field-strength elements of smaller radii. The authors conclude that, "Although it is possible that a metastable, limited solid solution exists between thorite and $(OH)_4^{4-}$-dominant 'thorogummite' with intermediate compositions defined by $Th(SiO_4)_{1-x}(OH)_{4x}$, reported compositions indicate otherwise and it is suggested that the name 'thorogummite' be abandoned," (Piilonen et al. 2013).

Cottle (2014) suggested that thorite can be useful as a tool for dating various geological events because of the wide variety of environments in which it occurs, combined with its ability to accommodate large amounts of U and Th in its structure.

Huttonite $ThSiO_4$

Huttonite, the monoclinic dimorph of thorite, was first found as minute, colorless grains in sands at Gillespie's Beach and several other locales in New Zealand. In the original description of huttonite (Pabst and Hutton 1951) it was noted that huttonite is rarely metamict, whereas thorite is nearly always so. Huttonite is isostructural with monazite, which is also resistant to metamictization. Conversely, thorite is isostructural with zircon, and radioactive zircons are often metamict, leading Pabst (1952) to conclude that the huttonite crystal structure is inherently more stable than the thorite structure. Microprobe analysis of material from the type locale indicated that small amounts of phosphorus and uranium are always present (0.11 to 0.68 wt. % P_2O_5; 0.54 to 0.82 wt. % UO_2) along with 0.3 to 1.4 wt. % Y + REE oxides (Förster, Harlov, and Milke 2000).

Figure 111. Backscattered electron image of small huttonite grains in matrix from the Holiday mine, Hawthorne, Mineral Co., Nevada. In the BSE mode, areas of higher atomic number appear brighter than areas with lower atomic number, enhancing visual contrast between the huttonite grain in the center (arrow) and the surrounding silica-rich matrix. RJL898

Huttonite is rather rare in nature compared to thorite, reflecting its restricted field of stability at temperatures and pressures that are normally found in the crust of the earth. Reported occurrences include: in granitic pegmatite at Bogatynia, Poland; in nepheline syenite at Brevik, Norway; in granitic pegmatite in the Santa Maria de Itabira district, Minas Gerais, Brazil; the Lovozero complex, Kola Peninsula, Russia; and at the Holliday mine, Mineral Co., Nevada (Förster, Harlov, and Milke 2000).

Huttonite forms solid solution series with the other members of the monazite group, monazite $[(Ce,La,Nd)PO_4]$ and brabantite $[Ca_{0.5}Th_{0.5}(PO_4)]$. Förster and Harlov (1999) present examples of Th-rich ("huttonitic") monazites, including one from the Kirchberg granite, Germany, which lies nearly at the boundary between monazite and huttonite (41.8 wt. % ThO_2).

Coffinite $USiO_4$

Coffinite was described from the La Sal No. 2 mine, Mesa Co., Colorado, by Stieff, Stern, and Sherwood (1956), who listed occurrences at several dozen uranium deposits on the Colorado Plateau and at several peripheral deposits. They also noted occurrences of the mineral in association with uraninite in a vein deposit in the Copper King mine, Larimer Co., Colorado, and associations with asphaltic pellets from sediments in Oklahoma and Texas. The material is black, sometimes with an adamantine luster "resembling that of a high-rank coal."

Coffinite and "thorogummite" were long thought to be isostructural with thorite but with some (SiO_4) replaced by $(OH)_4$, leading to the following formulas: coffinite = $U(SiO_4)_{1-x}(OH)_{4x}$ and thorogummite = $Th(SiO_4)_{1-x}(OH)_{4x}$. However, infrared data suggest that coffinite contains molecular water, implying the formula $U(SiO_4) \cdot nH_2O$. The situation is complicated by the fact that most coffinite is extremely fine-grained, metamict, possibly impure, and/or altered by ground water, so analytical work on these minerals remains challenging (Finch and Murakami 1999).

Understanding the nature of coffinite and its formation is crucial to predicting the long-term behavior of spent nuclear fuel and other high-level waste forms in geological repositories. It is also important from the perspective of the ore geologist: why is coffinite so common in nature and yet so hard to synthesize in the laboratory? Pointeau et al. (2009) noted that their synthetic coffinite nanoparticles were always associated with some UO_2 particles, and that the formation of coffinite is very sensitive to several experimental variables. Costin et al. (2011) reviewed the methods that had been reported for synthesizing $USiO_4$ and the difficulty in reproducing the results. Based on their own synthesis of uranothorite solid solutions, $Th_{1-x}U_xSiO_4$, they suggest that the difficulty lies in a kinetic hindrance in the "coffinitization" reaction. In a related report, Szenknect et al. (2013) note that uranothorite solid solutions with x>0.26, and thus coffinite as well, are less stable than the mixture of binary oxides, which is consistent with qualitative evidence from petrographic studies of uranium ore deposits.

Navrotsky, Shvareva, and Guo (2013) note that the enthalpies of formation of thorite and coffinite from UO_2 and SiO_2 are positive; i.e., these minerals are metastable at room temperature but may persist from high temperatures. Coffinite is often found in reducing, low-temperature environments rich in organic matter, suggesting formation from earlier U^{6+} phases, and in these deposits the coffinite is very fine-grained and often hydrated. By contrast, thorite typically forms at much higher temperatures. The formation of coffinite associated with organic matter in the Grants region, New Mexico, was studied by Deditius, Utsonomiya, and Ewing (2008). They note, "The variable, but high, totals of the chemical analyses (92.8 - 99.7 wt. %) suggest that coffinite contains molecular H_2O, which is not an essential component of coffinite structure. The amount of H_2O varies between 0 and 2 molecules per formula unit, and a minor amount of $(OH)^-$ may be accommodated into the structure . . . The U-Pb chemical ages limit the age of coffinite precipitation to between 36.6 and 0.7 Ma, which indicates continuous precipitation of coffinite since the mid-Tertiary. These results suggest that organic matter, which has preserved reducing conditions even in the presence of oxidizing fluids, over this long period of approximately 30 million years, plays an important role in the sustained presence of coffinite in these deposits."

Recent thermodynamic studies have shed more light on this problem. Guo et al. (2015) note that coffinite slowly decomposes to U_3O_8 and SiO_2 starting around 450°C in air. Its poor thermal stability explains why coffinite can't be directly synthesized from uraninite and quartz, but can be made by low-temperature precipitation in aqueous and hydrothermal environments. Calorimetric data show that coffinite can form from SiO_2-rich aqueous solutions in contact with dissolved U-containing species in a reducing environment. Thus, coffinitization involves the dissolution of UO_2, forming hexavalent uranium intermediates (i.e., uranyl species), followed by reaction with SiO_2-rich fluids.

Nearly all chemical analyses of coffinite report some phosphorus; samples found at the site of the ancient natural fission reactor at Bangombé, Gabon, had as much as 8.5 wt % P_2O_5, which would represent a replacement of roughly 45% of the Si atoms with P atoms in the structure. The material also contained Ca, REE, and Pb partially replacing the U (Janeczek and Ewing 1996; Deditius et al. 2009). Similarly high concentrations of P have been reported in Th- and Y-rich coffinite in the reefs of the Witwatersrand, South Africa. There, textural relations indicate that the coffinite was produced by direct transformation of detrital grains of uraninite via *in-situ* uptake of silica in an alkaline and reducing environment (Smits 1989). Coarse-grained coffinite from the Grants district, New Mexico, was found to contain some Ca, Y, REE, Pb, and P (Hansley and Fitzpatrick 1989).

Although coffinite never forms attractive specimens for the collector, it is an extremely important component of the Colorado Plateau deposits and is second only to uraninite as a uranium ore mineral.

Figure 112. Thumbnail sized mass of coffinite in sandstone from the San Rafael Swell, Emery Co., Utah, showing the typical dull appearance of coffinite in the hand specimen. RJL3141

Figure 113. A closer view of the specimen in figure 112 showing the black coal-like coffinite cementing the quartz sand particles together.

SECONDARY MINERALS

The taxonomy of secondary U minerals, i.e., those that generally contain the uranyl ion, is based primarily on the topology on which the uranyl ions are arranged and interconnected. For this reason, most of the structure diagrams will be rendered as polyhedral models, rather than ball and stick models, so that the topology will be more evident. Cations other than $(UO_2)^{2+}$ may be shown either as individual atoms (balls, or balls with "wire bonds") or as their corresponding coordination polyhedra, depending on which views make the key structural themes easiest to visualize. Wherever possible, the same color is used for a particular element in all of the diagrams.

A few words are in order to help the reader better understand the crystal structure diagrams and relate them to physical structures and to the descriptions found in the literature. First, the uranyl ion *Ur* consists of two oxygen ions firmly attached to a U^{6+} ion to form a more or less linear group as shown in Figure 114.

The uranyl ion is then typically coordinated to a number of additional oxygens or other ligands φ (OH^- or H_2O, for example) to form a polyhedron of various shapes with a uranium at the center. The polyhedron is typically a bipyramid with square, pentagonal, or hexagonal cross-section. In the scientific literature, these may be designated as follows: UO_6, $U\varphi_6$, or $Ur\varphi_4$ (square); UO_7, $U\varphi_7$, or $Ur\varphi_5$ (pentagonal); and UO_8, $U\varphi_8$, or $Ur\varphi_6$ (hexagonal). In all cases, the bipyramid is somewhat flattened because the uranyl oxygens forming the "poles" of the bipyramid are more tightly bonded to the central U, whereas the other ligands that form the "equator" are more weakly bonded and those bonds are therefore longer.

As with any system of classifying complex individuals into groups, there will always be room for debate as to which species belong in which group. The IMA has clear guidelines for defining a valid species, but membership in a group is somewhat more subjective. In the sections that follow, an attempt has been made to treat the minerals in logical groups based as much as possible on structural considerations, but it will be clear that in some cases the relationships between particular species are closer than in others. It is best to keep in mind that all of these "pigeonholes" are an artificial construct of man's desire to organize the natural world, and much like political borders, are completely irrelevant to Nature herself.

Figure 114. The uranyl ion, represented as it would appear in a space-filling model. The U^{6+} ion is pale blue-gray, and the O^{2-} ions are dark red. The entire cluster is treated as a divalent cation, $(UO_2)^{2+}$.

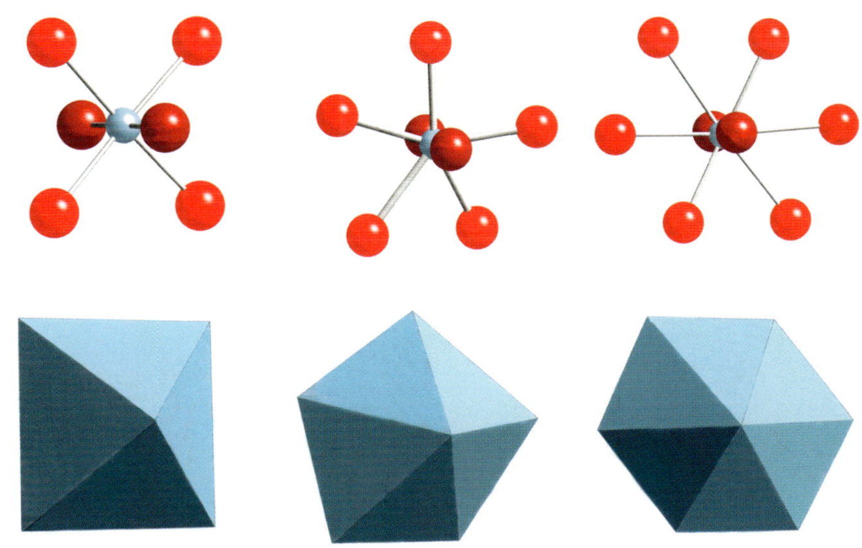

Figure 115. Three coordination structures that are known in secondary uranium minerals, viewed just off the axis of the linear $(UO_2)^{2+}$ ion. In the ball and stick models (top) the uranyl oxygens are dark red and the other oxygens are bright red. In the polyhedral models (bottom) the entire cluster is shown as a solid polyhedron where each corner represents the center of a surrounding oxygen ion. [See Burns (1999a) for a more detailed discussion of these structural issues.]

OXIDES AND HYDROXIDES

Secondary uranyl oxides may be conveniently grouped into three categories: those that contain no metal ions other than uranium, those that contain uranium and lead, and those that contain uranium and other metals. All of these minerals contain water and/or hydroxyl groups in their formula.

Table 43. Secondary uranyl oxides

Mineral	Formula	Symmetry
Uranyl oxides		
Ianthinite	$UO_2 \cdot 5UO_3 \cdot 10H_2O$	orth
Schoepite	$[(UO_2)_3O_2(OH)_{12}] \cdot 12H_2O$	orth
Metaschoepite	$UO_3 \cdot nH_2O$ (n<2)	orth
Paraschoepite	$UO_3 \cdot 2H_2O$ (?)	orth
Paulscherrerite	$UO_2(OH)_2$	mon
Heisenbergite	$UO_2(OH)_2 \cdot H_2O$	orth
Studtite	$UO_4 \cdot 4H_2O$	mon
Metastudtite	$UO_4 \cdot 2H_2O$	orth
Uranyl oxides containing lead		
Curite	$Pb_2U^{6+}{}_5O_{17} \cdot 4H_2O$	orth
Fourmarierite	$PbU^{6+}{}_4O_{13} \cdot 4H_2O$	orth
Masuyite	$Pb(UO_2)_3O_3(OH)_2 \cdot 3H_2O$	mon
Richetite	$(Fe,Mg)_xPb_{8.57}[(UO_2)_{18}O_{18}(OH)_{12} \cdot 41H_2O$	tric
Sayrite	$Pb_2(UO_2)_5O_6(OH)_2 \cdot 4H_2O$	mon
Spriggite	$Pb_3[(UO_2)_6O_8(OH)_2].3H_2O$	mon
Vandendriesscheite	$Pb_{1.5}(UO_2)_{10}O_6(OH)_{11} \cdot 11H_2O$	orth
Metavandendriesscheite	$PbU_7O_{22} \cdot nH_2O$ (n < 12)	orth
Wölsendorfite	$Pb_7(UO_2)_{14}O_{19}(OH)_4 \cdot 12H_2O$	orth
Uranyl oxides containing other metals		
Agrinierite	$K_2(Ca_{0.65}Sr_{0.35})[(UO_2)_2O_3(OH)_2]_2 \cdot 5H_2O$	orth
Bauranoite	$BaU_2O_7 \cdot 4\text{-}5H_2O$	?
Becquerelite	$Ca(UO_2)_6O_4(OH)_6 \cdot 8H_2O$	orth
Billietite	$Ba[(UO_2)_6O_4(OH)_6] \cdot 8H_2O$	orth
Calciouranoite	$(Ca,Ba,Pb)U_2O_7 \cdot 5H_2O$	?
Metacalciouranoite	$(Ca,Ba,Pb)U_2O_7 \cdot 2H_2O$	?
Clarkeite	$(Na,K,Ca,Pb)[(UO_2)O(OH)] \cdot H_2O$	hex
Compreignacite	$K_2[(UO_2)_3O_2(OH)_3]_2 \cdot 7H_2O$	orth
Protasite	$Ba[(UO_2)_3O_3(OH)_2] \cdot 3H_2O$	mon
Rameauite	$K_2CaU^{6+}{}_6O_{20} \cdot 9H_2O$	mon
Uranosphaerite	$Bi(UO_2)O_2OH$	mon
Vandenbrandeite	$Cu(UO_2)(OH)_4$	tric

Hydrous uranyl oxides

Secondary hydrated uranyl oxides that contain no other metal ions include: heisenbergite, ianthinite, schoepite, metaschoepite, paraschoepite, paulscherrerite, studtite, and metastudtite. These minerals tend to form on or very close to uraninite, typically as surface alterations.

Ianthinite $UO_2 5UO_3 10H_2O$

Ianthinite was described by Schoep (1926) from Kasolo, Congo, where it forms dark violet to black, acicular crystals on uraninite, associated with becquerelite and schoepite. Crystals have a submetallic luster and micaceous cleavage on (100), and tend to alter on the edges to yellow. The name, which Schoep spelled "janthiniet," is derived from the Greek *ιανθινος* (= violet colored), is only stable under somewhat reducing conditions, because it contains some U^{4+}, which will continue to oxidize under normal atmospheric conditions. It forms minute purple-black laths when fresh; on exposure to air it gradually turns brown and eventually changes completely to yellow schoepite.

In addition to the TL, ianthinite has been reported from: Wölsendorf and Menzenschwand, Germany; several localities in the Colorado Plateau region; and numerous deposits in France.

Guillemin and Protas (1959) suggested that material

Figure 116. Ianthinite needles, which have nearly completely altered to schoepite, on uraninite from Shinkolobwe, Congo. RJL686

described by Frondel and Cuttitta (1954) and by Bignand is actually a different mineral (a uranyl carbonate), which they named wyartite. The paper by Frondel and Cuttitta illustrates the difficulties encountered by early researchers attempting to sort out the various phases and their decomposition, oxidation, or dehydration products.

Ianthinite is orthorhombic with a=7.178, b=11.473, and

c=30.39 Å. The structure contains both U^{4+} and U^{6+}. The U^{6+} cations are present as pentagonal bipyramids. The U^{4+} cations are coordinated by O^{2-}, OH^- and H_2O in a distorted octahedral arrangement. The U^{4+} and U^{6+} polyhedra share edges to form two symmetrically distinct sheets at $z \approx 0.0$ and $z \approx 0.25$, which are parallel to (001). The sheets have the β-U_3O_8 topology and are linked to one another only by hydrogen bonding to the interlayer H_2O groups (Burns et al. 1997).

Schoepite $[(UO_2)_3O_2(OH)_{12}](H_2O)_{12}$
Metaschoepite $UO_3.nH_2O$ (n<2)
Paraschoepite $UO_3.2H_2O$ (?)

Schoepite was described from Kasolo, Congo, by Walker (1923), who had received a suite of minerals representing the various species that had recently been described by Schoep (1921, 1922). This material contained crystals that had distinctly different optical and crystallographic properties from the known associates. Although insufficient material was available for chemical analysis, Walker based the description on optical and goniometric measurements. He states: "The mineral in question closely resembles Schoep's becquerelite, being sulphur yellow in color, and adamantine in lustre. The crystals are seldom more than one and a half mm. in length, and about one mm. in diameter. They possess one very perfect cleavage (001). When examined under the microscope, most of the cleavage plates are found to be transparent, lemon yellow in color, and exhibit no pleochroism." Because some gas bubbles were evolved on dissolution in acid (perhaps from admixed rutherfordine?), Walker assumed the mineral was a carbonate.

Schoepite forms golden-yellow, tabular pseudohexagonal crystals and fine-grained crusts. It is fairly abundant at several locales in Congo including Kasolo (TL), Musonoi, and Shinkolobwe, where it forms sharp pseudomorphs after cubic uraninite crystals. It occasionally replaces other minerals as well. A frequently seen assemblage is uraninite-schoepite-curite (Deliens, Piret, and Comblain 1981). A very colorful association seen at Musonoi is cuprosklodowskite-rutherfordine-schoepite.

Some other locales include: Wölsendorf, Germany; Margnac mine, France; several sites on the Colorado Plateau; and as pseudomorphs after uraninite crystals at Beryl Mountain, New Hampshire.

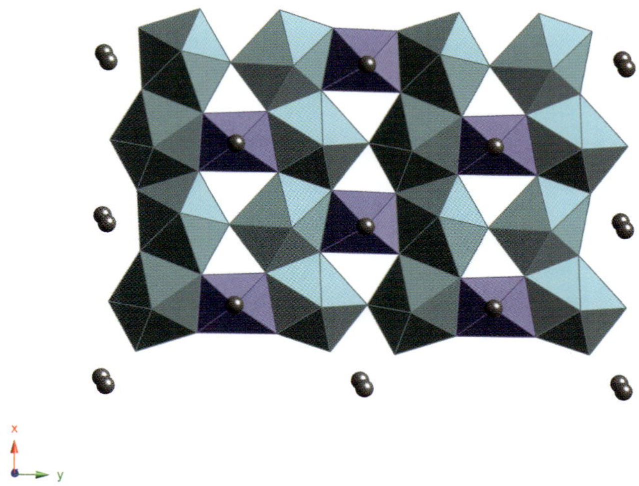

Figure 117. The sheet structure of ianthinite, showing U^{6+} pentagonal bipyramids (blue-gray) and U^{4+} distorted octahedra (lavender). Water is shown in gray.

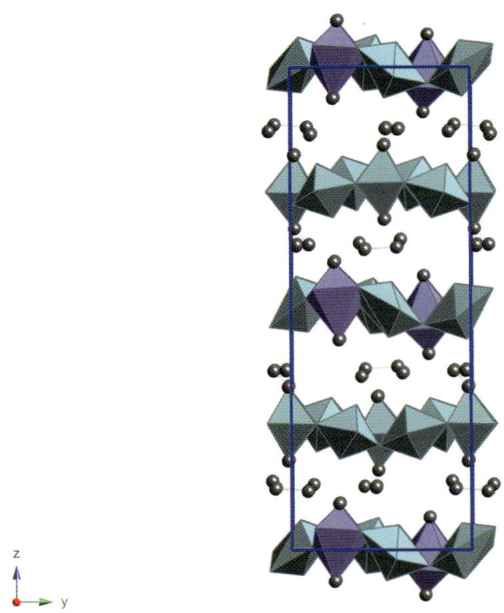

Figure 118. The unit cell of ianthinite viewed along [100], parallel to the sheets. Adjacent sheets are linked via hydrogen bonds involving the water molecules in the interlayer.

Figure 119. Yellow schoepite as an alteration product, filling cracks in massive uraninite, from Shinkolobwe, Congo. RJL353

Schoepite, metaschoepite, and paraschoepite are closely related compounds differing only in the water content, with schoepite having the highest hydration. When exposed to air, schoepite spontaneously dehydrates, gradually forming a mixture of two or perhaps all three phases. Detailed crystallographic analysis (Christ and Clark 1960; Christ 1965) identified three phases as schoepite I, II, and III. Schoepite I corresponds to unaltered or fully hydrated material, such as the amber core of a crystal having a golden-yellow alteration rim. Schoepite III corresponds closely to the paraschoepite of Schoep and Stradiot (1947). The exact relationship between these species remains unclear and it has been suggested that another phase, "dehydrated schoepite," is also common in nature and that the validity of paraschoepite might be questionable (Finch and Murakami 1999). The name epi-ianthinite (Schoep and Stradiot 1947) was discredited by the IMA in 1962 as being a synonym of schoepite. "Dehydrated schoepite" has since been described under the name paulscherrerite (see entry below).

Schoepite is orthorhombic with a=14.337, b=16.813, and c=14.731 Å. The structure contains charge-neutral sheets of composition $[(UO_2)_8O_2(OH)_{12}]$ alternating with almost planar sheets of interlayer H_2O groups. Adjacent sheets are linked together via hydrogen bonding involving the H_2O (Finch et al. 1996).

Metaschoepite is orthorhombic with a=13.99, b=16.72, and c=14.73 Å. Thus, the transformation of schoepite to metaschoepite involves a 2% decrease in the a cell dimension, a slight decrease in b, and little or no change in c. Finch, Hawthorne, and Ewing (1998) suggest that, "Differences in unit-cell volumes induce strain in crystals in which the transformation to metaschoepite is incomplete, and stored strain energy may be sufficient to rapidly drive the transformation of the remaining schoepite to 'dehydrated schoepite' [a 6.86, b 4.26, c 10.20 Å, *Abcm* (?)] when partly altered crystals are exposed to an external stress (e.g., heat, sunlight or mechanical pressure). Metaschoepite is apparently stable in air; canary yellow altered crystals commonly consist of a polycrystalline mixture of 'dehydrated schoepite' and metaschoepite. The alteration of schoepite to 'dehydrated schoepite' occurs in three steps: (1) loss of all interlayer H_2O from schoepite, causing collapse of the layers, (2) atomic rearrangement within the structural sheets to a configuration that may be similar to that of metaschoepite, and (3) further re-arrangement to a defect α-$UO_2(OH)_2$-type sheet. The complete reaction is $[(UO_2)_8O_2(OH)_{12}](H_2O)_{12} \rightarrow 8[(UO_2)O_{0.25}(OH)_{1.5}] + 12H_2O$."

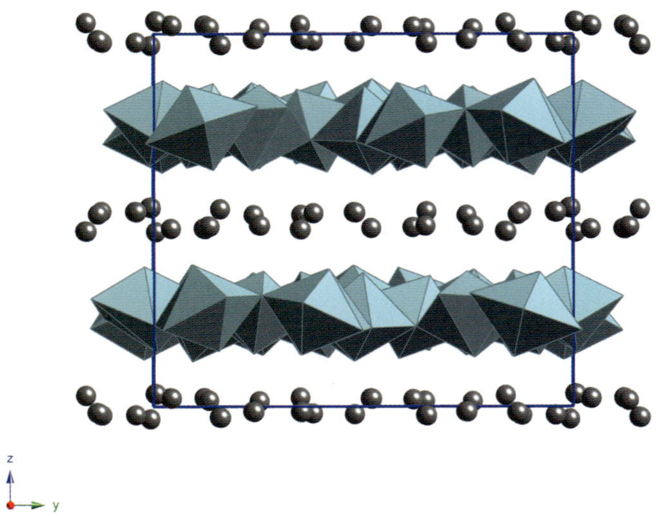

Figure 120. The structure of schoepite, viewed along [100] showing the uranyl sheets and the interlayer H_2O groups (gray) that link the sheets together by hydrogen bonds.

Paulscherrerite $UO_2(OH)_2$

Paulscherrerite was described from the Number 2 Workings at Radium Ridge, Northern Flinders Ranges, South Australia, where it occurs as a dehydration product of metaschoepite. It is canary yellow and is not fluorescent. Individual crystallites are extremely small (500 nm). Along with weeksite, metaschoepite, Fe-Mn-oxyhydroxides, and occasionally spriggite, paulscherrerite is the main mineral replacing an unknown earlier U-Pb-Mn(?)-bearing phase. It also forms pseudomorphs after platy metaschoepite in cavities in late epithermal quartz.

The structure is related to that of synthetic orthorhombic α-$UO_2(OH)_2$, but XRD results suggest monoclinic symmetry in the type material, with a=4.288, b=10.271, c=6.885 Å, and β=90.39°. Before the formal mineral description, "dehydrated schoepite" had been reported as an early weathering product of uraninite or pitchblende at several locales, including: Shinkolobwe, Congo; the Nopal I deposit, Mexico; and the granitic pegmatites of New Hampshire. These should now be regarded as additional occurrences of paulscherrerite (Brugger et al. 2011).

Heisenbergite $UO_2(OH)_2.H_2O$

Heisenbergite was described from the Menzenschwand U deposit, Southern Black Forest, Germany, where it occurs as a rare secondary mineral associated with quartz, barite, uraninite, billietite, studtite, rutherfordine, joliotite, geothite, and hematite. The phase had been previously noted, but not formally described, at El Sharana, South Alligator district, Northern Territory, Australia; and at Jachymov, Czech Republic. At the TL, it forms anhedral particles and prismatic crystals to ~15 μm long × 2 μm thick. Crystals are yellow, yellow-brown, or orange-brown, and are not fluorescent.

Although a full structure refinement wasn't possible, based on powder XRD the structure is orthorhombic, with a=113.10, b=13.76, and c=14.50 Å. The simplified formula is $UO_2(OH)_2 \cdot 2H_2O$ or $UO_3 \cdot 2H_2O$. Based on the Raman spectrum, it is suggested that there are two distinct types of U coordination polyhedra in the structure (Walenta and Theye 2012).

Studtite $UO_4.4H_2O$
Metastudtite $UO_4.2H_2O$

Studtite was described from the oxidation zone at Shinkolobwe, Congo, by Vaes (1947) who considered it to be a carbonate. It forms aggregates and crusts of minute, pale yellow, flexible needles associated with other secondary minerals. It is not fluorescent. Walenta (1974) investigated studtite from Menzenschwand, Germany, and concluded that the mineral was not a carbonate, but rather a hydrous uranium oxide identical with a synthetic compound that was regarded as a peroxide hydrate.

At Shinkolobwe, studtite is associated with uranophane, becquerelite, billietite, and wölsendorfite. At Swambo, it is accompanied by other hydrous uranyl oxides, and at Kobokobo it occurs with an aluminum uranyl phosphate (Deliens, Piret, and Comblain 1981). Studtite was considered fairly rare; however, it has since been identified at many localities including Lodeve, France, and the Krunkelbach deposit, near Menzenschwand, Germany.

Metastudtite, also reported from Shinkolobwe, Congo, is presumably a dehydration product of studtite.

Studtite is monoclinic with a=14.069, b=6.721, c=8.428 Å, and β=123.356°. The structure of studtite contains one symmetrically distinct U^{6+} cation and four O atoms, two of which occur as H_2O groups. The O-O bond-length in the peroxide group is 1.46(1) Å. The O-O bonds of two peroxide groups form two equatorial edges of each distorted uranyl hexagonal bipyramid. Uranyl polyhedra are polymerized into chains running along [001] by sharing peroxide groups. Adjacent chains are linked by hydrogen bonds extending to and from an interstitial H_2O group. Studtite and metastudtite are especially interesting to geochemists because they are

Figure 121. Studtite (yellow) with wölsendorfite (orange-red) on massive uraninite from Shinkolobwe, Congo. RJL354

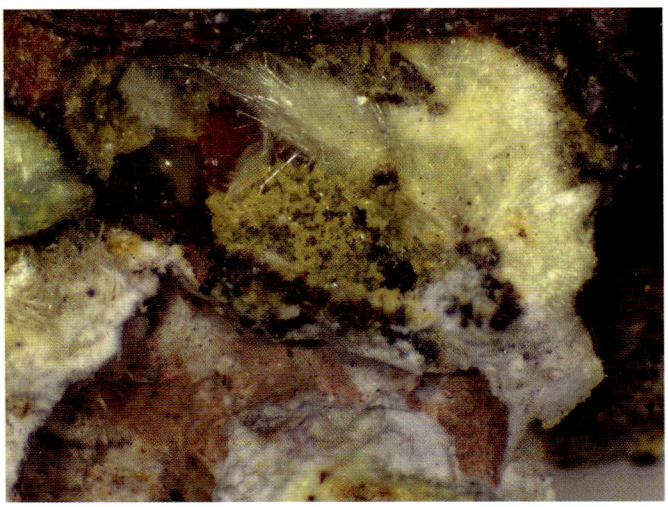

Figure 122. Studtite (cream-white fibrous aggregates) with tiny golden-yellow billietite crystals, from Menzenschwand, Germany. RJL696

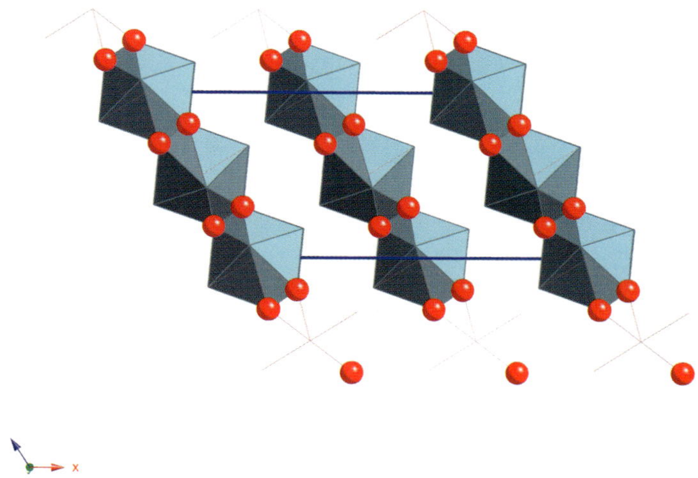

Figure 123. The uranyl peroxide chains of studtite, lying parallel to [001]. Red atoms indicate the oxygens in the peroxide groups; note the shorter distance between these oxygens compared to the other equatorial oxygens in the polyhedra. Each peroxide group is shared between two adjacent polyhedra.

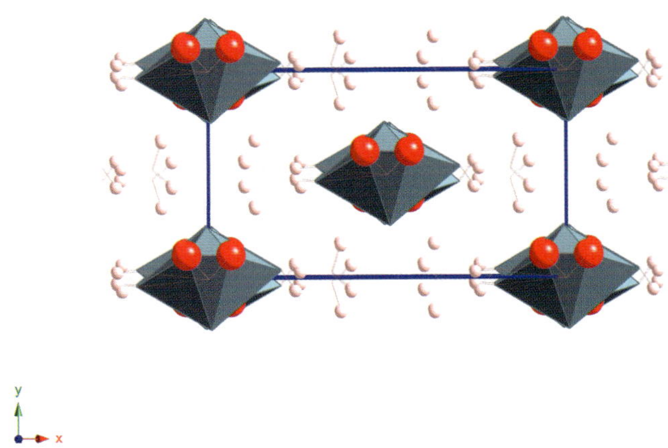

Figure 124. Crystal structure of studtite, viewed along [001] so the chains are seen end-on. Pink atoms with wire bonds represent hydrogens in the interstitial H_2O groups. These provide the hydrogen bonds that link adjacent chains together.

the only minerals believed to contain the peroxide (O-O) group. The structural formula for studtite may therefore be expressed as $[(UO_2)(O_2)(H_2O)_2](H_2O)_2$. The compound has been found both on natural uranium and on highly radioactive spent nuclear fuel that had been reacted with water for 1.5 years at 25°C. It is also noted on the surface of Chernobyl "lava" formed at the damaged Chernobyl Unit No. 4 reactor, where it formed through reaction of the fuel-containing masses of "lava" with water and air (Burakov, Strykanova, and Anderson 1996). However, the compound doesn't form in experiments using uranium oxide containing depleted uranium (i.e., with very little fissile ^{235}U isotope), leading to the conclusion that the source of peroxide is radiolysis of water by high-energy alpha particles. Thus studtite and metastudtite are the only known minerals that seem to need a radioactive environment in order to form (Burns and Hughes 2003b).

Hydrous uranyl oxides containing lead

This group includes curite, fourmarierite, masuyite, richetite, sayrite, spriggite, vandendriesscheite, metavandendriesscheite, and wölsendorfite. These colorful species have great appeal for the collector. Some of the minerals are fairly common, particularly masuyite and vandendriesscheite; others such as richetite and sayrite are quite rare. Considering them in order of increasing Pb:U ratio reflects the process through which the minerals evolve. As more uranium is converted to soluble uranyl ions, which move into the ground water, the solid phases left behind become increasingly enriched in lead, which is relatively less soluble: vandendriesscheite → fourmarierite, richetite → masuyite → curite, sayrite → wölsendorfite, spriggite.

These minerals are, therefore, especially important to our understanding of the oxidation zones of geologically old uranium deposits in which substantial amounts of radiogenic lead are present. At the same time, detailed understanding of their crystal chemistry has lagged behind many other mineral groups because of the inherent difficulty in determining their structures. They rarely form suitably large crystals for structure analysis, they are generally twinned, often possess pseudosymmetry and extremely large unit cells (as will be clear from the structure diagrams), and are strong X-ray absorbers (Burns and Hanchar 1999).

The Pb^{2+}-containing hydrous uranyl oxides generally have a lower solubility than uranyl-hydroxyhydrate minerals containing other metal cations (Finch & Murakami 1999). Schindler et al. (2006) studied the formation of etch pits during the dissolution of synthetic curite crystals in order to gain insights into the dissolution of hydrous uranyl oxides. Among other things, they noted that Sr^{2+} in the solution at pH 2.7 changes the morphology of etch pits, suggesting that the presence of radionuclides in the surrounding environment can influence the overall dissolution rate.

Christ and Clark (1960) reported a systematic study of becquerelite, billietite, fourmarierite, masuyite, schoepite, and vandendriesscheite. Tabulating the physical properties of color, habit, twinning, and cleavage, they noted that these properties vary only slightly from one species to the next: "It is evident that identification of any one mineral on the basis of these physical properties can only be tentative and X-ray and optical studies are also needed for definitive characterization." Using X-ray precession techniques, they concluded that all of the structures are based on a $UO_2(OH)_2$ sheet, parallel to the perfect (001) cleavage, with water of hydration and any cations occupying interlayer positions.

Deliens (1977) focusing specifically on Pb-U oxides, presented new powder XRD data for vandendriesscheite, fourmarierite, masuyite, and wölsendorfite. The results were mostly in good agreement with those of Christ and Clark (1960).

Curite $Pb_2U^{6+}{}_5O_{17}.4H_2O$

Curite was described from Kasolo Hill, Congo, where it was first noted in three habits: 1. reddish brown acicular crystals (yellow in transmitted light) implanted on torbernite; 2. orange, sugary crystalline aggregates; and, 3. compact, earthy orange masses (Schoep 1921).

Curite is plentiful at Shinkolobwe, Congo. Other reported locales include: at Menzenschwand and Wölsendorf, Germany; La Crouzille, Haute-Viene, France; in the South Alligator district, Northern Territory, Australia; and Malakialina, Madagascar. It is a common component in *gummite*, and occasionally forms pseudomorphic replacements of uraninite, as shown earlier in figure 17.

Curite is orthorhombic with a=12.53-12.58, b=13.01-13.03, and c=8.39-8.40 Å, based on structure refinements of fourteen single crystals from various localities in the

Figure 125. Massive orange curite from Kasolo, Congo, ~5 cm wide and extremely "hot" (~20,000 cpm at 1 cm). This old-timer is accompanied by several labels from old German collections, including that of Wilhelm Maucher (ca. 1936 or earlier). RJL4533

Figure 126. Red-orange curite needles with green torbernite from Shinkolobwe, Congo. RJL352

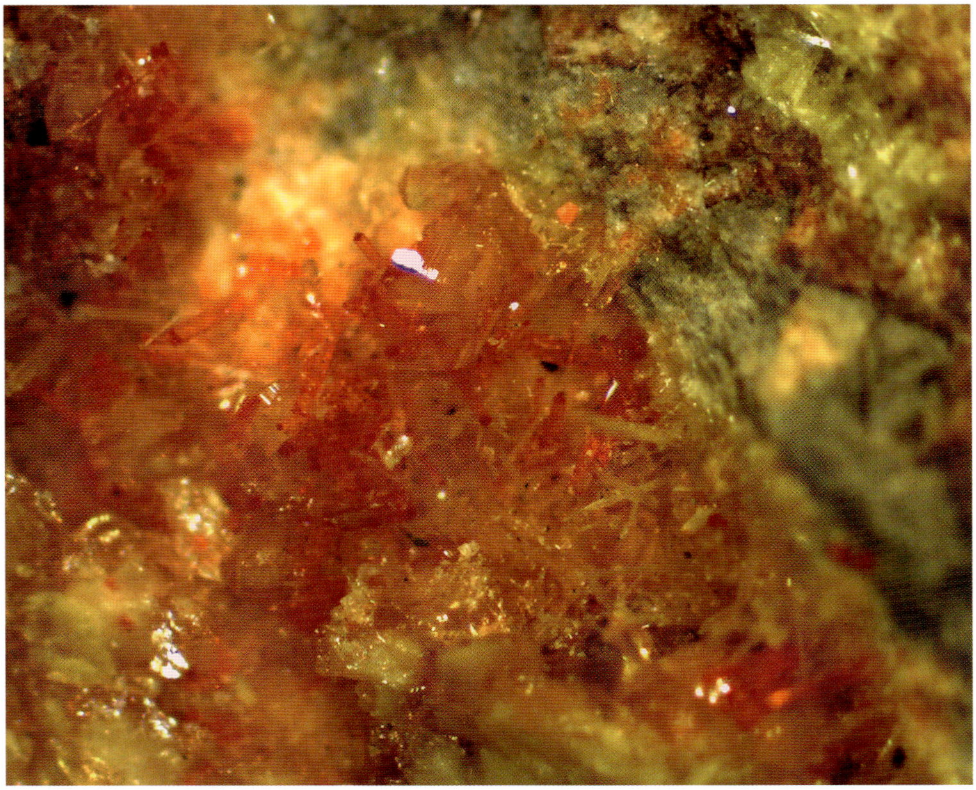

Figure 127. Orange curite with yellow acicular swamboite, from Swambo, Congo. RJL904

Figure 128. Glassy orange curite needles, ~1 mm long, associated with equant yellow soddyite crystals, from Swambo, Congo. RJL3899

Congo. Li and Burns (2000a) note that, "The structure contains uranyl square bipyramids and uranyl pentagonal bipyramids that share edges and corners to form sheets oriented parallel to (100). There are two symmetrically distinct Pb^{2+} cations and one H_2O group located in the interlayer. On the basis of the structure refinements, the site occupancy of Pb(1) is slightly deficient, ranging from 89 to 100%, whereas the Pb(2) site occupancy ranges from 57 to 63%. . . . the current study indicates that Pb variability in curite is limited, and supports minor variation in the hydroxyl content of the sheet of uranyl polyhedra as the charge balancing mechanism. On the basis of the structure refinements, the structural formula for curite may be written as $Pb_{3+x}(H_2O)_2[(UO_2)_4O_{4+x}(OH)_{3-x}]_2$, Z=2, with the constituents of the sheets of uranyl polyhedra enclosed in square braces."

Figure 129. The uranyl sheet in curite, viewed along [100], showing uranyl square bipyramids and uranyl pentagonal bipyramids that share edges and corners to form the sheet.

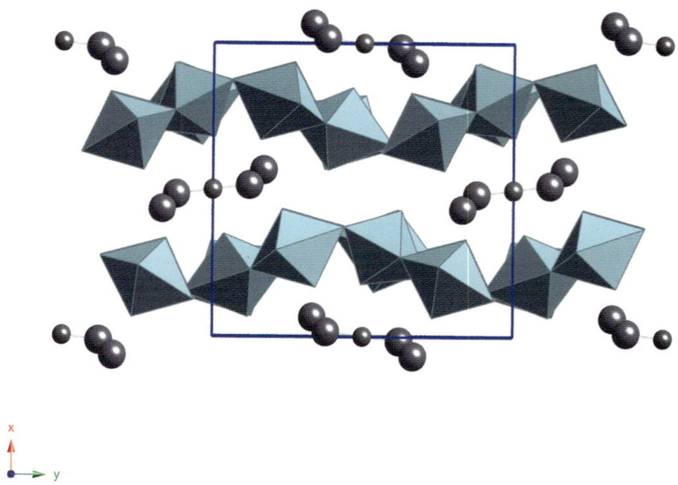

Figure 130. The structure of curite viewed along [001]. The interlayer contains Pb^{2+} (large gray atoms) and H_2O groups (small gray).

Fourmarierite $PbU^{6+}{}_4O_{13}.4H_2O$

Fourmarierite was described from Shinkolobwe, Congo, where it forms small red crystals, associated with torbernite, kasolite, and curite. Buttgenbach (1924) compared the red color to that of vanadinite, and drew an idealized pseudohexagonal crystal. He noted that the mineral contains PbO, UO_3, and H_2O, but didn't provide a quantitative analysis or proposed formula. Based on goniometric measurements, he deduced that it is orthorhombic. Brasseur (1948) reported unit cell dimensions, using XRD, but noted that some uncertainty remained as to the correct formula.

Fourmarierite typically forms reddish-orange to brown tabular pseudohexagonal crystals up to 2 mm across, as well as elongated to acicular crystals and dense crusts. It is a common component of *gummite*, e.g., at Palermo mine, New Hampshire, and the Spruce Pine district, North Carolina. It is sometimes found in uraniferous mineralized fossil wood, e.g., at Monument No. 2 mine, Arizona. Fine specimens are found at Shinkolobwe, Congo (TL); other locales include: the Eldorado mine, Great Bear Lake, Northwest Territory, Canada; Mounana, Gabon; and Wölsendorf, Germany.

Fourmarierite is orthorhombic; Piret (1985) reported XRD data and a revised structural formula. Li and Burns (2000b) refined the structures of twelve crystals from various localities in the Congo and one synthetic crystal, and give the cell parameters as a=14.00-14.03, b=16.40-16.48, and c=14.32-14.38. They note: "The structure of fourmarierite contains uranyl pentagonal bipyramids that share edges and corners to form sheets oriented parallel to (001). There are two distinct Pb^{2+} cations and eight H_2O groups located in the

Figure 131. Thumbnail-sized patch of orange fourmarierite on ore from Shinkolobwe, Congo RJL967

Figure 132. Tabular, red-orange fourmarierite microcrystals with other secondary minerals from Shinkolobwe, Congo. FOV is ~5 mm wide. RJL1144

Figure 133. SEM image of the previous sample, showing the tabular habit of the fourmarierite crystals. RJL1144

Figure 134. A thumbnail-sized mass of uraninite with a thin coating of orange fourmarierite, from Keystone, South Dakota. RJL4178

interlayer. On the basis of the structure refinements, the site occupancy of Pb(1) in the natural crystals is typically deficient, ranging from 71 to 100%, whereas the site occupancy of Pb(2) is ~100% in the crystals studied. . . . The substitution $O^{2-} \leftrightarrow (OH)^-$ at two anion sites [O(12) and O(15)] in the sheet of uranyl polyhedra provides the charge-balance mechanism associated with interlayer cation variation. The structural formula for fourmarierite may be written as $Pb_{1-x}[(UO_2)_4O_{3-2x}(OH)_{4+2x}](H_2O)_4$, Z=8, with the constituents of the sheet of uranyl polyhedra enclosed in square braces."

Figure 135. The uranyl sheet of fourmarierite, viewed along [001].

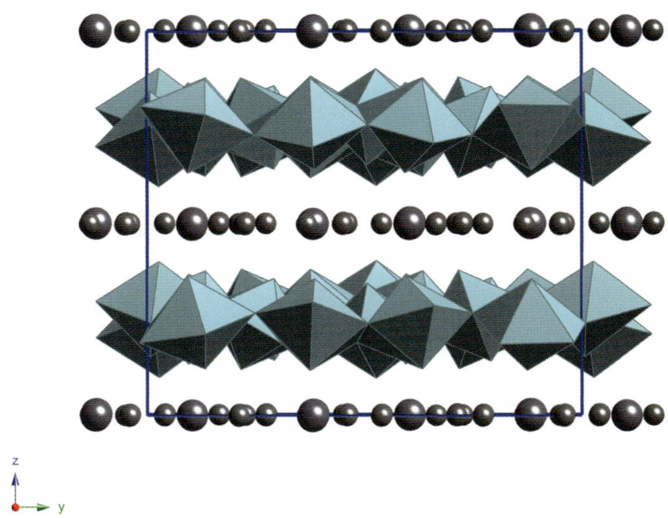

Figure 136. The fourmarierite structure viewed along [100], showing Pb^{2+} (large gray) and H_2O (small gray) in the interlayer.

Masuyite $Pb(UO_2)_3O_3(OH)_2.3H_2O$
Masuyite was described from Shinkolobwe, Congo, where it forms small (0.1 mm), pseudohexagonal, orange-red scales. It occurs in small geodes in pitchblende (Vaes 1947). Masuyite crystals are apparently always twinned. In addition to the TL at Shinkolobwe, it is found at the Kamoto mine, Congo, lining cavities in uraninite; other locales include Rabejac, France, and the South Alligator Valley, Northern Territory, Australia.

Burns and Hanchar (1999) reviewed the complicated history of masuyite and the lack of agreement among various published accounts, and suggest that two, or possibly three, different species have been referred to as masuyite in the literature. Compositions reported by different authors gave lead:uranium ratios ranging from 4:9 to 1:3, and from microanalytical studies it was clear that "masuyites" may be zoned, included, or otherwise inhomogeneous.

Masuyite is monoclinic with a=12.241, b=7.008, c=6.983 Å, and β=90.402°. The structure contains three symmetrically distinct U^{6+} positions, forming pentagonal bipyramids that share edges to form α-U_3O_8-type sheets that lie parallel to (010). The interlayer contains two distinct Pb^{2+} sites as well as three H_2O groups. The Pb(1) site is close to fully occupied and is coordinated by seven atoms of O that are contained in the sheets of uranyl polyhedra, and three H_2O groups. The Pb(2) site is only ~12% occupied and is coordinated by six atoms of O from the sheets of uranyl polyhedra and three H_2O groups. The structure of masuyite is closely related to that of protasite, but has an additional cation site in the interlayer (Burns and Hanchar 1999).

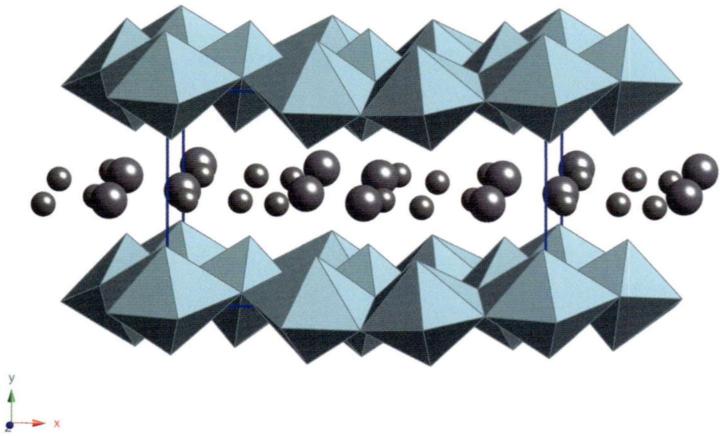

Figure 138. The structure of masuyite, viewed just off [001], showing uranyl sheets with Pb^{2+} (large gray) and H_2O (small gray) in the interlayer.

Figure 137. Orange masuyite with other secondary minerals on uraninite from Rabbit Lake, Saskatchewan. Sample is ~3 cm wide. RJL4567

Richetite $(Fe,Mg)_xPb_{8.57}[(UO_2)_{18}O_{18}(OH)_{12}.41H_2O$
Richetite was described from Shinkolobwe, Congo, where it forms minute (100 µm), black to brown, pseudohexagonal platy crystals on needles of uranophane. Crystals are commonly twinned and exhibit a perfect cleavage on (001) (Vaes 1947).

Richetite is triclinic with a=20.939, b=12.100, c=16.345 Å, α=103.87, β=115.37, and γ=90.27°. Piret and Deliens (1984) reported that microprobe analysis indicated the formula $PbO.4UO_3.4H_2O$ and that the structure is triclinic based on powder XRD data. Burns (1998a) reported a complete structure refinement, which showed that small amounts of other cations (Fe^{3+} and Mg^{2+}) can occupy certain interlayer sites: "The structure contains 36 unique U^{6+} positions, . . . forming pentagonal bipyramids. The uranyl polyhedra share edges to form symmetrically distinct but topologically identical α-U_3O_8-type sheets at z=0.25 and z=0.75. Although α-U_3O_8-type sheets of uranyl polyhedra occur in several structures, the richetite sheets are unique in their arrangement of OH^- anions. There are 13 partially occupied unique Pb^{2+} sites, two octahedrally coordinated M sites that may contain Fe^{3+} or other cations, and 41 unique H_2O groups in two distinct interlayers at z=O and z=0.5. Both the Pb^{2+} and M cations link to uranyl-ion O-atoms from adjacent sheets, and thus provide linkage of the sheets to the interlayer constituents. An extensive network of H bonds provides additional linkage."

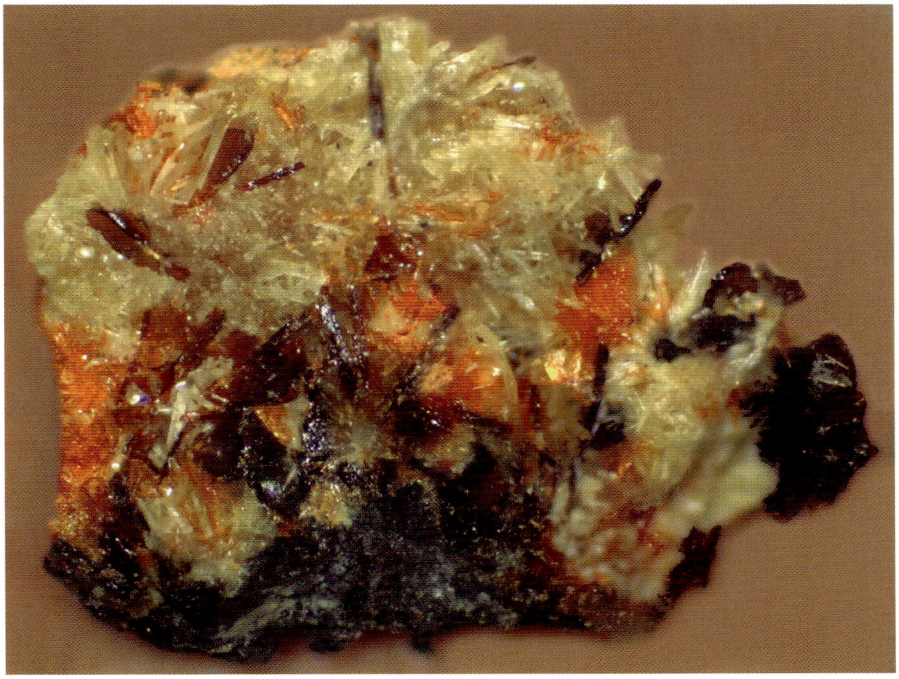

Figure 139. Richetite from the TL at Shinkolobwe, Congo, forming minute brown, platy crystals with acicular uranophane and other secondary minerals. FOV is ~3 mm. RJL676

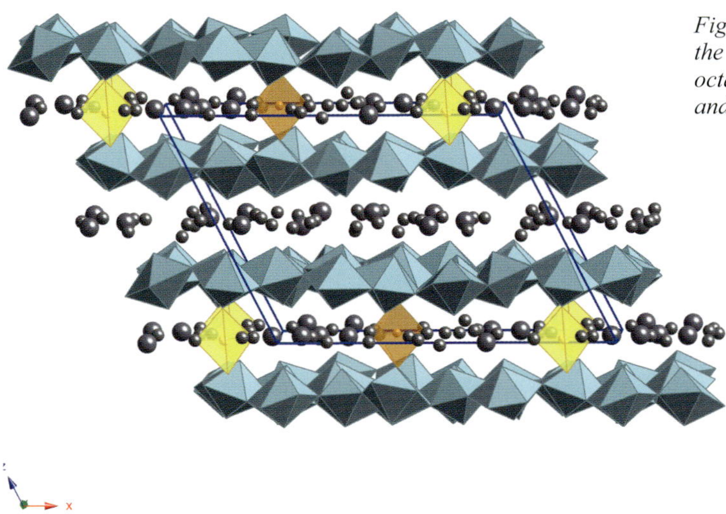

Figure 140. The structure of richetite, viewed just off [010] to show the uranyl sheets linked via interlayer species Pb (large gray), octahedrally coordinated M cations Fe (brown) and Mg (yellow), and hydrogen bonding involving the H_2O groups (small gray).

Sayrite $Pb_2(UO_2)_5O_6(OH)_2.4H_2O$
Sayrite was described from Shinkolobwe, Congo, where it forms small, yellow-orange to red-orange, isolated prismatic crystals. The crystals are elongated and flattened, with maximum dimensions of 0.6 × 0.3 × 0.1 mm. Sayrite is not fluorescent. Associated minerals include uranophane, becquerelite, masuyite, and richetite (Piret et al. 1983).

The mineral is quite rare at the TL. It can be confused with masuyite because of their similar color and habit, but the two can be distinguished in the microscope by the different terminal angles on the crystals: sayrite is 108°, whereas masuyite is ~120° (Deliens, Piret, and Comblain 1984).

Sayrite is monoclinic with a=10.704, b=6.960, c=14.553 Å, and β=116.81°. The structure comprises layers of composition $[(UO_2)_5O_6(OH)_2]$ that are linked via Pb^{2+} and H_2O groups in the interlayer (Piret et al. 1983).

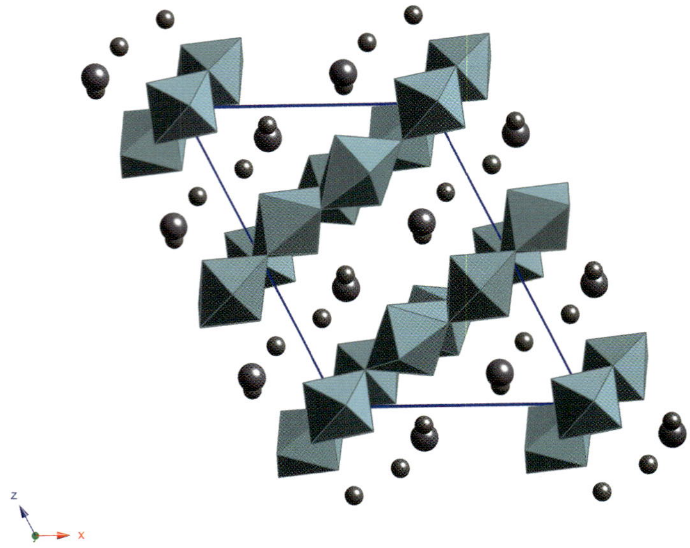

Figure 141. The structure of sayrite viewed just off [010], showing the arrangement of the uranyl sheets and interlayer Pb^{2+} (large gray) and H_2O (small gray).

Spriggite $Pb_3[(UO_2)_6O_8(OH)_2](H_2O)_3$
Spriggite was described from the Number 2 workings on Radium Ridge near Mt. Painter, South Australia, where it forms bright orange, cm-sized aggregates of randomly oriented prismatic crystals. Individual crystals range up to ~150 μm long × 40 μm across. Penetration twins forming six-legged stars are common. Associated minerals include uranophane-β, soddyite, kasolite, Ce-rich françoisite-(Nd), metatorbernite, billietite, Ba-rich boltwoodite, schoepite, metaschoepite, and weeksite. The assemblage is the product of supergene alteration of U-Nb-REE-bearing hydrothermal hematite breccia (Brugger, Ansermet and Pring 2003; Brugger et al. 2004).

Spriggite is monoclinic with a=28.356, b=11.990, c=13.998 Å, and β=104.24°. The structure consists of $[(UO_2)_6O_8(OH)_2]^{6+}$ sheets of uranyl polyhedra with the β-U_3O_8 anion topology, and Pb^{2+} and H_2O groups in the interlayer. Billietite and spriggite contain only hexavalent U in the uranyl sheets, whereas the similar sheets in β-U_3O_8 contain U^{5+} and U^{6+}, and those in ianthinite contain U^{4+} and U^{6+} (Brugger et al. 2004).

Figure 142. The uranyl sheet in spriggite, which has the β-U_3O_8 anion topology.

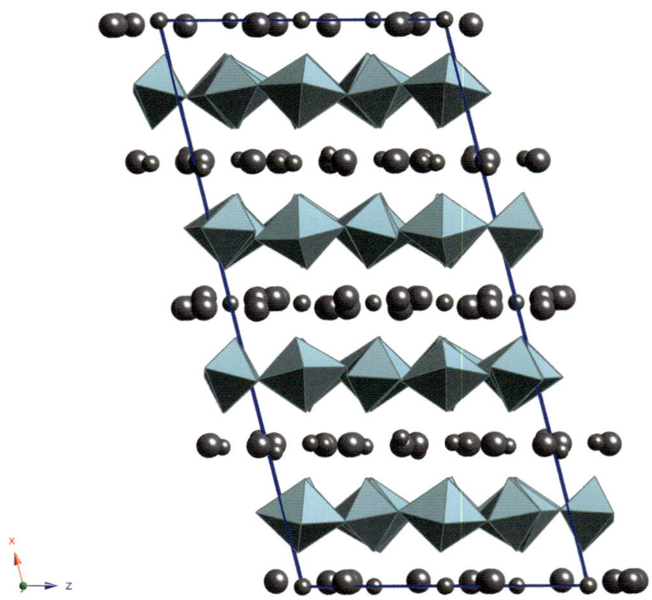

Figure 143. The spriggite structure viewed along [010], showing the uranyl sheets with Pb (large gray) and H_2O (small gray) in the interlayer.

Vandendriesscheite $Pb_{1.5}(UO_2)_{10}O_6(OH)_{11}.11H_2O$
Metavandendriesscheite $PbU_7O_{22}.nH_2O$ (n < 12)
Vandendriesscheite was described from Shinkolobwe, Congo, where it is associated with fourmarierite and rutherfordine. It forms yellow-orange to orange pseudohexagonal crystals (up to several mm), subparallel aggregates, and dense microcrystalline intergrowths with metavandendriesscheite (Vaes 1947). The mineral appears to spontaneously dehydrate in air to some degree, like schoepite, so most if not all specimens likely contain a mixture of vandendriesscheite and metavandendriesscheite.

It is the most common Pb-U oxide, and has been reported from many locales, including: Shinkolobwe, Congo (TL); as a component of *gummite* at the Ruggles pegmatite, New Hampshire, and the Spruce Pine district, North Carolina; the Eldorado mine, Great Bear Lake, Northwest Territory, Canada; Les Bois Noirs, France; and Menzenschwand, Germany.

Vandendriesscheite is orthorhombic with *a*=14.117, *b*=41.378, and *c*=14.535 Å. The structure consists of uranyl sheets that have a unique and extremely complex topology; the sheets are structurally intermediate between those of schoepite and becquerelite. Adjacent sheets are linked via bonds to the interlayer Pb^{2+} and H_2O groups. There is an extensive network of hydrogen bonds (Burns 1997).

Figure 144. Orange vandendriesscheite crystals associated with tabular red-orange fourmarierite and pale yellow rutherfordine(?), on massive uraninite, from the TL at Shinkolobwe, Congo. RJL1144

Figure 145. SEM photo of the sample in the previous figure, emphasizing the blocky crystal habit. RJL1144

Figure 146. Orange microcrystalline to earthy coatings of vandendriesscheite, associated with other secondary minerals, as alteration products on uraninite from Les Bois Noirs, Loire, France. RJL574

Figure 147. The uranyl sheet in vandendriesscheite, viewed along [001].

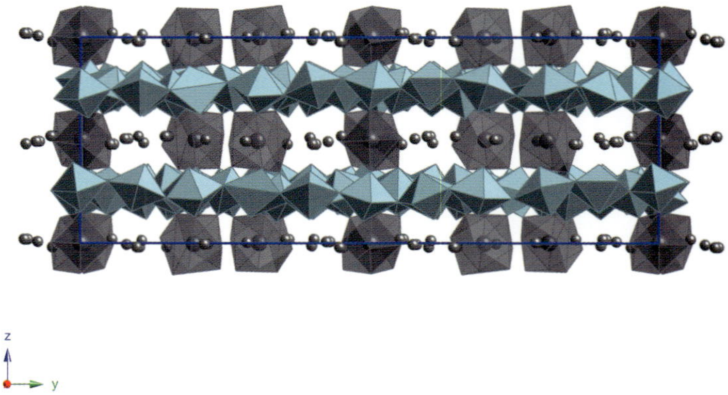

Figure 148. The vandendriesscheite structure viewed along [100], showing the Pb polyhedra (translucent gray) and H_2O groups (gray atoms) that link the sheets together.

Wölsendorfite $(Pb,Ba)_7(UO_2)_{14}O_{19}(OH)_4.12H_2O$
Wölsendorfite was described from Wölsendorf, Bavaria, Germany, where it forms red crystalline masses in fissures in fluorite (Protas 1957). It can form resinous masses, bladed to tabular pseudohexagonal crystals, and fibrous to acicular aggregates. It had previously been confused with fourmarierite.

A number of occurrences have been noted, including: orange-red crystalline incrustations on pitchblende at Great Bear Lake, Canada; orange-red spherulitic aggregates with pitchblende at Kersegalec, Lignol, France; as carmine-red nodules and crystals to 4 mm at Shinkolobwe, Congo; in pegmatite at the Corrego do Urucum mine, Minas Gerais, Brazil; the Koongarra uranium deposit, Northern Territory, Australia; and red crystals at Horni Hoštice, Czech Republic. Coatings and aggregates of Ba-rich wölsendorfite occur in rhyolite at Randbøldal, northern East Greenland (Beddoe-Stephens and Secher 1982).

Wölsendorfite is orthorhombic with a=14.131, b=13.885, and c = 55.969 Å. The structure contains two unique $U\varphi_6$ square bipyramids and six unique $U\varphi_7$ pentagonal bipyramids that link by the sharing of equatorial corners and edges to form infinite sheets lying parallel to (100). The sheets have a primitive repeat distance of c = 55.969 Å, and are by far the most complex sheet of uranyl polyhedra yet observed. The interlayer between the uranyl sheets contains Pb^{2+} and Ba cations, as well as H_2O groups that are either bonded to the interlayer cations or are held in the structure by hydrogen bonding only. There are eight unique cation positions in the interlayer, which are coordinated by six to ten anions. The structure of wölsendorfite is remarkable both in the complexity of the sheets of uranyl polyhedra and the connectivity of the interlayer. The wölsendorfite sheet is composed of slabs of the simpler α-U_3O_8 and β-U_3O_8-type sheets (Burns 1999).

Figure 149. Veins of orange wölsendorfite with other secondary minerals replacing massive uraninite from Shinkolobwe, Congo. RJL354

Figure 150. The uranyl sheet structure of wölsendorfite, viewed along [100].

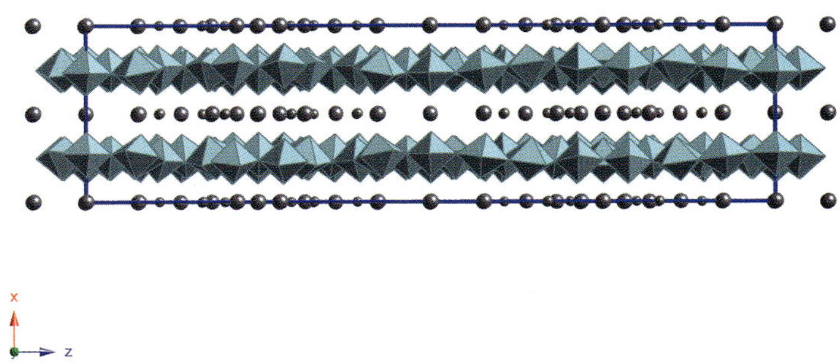

Figure 151. The wölsendorfite structure, viewed along [010], showing Pb^{2+} (large gray) and H_2O (small gray) in the interlayer. Note the enormous unit cell, with Z=8 and a volume of nearly 11,000 Å³!

Hydrous uranyl oxides containing other metals
Secondary uranyl oxides containing other metal ions include those containing alkali or alkaline earth metals such as K, Sr, Ca, and Ba, or transition metals such as Bi and Cu. Minerals in this category include agrinierite, bauranoite, becquerelite, billietite, calciouranoite and metacalciouranoite, compreignacite, protasite, rameauite, uranosphaerite, and vandenbrandeite.

Agrinierite $K_2(Ca_{0.65}Sr_{0.35})[(UO_2)_2O_3(OH)_2]_2.5H_2O$
Agrinierite was described from Margnac, France, where it occurs with uranophane in small cavities in *gummite*. It forms orange crystals, tabular on {001} with pseudohexagonal section bounded by the trace of (010) and showing sector twinning like that of billietite. There is good cleavage on {001} (Cesbron et al. 1972).

Agrinierite is orthorhombic with *a*=14.094, *b*=14.127, and *c*=24.106 Å. The structure comprises α-U_3O_8 sheets created by linking uranyl polyhedra by sharing oxygens along their equatorial edges; this topology is also seen in billietite, protasite, becquerelite, richetite, compreignacite, and masuyite. The arrangement of OH^- within the sheets varies among these minerals; the arrangement in agrinierite is the same as that in protasite. The interlayer is occupied by Ca, Sr, K, and H_2O. The authors note, "The inclusion of Sr in the structure of agrinierite suggests that the release of radioactive ^{90}Sr may be impacted by incorporation into this phase if it forms in a geological repository for nuclear waste," (Cahill and Burns 2000).

Figure 152. Brilliant orange coating of agrinierite on massive uraninite, ~2 × 3 cm, from the TL at Margnac, France. RJL419

Figure 153. Detail of the previous sample, showing tiny orange, tabular agrinierite crystals, associated with a yellow phase (possibly uranophane). RJL419

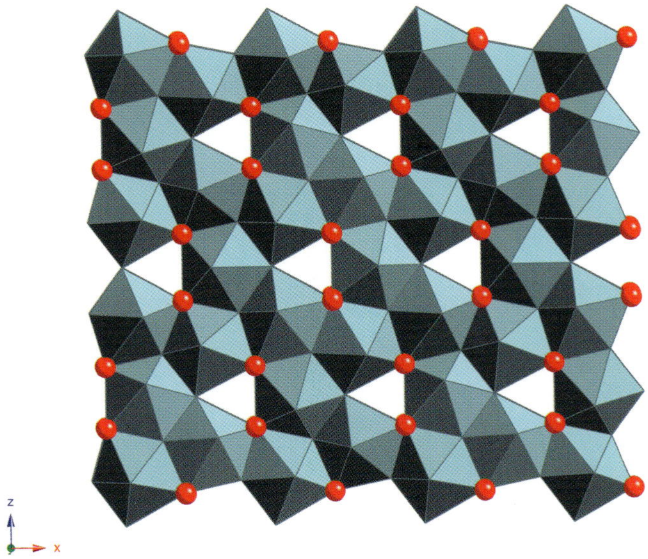

Figure 154. Crystal structure of agrinierite, viewed along [010], showing the α-U_3O_8 sheet. Red atoms represent OH^- oxygens. This is the same sheet topology and OH^- arrangement as in protasite.

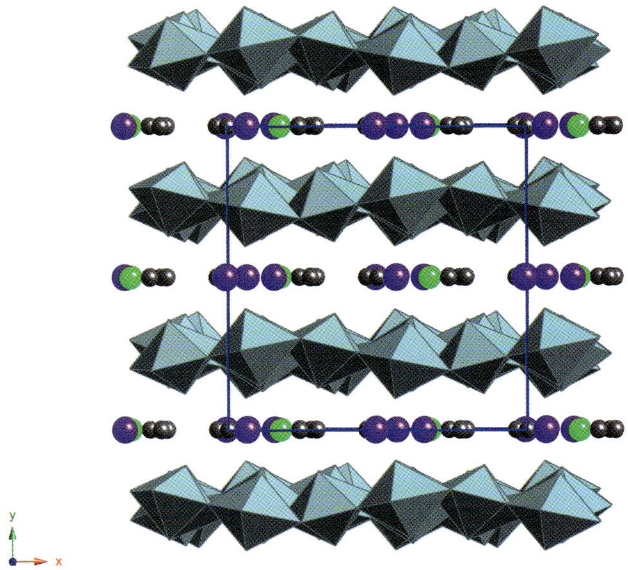

Figure 155. Structure of agrinierite, viewed along [100], showing the sheets edge-on, with K (purple), Sr (green), and H_2O (gray) in the interlayer.

Bauranoite $BaU_2O_7.4\text{-}5H_2O$
Bauranoite was described by Rogova et al. (1973) from the oxidation zone of a U-Mo deposit, which Pekov (1998) has identified as the Oktyabr'skoe deposit, Strel'tsovskoye ore field, near Krasnokamensk, Eastern Transbaikal region, Russia. It forms fine-grained, dense, reddish brown aggregates replacing pitchblende or nasturan and being replaced by uranophane.

The powder XRD pattern shows some resemblance to those of metacalciouranoite and wölsendorfite. Belova et al. (1985) provide a further discussion of the relationships of these minerals.

Becquerelite $Ca(UO_2)_6O_4(OH)_6.8H_2O$
Becquerelite was described from Kasolo, Congo, where it forms tabular to prismatic or lath-like crystals, and fine grained crusts in various shades of golden- to lemon-yellow, yellow-orange, and brownish yellow, in cavities and fissures in pitchblende (Schoep 1922b). The prismatic crystals are tabular on (001) and elongated on [010]. Cleavage is perfect on {001} and good on {010}. Associated minerals include billietite, curite, wölsendorfite, fourmarierite, schoepite, studtite, and rutherfordine. Microcrystalline becquerelite and schoepite occasionally form pseudomorphs of cubic uraninite crystals (Deliens, Piret, and Comblain 1981).

It has been identified at many locales, including: Shinkolobwe, Congo; Wölsendorf, Germany; the Eldorado mine, Canada; the Margnac II mine and elsewhere in France; and at numerous sites in the Colorado Plateau region.

The true composition of becquerelite took some time to be established. Schoep (1922b) considered it to be a hydrous

Figure 156. Amber becquerelite crystals to ~2 mm long, filling a cavity ~2 cm wide in massive uraninite, from Shinkolobwe, Congo. RJL801

uranyl oxide, and the Pb originally reported was assumed to be an impurity. Schoep and Stradiot (1948) described a Pb-free variety, and suggested that substitution of Pb for U might occur. Frondel and Cuttitta (1953) gave the formula as $7UO_2 \cdot 11H_2O$, but noted that becquerelite is isostructural with billietite ($BaO \cdot 6UO_2 \cdot 11H_2O$). Piret-Meunier and Piret (1982) presented new crystal structure data and a revised formula, which made it clear that becquerelite is the Ca analog of billietite, as suggested by Protas (1957a). Pagoaga, Appleman, and Stewart (1987) presented a detailed comparison of the structures of becquerelite, billietite, and protasite.

Becquerelite is orthorhombic, with a=13.853, b=12.393, and c=14.930 Å. The structure contains α-U_3O_8-type sheets, with a single symmetrically distinct Ca cation and eight symmetrically distinct H_2O groups in the interlayer (Burns and Li 2002).

Figure 157. Yellow becquerelite forming a crust on sandstone from the Happy Jack mine, Utah. Specimen is 8 cm wide. RJL2641

Figure 158. Detail of the becquerelite specimen in the previous figure. FOV is ~ 3 mm. RJL2641

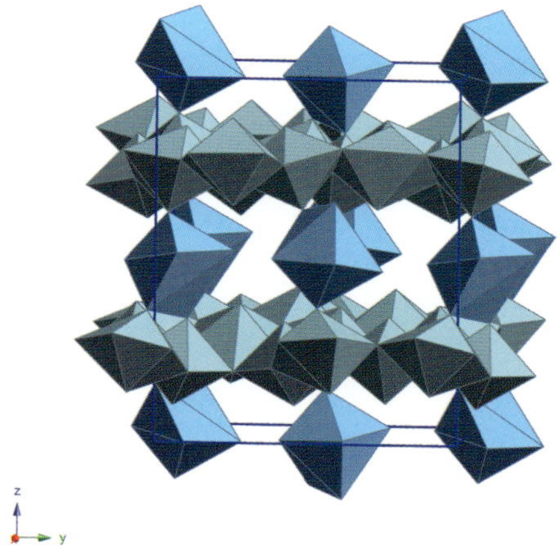

Figure 159. The structure of becquerelite, showing how the U_3O_8 sheets (blue-gray) are bonded together via Ca (blue polyhedra).

Billietite $Ba(UO_2)_6O_4(OH)_6.8H_2O$

Billietite was described from Shinkolobwe, Congo, (Vaes 1947) where it forms small (up to 2 mm) amber-yellow, tabular crystals that resemble becquerelite. Crystals are orthorhombic, pseudohexagonal, and usually twinned on (111). There is a perfect cleavage on (010). The crystals are transparent to translucent, strongly pleochroic (α=pale yellow to colorless, β and γ=golden-yellow), with adamantine luster, and are not fluorescent. It is a somewhat rare product of early-stage alteration of uraninite at Shinkolobwe; at Swambo it is associated with soddyite, and at Luiswishi it is associated with cuprosklodowskite, vandenbrandeite, and malachite (Deliens, Piret, and Comblain 1981).

Other notable locales include: Menzenschwand, Germany; and at Margnac II mine and Rabejac, France.

Billietite is orthorhombic with a=12.094, b=30.211, and c=7.156 Å. The crystal structure consists of $[(UO_2)_6O_4(OH)_6]$ sheets, linked by interstitial Ba atoms and H_2O groups. There

Figure 161. Amber billietite crystals with pale yellow spherical aggregates of uranophane(?), from the Krunkelbach deposit. FOV is ~2 mm wide. RJL582

Figure 160. Earthy yellow film of billietite on matrix, from the Krunkelbach deposit, Germany. RJL2702

Figure 162. Amber to yellow billietite crystals with dark green vandenbrandeite, apple-green cuprosklodowskite, and other secondary phases typical of the complex mineral assemblage at Musonoi, Congo. RJL520

are two crystallographically distinct structural sheets and eight interlayer H_2O groups. Four H_2O are bonded to the interlayer Ba atoms, and the remaining four H_2O are not bonded to a cation, but are held in the structure by a network of hydrogen bonds (Finch et al. 2006).

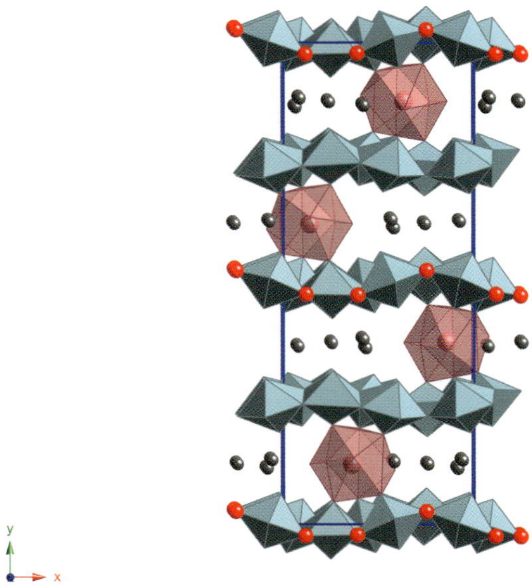

Figure 163. Crystal structure of billietite, viewed along [001], showing two structurally distinct sheets of uranyl polyhedra alternating along [010] and linked together via Ba (magenta polyhedra). OH groups are red, and interlayer H_2O groups are gray. The unit cell is rather large, with Z=4 and a volume of 2615Å^3.

Calciouranoite $(Ca,Ba,Pb)U_2O_7.5H_2O$
Metacalciouranoite $(Ca,Ba,Pb)U_2O_7{\cdot}2H_2O$
Calciouranoite was described by Rogova et al. (1974) from deep levels of the oxidation zone at a locale that Pekov (1998) has identified as the Oktyabr'skoe deposit, Strel'tsovskoye ore field, near Krasnokamensk, Eastern Transbaikal region, Russia. There, brown and orange-brown massive calciouranoite and metacalciouranoite were found replacing pitchblende. The powder XRD pattern shows some similarities to those of bauranoite and wölsendorfite; see also Belova et al. (1985).

A K-Sr-containing calciouranoite was reported by Fedorov, Ryzhov, and Lyubomilova (1981). Analysis yielded the empirical formula $(Ca_{.577}K_{.183}Sr_{.163}Pb_{.058}Na_{.019})O{\cdot}2.12UO_2{\cdot}3.89H_2O$, or ideally $(Ca,K,Sr,)O{\cdot}2UO_2{\cdot}4H_2O$.

Clarkeite $(Na,K,Ca,Pb)(UO_2)O(OH).0\text{-}1H_2O$
Clarkeite was described from the Spruce Pine district, North Carolina, where it forms pseudomorphic replacements of uraninite, often intergrown with uranophane and *gummite.* Ross, Henderson, and Posnjak (1931) observed, "The uranium minerals of Spruce Pine make strikingly beautiful specimens. In some there is a central core of black, sub-metallic uraninite. This is surrounded by a zone of dark reddish brown clarkeite; this by brilliant orange-red gummite; and the whole by an outer zone of bright yellow uranophane. The various zones vary in width but in the larger specimens they may be several centimeters wide. ... Clarkeite is massive, without cleavage, has a conchoidal fracture and a hardness of 4 to 4.5. The specific gravity is 6.39. The color is very dark brown in the hand specimen and reddish brown to orange brown in microscopic grains. The luster is slightly waxy and the streak is lighter in color than the mineral and more of a yellowish brown." Their proposed general formula was $RO{\cdot}3UO_3{\cdot}3H_2O$, where R represents alkalis and alkaline earths, predominantly Na; this formula was based on wet-chemical analysis on carefully hand-picked grains.

Gruner (1954) showed that clarkeite is isostructural with several Na or Ca uranyl oxide phases. Frondel and Meyrowitz (1956) described a clarkeite from the Ajmer district, Rajputana, India, where it forms a chocolate-brown, microcrystalline alteration product of uraninite in pegmatite. They proposed the formula $(Na,Ca,Pb,Th,H_2O)_2U_2(O,OH)_7$.

Clarkeite is hexagonal with *a*=3.954 and *c*=17.73 Å. The structure consists of $[(UO_2)(O,OH)_2]$ sheets that are bonded to each other through interlayer cations and H_2O groups. Finch and Ewing (1997) give the empirical formula for material from the Fanny Gouge mine, Spruce Pine district, North Carolina, as: $\{Na_{.733}K_{.029}Ca_{.021}Sr_{.009}Y_{.024}Th_{.006}Pb_{.058}\}_{\Sigma.880}[(UO_2)_{.942}O_{.918}(OH)_{1.082}](H_2O)_{.069}$ where Na predominates and the Pb is radiogenic. They give the general formula for clarkeite as $\{(Na,K)_pM^{2+}_qM^{3+}_rM^{4+}_sPb_x\}[(UO_2)_{1-x}O_{1-y}(OH)_{1+y}](H_2O)_z$, where Na >> K and p > (q + r + s). The number of O^{2-} ions and OH groups in the structural unit is determined by the net charge of the interlayer cations (except Pb): y = 1 - (p + 2q + 3r + 4s). This suggests that the ideal formula for ideal end-member clarkeite is $Na[(UO_2)O(OH)](H_2O)_{0\text{-}1}$. The structural sheets are destabilized as U decays to Pb (increasing x), and Pb enters vacant interlayer cation sites. Clarkeite eventually recrystallizes to lead uranyl oxide hydrates such as wölsendorfite or curite.

Compreignacite $K_2[(UO_2)_3O_2(OH)_3]_2.7H_2O$
Compreignacite was described from the Margnac deposit, near Compreignac, Haute-Vienne, France, where it formed as an oxidation product of pitchblende ore. It was quite rare, having been found on only a few samples. It formed tiny yellow crystals, flattened on (001), and frequently twinned on (110) producing tabular crystals with a pseudorhombic outline. Repeated twinning on (110) produces pseudohexagonal crystals similar to aragonite. The structure is analogous to that of billietite (Protas 1964; Granger and Protas 1965).

Figure 164. Resinous brownish clarkeite associated with uranophane and gummite, as an alteration product of uraninite, in pegmatite from North Grafton, New Hampshire. RJL595

Figure 165. Compreignacite forming an earthy, bright yellow crust on granite from Les Bois Noirs, France. FOV is ~4 mm wide. RJL1204

Compreignacite has also been reported from Cornwall, England. Yellow translucent spherules up to 0.1 mm on mineralized hornfels were found on spoil heaps at Wheal Edward, associated with cuprosklodowskite, uranophane, brochantite, connellite, and calcite. In a backfilled stope at the nearby West Wheal Owles, it forms thin crusts of transparent, bright to golden yellow, prismatic crystals (≤1 mm long). There, it is intimately associated with later-formed cuprosklodowskite and uranophane. The matrix is mafic hornfels containing inclusions of pink K-feldspar, the decomposition of which is presumed to supply the K^+ to form compreignacite (Elton, Hooper, and Ryback 1994). It has also been reported from Menzenschwand and Ronneburg, Germany; and from pegmatite at the Palermo No. 1 mine, Grafton Co., New Hampshire.

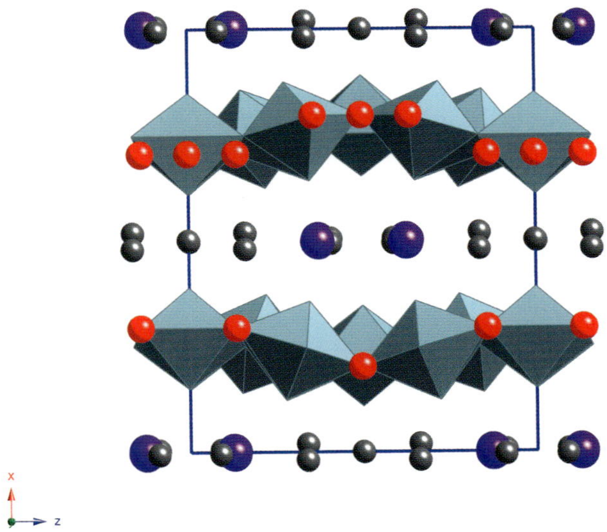

Figure 166. Structure of compreignacite, viewed along [010], showing the α-U_3O_8-type sheets (OH are shown in red), with potassium (purple) and H_2O (gray) in the interlayer.

Compreignacite is orthorhombic with a=14.859, b=7.715, and c=12.187 Å. The structure contains uranyl sheets like those found in becquerelite, billietite, protasite, richetite, and α-U_3O_8. Potassium and water occupy the interlayer. Each K cation is coordinated by four O atoms of the adjacent sheets of uranyl polyhedra as well as three H_2O groups. Two of the symmetrically distinct H_2O groups in the structure are bonded to K, and the other is held in the structure only by H bonds. The K polyhedra share a face, forming a dimer with the composition $K_2O_6(H_2O)_4$. Additional linkages between the interlayer constituents and the sheets of uranyl polyhedra are provided by H bonds (Burns 1998b).

Protasite $Ba[(UO_2)_3O_3(OH)_2].3H_2O$

Protasite was described from Shinkolobwe, Congo, where it forms bright orange, pseudohexagonal platelets to 0.5 mm long × 0.1 mm thick, flattened on (010). The type specimen is associated with uraninite and uranophane on an unidentified fine-grained rock. It strongly resembles fourmarierite in color and habit; the two species can be distinguished by fourmarierite's tendency to form parallel aggregates of blocky crystals versus protasite's separate, thinner platelets, which are striated because of twinning (Pagoaga, Appleman, and Stewart 1986).

Protasite is monoclinic with a=12.295, b=7.2221, c=6.956 Å, and β=90.401°. The structure contains α-U_3O_8-type sheets, linked to one another via interlayer Ba ions. All interlayer H_2O molecules are coordinated to the Ba, giving protasite the simplest structure of the hydrous uranyl oxides (Pagoaga, Appleman, and Stewart 1987).

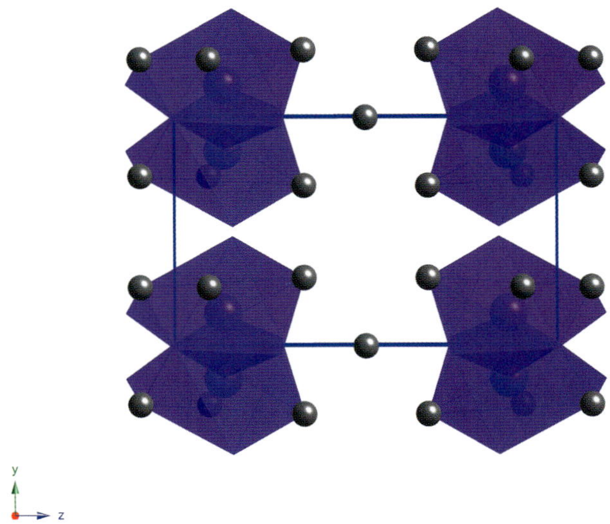

Figure 167. The interlayer contents of compreignacite, viewed normal to the sheet, to show the $K_2O_6(H_2O)_4$ dimers (translucent purple polyhedra) and the unattached H_2O groups (gray) that are held in place by hydrogen bonding

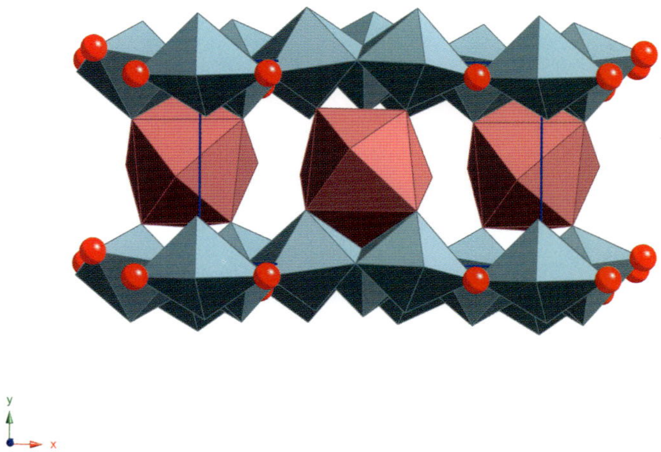

Figure 168. Structure of protasite viewed along [010], showing the α-U_3O_8-type sheets (OH are shown in red), linked via Ba (magenta polyhedra) in the interlayer. Interlayer water groups are not shown.

Rameauite $K_2CaU^{6+}{}_6O_{20}.9H_2O$
Rameauite was described from the Margnac mine, Haute-Vienne, France, where it forms small (up to 1 mm), orange crystals associated with calcite and uranophane on a block of pitchblende. The crystals are twinned on {100}, slightly flattened on {010} and elongated on [001], with a pseudohexagonal section. TGA and DTA indicated water loss peaks at 95, 130, and 170°C (Cesbron et al. 1972).

It has also been noted at the Orphan mine, Grand Canyon, Coconino Co., Arizona; and at Shinkolobwe, Congo.

Rameauite is monoclinic with a=13.97, b=14.26, c=14.22 Å, and β=121°1' (±5'). A complete structure refinement hasn't been reported to date.

Uranosphaerite $Bi(UO_2)O_2OH$
Uranosphaerite was described from the Weisser Hirsch mine, near Schneeberg, Saxony, Germany, where it formed as an oxidation product of uraninite and native bismuth (Weisbach 1873). It forms minute, acute crystals elongated along [001], in globular aggregates to ~0.5 mm. Color ranges from yellow-orange to reddish-orange, with greasy to dull luster.

Other reported locales include: the Clara mine, Germany; and in the Czech Republic at Kladska, near Marianske Lazne, and at Ryžovište, near Harrachov.

Uranosphaerite is monoclinic with a=7.559, b=7.811, c=7.693 Å, and β=92.88°. The structure consists of sheets formed by the sharing of edges between dimers of uranyl pentagonal bipyramids and $Bi(O,OH)_4$ polyhedra. Adjacent sheets are linked via weak Bi-O bonds and weak hydrogen bonding. The sheet is based on the anion topology of the francevillite sheet (Hughes, Burns, and Kolitsch 2003).

Figure 169. Reddish veins of rameauite in massive uraninite, from Shinkolobwe, Congo. Sample is ~3 mm wide. RJL636

Vandenbrandeite $Cu(UO_2)(OH)_4$
Vandenbrandeite was described from Kalongwe, Congo, where it forms deep green to nearly black tabular crystals up to 1 cm long, and parallel groups of crystals in which the (001) faces predominate. There is a perfect cleavage on (001) and very thin cleavage fragments are translucent. The mineral assemblage at Kalongwe differs from that of Shinkolobwe

Figure 170. Crystal structure of uranosphaerite, viewed normal to the sheet, showing edge-sharing of dimers of uranyl pentagonal bipyramids (blue-gray) and dimers of $Bi(O,OH)_4$ polyhedra (purple).

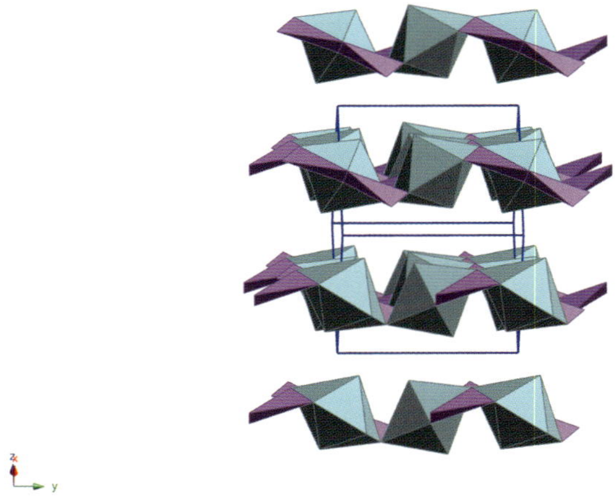

Figure 171. Structure of uranosphaerite viewed parallel to the sheets (approximately diagonal to the unit cell).

Figure 172. Dark green, somewhat curved vandenbrandeite crystals, ~1 mm long, in massive vandenbrandeite from Musonoi, Congo. RJL886

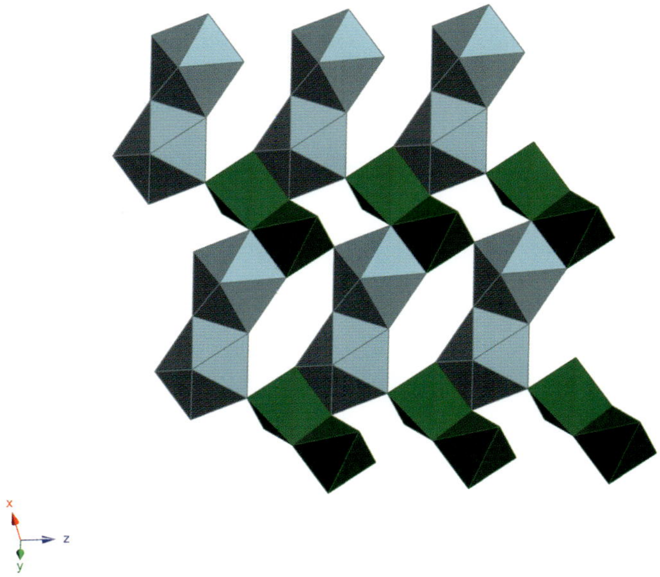

Figure 173. The copper-uranyl oxide sheet in vandenbrandeite, showing the linkage between Cu(O,OH)$_5$ square pyramids (dark green) and uranyl bipyramids (blue-gray).

in that the uraninite is associated with copper sulfides, which are rare or absent at Shinkolobwe (Schoep 1932).

At the TL, vandenbrandeite is associated with pitchblende, chalcopyrite, chalcocite, kasolite, sklodowskite, and malachite (with which it can sometimes be confused. Other U-Cu minerals that are typically associated include torbernite and cuprosklodowskite (Deliens, Piret, and Comblain 1981). It has also been found at Kambove, Luiswishi, Musonoi, Shinkolobwe, and Swambo in the Congo; also at Rabejac, France.

Vandenbrandeite is triclinic with a=7.85, b=5.45, c=6.10 Å, α=91°24', β=102°, and γ=89°12'. Milne and Nuffield (1951) reported structural data based on powder XRD; they discussed the problem of determining the best choice of cell orientation in triclinic minerals and noted that the cell should be chosen to best describe the mineral, taking into account important morphological features. Among hydrous uranyl oxide minerals, vandenbrandeite is the only one besides uranosphaerite that contains non-uranyl polyhedra in the uranyl sheets. The vandenbrandeite sheet contains dimers of edge-sharing uranyl pentagonal bipyramids, as well as dimers of Cu(O,OH)$_5$ square pyramids formed by the sharing of equatorial edges. The apical ligands of adjacent Cu(O,OH)$_5$ pyramids within each dimer point in opposite directions. Sheets are linked directly through the apical ligands of the Cu(O,OH)$_5$ square pyramids, which are uranyl oxygen atoms of adjacent sheets (Rosenzweig and Ryan 1977; Hughes, Burns, and Kolitsch 2003).

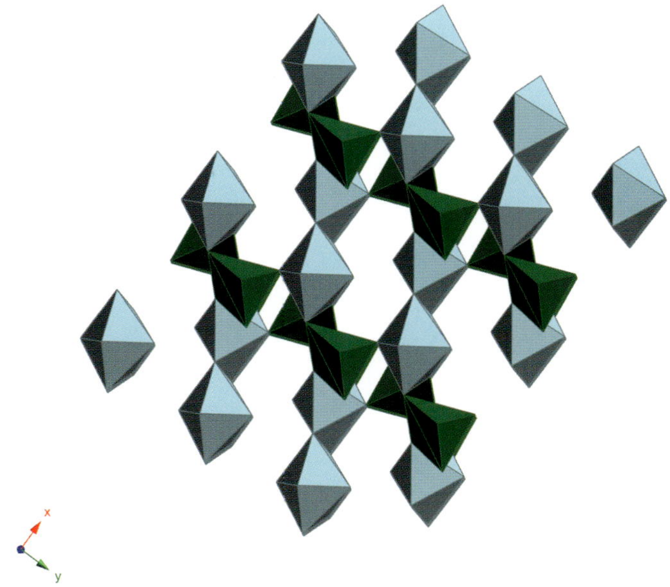

Figure 174. The vandenbrandeite structure viewed along [001], showing the linkage of adjacent sheets by the copper polyhedra.

CARBONATES

The uranyl carbonates may be conveniently placed into four groupings: uranyl monocarbonates; uranyl di- and tricarbonates that contain other metal ions; uranyl carbonates that contain rare earths; and other uranyl carbonates, including some that contain other anions. Researchers have shown that in aqueous solutions, the U^{6+} ion can form complexes with the general formula $[(UO_2)_x(CO_3)_y(OH)_z(H_2O)_q]$. The uranyl tricarbonates tend to form in water having a relatively high pH, and their structures are based on isolated uranyl tricarbonate clusters. Minerals with $U:CO_3$ ratios other than 1:3 can form at lower pH and higher pCO_2 and can have other structural features, such as uranyl carbonate sheets. Thus, the uranyl carbonates are a natural group for researchers to study in order to better understand the relationship between mineral paragenesis and crystal structure (Li, Burns, and Gault 2000).

Table 44. Uranyl carbonates

Species	Formula	Symmetry
Monocarbonates		
Blatonite	$UO_2CO_3 \cdot H_2O$	hex or trig
Joliotite	$UO_2CO_3 \cdot nH_2O$	orth
Oswaldpeetersite	$(UO_2)_2CO_3(OH)_2 \cdot 4H_2O$	mon
Rutherfordine	$UO_2(CO_3)$	orth
Di- and Tricarbonates		
Agricolaite	$K_4(UO_2)(CO_3)_3$	mon
Andersonite	$Na_2Ca(UO_2)(CO_3)_3 \cdot 6H_2O$	trig
Bayleyite	$Mg_2(UO_2)(CO_3)_3 \cdot 18H_2O$	mon
Čejkaite	$Na_4(UO_2)(CO_3)_3$	mon
Grimselite	$K_3Na(UO_2)(CO_3)_3 \cdot (H_2O)$	hex
Liebigite	$Ca_2(UO_2)(CO_3)_3 \cdot 11H_2O$	orth
Linekite	$K_2Ca_3[(UO_2)(CO_3)]_2 \cdot 7H_2O$	orth
Rabbittite	$Ca_3Mg_3(UO_2)_2(CO_3)_6(OH)_4 \cdot 18H_2O$	mon
Swartzite	$CaMg(UO_2)(CO_3)_3 \cdot 12H_2O$	mon
Widenmannite	$Pb_2(UO_2)(CO_3)_3$	orth
Zellerite	$Ca(UO_2)(CO_3)_2 \cdot 5H_2O$	orth
Metazellerite	$Ca(UO_2)(CO_3)_2 \cdot 3H_2O$	orth
Znucalite	$CaZn_{11}(UO_2)(CO_3)_3(OH)_{20} \cdot 4H_2O$	orth
Carbonates with REE		
Astrocyanite-(Ce)	$Cu_2(Ce, Nd, La)_2(UO_2)(CO_3)_5(OH)_2 \cdot 1.5 H_2O$	hex
Bijvoetite-(Y)	$(Y,REE)_2(UO_2)_4(CO_3)_4(OH)_6 \cdot 11H_2O$	mon
Kamotoite-(Y)	$Y_2(UO_2)_4(CO_3)_3(OH)_8 \cdot 10\text{-}12H_2O$	mon
Lepersonnite-(Gd)	$Ca(Gd,Dy)_2(UO_2)_{24}(CO_3)_6Si_4O_{12} \cdot 60H_2O$	orth
Shabaite-(Nd)	$Ca(Nd,Sm,Y)_2(UO_2)(CO_3)_4(OH)_6.6H_2O$	mon
Other uranyl carbonates		
Albrechtschraufite	$Ca_4Mg(UO_2)_2(CO_3)_6F_2$	tric
Fontanite	$Ca[(UO_2)_3(CO_3)_2O_2](H_2O)_6$	mon
Roubaultite	$Cu_2(UO_2)_3(CO_3)_2O_2(OH)_2 \cdot 4H_2O$	tric
Schröckingerite	$NaCa_3(UO_2)(CO_3)_3(SO_4)F \cdot 10H_2O$	tric
Sharpite	$Ca(UO_2)_6(CO_3)_5(OH)_4 \cdot 6H_2O$	orth
Urancalcarite	$Ca(UO_2)_3(CO_3)(OH)_6 \cdot 3H_2O$	orth
Voglite	$Ca_2Cu^{2+}(UO_2)(CO_3)_4 \cdot 6H_2O$	?
Wyartite	$CaU^{5+}(UO_2)_2(CO_3)O_4(OH) \cdot 7H_2O$	orth

Uranyl monocarbonates

Uranyl monocarbonates include three structurally distinct species that contain no metals other than U and differ from one another by their degree of hydration: blatonite ($\cdot H_2O$), joliotite ($\cdot 2H_2O$), and rutherfordine (anhydrous). Interestingly, when joliotite was dehydrated by heating for two hours at 100°C, changes in the optical and crystallographic properties were noted, but the dehydrated material was distinct from blatonite, which thus cannot be viewed as a *meta-* form of joliotite (Vochten and Deliens 1998). The hydrated basic uranyl carbonate, oswaldpeetersite, is included here as well; it contains OH^- groups in addition to a carbonate group, but it contains no metals other than U.

Blatonite $UO_2CO_3.H_2O$

Blatonite was described from material collected by Patrick Haynes at the Jomac mine, Brown's Rim, San Juan Co., Utah. At the TL it occurs in seams of gypsum that lie along bedding planes in a layer of siltstone contained within the Triassic Shinarump conglomerate. The mineral typically forms asbestiform aggregates or bundles of very thin needles; the bundles are about 0.1 mm wide and 1 mm long. It is translucent, canary yellow, with a silky luster, and is strongly fluorescent in UV. Blatonite will dissolve with effervescence in dilute acids, including acetic acid. Associated uranium minerals include boltwoodite, coconinoite, metazeunerite, and rutherfordine; other minerals noted from the mine are azurite, brochantite, carbonatecyanotrichite, malachite, Mn-rich smithsonite, and small amounts of sulfide minerals (Vochten and Deliens 1998).

The Jomac mine was worked off and on from about 1950 to 1985; it has since been sealed and reclaimed and the area is now part of the Glen Canyon National Recreation Area, so further collecting is not permitted.

Figure 175. A typical example of siltstone with thin white seams of gypsum, from the blatonite TL at the Jomac mine, San Juan Co., Utah. RJL2722

Figure 176. Detail of the sample in the previous photo, showing the minute pale yellow fibers of blatonite associated with gypsum on matrix, from the type locale at the Jomac mine, San Juan Co., Utah. FOV is ~4 mm. RJL2722

Figure 177. Yellow mass of blatonite on gypsum, collected by Patrick Haynes at the Jomac mine, Utah, in 1989. FOV is ~4 mm. RJL3144

The minutely fibrous habit has prevented single crystal structure refinement. Based on powder XRD, the structure is reported to be hexagonal or trigonal (space group unknown), with a=15.79 and c=23.93 Å (Vochten and Deliens 1998).

Joliotite $UO_2CO_3.nH_2O$

Joliotite was described from the Krunklebach Valley uranium deposit, Menzenschwand, Germany. The species is rare at the TL, where it forms lemon yellow crusts on limonite and barite, in association with other secondary species, including billietite, rutherfordine, studtite, and an unknown U oxide. The minute (100 µm) spherulitic aggregates are made up of scaly, orthorhombic crystals flattened on (100). It is weakly fluorescent. The structure is orthorhombic with a=8.16, b=10.35, and c=6.32 Å (Walenta 1976).

Joliotite has also been reported from the Müllenbach uranium deposit, Baden-Baden, Germany.

Oswaldpeetersite $(UO_2)_2CO_3(OH)_2.4H_2O$

Oswaldpeetersite, a basic uranyl carbonate, was described from the Jomac mine, San Juan Co., Utah, based on material collected by Patrick Haynes. It occurs in a setting like that of blatonite (discussed above), associated with gypsum along bedding planes of a siltstone layer contained within the

Shinarump conglomerate. The mineral forms minute (100 × 10 × 2 μm), lath-like to acicular, transparent, canary yellow crystals. It has a vitreous luster and is not fluorescent (Vochten, Deliens, and Medenbach 2001).

Based on powder XRD refinement, the structure is monoclinic with a=4.143, b=14.098, c=18.374 Å, and β=103.62°.

Rutherfordine $UO_2(CO_3)$
Rutherfordine, the anhydrous uranyl carbonate, was first described by Marckwald (1906) from Morogoro, Tanzania. In 1947, Vaes described (as *diderichite*) a mineral from Shinkolobwe that was later shown to be the same as rutherfordine (Frondel and Meyrowitz 1956). It typically forms off-white, powdery or earthy aggregates of minute fibrous or lath-like crystals flattened on (100) and elongated along [001] (Frondel 1958).

Rutherfordine often forms the outermost alteration layer on uraninite that is pseudomorphously replaced by *gummite* and uranyl silicates. It is also a component of Colorado Plateau type ores. It is frequently seen at Shinkolobwe and Musonoi, Congo, in association with cuprosklodowskite, schoepite, and other secondary minerals. In these specimens the rutherfordine often forms visually distinctive pale yellowish to cream-colored bundles of acicular crystals. Other reported locales include: Palermo No. 3 mine, New Hampshire; Newry, Maine; the Rabejac deposit, France; and Menzenschwand, Germany.

Rutherfordine is orthorhombic with a=4.840, b=9.273, and c=4.298 Å. Clark and Christ (1956) determined the structure using powder XRD and also determined the optical properties using a sample from Shinkolobwe that contained unusually large, lath-like single crystals up to 2–3 mm long. A full structure refinement was reported by Finch et al. (1999): The uranyl dipyramids are linked by sharing pairs of edges to form chains parallel to [001]; the chains are joined in the [100] direction by sharing vertices, forming a sheet of hexagonal bipyramids. The CO_3 group shares two edges with neighboring (UO_8) polyhedra, and all the CO_3 groups in the sheet point in the same direction. These sheets are then stacked along [010] so that neighboring sheets mesh together. However, there are no shared oxygens to directly link the sheets, and they are held together only by van der Waals forces.

Figure 178. Pale yellow rutherfordine filling interstices in a cuprosklodowskite nodule from Musonoi, Congo. Tiny golden yellow crystals are possibly schoepite. Field of view is about 2 cm wide. RJL 2571

Figure 179. Detail of rutherfordine in cuprosklodowskite nodule with golden yellow schoepite(?). RJL 2571

Figure 180. Cream-yellow crust of rutherfordine on altered uraninite, with bright orange vandendriesscheite and red-orange fourmarierite. Specimen is from Shinkolobwe, Congo. RJL1144

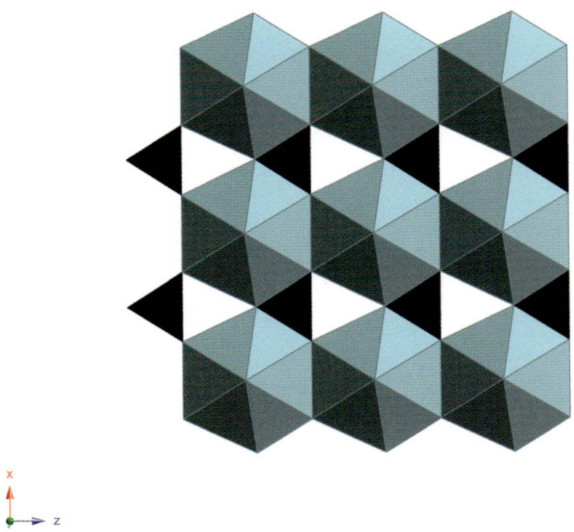

Figure 181. The uranyl carbonate sheet in rutherfordine. CO_3 groups (black triangles) share two edges with neighboring uranyl polyhedra and all point in the same direction.

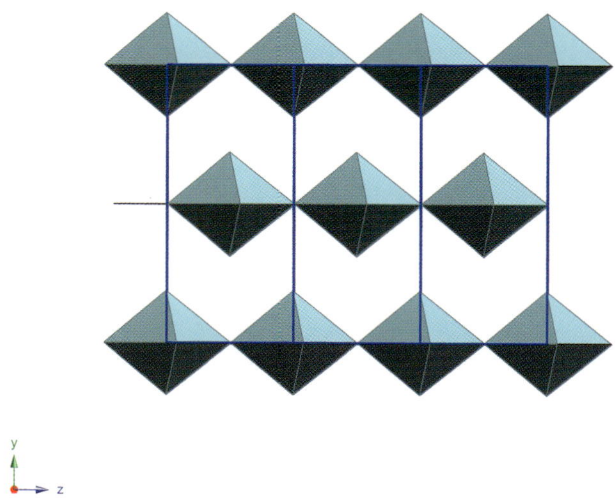

Figure 182. The structure of rutherfordine, viewed along [100], showing how the $UO_2(CO_3)$ sheets are stacked along [010]. There are no direct bonds, and the sheets are held together by van der Waals forces. Three unit cells are shown.

Uranyl dicarbonates and tricarbonates

Uranyl dicarbonates include zellerite and metazellerite. Uranyl tricarbonates include agricolaite, andersonite, bayleyite, čejkaite, grimselite, liebigite, linekite, rabbittite, swartzite, widenmannite, and znucalite. Their structures are based on uranyl tricarbonate clusters that are interconnected via bonds from apical uranyl oxygens to low-valence cations. Some species are anhydrous, e.g., čejkaite, whereas others are hydrated.

Agricolaite $K_4(UO_2)(CO_3)_3$

Agricolaite was described from the abandoned Giftkiesstollen adit at Jachymov, Czech Republic, where it occurs in vugs of ankerite gangue in gneisses. It forms isolated, equant, irregular yellow grains up to 0.3 mm, with pale yellow streak and vitreous luster. It is not fluorescent. Associated minerals include aragonite, posnjakite, malachite, rutherfordine, and "pseudo-voglite" (Skála et al. 2011).

Agricolaite is monoclinic with a=10.238, b=9.193, c=12.211 Å, and β=95.108°. The structure is identical to that of synthetic $K_4(UO_2)(CO_3)_3$ and involves separate $UO_2(CO_3)_3$ groups arranged in layers parallel to (100), and two crystallographically non-equivalent sites occupied by K^+.

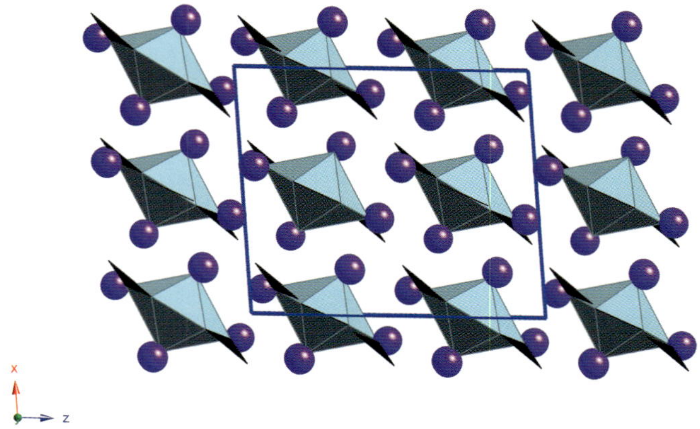

Figure 183. The structure of agricolaite, viewed along [010], showing isolated uranyl carbonate groups, which are linked together via bonds between the K^+ (purple) and oxygens on the uranyl polyhedra.

Andersonite $Na_2Ca(UO_2)(CO_3)_3.6H_2O$

Andersonite is one of several species described by Axelrod et al. (1951) from the Hillside mine, Yavapai Co., Arizona, where it forms minute clusters of pseudocubic crystals associated with gypsum, bayleyite, schröckingerite, and swartzite. It is water soluble and consequently it is often found as an efflorescence on mine walls in fairly arid settings. It is strongly fluorescent bright green. The authors noted that, "The rare andersonite is easily overlooked; superficially, it resembles swartzite, but when the two minerals are seen together, the distinctive characters are apparent." These are compared in the following table.

Figure 184. Bright green massive to platy veins of andersonite in sandstone, from the Atomic King mine, Utah, shown here in visible light. Sample is about 4 cm across. RJL767

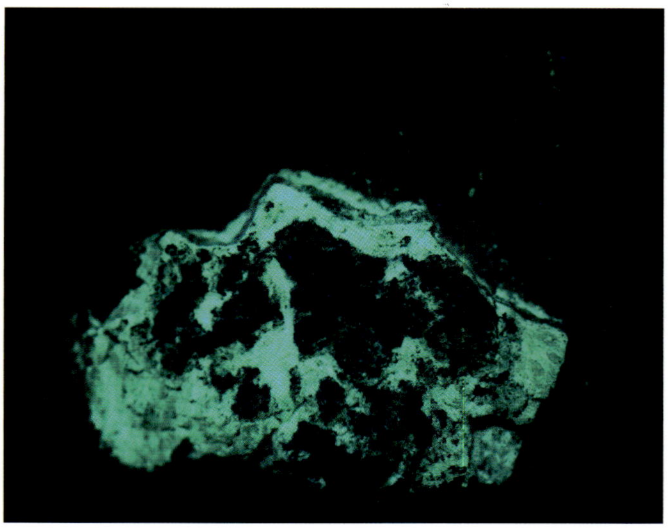

Figure 185. The andersonite specimen from the preceding figure, shown here in LWUV light. Note that the fluorescence is whitish green, rather than the yellowish green often seen in uranyl compounds. RJL767

Figure 186. Scanning electron micrograph of andersonite from the Atomic King mine, Utah, showing platy crystals in a parallel arrangement. FOV is about 1 mm wide. RJL767

Table 45. Comparison of andersonite, swartzite, and bayleyite[a]

Species	Andersonite	Swartzite	Bayleyite
Color	bright yellow-green	green	yellow
Habit	pseudocubic	prismatic	prismatic
Fluorescence	bright whitish green	bright yellowish green	weak

[a]Data from Axelrod et al. (1951)

At the Atomic King No. 2 mine, near Moab, Utah, thick crusts of andersonite were mined as ore and crystals up to 1 cm were found. Some other locales include: Jim Thorpe, Pennsylvania; Jachymov, Czech Republic; Ronneburg, Germany; and the Geevor mine, Cornwall, England.

Andersonite is trigonal with *a*=18.009 and *c*=23.838 Å. The crystal structure consists of isolated $[(UO_2)(CO_3)_3]$ clusters, in which the six equatorial oxygens belong to three CO_3 groups. The clusters are linked together by Na and Ca having coordinations of 6 and 7 respectively (Coda, Della Giusta, and Tazzoli 1981).

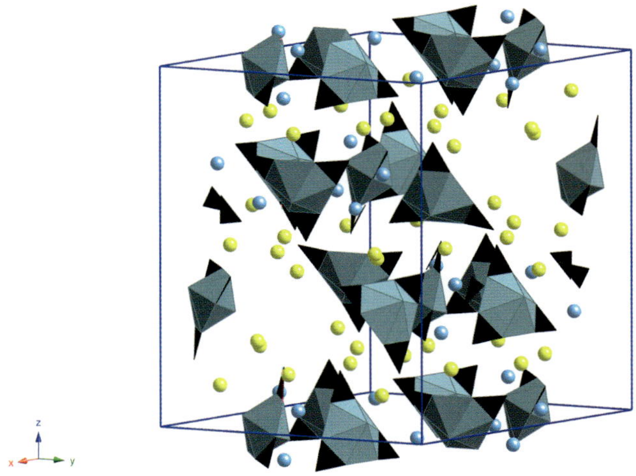

Figure 187. The structure of andersonite, a typical uranyl tricarbonate, in which the dominant structural theme comprises isolated $[UO_2(CO_3)_3]$ clusters. Black triangles represent the CO_3 groups attached to the equators of the uranyl groups (blue-gray polyhedra). Note that the uranyl polyhedra appear flattened because the two apical oxygens that form the $(UO_2)^{2+}$ ion are much closer to the U than are the equatorial oxygens, as discussed earlier. The clusters are bonded via coordination to Ca (light blue) and Na (yellow). For simplicity, water molecules are not shown.

Bayleyite $Mg_2(UO_2)(CO_3)_3.18H_2O$

Bayleyite was described from the Hillside mine, Yavapai Co., Arizona, where it occurs with andersonite as an efflorescence on rock surfaces. It forms yellow crusts and small prismatic crystals that are weakly fluorescent under UV (Axelrod et al. 1951; Frondel 1958). The preceding table above gives some guidance for distinguishing these phases from each other. Other locales include: the Pumpkin Buttes area of the Powder River Basin, Wyoming; the Ambrosia Lake district, New Mexico; numerous mines in Utah; and a mine tunnel at Aznegour, Morocco.

Bayleyite is monoclinic with a=26.560, b=15.256, c=6.505 Å, and β=92.90°. The crystal structure of synthetic bayleyite was reported by Mayer and Mereiter (1986). It consists of isolated $[(UO_2)(CO_3)_3]$ clusters that lie parallel to (001), isolated $[Mg(H_2O)_6]$ octahedra, and lattice H_2O molecules, all held together only by hydrogen bonding.

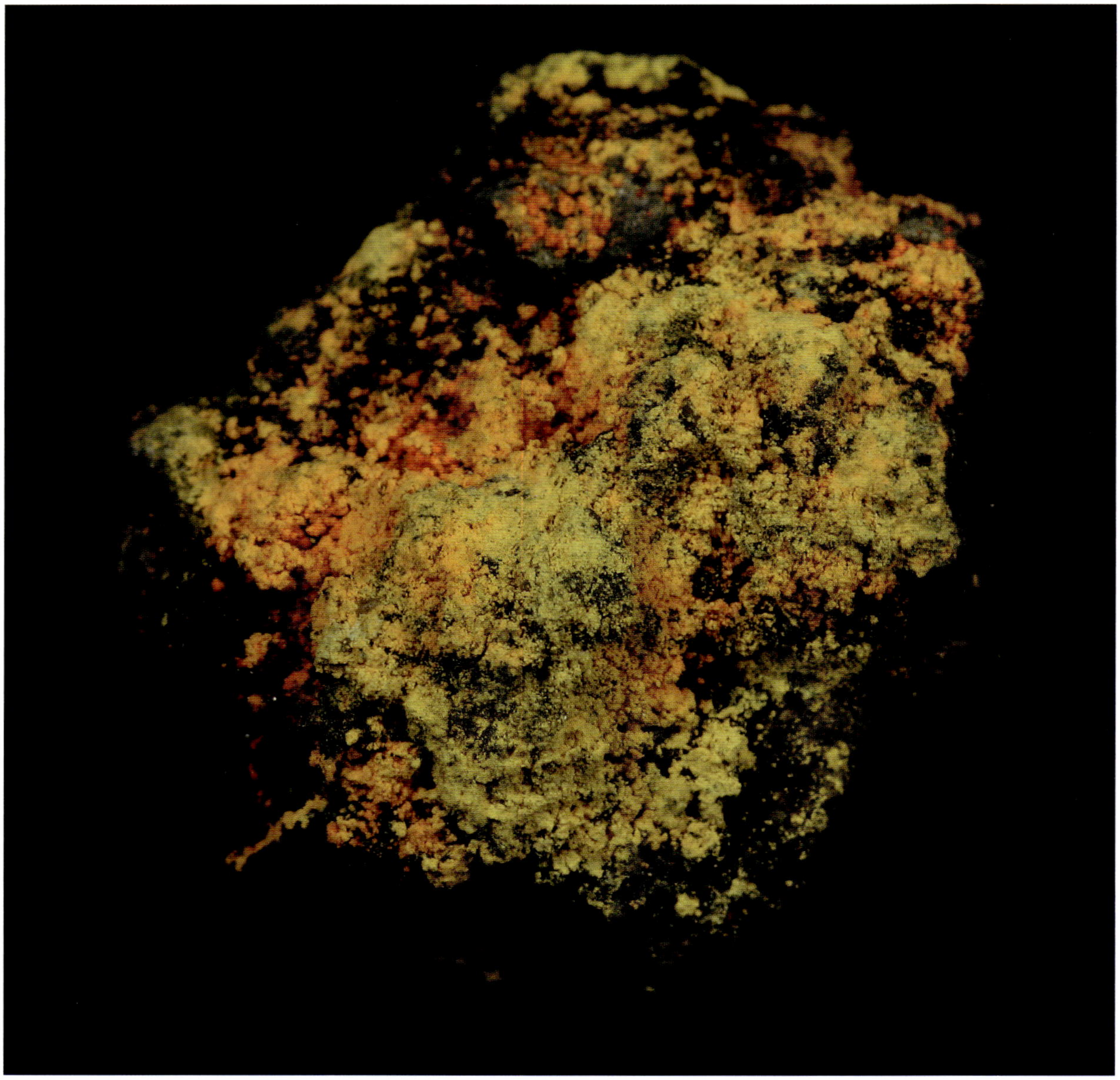

Figure 188. Bayleyite forming a bright yellow efflorescence on a thumbnail-sized mass of black ore from the Mi Vida mine, San Juan Co., Utah. RJL3057

Figure 189. Bright yellow bayleyite forming tiny clusters on greenish-yellow andersonite coating sandstone from the Ambrosia Lake district, Valencia Co., New Mexico. Specimen is about 5 cm wide. RJL3169

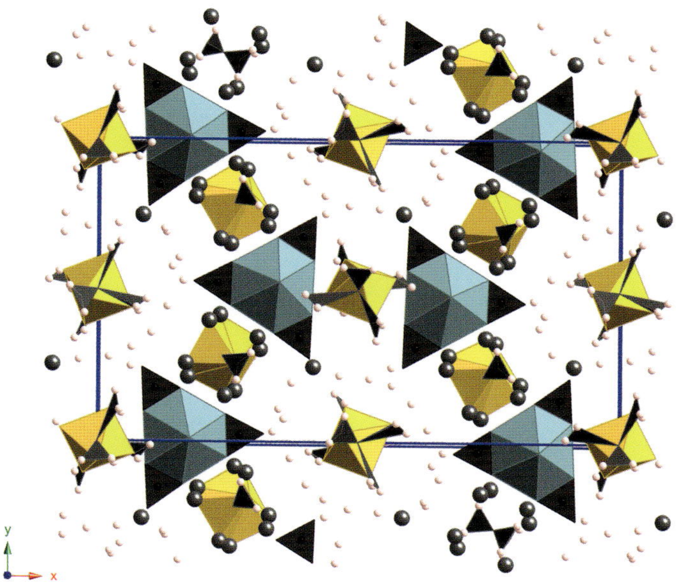

Figure 190. The structure of bayleyite, viewed along [001] showing isolated [$UO_2(CO_3)_3$] clusters, [$Mg(H_2O)_6$] polyhedra (gold) and some of the water oxygens (gray). Hydrogens are shown as pink atoms with wire bonds. Note that the Mg octahedra that appear to be doubled represent the two possible orientations of the disordered Mg(3) octahedra.

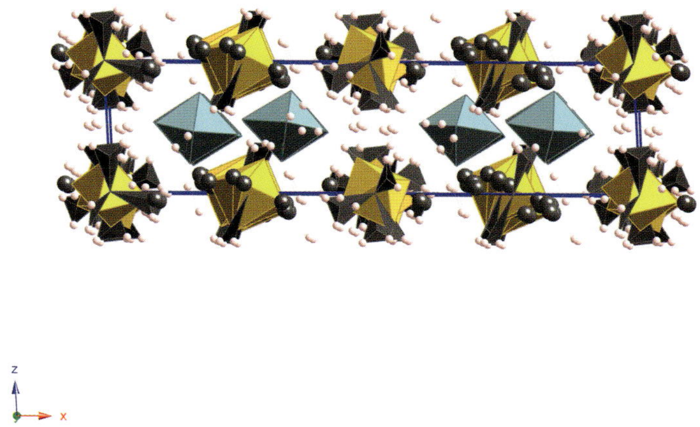

Figure 191. The bayleyite structure viewed along [010], showing that there are no shared oxygens and the polyhedra are linked only via hydrogen bonding.

Čejkaite $Na_4(UO_2)(CO_3)_3$

Čejkaite is an extremely rare species among several dozen new minerals discovered in the course of work on the secondary minerals in the Jachymov ore district (Ondruš et al. 2003, 2003a). It forms a thin, earthy efflorescence over a calcite vein associated with disintegrated uraninite. The individual crystallites are minute, typically 0.2 to 0.6 μm, with an indistinct hexagonal morphology observed by TEM. The earthy sprays are pale yellow to beige with a vitreous luster. Čejkaite is weakly fluorescent yellow to yellow-green in both SW and LWUV.

The type material was found on a single specimen from the Geschieber vein in the Svornost mine, at Jachymov. The čejkaite crust covers an area about 2 × 5 cm; associated species include andersonite and schröckingerite, although the čejkaite is not in direct contact with any other secondary phase. It is postulated that čejkaite formed recently under fairly dry conditions with rapid precipitation, probably from a change in the pH of sodium carbonate solutions. In contrast to the more common uranyl carbonates at Jachymov, gypsum is absent from the čejkaite mineral assemblage. The rarity of čejkaite is attributed to its inherent instability under even slightly acidic conditions: "The formation of the $[UO_2(CO_3)_3]^{4-}$ complex, as well as its mobility, is limited to a relatively narrow range of pH values. The $[UO_2(CO_3)_3]^{4-}$ anion is stable in the pH range 6.5 to 11.5, which indicates that čejkaite is not stable in an acidic environment as typically found at the Jáchymov ore district" (Ondruš et al. 2003).

Interestingly, čejkaite occurs at the Apex mine, Lander Co., Nevada, but there it is associated with earlier hydroglauberite, $Na_{10}Ca_3(SO_4)_8{\cdot}6H_2O$, so at least in some circumstances it can coexist with some sulfates (Jensen and Hermann 2012). Analytical work on the material from this locality was done by A. J. Nikischer, who provides the following guidance: "Visually, [čejkaite] is similar to andersonite at the locality (although it is usually much paler yellow to nearly white crusty masses there), with the andersonite being brilliantly fluorescent. Čejkaite is occasionally associated

Figure 192. Tiny sample of čejkaite forming a thin crust on matrix from the TL in the Svornost mine; sample obtained from Jiri Čejka himself. FOV is ~3 mm wide. RJL3675

Figure 193. Pale yellow patch of čejkaite, ~2 mm wide, on white hydroglauberite from the Apex mine, Lander Co., Nevada. RJL4181

with the andersonite crusts. If your material is *not* brightly fluorescent, it is likely the čejkaite, which has only modest response to UV; if intensely fluorescent, it is probably andersonite (which we also confirmed at the locality). Other potentially yellow to yellow-green crystalline minerals found at Apex include meta-autunite, metatorbernite, rutherfordine, uranopilite, uranocircite and zippeite, all confirmed by XRD and chemistry as well. The most likely one to be mistaken for čejkaite (other than the andersonite) is uranopilite when it is in crusty masses."

Synthetic material of this composition had been known since the mid-1950s, and the structure had been solved by Li, Krivovichev, and Burns (2001) for material synthesized at 220°C, which was found to have trigonal symmetry. The type material, in contrast, was first reported to have triclinic (pseudohexagonal) symmetry, based on Rietveld refinement of powder XRD data; Ondruš et al. (2003) were able to synthesize this structure at room temperature and show that it reverts to a trigonal form when held under hydrothermal conditions at 135°C for several days. The high-temperature trigonal polytype has not as yet been reported in nature. More recently, a single-crystal structural refinement (Plašil et al. 2013b) indicated that the mineral is monoclinic with a=9.292, b=16.099, c=6.444 Å, and β=91.404°.

Figure 194. The structure of čejkaite, viewed along [001] to show coordination of the Na (yellow polyhedra) around the uranyl tricarbonate groups.

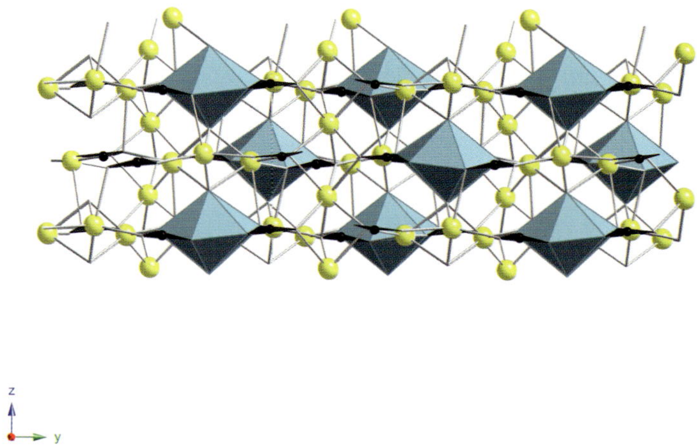

Figure 195. The čejkaite structure viewed along [100] showing linkage between the different polyhedra, with Na represented as yellow atoms with stick bonds.

Grimselite $K_3Na(UO_2)(CO_3)_3.H_2O$

Grimselite was described from a cable tunnel between Gerstenegg and Sommerloch, Grimsel area, Aar massif, Bern, Switzerland, in a place where granodiorite and aplite granite are cut by mineralized zones and fissure veins. The mineral forms crusts of fine-grained aggregates of generally anhedral grains, associated with schröckingerite, monohydrocalcite, and several unnamed new carbonate minerals. The crystals are yellow, with no distinct cleavage and a conchoidal fracture, and are not fluorescent (Walenta 1972).

Grimselite has been reported from Jachymov, Czech Republic, where greenish crystals to ~1 mm were found with čejkaite in old galleries at the 90 m level of the Rovnost shaft. It also occurs in the Krunkelbach Valley, Germany.

Grimselite is hexagonal; the structure was originally worked out using synthetic material by Li and Burns (2001), who showed that the structure contains the typical uranyl tricarbonate cluster. These clusters are oriented normal to [001] and are interconnected via bonds to Na and K polyhedra. Plášil et al. (2012c) reported a single-crystal structure refinement using natural crystals from Jachymov. For a crystal of nearly ideal composition (K_3), a=9.300 and c=8.294 Å. Anomalously Na-rich material ($K_{2.43}Na_{.57}$) has a=9.251 and c=8.179 Å, because the presence of Na in some of the KO_8 polyhedra reduces their volume.

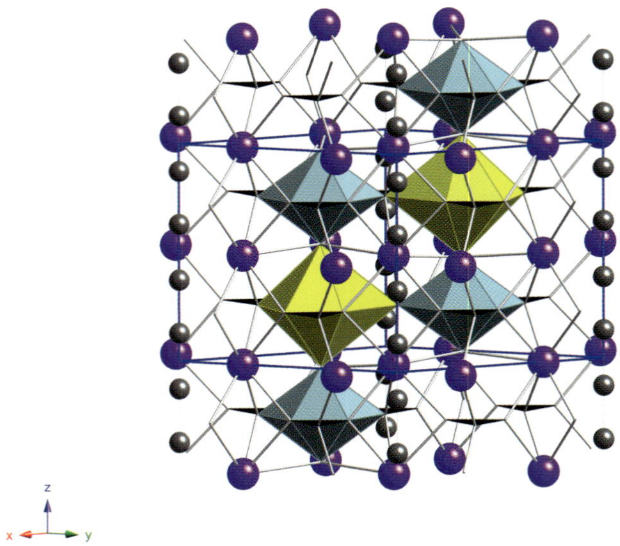

Figure 196. The structure of grimselite, showing uranyl tricarbonate clusters oriented normal to [001] and connected via bonds to K (purple atoms with stick bonds) and Na (yellow polyhedra). H_2O groups are gray.

Liebigite $Ca_2(UO_2)(CO_3)_3.11H_2O$

Liebigite was described from Edirne, near Adrianople, Turkey, where it formed coatings on pitchblende. Smith (1848) noted, "This mineral is not found crystallized, but occurs in the form of a mammillary concretion, having an apparent cleavage in one direction. It is of a beautiful apple-green color, transparent, with a vitreous fracture. Hardness between 2 and 2.5; sp. grav. not ascertained."

Liebigite usually forms bright green to yellowish-green, granular to scaly crusts and films; when it does form distinct crystals they are usually crude, with rounded edges and a waxy appearance. Good specimens are colorful in natural light and show brilliant bluish-green fluorescence under UV light. The classic American locality for liebigite is the Schwartzwalder mine, Jefferson Co., Colorado, but it is fairly widespread throughout the western US, including: the Mi Vida mine, Utah; the Pumpkin Buttes area, Wyoming; the Lucky Mc mine, Wyoming; the Hanosh mines, near Grants, New Mexico; and

Figure 197. Hand sample (~10 cm wide) of waxy yellow liebigite thickly covering schist from the Schwartzwalder mine, near Boulder, Colorado, shown here in visible light. RJL885

Figure 198. The liebigite specimen from the previous figure, shown here in SWUV light. RJL885

Figure 199. Small but very sharp liebigite crystals from the Schwartzwalder mine, Colorado. RJL264

Figure 200. Tabular, yellow to greenish liebigite crystals from the Midnite mine, Spokane, Washington. RJL2949

the Midnite mine, Washington. Other locales include: Jachymov, Czech Republic, where it is plentiful in the oxidized zone of uraninite veins and as an efflorescence on mine walls; Schneeberg, Germany; and Wheal Basset, Cornwall, England.

Liebigite is orthorhombic with *a*=16.70, *b*=17.56, and *c*=13.70 Å. The structure contains $[(UO_2)(CO_3)_3]$ clusters that are linked to two kinds of $CaO_4(H_2O)_4$ polyhedra and one kind of $CaO_3(H_2O)_4$ polyhedron to form puckered $Ca_2UO_2(CO_3)_3{\cdot}8H_2O$ layers parallel to (010). These layers are interconnected only by hydrogen bonds, both directly and via three additional interlayer water molecules (Mereiter 1982b).

Figure 201. Thumbnail-sized sample of liebigite covering matrix from the Pitch mine, Saguache Co., Colorado, showing the typical waxy, somewhat rounded crystals. RJL3063

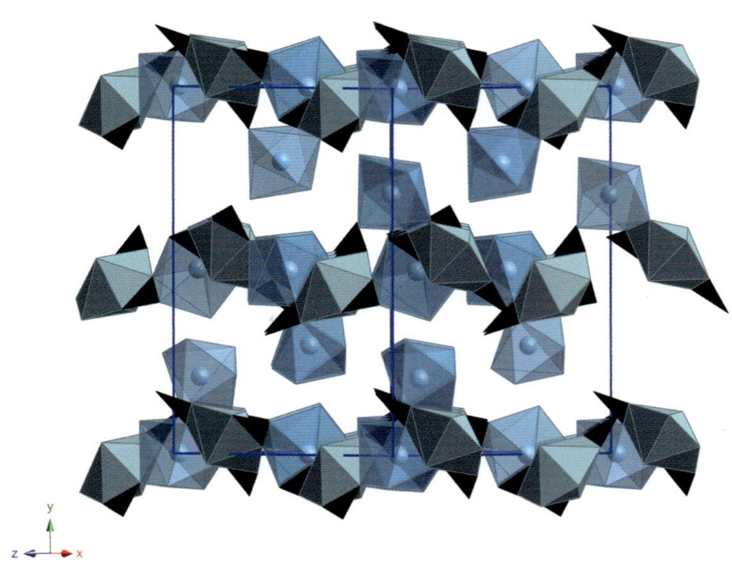

Figure 202. The structure of liebigite, viewed normal to [010], showing the uranyl carbonate groups linked by Ca polyhedra (translucent blue) to form puckered sheets. Note that there is no direct linkage between adjacent sheets.

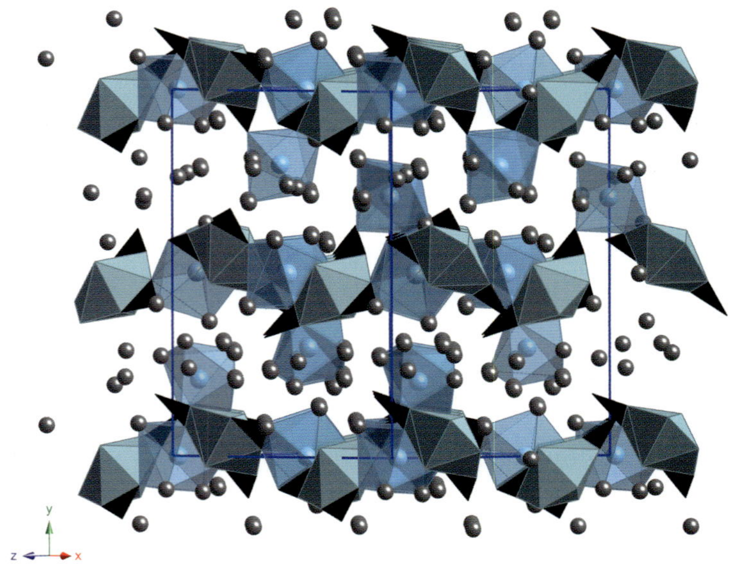

Figure 203. The same view as the previous figure, but with H_2O groups (gray) added. Some H_2O are directly coordinated to Ca, whereas others lie in the interlayer. Hydrogen bonding holds adjacent sheets together.

Linekite $K_2Ca_3[(UO_2)(CO_3)]_2.7H_2O$

Linekite was described from the Geschieber vein, Svornost mine, Jachymov, Czech Republic, where it forms transparent, yellowish-green tabular crystals, often forming stacked aggregates, associated with čejkaite, schröckingerite, and other secondary phases (Plášil et al. 2013f).

Linekite is the natural equivalent of a phase previously synthesized by Hughes Kubatko and Burns (2004), who predicted its likely occurrence in nature based on the ease with which it forms in the laboratory. It is orthorhombic with a=17.007, b=18.027, and c=18.337 Å. The structure consists of uranyl tricarbonate groups joined through bonds to K and Ca polyhedra.

Figure 204. Transparent yellow-green linekite forming clusters of small tabular crystals, from the TL at Jachymov, Czech Republic. Photo courtesy of Pavel Škacha

Figure 205. Tabular linekite crystals associated with bright yellow, spherical aggregates of a possible new species that is presently undergoing study, from the TL at Jachymov, Czech Republic. Photo courtesy of Pavel Škacha

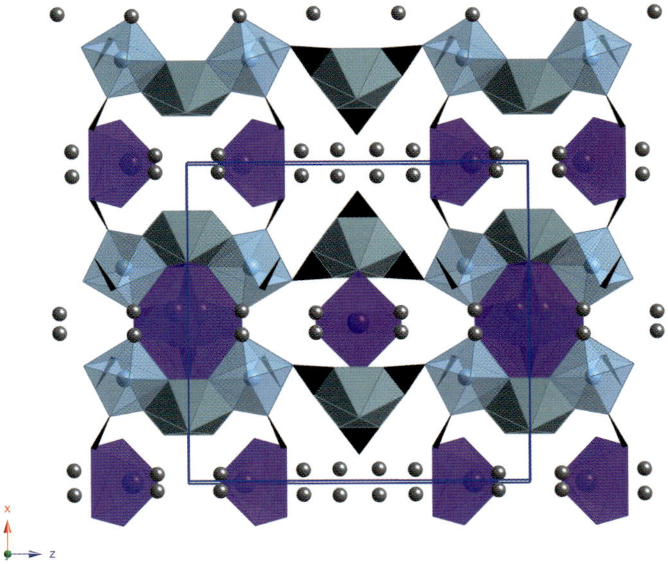

Figure 206. The structure of linekite based on synthetic material, viewed along [010], showing the linkage of uranyl tricarbonate groups by Ca (translucent blue) and K (translucent purple) polyhedra. Gray atoms represent water.

Rabbittite $Ca_3Mg_3(UO_2)_2(CO_3)_6(OH)_4.18H_2O$
Rabbittite was described from the Lucky Strike No. 2 mine, San Rafael district, Emery Co., Utah, where it formed pale green, acicular to fibrous efflorescences on a pillar of high-grade ore near the mine portal. The ore contains pitchblende, which is only partially oxidized because the deposit is somewhat protected by the overlying formations. Associated minerals included fourmarierite, zippeite, gypsum, and several cobalt-containing species. Rabbittite is slowly soluble in cold water and effervesces in dilute HCl (Thompson, Weeks, and Sherwood 1955).

Other reported occurrences of rabbittite include: the Hideout No. 1 mine, White Canyon district, San Juan Co., Utah; and Jachymov, Czech Republic.

The structure was incompletely determined by XRD, owing to the minutely fibrous nature of the type material. Lattice parameters are reported as *a*=32.6, *b*=23.8, and *c*=9.45Å. Thompson, Weeks, and Sherwood (1955) note, "The mineral is considered monoclinic because of its optical properties. The β angle cannot be accurately determined on the present material, but it is probably close to 90°."

Swartzite $CaMg(UO_2)(CO_3)_3.12H_2O$
Swartzite was one of three new uranyl carbonates described from the Hillside mine, Yavapai Co., Arizona, where it formed small green clusters on granular gypsum, associated with andersonite, bayleyite, schröckingerite, and other species. At the time, the Hillside mine produced gold, silver, lead, and zinc; the source of U was not immediately evident, but shortly afterward samples of U-bearing ore were found at the 400-foot level of the mine (Axelrod et al. 1951). Comparative data to help distinguish swartzite from andersonite and bayleyite are given in Table 45 in the entry for andersonite.)

Swartzite has also been reported from: the Rifle mine, Garfield Co., Colorado; the Schwartzwalder mine, Jefferson Co., Colorado; and the Coral claim, San Juan Co., Utah.

Swartzite is monoclinic with *a*=11.080, *b*=14.634, *c*=6.439 Å, and β=99.43°. The crystal structure includes dinuclear groups $(Ca(H_2O)_6[UO_2(CO_3)_3])^{2-}$, where Ca forms $Ca(H_2O)_6O_2$ square antiprisms, which are edge-linked with UO_8 polyhedra. Isolated $Mg(H_2O)_6^{2+}$ octahedra and a system of hydrogen bonds link the structure together (Mereiter 1986a).

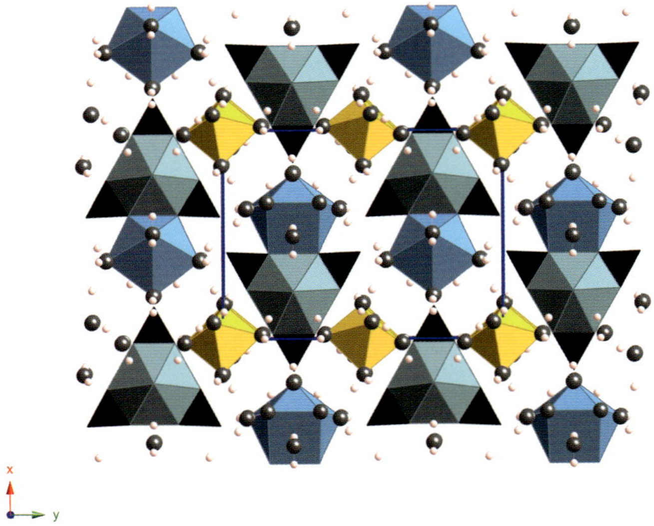

Figure 207. Crystal structure of swartzite, viewed along [001], showing $[UO_2(CO_3)_3]$ clusters (blue-gray polyhedra with black triangles), $Mg(H_2O)_6^{2+}$ (yellow polyhedra), and $Ca(H_2O)_6$ (blue polyhedra). Water molecules are shown as gray oxygen and pink hydrogens.

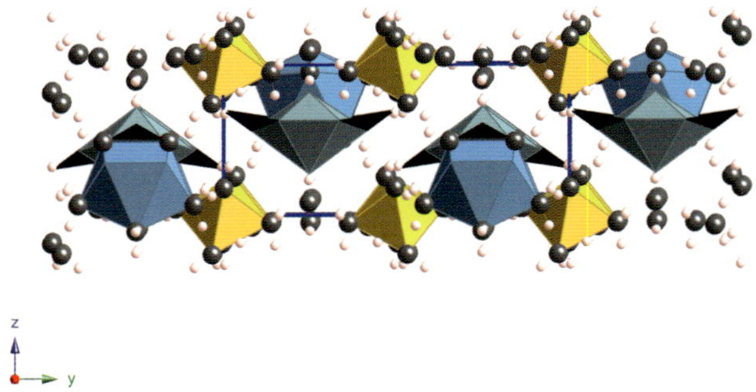

Figure 208. Crystal structure of swartzite, viewed along [100], showing that the $(Ca(H_2O)_6[UO_2(CO_3)_3])^{2-}$ groups are linked only via hydrogen bonding.

Widenmannite $Pb_2(UO_2)(CO_3)_3$

Widenmannite was described from the Michael mine, Weiler, Black Forest, Germany, where it formed radiating bundles of fragile yellow crystals, tabular on (010) and elongated on [001], with pearly to silky luster. It occurs as a secondary phase in cavities of dense quartz, associated with cerussite and altered galena; hügelite and halimondite were also associated. It is soluble with effervescence in dilute HNO_3 and is not fluorescent (Walenta and Wimmenauer 1961; Walenta 1976).

At Loe Warren zawn (cove), Botallack, Cornwall, England, it occurs in a localized patch of Pb-U-Cu mineralization. There, widenmannite is often associated with an unknown basic lead uranyl carbonate (U:Pb ratio ~1:4), which forms transparent to translucent hexagonal tabular crystals up to 100 μm across. The XRD pattern of this mineral is distinct from that of widenmannite, although the IR spectra of the two minerals are similar. Dewindtite is associated with widenmannite on a number of specimens and forms radiated groups and tufts of bright yellow doubly-terminated tapering "canoe-shaped" crystals (Elton and Hooper 1995).

Figure 209. Pale yellow, lath-like widenmannite crystals in radiating aggregates, from the Janska vein, Prokop mine, Pribram, Czech Republic. Collected by Pavel Škacha in 2003. FOV is ~5 mm wide. RJL3836

At an occurrence in the Janska vein, Prokop mine, Pribram, Czech Republic, two generations of widenmannite occur in several morphological types and mineral associations in hydrothermal veins. Based on alpha spectroscopy, the two generations have different ages, >220,000 and 118±12 y. The older "widenmannite I" forms fine-grained, gray to beige aggregates in the highly altered supergene part of the hydrothermal ore vein in association with pyromorphite, cerussite and goethite. The younger "widenmannite II" occurs as white, yellow or greenish-yellow thin tabular crystals up to 0.5 mm long in association with cerussite, anglesite, limonite, kasolite and an unnamed Pb-U-O phase Plášil et al. (2010).

The species has also been reported from the Needle's Eye mine, Scotland; and the Huron River uranium prospect, Michigan.

Widenmannite is orthorhombic with a=4.974, b=9.382, and c=8.954 Å. The structure was solved by Plášil et al. (2014a) using a combination of electron diffraction and synchrotron powder XRD. They note that, "The idealized structure formula of widenmannite is $Pb_2(OH)_2[(UO_2)(CO_3)_2]$, $Z = 2$. However, both data sets suggest that the widenmmanite structure is not that simple. There are two symmetrically independent, partly occupied U sites. . . . When the U(2) site is occupied, the $U(1)O_2$ group is absent, the two OH groups are substituted by O^{2-} and one Pb^{2+}-vacancy. The chemical formula of the real structure should be written as $Pb_{2-x}(OH)_{2-2x}[(UO_2)(CO_3)_2]$, where x is the probability of the substitution U(2) → U(1). The probability of occurrence of U(2) refines to x=0.074(15) from the powder-diffraction data and to

Figure 210. The uranyl carbonate chains in widenmannite, viewed along [010].

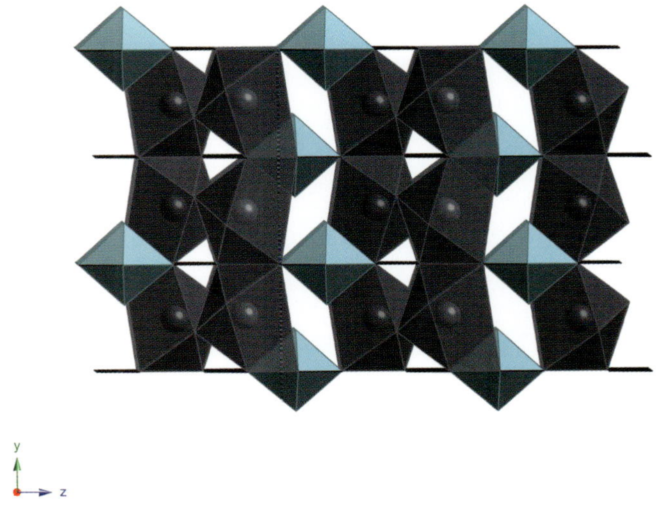

Figure 211. The widenmannite structure viewed along [001]. Uranyl carbonate chains lie between Pb-O sheets (translucent gray), connected by the Pb-O bonds.

x=0.176(4) from the electron-diffraction data. There is one Pb site (nearly fully occupied), which is coordinated by 11 anions (up to the distance of 3.5 Å), including O and OH^-. The shorter Pb-O bonds form a sheet structure, which is linked by the weaker bonds to the uranyl-carbonate chains to form a three-dimensional framework structure."

Zellerite $Ca(UO_2)(CO_3)_2 \cdot 5H_2O$
Metazellerite $Ca(UO_2)(CO_3)_2 \cdot 3H_2O$

Figure 212. Yellow zellerite forming a thin seam in sandstone ore from the White Canyon #1 mine, San Juan Co., Utah. RJL4565

Zellerite and metazellerite were described from the Lucky Mc mine, Fremont Co., Wyoming, where their formation was the result of surface oxidation of the uranium ore. Unoxidized ore at the Lucky Mc is a mixture of uraninite, coffinite, and iron sulfides that fill intergranular openings in arkose. Oxidation is facilitated by the porous structure of the host rock. During the early stages of oxidation, the primary ore develops seams of liebigite mixed with gypsum; this ore is rimmed by a rusty brown oxidation zone in which all of the primary uranium minerals and some of the iron sulfides have broken down. In this zone, zellerite is common, along with gypsum and limonite. It forms minute transparent yellow fibers (rarely over 2 mm long × 5 µm thick) aggregated into small "pincushion" clumps. Fully hydrated material is lemon-yellow with a dull luster and no discernable cleavage. It shows weak, patchy green fluorescence under both LW and SWUV.

Dehydration to metazellerite causes the unit cell volume to decrease by 186 $Å^3$, which was interpreted to imply a loss of 8 or 10 H_2O molecules, based on Z = 4 formula units per cell, leading to the formulas given above (Coleman, Ross, and Meyrowitz 1966).

Zellerite has also been reported from: the White Canyon district, San Juan Co., Utah; the Ambrosia Lake area, New Mexico; Lodève, France; and the Elias mine, Jachymov, Czech Republic. In all cases, the existence of accompanying metazellerite should be assumed.

Zellerite is orthorhombic with a=11.220, b=19.252, and c=4.933 Å. Metazellerite is orthorhombic with a=9.718, b=18.226, and c=4.965 Å. Complete structure refinements have not been reported as of this writing.

Figure 213. Earthy to fibrous, pale yellow zellerite, associated with voglite (green), liebigite (waxy yellow), and gypsum (colorless), from the White Canyon #1 mine, San Juan Co., Utah. FOV is ~4 mm. RJL4633

Znucalite $CaZn_{11}(UO_2)(CO_3)_3(OH)_{20}\cdot 4H_2O$
Znucalite was described from uraninite-bearing dump material from the Lill mine, Příbram, Czech Republic, where it had developed as an oxidation product (the mine had operated during the second half of the nineteenth century). Znucalite forms porous coatings of saucer-shaped aggregates to about 10 cm² consisting of microscopic plates (15 × 8 × 0.4 µm). The material is white, pale yellow, or grayish yellow, with a silky luster, and is fluorescent yellow-green in both SW and LWUV. It is soluble in acid, with effervescence. At the TL, znucalite is associated with gypsum, hydrozincite, aragonite, sphalerite, galena, and pyrite (Ondruš, Veselovsky, and Rybka 1990).

Znucalite was also reported from the oxidation zone of the Mas d'Alary uranium deposit near Lodève, France, where it forms 350-µm spherules (Chiappero and Sarp 1993). It has also been noted at: the St. Christoph mine, Erzgebirge district, Saxony, Germany; and the Sterling mine, Franklin district, New Jersey.

Znucalite is orthorhombic with a=10.72, b=25.16, and c=6.33 Å. A full structure refinement has not been reported.

Figure 214. A typical sample of znucalite from the TL at the Lill mine, Příbram, Czech Republic, where it has formed a thin, off-white crust on fracture surfaces, associated with colorless gypsum and other phases. RJL4626

Figure 215. Viewed in the microscope, znucalite forms pale yellow to cream-colored, globular aggregates on matrix, from the TL at the Lill mine, Příbram, Czech Republic. FOV is ~6 mm wide. RJL4566

Uranyl carbonates containing essential rare earth elements

Uranyl carbonates containing essential REE include astrocyanite-(Ce), bijvoetite-(Y), kamotoite-(Y), lepersonnite-(Gd), and shabaite-(Nd). Of these, bijvoetite-(Y) and lepersonnite-(Gd) were described from the oxidation zone at Shinkolobwe, Congo (Deliens, Piret, and Comblain 1984). The other three species were discovered during work at the Kamoto East extension, Congo (Deliens, Piret, and Van Der Meersche 1990).

Astrocyanite-(Ce) $Cu_2(Ce, Nd, La)_2(UO_2)(CO_3)_5(OH)_2 \cdot 1.5 H_2O$
Astrocyanite-(Ce) was described from Kamoto East, Cu-Co deposit, Congo, where it was formed as an oxidation product of uraninite. The mineral forms delicate 1-mm rosettes of platy crystals in various shades of pale blue, bright blue, to blue-green, as well as isolated transparent blue vitreous crystals; in clusters it is translucent to opaque with an earthy luster. It dissolves with effervescence in dilute HCl. Associated minerals include kamotoite-(Y), françoisite-(Nd), shabaite-(Nd), masuyite, uranophane, and schuilingite (Deliens and Piret 1990; Deliens, Piret, and Van Der Meersche 1990).

Astrocyanite is hexagonal with a=14.96 and c=26.86 Å.

Figure 216. Bright blue astrocyanite-(Ce) forming a thin vein in a small chip of ore from Kamoto East, Congo. The yellow material is probably kamotoite. RJL4563

Bijvoetite-(Y) $[(Y,REE)_8(H_2O)_{25}(UO_2)_{16}O_8(OH)_8(CO_3)_{16}](H_2O)_{14}$
Bivoetite-(Y) was described from Shinkolobwe, Congo, where it occurs in association with lepersonnite-(Gd) and hydrous uranium oxides in the lower part of the oxidation zone. It is extremely rare and the description was based on a single specimen. The mineral forms minute tabular crystals with a very good cleavage in the plane of the tablet. The crystals are transparent sulfur-yellow to translucent pale yellow and have a vitreous luster. It is not fluorescent (Deliens and Piret 1982a).

Bijvoetite-(Y) is monoclinic (pseudo-orthorhombic) with a=21.234, b=12.958, c=44.911 Å, and β=90.00°. The structure has an unusual combination of uranyl carbonate chains parallel to [100], cross-linked by irregular polyhedra to form (Y-REE)-bearing uranyl carbonate sheets parallel to (001). Adjacent sheets are connected by hydrogen bonds only (Li, Burns, and Gault 2000).

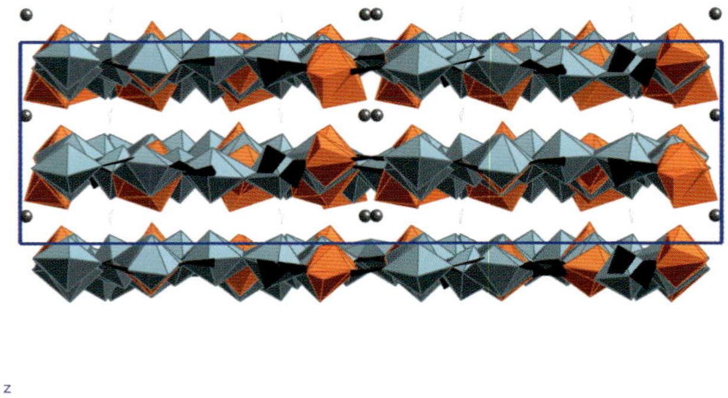

Figure 217. Polyhedral model of the structure of bijvoetite-(Y) projected along [100] to emphasize the sheet structure. Blue-gray polyhedra represent U; Y,REE are shown as orange polyhedra; and small gray atoms are H_2O groups. Note the extremely large unit cell, with a volume of 12,357 Å³.

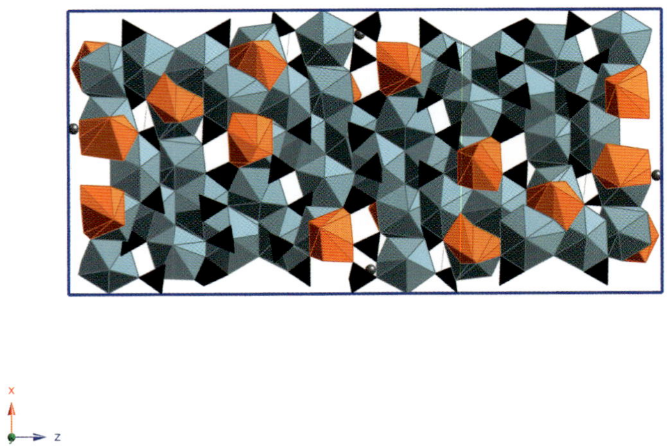

Figure 218. The sheet in bijvoetite-(Y) projected along [010]. Coordination polyhedra are represented in the same colors as the previous figure, and CO_3 groups are shown as black triangles.

Kamotoite-(Y) $Y_2(UO_2)_4(CO_3)_3(OH)_8 \cdot 10\text{-}12H_2O$
Kamotoite-(Y) was described from the Kamoto East mine, Congo. The type material contains several percent each of Nd, Sm, Gd, and Dy in partial substitution for Y. Two somewhat distinct habits are seen: transparent, bright yellow bladed crystals (to about 5 mm) flattened on (010) and elongated on [100] with a sharp point bounded by {100} and {101} forming an angle of 35°; alternatively, the mineral forms flattened fan-like aggregates. The mineral effervesces strongly in dilute HCl, and the effervescence and crystal habit help to distinguish it from other yellow secondary minerals that occur at Kamoto East. Dehydration causes the crystals to become opaque and the color becomes less intense (Deliens and Pieret 1986).

Kamotoite-(Y) is one of the more abundant secondary minerals in uraniferous pockets at Kamoto East. It can form cm-sized layers on altered uraninite in the supergene zone. Associated minerals include pale yellow rosettes of shabaite-(Nd), minute acicular crystals and spherical nodules of uranophane, blue rosettes of astrocyanite-(Ce), and transparent blue-green crystals of schuilingite-(Nd).

It has also been found sparingly at the Gole quarry, near Madawaska, Ontario, Canada (Back et al. 2007). Kamotoite-(Y) is monoclinic with a=21.22, b=12.93, c=12.39 Å, and β=115.3°.

Figure 219. Kamotoite-(Y) as a drusy coating of transparent microcrystals on massive uraninite, from Kamoto East, Congo. RJL2569

Figure 220. Detail of the kamotoite-(Y) crystals in the previous figure. RJL2569

Figure 221. Alternate habit of kamotoite-(Y) showing a flattened fanlike spray of opaque (partially dehydrated) yellow lath-like crystals. RJL1372

Lepersonnite-(Gd) $Ca(Gd,Dy)_2(UO_2)_{24}(CO_3)_6Si_4O_{12} \cdot 60H_2O$
Lepersonnite-(Gd) was described from Shinkolobwe, Congo, where it forms crusts and isolated spherules (to 5 mm across) of radiating acicular crystals. The crystals are bright yellow, transparent to translucent, and are not fluorescent. The type material was described from a black crust in the lower portion of the oxidation zone at Shinkolobwe, where lepersonnite-(Gd) was associated with bijvoetite-(Y), sklodowskite, uranophane, curite, becquerelite, and rutherfordine. During the study that led to the discovery of lepersonnite-(Gd), the authors found that about half of the supposed "studtite" found there was actually lepersonnite-(Gd); the two species have very similar appearance (Deliens and Piret 1982).

Lepersonnite-(Gd) is orthorhombic with a=16.23, b=38.75, and c=11.73 Å.

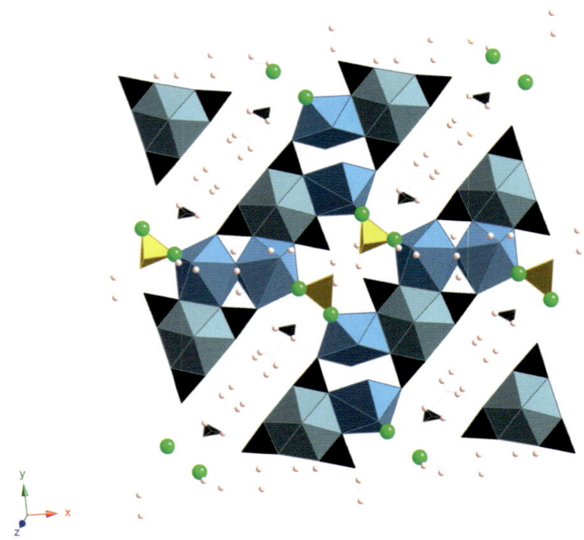

Figure 222. The sheet structure of albrechtschraufite with composition $MgCaF_2[(UO_2)(CO_3)_3] \cdot 8H_2O$

Shabaite-(Nd) $Ca(Nd,Sm,Y)_2(UO_2)(CO_3)_4(OH)_6 \cdot 6H_2O$
Shabaite-(Nd) was described from the Kamoto East mine, Shaba province, Congo, where it forms pale yellow micaceous flakes and rosettes to about 5 mm diameter. It is translucent to opaque with a pearly luster. Associated minerals include uraninite, kamotoite-(Y), schuilingite-(Nd), and uranophane (Deliens and Piret 1989).

Shabaite-(Nd) is monoclinic with a=9.208, b=32.09, c=8.335 Å, and β=90.3°.

Other uranyl carbonates
Other uranyl carbonates include albrechtschraufite, fontanite, roubaultite, schröckingerite, sharpite, urancalcarite, voglite, and wyartite. Of these, wyartite is notable as the first mineral having pentavalent uranium (U^{5+}) in its structure (Burns and Finch 1999a).

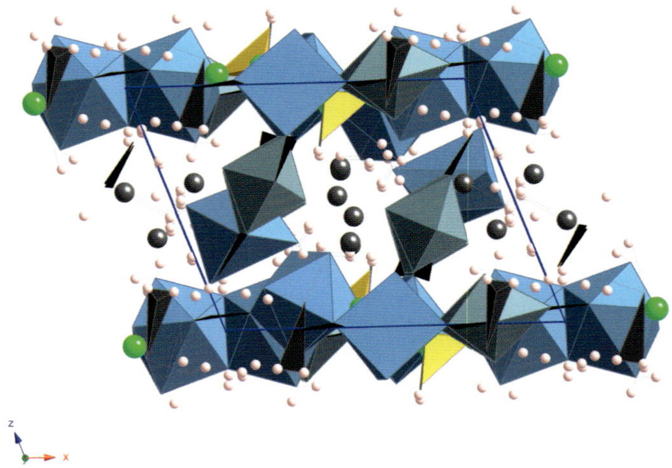

Figure 223. The structure of albrechtschraufite viewed along [010] to show the interconnection of layers by $Ca[(UO_2)(CO_3)_3] \cdot 5H_2O$ units.

Albrechtschraufite $Ca_4Mg(UO_2)_2(CO_3)_6F_2 \cdot 17H_2O$
Albrechtschraufite was described from a sample detected in small amounts with schröckingerite on a museum specimen of uranium ore from Jachymov, Czech Republic. The particular specimen was apparently part of the original material studied by Albrecht Schrauf in his description of schröckingerite as a new mineral (Schrauf 1873). The specimen consists of a matrix of pitchblende and red dolomite covered by scattered globular aggregates of schröckingerite, and in one place an incrustation of yellow crystals that proved to be a new mineral. This species forms transparent yellow-green crystals with a vitreous luster and pale bluish-green fluorescence under both LW and SWUV. The grain-like subhedral crystals (up to 0.2 mm) resemble liebigite.

Albrechtschraufite is triclinic with a=13.569, b=13.419, c=11.622 Å, α=115.82, β=107.61, and γ=92.84°. The structure has layers of composition $MgCaF_2[(UO_2)(CO_3)_3] \cdot 8H_2O$, parallel to (001), which are interconnected into a framework by $Ca[(UO_2)(CO_3)_3] \cdot 5H_2O$ units (Mereiter 1984, 2013).

Fontanite $Ca[(UO_2)_3(CO_3)_2O_2](H_2O)_6$
Fontanite was described from the oxidized zone of the Rabejac uranium deposit, near Lodeve, Herault, France, where the primary minerals include uraninite (var. *pitchblende*), uraniferous carbon, and coffinite. The mineral forms bright yellow rectangular lath-shaped crystals (up to 1.2 × 0.15 × 0.05 mm) in radiating aggregates up to about 4 mm in diameter. The crystal habit in some ways resembles uranophane-β. The crystals are transparent, with a vitreous luster, and exhibit light green fluorescence under LWUV light. Associated secondary minerals include billietite and uranophane (Deliens and Piret 1992).

Fontanite is monoclinic with a=6.968, b=17.276, c=15.377 Å, and β=90.064°. The structure contains uranyl carbonate sheets having the composition $[(UO_2)_3(CO_3)_2O_2]^{2-}$. The Ca^{2+} ion is located between the sheets, and is coordinated by six H_2O groups and by two O atoms of uranyl ions of adjacent sheets. The uranyl carbonate sheet in fontanite is novel but is based on the phosphuranylite anion topology that forms the basis for many uranyl phosphate, sulfate, and selenite sheet structures in uranium minerals (Hughes and Burns 2003a).

Figure 224. The structure of fontanite, viewed along [010], in which the dominant structural theme comprises chains of uranyl polyhedra, joined by carbonate groups (black triangles) into sheets.

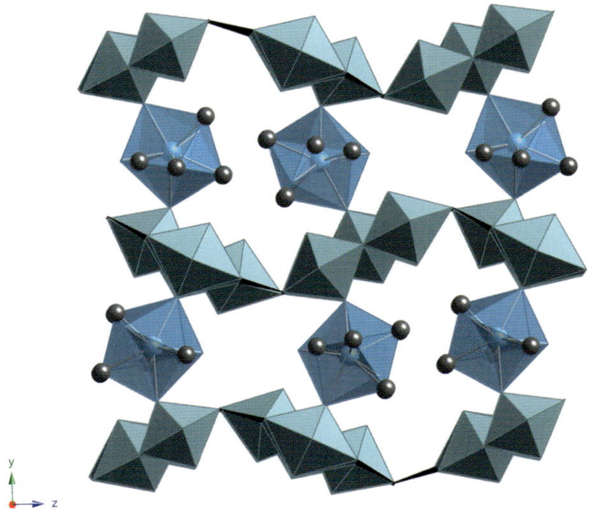

Figure 225. Crystal structure of fontanite, viewed along [100], showing Ca (blue polyhedra) bridging between the uranyl carbonate sheets. H_2O groups are gray.

Roubaultite $Cu_2(UO_2)_3(CO_3)_2O_2(OH)_2{\cdot}4H_2O$
Roubaultite was described from the oxidation zone at Shinkolobwe, where it is extremely rare. It forms small rosettes, to ~3 mm, of apple-green crystals flattened on (100). The type material occurred in a block of primary uraninite, associated with becquerelite, vandenbrandeite, soddyite, and heterogenite. Its brighter green color distinguishes it from vandenbrandeite. It was first thought to be a copper uranyl hydrate (Cesbron, Pierrot, and Verbeek 1970); however, later work (Ginderow and Cesbron 1985) determined that the mineral is actually a carbonate, with the formula given above.

Roubaultite is triclinic with a=7.767, b=6.924, c=7.850 Å, α=92.16, β=90.89, and γ=93.48°. The crystal structure contains chains of distorted $[Cu(O,OH,H_2O)_6]$ octahedra alternating with chains of $[UO_8]$ and $[UO_6(OH)]$ hexagonal and pentagonal bipyramids. All of the chains are parallel to [010] and are linked together via hydrogen bonding to form sheets parallel to (010). In addition to the crystal structure determination, thermogravimetric tests showed two major losses on heating, totaling 15%. These were interpreted as five molecules of water (7.54%) and two of CO_2 (7.36%).

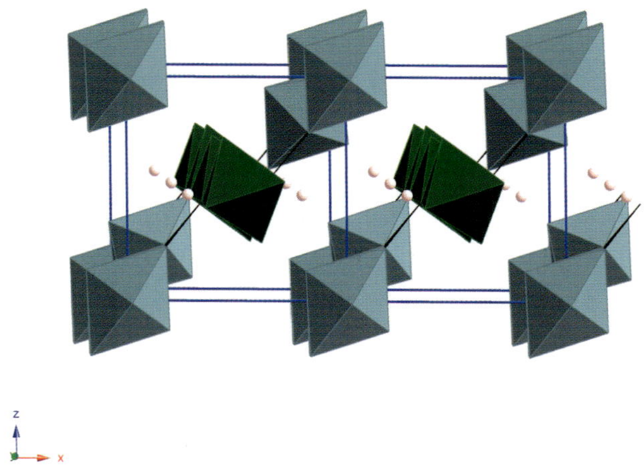

Figure 226. The structure of roubaultite, viewed just off [010], showing the relationship between the uranyl polyhedra (blue-gray), carbonate groups (black triangles), and octahedrally coordinated Cu (dark green). Pink atoms represent H in the water groups. Two unit cells are shown. Note that the structure is triclinic, but all three angles are close to 90° (α=92.16°; β=90.89°; γ=93.48°).

Schröckingerite $NaCa_3(UO_2)(CO_3)_3(SO_4)F \cdot 10H_2O$
Schröckingerite (schroeckingerite) was described by Schrauf (1873) from Jachymov, Czech Republic; material found in Wyoming and described as "dakeite" by Larsen and Gonyer (1937) was shown by Nováček (1939) to be identical to schröckingerite. Later analyses showed that the mineral contains sufficient fluorine to include F as an essential part of the formula (Jaffe, Sherwood, and Peterson 1948). It is pale greenish yellow with vitreous to pearly luster and shows bright yellow-green fluorescence under UV. Distinct crystals are rare; the common habit is crusts and globular aggregates of minute scaly grains. The water content can vary zeolitically to some degree with humidity, with little change in the crystal structure (Smith 1959); the fluorescence of dehydrated material is somewhat weaker than before dehydration (Hurlbut 1954).

Schröckingerite is plentiful along Lost Creek, near Wamsutter, Sweetwater Co., Wyoming, where it forms small concretionary aggregates in clays and silts containing gypsum. The deposit is a very recent near surface caliche-type formation (Frondel 1958). The mineral is widespread at sites in the western US, including: the Hillside mine, Arizona; as encrustations with gypsum and andersonite on mine walls in the Ambrosia Lake district, New Mexico (Barczak 1966); the Poison Basin area, Carbon Co., Wyo-

Figure 227. Pale yellow pearly schröckingerite forming a pea-sized nodule in clay from Sweetwater Co, Wyoming, shown here in natural light. RJL570

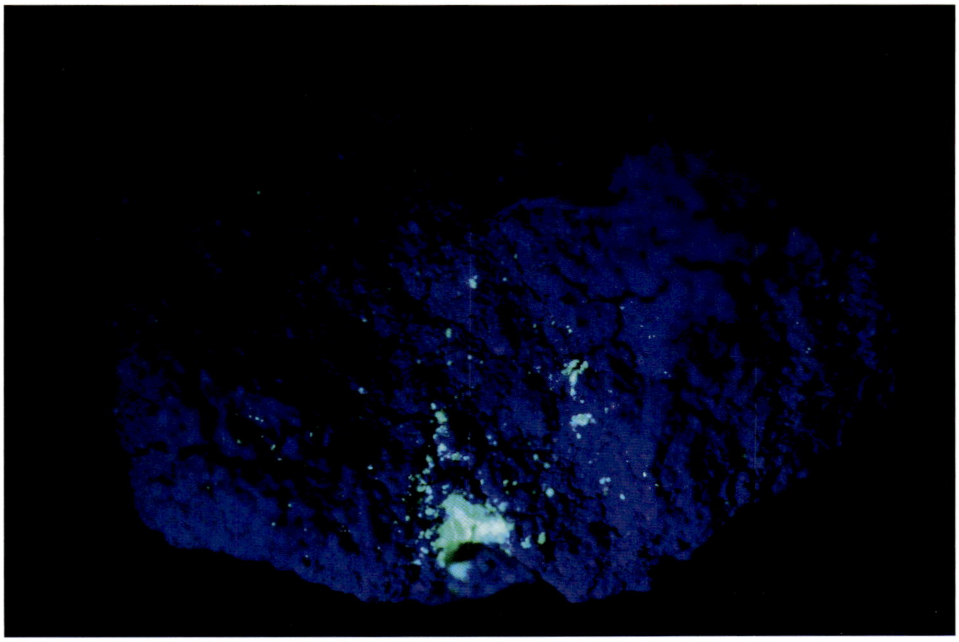

Figure 228. The sample of schröckingerite in the previous figure, shown in LWUV light. RJL570

Figure 229. Pale yellow plates and scales of schröckingerite on a sulfide-rich matrix from the Zeehan mining district, Tasmania, shown here in natural light. RJL689

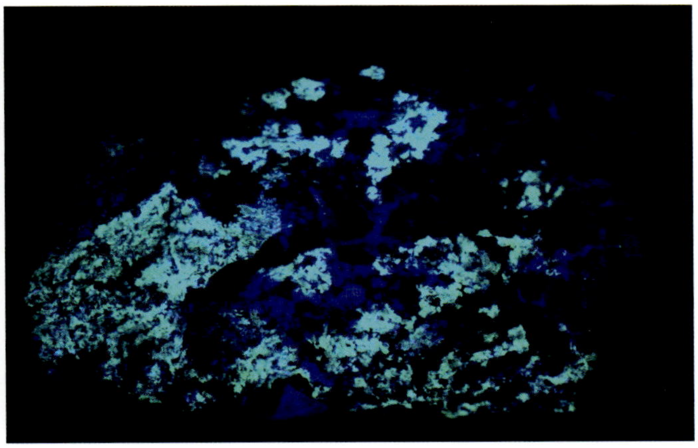

Figure 230. The sample of schröckingerite in the preceding figure, shown in LWUV light. RJL689

ming; Atomic King #2 mine, Cane Wash, San Juan Co., and numerous other districts in Utah. Worldwide occurrences include: Jachymov, Czech Republic (TL); Johanngeorgenstadt, Germany; the La Soberania mine, Mendoza Province, Argentina (Hurlbut 1954); and the Zeehan district, Tasmania, Australia.

Schröckingerite is triclinic (pseudotrigonal) with a=9.634, b=9.635, c=14.391 Å, α=91.41, β=92.33, and γ=120.26°. In the structure, three symmetrically unique $Ca\varphi_7$ polyhedra link by sharing three vertices to form a $Ca_3\varphi_{18}$ trimer. The trimers combine with uranyl tricarbonate clusters to form sheets; these are linked to SO_4 tetrahedra on one side, and adjacent sheets link through hydrogen bonds to H_2O groups in the interlayer (Mereiter 1986b; Li, Krivovichev and Burns 2001)

Figure 231. The schröckingerite structure viewed along [001] (normal to the sheet) to illustrate the linkage between the $(UO_2)(CO_3)_3$ clusters and the Ca coordination polyhedra. Bright green atoms are fluorine.

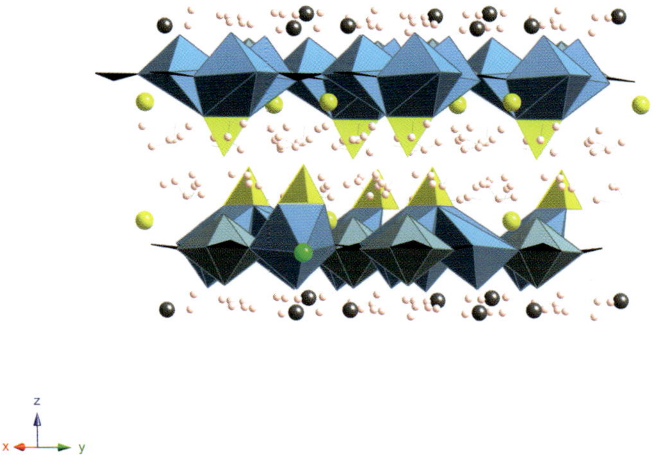

Figure 232. The schröckingerite structure viewed normal to [001] (parallel to the sheets) to show the positions of the SO_4 (yellow tetrahedra), Na (yellow atoms), and F (bright green). Gray atoms represent water in two distinctive sites; the remaining water is shown as H (pink atoms) with wire bonds to their corresponding oxygens (not shown). Adjacent sheets are connected only via hydrogen bonding.

Sharpite $Ca(UO_2)_6(CO_3)_5(OH)_4 \cdot 6H_2O$
Sharpite was described from Shinkolobwe, Congo, by Mélon (1938), who named it in honor of Major R. R. Sharp, who had discovered the Shinkolobwe deposit in 1915. At the TL, sharpite forms fibroid masses of clear olive-green to yellow-beige acicular crystals; the material is frequently darkened by hydrous cobalt oxides. The mineral is very rare at Shinkolobwe, and is associated with wölsendorfite, curite, becquerelite, and uranophane (Deliens, Piret, and Comblain 1981).

Sharpite has also been noted from: the Ellweiler uranium deposit, Rhineland-Palatinate, Germany; Le Brugeaud mine, Haute-Vienne, and the Kruth uranium deposit, Alsace, France.

Čejka, Mrazek, and Urbanec (1984) presented detailed analyses of sharpite using chemical, optical, infrared absorption, thermal decomposition, and X-ray diffraction; they confirmed the previously proposed formula, but noted that the density is 4.45, rather than the value of 3.30 previously reported. They report that the structure is orthorhombic with *a*=21.99, *b*=15.63, and *c*=4.487 Å.

Urancalcarite $Ca(UO_2)_3(CO_3)(OH)_6 \cdot 3H_2O$
Urancalcarite was described from Shinkolobwe, Congo, where it forms bright yellow acicular crystals in radiating aggregates to about 4 mm. Fibers are elongated on [001] and flattened on (100). It is found on uraninite in association with masuyite, uranophane, and wyartite (Deliens and Piret 1984a).

The species has also been reported from: Rabejac, Lodeve, France; Jachymov, Czech Republic; and the Bjertnes pegmatite, Buskerud, Norway.

Urancalcarite is orthorhombic with *a*=15.42, *b*=16.08, and *c*=6.970 Å (Deliens and Piret 1984a).

Voglite $Ca_2Cu^{2+}(UO_2)(CO_3)_4 \cdot 6H_2O$
Voglite was described by Vogl (1853) as "Uran-Kalk-Kupfer-Carbonat" based on material found at the Elias mine, Jachymov, Czech Republic. In an unusual addendum to Vogl's paper, Haidinger proposed the name voglite as preferable to a name based on the chemical composition. The mineral was associated with liebigite as an alteration product of uraninite. It typically forms emerald- to grass-green aggregates of microscopic scaly crystals. It resembles liebigite, with which it is sometimes found, but voglite is not fluorescent. (Copper tends to quench the UV fluorescence of uranyl minerals.)

Voglite is comparatively rare; in addition to the TL, documented localities include: Frey Point and the White Canyon No. 1 mine, San Juan Co., Utah; and the Red Mesa district, Navajo Co., Arizona.

Figure 233. Platy, emerald-green voglite forming a seam about 1 mm thick in matrix from the White Canyon district, San Juan Co., Utah. RJL756

Figure 234. Layer of yellow liebigite about 3 mm thick growing on a thin seam of green voglite from the White Canyon district, San Juan Co., Utah. RJL755

Wyartite $CaU^{5+}(UO_2)_2(CO_3)O_4(OH)\cdot 7H_2O$
Wyartite was described from Shinkolobwe, Congo, by Guillemin and Protas (1959), who pointed out that some samples previously thought to be ianthinite were in fact this new species. The appearance of wyartite, black to dark violet when fresh, was the likely source of confusion. Alteration causes the color to change to yellowish brown or greenish brown. The mineral forms as an alteration product of a previous red alteration product of uraninite (presumably wölsendorfite) (Guillemin and Protas 1959; Deliens, Piret, and Comblain 1981). Wyartite has also been reported from: Les Bois-Noirs mine, Rhone-Alpes, France; and the Ranger mine, Jabiru, Northern Territory, Australia.

Wyartite is orthorhombic, with *a*=11.271, *b*=7.106, and *c*=20.807 Å. It was believed early on that wyartite (like ianthinite) contained both U^{4+} and U^{6+}, and it was observed that wyartite slowly transforms in air to a phase referred to as wyartite II (Clark 1960). On long-term exposure to air, wyartite alters to schoepite or other fully oxidized phases, which no doubt contributes to its overall rarity at Shinkolobwe. Recent structural analysis of wyartite has shown that the structure contains essential U^{5+}. The structure has three unique U positions, two of which contain U^{6+} and represent uranyl ions with the normal pentagonal-bipyramidal coordination. In the other U position, seven anions coordinate the uranium but no uranyl ion is present. Based on the geometry of the polyhedron, the sum of bond valences at this site, and the requirement of electroneutrality, it was concluded that this site contains U^{5+}. Overall, the $[U(O,OH,H_2O)_7]$ polyhedra share edges and corners to form a sheet in which a CO_3 group shares an edge with the U^{5+} polyhedron. Structural units are linked together through hydrogen bonding (Burns and Finch 1999a).

Hawthorne, Finch, and Ewing (2006) described "dehydrated wyartite" from Shinkolobwe, with a similar formula but only $3H_2O$. They concluded that, "The principal difference between the structures of wyartite and dehydrated wyartite is that the sheets in the former are linked only through interstitial hydrogen bonding, whereas the sheets in the latter are linked directly by Ca-O bonds (plus interstitial hydrogen-bonds)."

Figure 235. The sheet structure of wyartite, viewed along [001]. For clarity, the polyhedra that contain U^{5+} are shown in lavender and normal uranyl polyhedra are blue-gray. CO_3 groups (black) are viewed nearly edge-on.

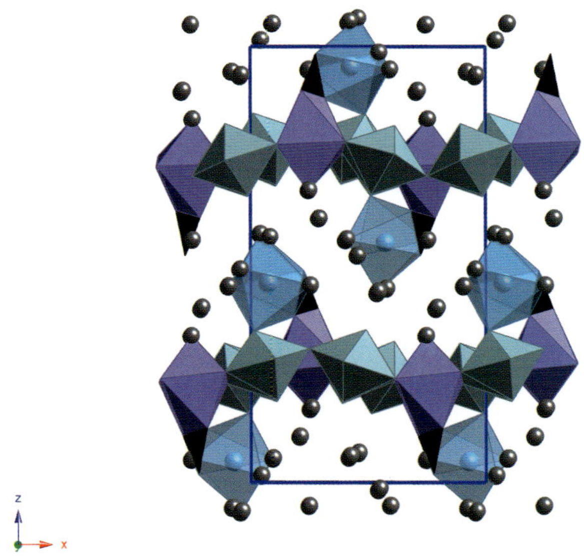

Figure 236. The wyartite cell viewed along [010], showing that the Ca polyhedra (translucent blue) don't bridge from one sheet to the next; adjacent sheets are connected only through hydrogen bonds involving the H_2O groups (gray).

SULFATES

Uranyl sulfates may be divided into: 1. the zippeite group, which includes cobaltzippeite, magnesiozippeite, marecottite, natrozippeite, nickelzippeite, pseudojohannite, zinczippeite, and zippeite; and 2. other sulfates, which includes adolfpateraite, běhounekite, belakovskiite, beshtauite, bluelizardite, bobcookite, deliensite, fermiite, jachymovite, johannite, mathesiusite, meisserite, oppenheimerite, plášilite, rabejacite, uranopilite, metauranopilite, and wetherillite.

Our knowledge of the uranyl sulfates is rapidly evolving, and their complexity has been made clear with the recent discovery of a number of new species with interesting structures. In connection with the description of meisserite, Plášil et al. (2013e) make a very important point: "Prior to the discovery of meisserite and several other new Na-bearing uranyl-sulfate minerals at the Blue Lizard mine, natrozippeite . . . was the only known Na-bearing uranyl-sulfate mineral. That made the identification of natrozippeite in the geological material seemingly 'straightforward', often based only on qualitative EDS analyses . . . With the discovery of meisserite, as well as the other new Na-bearing uranyl sulfates, the identification of Na-bearing uranyl sulfates in natural environments will clearly involve more demanding analytical study, and earlier identifications, based primarily on composition, require reassessment."

Table 46. Uranyl sulfates

Species	Formula	Symmetry
Zippeite group		
Cobaltzippeite	$Co(H_2O)_{3.5}[(UO_2)_2(SO_4)O_2]$	mon
Magnesiozippeite	$Mg(H_2O)_{3.5}[(UO_2)_2(SO_4)O_2]$	mon
Marecottite	$Mg_3(H_2O)_{18}[(UO_2)_8(SO_4)_4O_6]\cdot 10H_2O$	tric
Natrozippeite	$Na_5(H_2O)_{12}[(UO_2)_8(SO_4)_4O_5(OH)_3]$	mon
Nickelzippeite	$Ni(H_2O)_{3.5}[(UO_2)_2(SO_4)O_2]$	mon
Pseudojohannite	$Cu_3(OH)_2[(UO_2)_2O_4(SO_4)_2]\cdot 12H_2O$	tric
Sejkoraite-(Y)	$(Y,Dy)_{2.22}H^{+}_{.34}[(UO_2)_8O_{88}O_7OH(SO_4)_4(OH)\cdot 26H_2O$	tric
Zinczippeite	$Zn(H_2O)_{3.5}[(UO_2)_2(SO_4)O_2]$	mon
Zippeite	$K_3(H_2O)_3[(UO_2)_4(SO_4)_2O_3(OH)]$	mon
Other sulfates		
Adolfpateraite	$K(UO_2)(SO_4)(OH)\cdot H_2O$	mon
Běhounekite	$U(SO_4)_2\cdot 4H_2O$	orth
Belakovskiite	$Na_7(UO_2)(SO_4)_4(SO_3OH)(H_2O)_3$	tric
Beshtauite	$(NH_4)_2(UO_2)(SO_4)_2\cdot 2H_2O$	mon
Bluelizardite	$Na_7(UO_2)(SO_4)_4Cl(H_2O)_2$	mon
Bobcookite	$NaAl(UO_2)_2(SO_4)_4\cdot 18H_2O$	tric
Deliensite	$Fe(UO_2)_2(SO_4)_2(OH)_2\cdot 7H_2O$	orth
Fermiite	$Na_4(UO_2)(SO_4)_3\cdot 3H_2O$	orth
Jachymovite	$(UO_2)_8(SO_4)(OH)_{14}\cdot 13H_2O$	mon
Johannite	$Cu(UO_2)_2(SO_4)_2(OH)_2\cdot 8H_2O$	tric
Leydetite	$Fe(UO_2)(SO_4)_2\cdot 11H_2O$	mon
Mathesiusite	$K_5(UO_2)_4(SO_4)_4(VO_5)\cdot 4H_2O$	tetr
Meisserite	$Na_5(UO_2)(SO_4)_3(SO_3OH)(H_2O)$	tric
Oppenheimerite	$Na_2(UO_2)(SO_4)_2\cdot 3H_2O$	tric
Plášilite	$Na(UO_2)(SO_4)(OH)\cdot 2H_2O$	mon
Rabejacite	$Ca(UO_2)_4(SO_4)_2(OH)_6\cdot 6H_2O$	orth
Uranopilite	$(UO_2)_6(SO_4)(OH)_{10}\cdot 12H_2O$	tric
Metauranopilite	$(UO_2)_6(SO_4)(OH)_{10}\cdot 6H_2O$?	?
Wetherillite	$Na_2(Mg,Zn)(UO_2)_2(SO_4)_4\cdot 18H_2O$	mon

Zippeite Group

The first naturally occurring uranium sulfates were studied by J. F. John in 1821, who analyzed an emerald-green "uranvitriol" and a yellow earthy "basisches schwefelsaueres uranoxyd," both from Jachymov, Czech Republic. The green mineral was formally described as johannite by Haidinger (1830), who provided a crystallographic analysis. Haidinger (1845, p. 510) applied the name zippeite to the yellow earthy material, but did not provide descriptive data.

The minerals of the zippeite group are monoclinic or triclinic uranyl sulfates defined by a characteristic topology of the uranyl sulfate layer (Brugger, Burns, and Meissner 2003; Burns, Deely, and Hayden 2003). The interlayer may contain monovalent, divalent, or trivalent cations, giving rise to general formulas of the following types:
$\mathbf{A}^{+}_{x}(H_2O)_y[(UO_2)_a(SO_4)_bO_c(OH)_d]$ where **A** = K or Na;
$\mathbf{B}^{2+}(H_2O)_{3.5}[(UO_2)_2(SO_4)O_2]$ where **B** = Co, Mg, Ni, Zn, or Cu;
$(Y,REE)^{3+}_{3}(OH)_2[(UO_2)_8O_7OH(SO_4)_4](H_2O)_{24}$.

Members of the zippeite group are fairly widespread but for many years they remained poorly understood, partly because the material is commonly extremely fine powder with admixed impurities, and partly because the minerals were long believed to be simply hydrated uranyl sulfates. Detailed studies of synthetic zippeite-like phases showed that potassium was an essential constituent, and re-analysis of natural materials confirmed that zippeite is actually a potassium uranyl sulfate. Tandem studies of synthetic and natural materials led to the formal recognition of natrozippeite, cobaltzippeite, magnesiozippeite, nickelzippeite, and zinczippeite (Frondel et al. 1976). Detailed structural analysis on synthetic analogues of the zippeite minerals showed them to be monoclinic.

Interestingly, Frondel et al. (1976) noted that extensive solid solutions exist between the members with divalent

Figure 237. The zippeite-type uranyl sulfate sheet, in this case that of zippeite, viewed along [010], showing double chains of uranyl bipyramids (blue-gray) cross-linked by SO_4 groups (yellow tetrahedra).

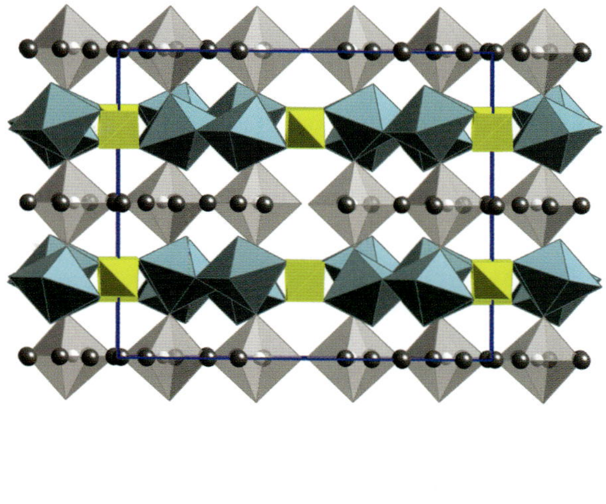

Figure 239. The zinczippeite structure viewed parallel to the sheets, showing how divalent interlayer ions such as Zn^{2+} (translucent pale gray octahedra) are coordinated to oxygens on the uranyl polyhedra and also to interlayer water molecules (gray balls).

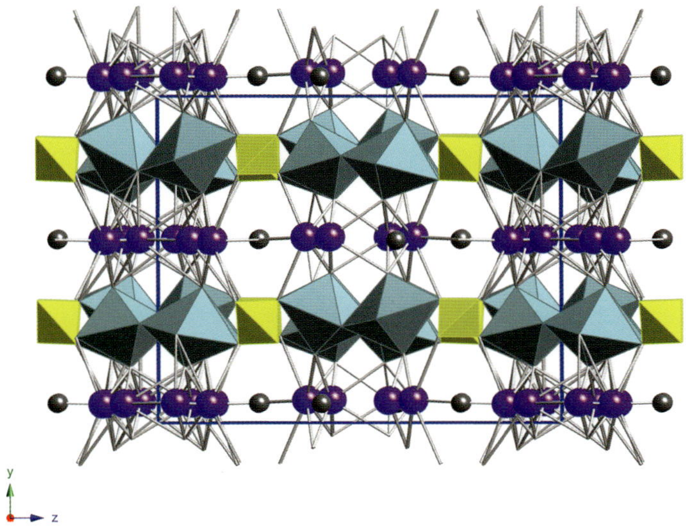

Figure 238. The zippeite structure viewed parallel to the sheets, with interlayer water (gray) and K^+ (purple ions) with stick bonds to show how monovalent interlayer ions link the sheets together.

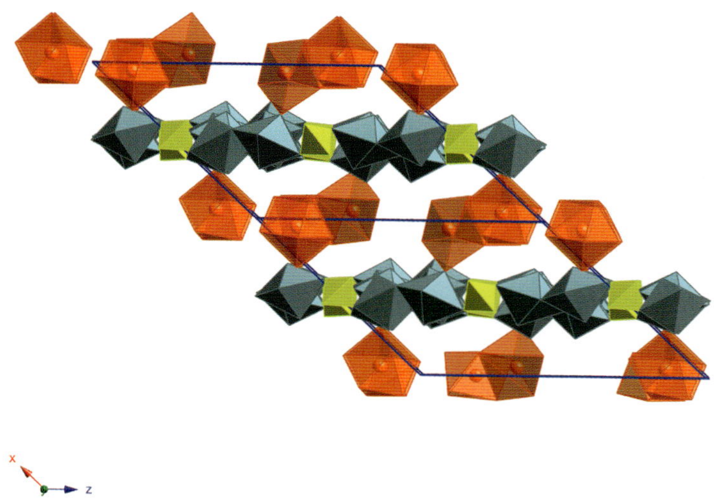

Figure 240. The structure of sejkoraite-(Y), viewed along [010] to show the different structure that arises with a trivalent interlayer ion, Y^{3+} (orange polyhedra). Two unit cells are shown.

interlayer cations, but not between the K- and Na-dominant members, zippeite and natrozippeite.

A characteristic feature of all these minerals is a sheet that consists of edge-sharing uranyl groups (pentagonal bipyramids) that are cross-linked by sharing the vertices of sulfate tetrahedra. It is interesting to note that the substitution $O^{2-} \leftrightarrow OH^-$ within the uranyl sheet allows for different sheet charges and thus different interlayer cations. The detailed sheet composition varies from one species to the next, as do the interlayer configurations, which are influenced by the bonding requirements of the sheets and the valence of the interlayer ions (Burns, Deely, and Hayden 2003). Marecottite is triclinic (Brugger, Burns, and Meissner 2003), as is pseudojohannite (Brugger et al. 2006); however, they share many structural similarities with the zippeites and are treated here as members of the group.

The crystal structures of cobaltzippeite, magnesiozippeite, and zinczippeite are built along a common theme in which the **B** ions (**B** = Co, Mg, Zn) occupy two symmetrically distinct sites in the interlayer, each of which is coordinated by six ligands (oxygen or water) forming a distorted octahedron. Each **B** is coordinated by two O atoms belonging to uranyl polyhedra of both neighboring sheets, so the **B** sites form a direct bridge between the uranyl sulfate sheets (Burns, Deely, and Hayden 2003).

All of the zippeites are secondary minerals that form in the oxidized zone of uraninite deposits that contain sulfide minerals. They often form efflorescences on mine walls, and are commonly associated with gypsum, johannite, and uranopilite. Natrozippeite is typical of sandstone-type deposits, whereas zippeite is more characteristic of weathered uraninite veins in igneous or metamorphic rocks (Frondel et al. 1976). Collectors are cautioned that members of the group cannot be distinguished from one another reliably by visual examination, and many "zippeite" specimens contain more than one member of the group in close association. Collectors should also be aware that the IMA recently adopted new rules for mineral nomenclature that, among other things, eliminated two-word names. Thus when searching older literature, one must include the now-obsolete versions of the names; in the entries below, the obsolete names are *italicized* in parentheses.

Figure 241. Cream-yellow specks of cobaltzippeite in a sandstone fragment ~12 mm wide, from the TL at the Happy Jack mine, Utah. RJL3838

Cobaltzippeite $Co(H_2O)_{3.5}[(UO_2)_2(SO_4)O_2]$
Yellowish crusts of zippeites containing cobalt, nickel, and other metals over a range of compositions were found abundantly as coatings on the walls at the Happy Jack mine, Emery Co., Utah. One of these species is cobaltzippeite (*cobalt zippeite*). Associated minerals included: natrozippeite, uranopilite, johannite, and zeunerite, along with other secondary minerals such as antlerite, erythrite, chalcanthite, and gypsum. The Co and Ni were derived from gersdorffite and other Co-Ni sulfosalts, which were locally abundant in the uranium ore (Frondel et al. 1976).

Figure 242. A complex incrustation of secondary minerals from Les Broses, near Grury, Saone-et-Loire, France. Magnesiozippeite is one of the components of the pale beige material (J. Vajdak XRD analysis). RJL3835

Magnesiozippeite $Mg(H_2O)_{3.5}[(UO_2)_2(SO_4)O_2]$
Magnesiozippeite (formerly *magnesium zippeite*) has been found in small amounts at the Lucky Strike No. 2 mine, Emery Co., Utah, in association with natrozippeite, gypsum, bieberite, cobaltocalcite, and rabbittite. It is possible that a complete solid solution series extends between the Mg-, Co-, and Ni-zippeite end members (Frondel et al. 1976). The mineral has also been reported from the Evangelista vein, Jachymov, Czech Republic; and from the La Creusaz uranium prospect, Valais, Switzerland.

Figure 243. Tiny orange aggregates of microscopic marecottite crystals on matrix from the TL at La Creusaz. RJL4400

Marecottite $Mg_3(H_2O)_{18}[(UO_2)_8(SO_4)_4O_6]\cdot 10H_2O$
Marecottite was described from the La Creusaz uranium prospect near Les Marecottes, in the Western Swiss Alps (Brugger, Burns, and Meisser 2003). The mineral formed on stockpiled ore in an exploration tunnel where it had been left since 1981 and exposed to acid mine drainage waters. Marecottite forms rosettes of minute, diamond-shaped tabular crystals up to about 0.5 mm long. The crystals are transparent yellow-orange with vitreous luster, and are not fluorescent.

In addition to its occurrence at the TL, marecottite is found at Jachymov, Czech Republic, where it had been provisionally described as "pseudo-zippeite(Mg)" in a 1997 review of the locality. At Jachymov, marecottite tends to form yellow powdery coatings or earthy aggregates, rather than distinct crystals, usually admixed with zippeite, natrozippeite, uranopilite, jachymovite, and other species (Ondrus 2003a). It also coexists with magnesiozippeite at the Lucky Strike #2 mine, Emery Co., Utah.

The crystal structure of marecottite differs somewhat from the zippeites. It contains the same zippeite-type uranyl sulfate sheet, but the interlayer configuration is different. Specifically, there are three **M**$^{2+}$ cations (ideally all Mg) and twenty-eight symmetrically distinct H_2O groups. The **M**$^{2+}$ ions are completely coordinated by eighteen of the water groups, and are thus isolated and don't form bridges between the uranyl sulfate sheets as they do in magnesiozippeite. Overall, the structure is held together by hydrogen bonding (Brugger, Burns, and Meisser 2003).

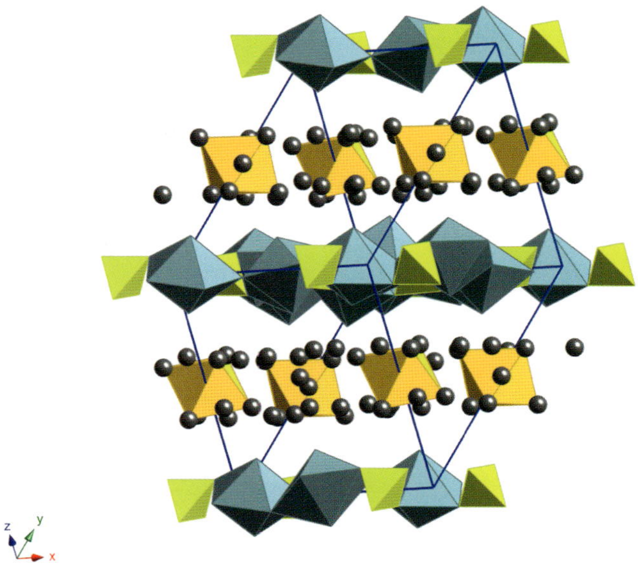

Figure 244. Crystal structure of marecottite viewed parallel to the uranyl sulfate sheets, showing Mg (yellow octahedra) coordinated by water (gray). Note that the interlayer ions are isolated and don't share any oxygens with the uranyl sheets.

Natrozippeite $Na_5(H_2O)_{12}[(UO_2)_8(SO_4)_4O_5(OH)_3]$
Natrozippeite (*sodium zippeite*) is the commonest member of the group, according to Frondel et al. (1976), who recognized it as a distinct species in the early 1950s based on optical and X-ray measurements. They state that the "zippeite" and zippeite-like minerals listed in many AEC and USGS reports from the 1950s for numerous locales on the Colorado Plateau and elsewhere " . . . probably are sodium zippeite in large part. Some of the earlier reported optical and X-ray data, however, cannot be reconciled with any of the present descriptions."

This species has been documented from numerous mines in Utah, including: Happy Jack, Delta, Atomic King, Oyler, Lucky Strike No. 2, Parco No. 23, Soda Roll, and W. M. mines. In Arizona it is reported from the Sue mine and the Hillside mine; in New Mexico it occurs at the Jackpile mine. Natrozippeite is also found at Jachymov, and at the Bukov mine, Czech Republic; and at the Geevor mine, Cornwall, England. Recently, extremely well-crystallized material was found at Green River, Emery Co., Utah. There, bright yellow needles of natrozippeite occur as thick coatings with powdery zippeite.

The structure of natrozippeite has a complex interlayer containing five Na ions and twelve H_2O groups, eight of

Figure 245. A thick, bright yellow crust of natrozippeite with zippeite, on sandstone from Green River, Utah. RJL2662

Figure 246. Yellow natrozippeite and colorless gypsum, from Bukov mine, Rožna deposit, Moravia, Czech Republic. RJL3904

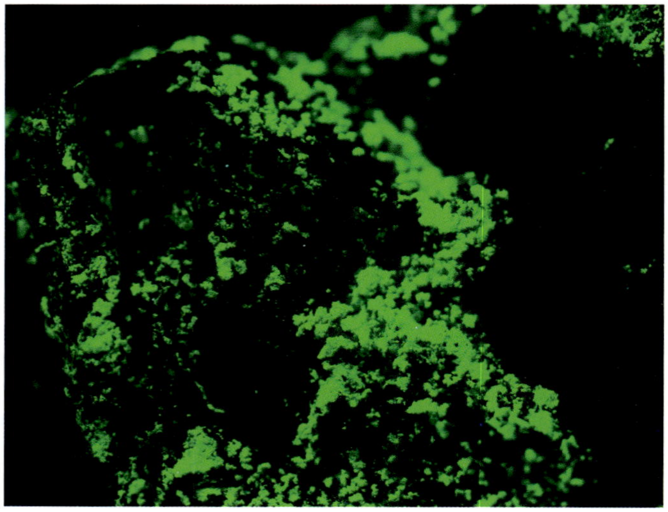

Figure 247. The natrozippeite sample from the previous photo, shown here in SWUV. RJL3904

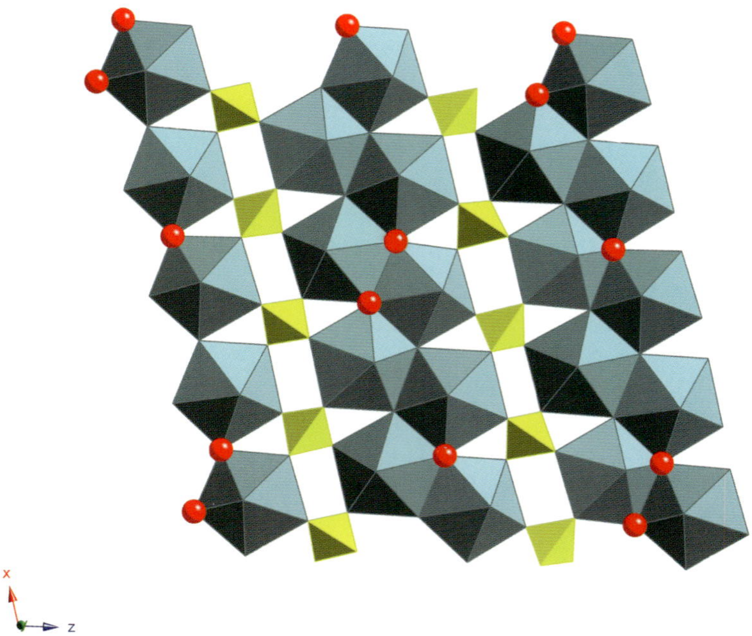

Figure 248. Crystal structure of natrozippeite, viewed along [010], showing the uranyl sulfate sheet, with OH^- ions indicated in red.

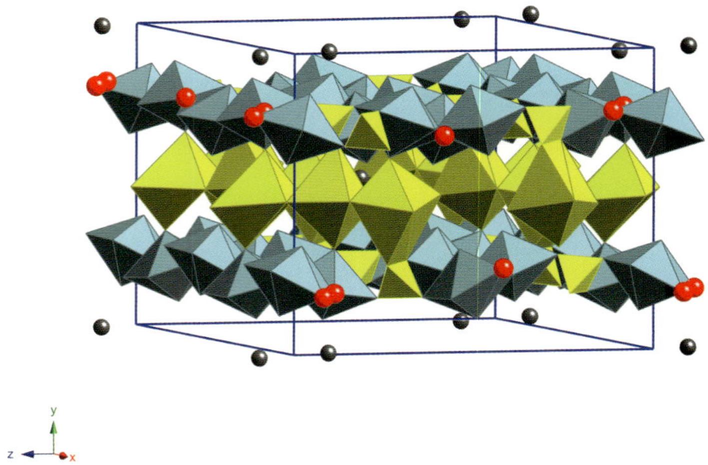

Figure 249. The natrozippeite structure viewed oblique to the axes to emphasize the Na coordination polyhedra in the interlayer.

which are shared between two Na ions and four of which are bonded to only one Na ion. The Na ions are coordinated by either six or seven ligands, so their coordination polyhedra are either octahedra or distorted pentagonal bipyramids, respectively. There is also a significant amount of hydrogen bonding in the interlayer (Burns, Deely, and Hayden 2003). The uranyl sulfate sheet contains some OH groups substituting for O in the uranyl dipyramids.

Nickelzippeite $Ni(H_2O)_{3.5}[(UO_2)_2(SO_4)O_2]$
Nickelzippeite (*nickel zippeite)* forms yellow-orange to yellow-brown aggregates and occasionally thin tabular crystals. In addition to its occurrence with cobaltzippeite at the Happy Jack mine, it has also been reported from the Hillside mine, Yavapai Co., Arizona, with zippeite and natrozippeite; from the Czech Republic at Jachymov, and at the Rožna deposit, Moravia; in fine botryoidal aggregates at the Beerwald mine, near Ronneburg, Germany; and the Eldorado mine, Great Bear Lake, Northwest Territories, Canada.

Figure 250. Orange, powdery nickelzippeite with other secondary minerals, from the Happy Jack mine, White Canyon, Emery Co., Utah. RJL1383

Figure 251. Orange nickelzippeite with bright yellow natrozippeite and colorless gypsum, covering an area ~1 × 2 cm, from the Bukov mine, Rožna deposit, Moravia, Czech Republic. RJL3903

Pseudojohannite $Cu_3(OH)_2[(UO_2)_2O_4(SO_4)_2]\cdot 12H_2O$
Pseudojohannite was described from Jachymov, Czech Republic (Ondrus et al. 2003a) and later from Musonoi, near Kolwezi, Congo, and from the La Creusaz uranium prospect in the Swiss Alps (Brugger et al. 2006). It is a product of the interaction of acid sulfate mine drainage water with either uraninite (Jachymov and La Crusaz) or uranyl silicates (Musonoi). It typically forms green aggregates of minute (< 25μm) irregularly shaped crystals; it is not fluorescent.

At the TL, only one sample (from the Rovnost shaft) is known to contain pseudojohannite. In that sample, the gray-olive pseudojohannite grew directly on heavily weathered uraninite containing pyrite, tennantite, and chalcopyrite, in association with johanite, uranopilite, and gypsum (Ondrus et al. 2003a). Individual crystals in the 1–10 μm range form 5-mm aggregates and irregular coatings on the ore.

At the Musonoi mine, light green pseudojohannite is associated with yellow uranopilite and orange-yellow metaschoepite on quartz matrix containing crosscutting veins of kasolite, uranophane-α, and cuprosklodowskite. Here, the crystals range up to 25 μm long and are flattened parallel to the cleavage plane.

At La Creusaz, pseudojohannite is associated with gypsum, ktenasite, marecottite, uranopilite, and other secondary minerals encrusting primary ore consisting mostly of uraninite and quartz. Here, poorly formed crystals about 10 μm long form moss-green, earthy aggregates to about 3 mm.

At the Widowmaker mine, White Canyon, Utah, it occurs with johannite and natrozippeite in rock consisting of silt containing lamellar coal with disseminated uraninite and chalcocite. There, it forms rich crystalline aggregates of light greenish crystals up to ~40 μm long (Plášil et al. 2012).

Figure 253. Crystals of pseudojohannite from White Canyon, Utah, having typical crystal habit of zippeite group minerals. Backscattered electron image by N. Meisser. Reproduced with the kind permission of the Mineralogical Society of America from a paper by Plášil et al. (2012)

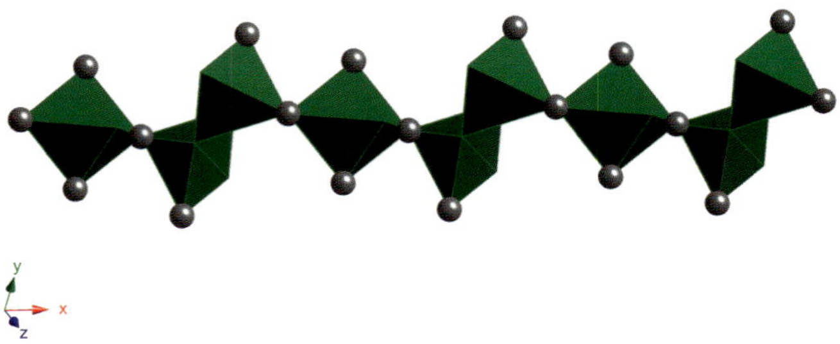

Figure 254. The arrangement of copper in the interlayer of pseudojohannite, in which edge-sharing $Cu(OH)_2(H_2O)_3$ square pyramids (dark green) are linked by corner-sharing to octahedrally coordinated Cu, forming chains running parallel to [100].

Figure 252. Two small pseudojohannite rosettes (the larger is ~0.3 mm wide) from the TL at Jachymov, Czech Republic. RJL3067

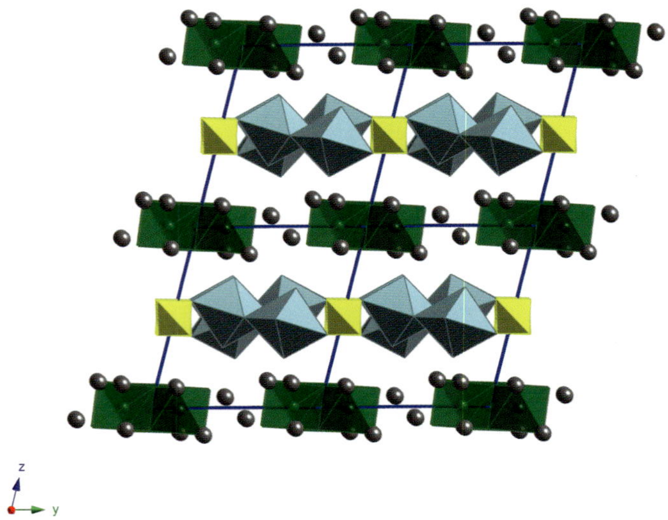

Figure 255. The overall structure of pseudojohannite, viewed parallel to the uranyl sulfate sheets; the Cu chains are seen end-on. Water molecules are gray. Four unit cells are shown.

Pseudojohannite is triclinic, with *a*=8.6744, *b*=8.8692, *c*=10.0090 Å, α=72.105, β=70.544, and γ=76.035° (Plášil et al. 2012).

The structure was initially analyzed by synchrotron powder diffraction and determined to belong to the zippeite group. Because of the limitations of powder diffraction methods (compared to single-crystal techniques) it was not possible to precisely locate the Cu atoms within the interlayer. However, because of the large spacing between the uranyl sheets, it was assumed that the Cu polyhedra don't bridge between the two sheets. Pseudojohannite, with Cu:S = 1.6:1, has a larger number of cations in the interlayer versus marecottite (Mg:S = 3:4), suggesting that pseudojohannite has a unique interlayer topology in which some or all of the $Cu(O,H_2O,OH)_6$ octahedra are linked together (Brugger et al. 2006).

The unusual interlayer structure was confirmed by Plášil et al. (2012) in a single-crystal refinement using a crystal from the Widowmaker mine. The crystal structure has uranyl sulfate sheets of zippeite topology that do not contain any OH groups; these sheets are identical to those found in zippeites containing Mg^{2+}, Co^{2+}, and Zn^{2+}. The two Cu^{2+} sites in pseudojohannite are [5]- and [6]-coordinated by H_2O molecules and OH groups, forming chains of square pyramids and octahedra. This explains the large number of metal cations in the interlayer. The crystal structure of the pseudojohannite holotype specimen from Jáchymov was solved using Rietveld refinement of high-resolution powder diffraction data. Results indicated that the crystal structures of pseudojohannite from White Canyon and Jáchymov are identical.

Figure 256. Yellow to orange sejkoraite-(Y) crystals with gypsum and other phases, from the TL in the Rovnost mine, Jachymov, Czech Republic. Reproduced with the kind permission of the Mineralogical Society of America from a paper by Plášil et al. (2011a)

Sejkoraite-(Y) $(Y,Dy)_{2.22}H^{+}{}_{.34}[(UO_2)_8O_{88}O_7OH(SO_4)_4(OH)\cdot 26H_2O$

Sejkoraite-(Y) was described from the Červena vein, Rovnost mine, Jachymov ore district, Czech Republic, where it had formed on the surface of strongly altered gangue material that contained primary mineralizations of uraninite, tennantite, and chalcopyrite. Associated minerals include pseudojohannite, rabejacite, uranopilite, zippeite, and gypsum. It occurs as aggregates to ~1 mm, of small, well-formed yellow-orange to orange crystals, which are transparent to translucent, with vitreous to pearly luster. No UV fluorescence is observed (Plášil et al. 2011a).

Sejkoraite-(Y) is the first zippeite that contains trivalent ions in the interlayer, and the first that contains REE. Microprobe analysis of the type material gives the empirical formula: $(Y_{1.49}Dy_{.17}Gd_{.11}Er_{.07}Yb_{.05}Sm_{.02})_{\Sigma 1.90}H^{+}{}_{0.54}[(UO_2)_{8.19}O_7OH(SO_4)_{3.81}](H_2O)_{26}$. It is triclinic, with *a*=14.074, *b*=17.417, *c*=17.706 Å, α=75.93, β=128.00, and γ=74.419°. The crystal structure, figure 240, consists of uranyl sulfate sheets of zippeite anion topology, which alternate with an interlayer containing $Y^{3+}(H_2O)_n$ polyhedra and uncoordinated H_2O groups. Two yttrium atoms are linked to the sheet directly via uranyl oxygen atoms, and the remaining one is bonded by hydrogen bonds only (Plášil et al. 2011a).

Zinczippeite $Zn(H_2O)_{3.5}[(UO_2)_2(SO_4)O_2]$

Zinczippeite (*zinc zippeite*) was first mentioned as a likely new species (as a "zinc uranium sulfate") from the Hillside Mine, Yavapai Co., Arizona, where it occurred with other zippeite minerals as yellow, orange, or reddish-brown coatings on quartzose ore containing disseminated uraninite and sulfide minerals (Axelrod et al. 1951). The X-ray pattern of this material was shown to be identical to that of synthetic zinczippeite (Frondel et al. 1976). The major structural features of zinczippeite are like those of cobaltzippeite and magnesiozippeite (Burns, Deely, and Hayden 2003).

Zippeite $K_3(H_2O)_3[(UO_2)_4(SO_4)_2O_3(OH)]$

The composition of the original "zippeite" of J. F. John (see Haidinger 1845) wasn't identified specifically, so it isn't possible to know which member of the group (or mixture thereof) was intended. Nováček (1935) was unable to find any of the material studied by John, and presented some new analyses of "zippeite" samples. Although these analyses didn't report the presence of potassium, Frondel (1958) re-examined Nováček's neotype specimens and determined that the material was, in fact, a potassium uranyl sulfate. Frondel et al. (1976) proposed the designations for the various zippeites, with zippeite *sensu stricto* designating the K-dominant member and Jachymov representing the TL.

Typical of the group, zippeite forms deep golden yellow crusts and aggregates of very small crystals (typically < 0.5

mm); UV fluorescence is yellow-green to green. In addition to the type locale at Jachymov, Czech Republic, zippeite has been reported from: Great Bear Lake, Northwest Territories, Canada; in Colorado at the Telegraph, Wood, and Kirk mines, Gilpin Co., and the Diamond Joe and Remington mines, Clear Creek Co.; at the Hillside mine, Yavapai Co., Arizona; at the Sec. 35 mine, Grants district, Valencia Co., New Mexico; at the Gunnislake and South Terras mines, Cornwall, England; in Germany at several locales in Saxony including Schneeberg, Annaberg, and Johanngeorgenstadt, as well as at Menzenschwand, Schwartzenwald Mountains.

The structure of zippeite, like that of natrozippeite, contains some OH groups coordinated to the uranyl ions in the sheet. There are three symmetrically distinct water groups in the interlayer, each of which is bonded to at least one of the K ions. The K ions provide direct linkage between the uranyl sulfate sheets, and there is likely also some hydrogen bonding (Burns, Deely, and Hayden 2003).

Figure 257. Bright yellow zippeite forming soft, popcorn-like aggregates, with colorless gypsum crystals on black coffinite-impregnated sandstone, from the Sec. 35 mine, Grants district, New Mexico. RJL346

Figure 258. Earthy yellow zippeite forming a thin crust on 2 × 2 cm matrix, from Great Bear Lake, Canada. RJL2865

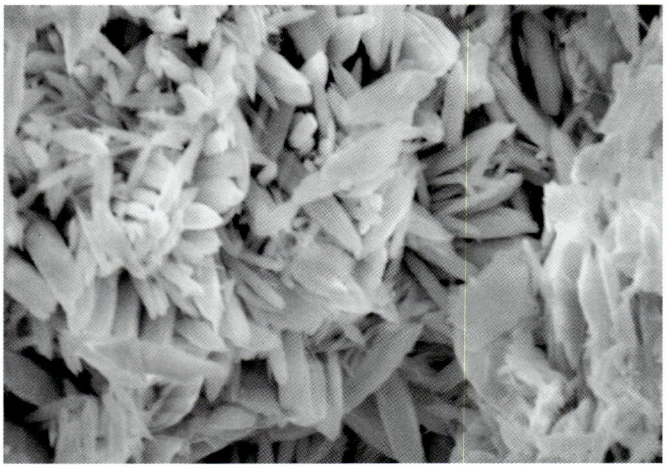

Figure 259. SEM photo showing individual zippeite crystals from Grants, New Mexico. FOV is ~50 µm wide. RJL750

Other Sulfates

Adolfpateraite $K(UO_2)(SO_4)(OH)(H_2O)$

Adolfpateraite was described from the supergene zone at the Svornost mine, Jachymov district, Czech Republic (Plášil et al. 2012a). It forms sulfur-yellow to greenish yellow crystalline aggregates up to 2 mm in diameter. Crystals are transparent to translucent with vitreous luster, and fluoresce green in LWUV. The mineral was found in an old gallery on the Geschieber vein, in association with gypsum, schoepite, čejkaite, and mathesiusite. It is believed to have formed fairly recently in the environment of the old mine works.

Adolfpateraite is monoclinic, with a=8.046, b=7.926, c=11.321 Å, and β=107.73°. Although it is a potassium uranyl sulfate, adolfpateraite is structurally unrelated to the zippeite group. The structure is based on a single chain of corner-sharing UO_7 bipyramids running parallel to [010]. SO_4 tetrahedra each share two corners with neighboring UO_7 groups. H_2O groups lie on the edges of the chains on the non-linking vertex of each UO_7 bipyramid. The K^+ ions and a single H_2O group lie between the chains, helping to link the chains together via hydrogen bonds (Plášil et al. 2012a).

Figure 260. Crystal aggregates of adolfpateraite (greenish prismatic) associated with more yellow, hedgehog-like crystal aggregates composed of the needle crystals of another new phase (later approved as mathesiusite) and gypsum (colorless) on gangue. FOV is 2.5 mm; photo by P. Škácha. Reproduced with the kind permission of the Mineralogical Society of America from a paper by Plášil et al. (2012a)

Figure 261. Prismatic crystals of adolfpateraite. Secondary electron image. Reproduced with the kind permission of the Mineralogical Society of America from a paper by Plášil et al. (2012a)

Figure 262. The uranyl sulfate chain of adolfpateraite, showing how each sulfate group shares a corner with each of two neighboring UO_7 bipyramids. The shared vertex of each bipyramid contains an OH^- group, and the non-linking vertex contains an H_2O group (not shown).

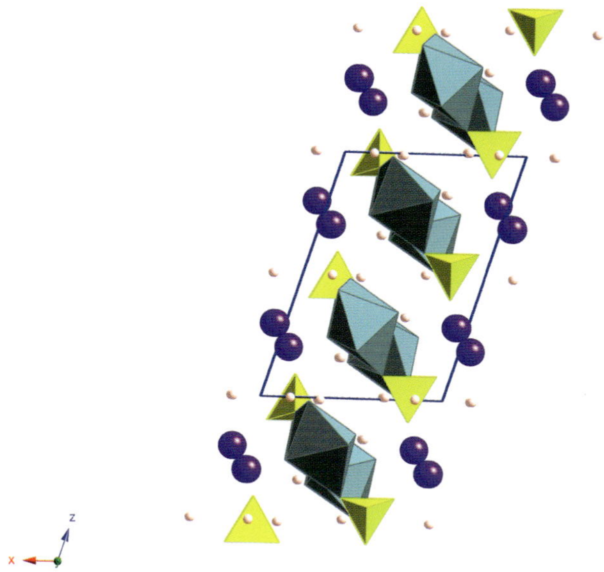

Figure 263. The structure viewed along [010] (i.e., parallel to the chains) showing K (large purple) and H (small pink atoms). The chains are cross-linked by the coordination of the K ions in the interlayer and via hydrogen bonds.

Běhounekite $U(SO_4)_2.4H_2O$

Běhounekite, the first natural sulfate of U^{4+}, was described from the Geschieber vein, Jachymov district, Czech Republic (Plášil et al. 2011b). It was found crystallized on the altered surface of native arsenic, associated with kaatialaite, arsenolite, claudetite, gypsum, and other phases. It forms short-prismatic to tabular green crystals, rarely up to 0.5 mm long. Crystals have a vitreous luster and very good cleavage on {100}; they are not fluorescent in either SW or LWUV.

Figure 264. Euhedral green and prismatic light green crystals of běhounekite on altered arsenic with arsenolite (partly in matrix) and an additional long prismatic (needle-like) phase which is not yet approved as a new mineral. FOV is 4 mm; photo by P. Škácha. Reproduced with the kind permission of the Mineralogical Society of Great Britain and Ireland from a paper by Plášil et al. (2011b)

Figure 265. SEM image of prismatic thin crystals of běhounekite. Reproduced with the kind permission of the Mineralogical Society of Great Britain and Ireland from a paper by Plášil et al. (2011b)

Běhounekite is orthorhombic with a=14.646, b=11.079, and c=5.691 Å. The crystal structure consists of sheets built up from [8]-coordinated uranium atoms and sulfate tetrahedra. The eight ligands include four oxygen atoms from the sulfate groups and four oxygen atoms from the H_2O molecules. Each uranium coordination polyhedron is connected via sulfate tetrahedra with other uranium polyhedra and through hydrogen bonds to the apices of sulfate tetrahedra. Běhounekite is structurally identical to the synthetic compound $U(SO_4)_2(H_2O)_4$ produced by P. Kierkegaard in 1956.

Figure 266. The structure of běhounekite viewed along [100] to show the uranium sulfate sheets.

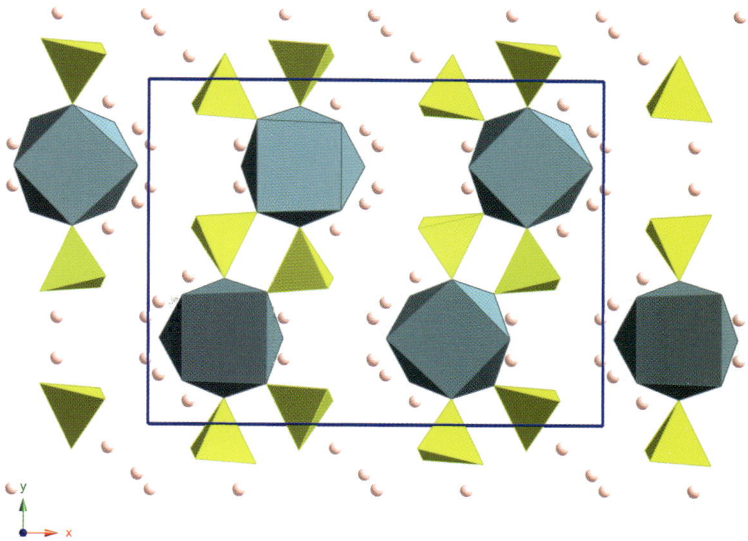

Figure 267. The structure viewed along [001] (i.e., parallel to the sheets). The U coordination polyhedra (blue-gray) are square Archimedean antiprisms, linked by SO_4 tetrahedra (yellow) and by hydrogen bonds between the sulfate groups and H atoms (pink) in the H_2O groups.

Belakovskiite $Na_7(UO_2)(SO_4)_4(SO_3OH)(H_2O)_3$
Belakovskiite was described from the Blue Lizard mine, Red Canyon, White Canyon district, San Juan Co., Utah, where it occurs on sandstone as a secondary alteration phase associated with blödite, ferrinatrite, kröhnkite, meisserite, and metavoltine. It forms very pale yellowish green hair-like fibers up to ~2 mm long and a few µm in diameter, elongated on [100] and slightly flattened on {021}.

The Blue Lizard mine was claimed in 1943 and began producing ore in 1951; high grade ore was discovered in 1954. The mine has been inactive since 1978. A number of new uranyl sulfates were recently discovered there, including meisserite, bluelizardite, and several more that are currently under study. They occur on tunnel walls as efflorescent crusts attributed to post-mining oxidation of primary ores (uraninite, pyrite, and chalcopyrite disseminated in organic matter) in the relatively humid underground environment.

Figure 268. Fibrous belakovskiite with light green meisserite crystals on blödite; FOV is 2 mm. Reproduced with the kind permission of the Mineralogical Society of Great Britain and Ireland from a paper by Kampf et al. (2014)

Belakovskiite is triclinic with a=5.4581, b=11.3288, c=18.4163 Å, α=104.787, β=90.093, and γ=96.768°. The structure contains $[(UO_2)(SO_4)_4(H_2O)]^{6-}$ polyhedral clusters

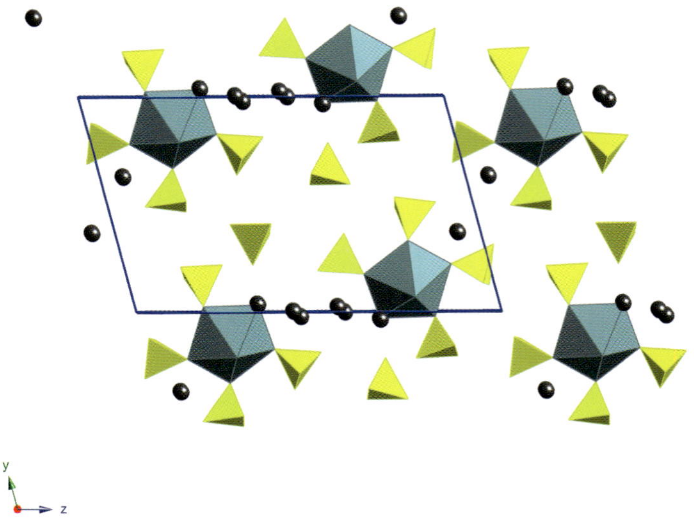

Figure 269. The structure of belakovskiite, viewed along [100], showing the uranyl sulfate clusters (polyhedra) and water (gray atoms).

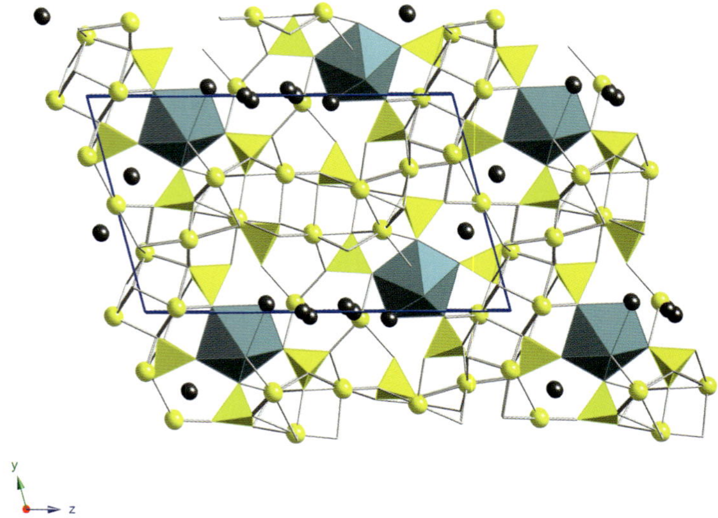

Figure 270. The same view but with Na atoms (yellow) added and Na-O stick bonds to show how the clusters are linked into a framework. The framework is further strengthened via hydrogen bonds to the water groups.

connected via an extensive network of Na-O bonds and hydrogen bonds involving eight Na sites, three other H_2O sites and an SO_3OH (hydrosulfate) group. The 3-D framework, thus defined, is unique among known uranyl sulfate structures (Kampf et al. 2014).

Beshtauite $(NH_4)_2(UO_2)(SO_4)_2 \cdot 2H_2O$

Beshtauite was described from the oxidation zone of the Beshtau U deposit, Mt. Beshtau, Northern Caucasus, Russia, where it is associated with rozenite, gypsum, lermontovite, and earlier marcasite, pyrite, halloysite, and opal. It forms well-developed short-prismatic crystals to ~0.1 × 0.15 × 0.2 mm in clusters and crusts up to 0.5 mm across on marcasite. Crystals are transparent, light green, with vitreous luster; fluorescence is strong yellow-green in both SW and LWUV. The authors of the study consider beshtauite to be an important indicator mineral because its presence is evidence of the transport of U^{6+} as mobile complexes of the uranyl ion with ammonia or polyamines (Pekov et al. 2014).

Beshtauite is monoclinic, with a=7.736, b=7.731, c=20.856 Å, and β=102.124°. The structure comprises $[UO_2(SO_4)_2(H_2O)]^{2-}$ layers consisting of corner-sharing $UO_6(H_2O)$ pentagonal bipyramids and SO_4 tetrahedra. These sheets are linked via hydrogen bonding involving interlayer NH^{4+} ions and H_2O groups. The IR spectrum closely resembles that of the analogous synthetic compound reported about thirty-five years ago.

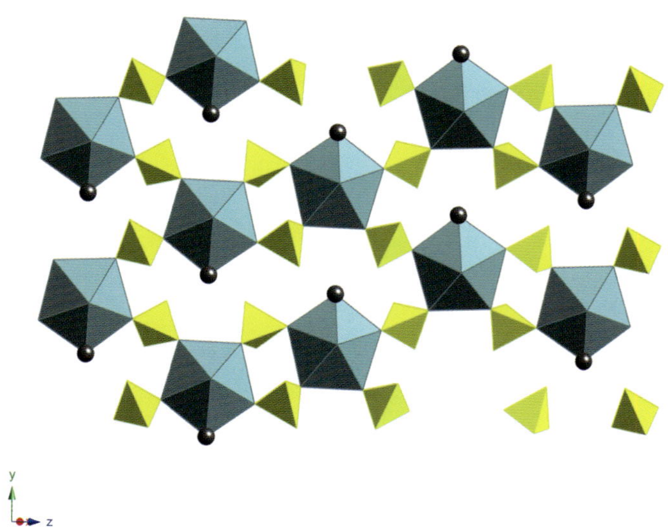

Figure 271. The uranyl sulfate sheet in beshtauite.

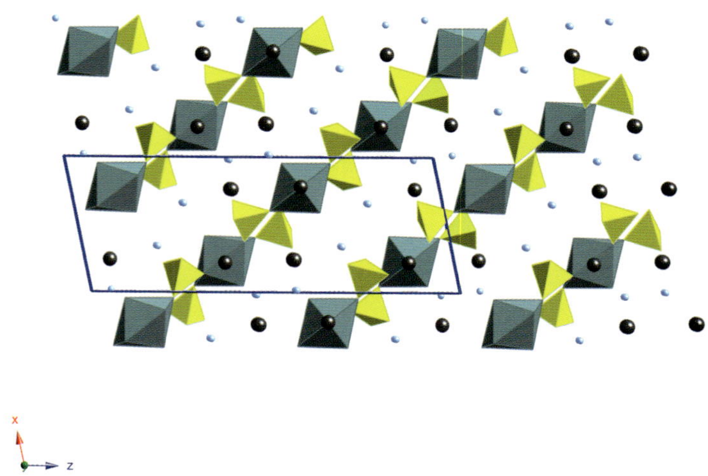

Figure 272. The beshtauite structure viewed along [010], parallel to the sheets. Ammonia N atoms are pale blue and water oxygens are gray.

Bluelizardite $Na_7(UO_2)(SO_4)_4Cl(H_2O)_2$
Bluelizardite was described from the Blue Lizard mine, White Canyon district, San Juan Co., Utah, where it occurs in sandstone in association with chalcanthite, copiapite, ferrinatrite, gypsum, kröhnkite, johannite, and several other new Na- and Mg-containing uranyl sulfates. The assemblage is the result of post-mining alteration of uraninite by SO_4-bearing acid mine drainage at approximately ambient temperatures (~15-20°C). The species forms bladed crystals up to 0.4 mm long, flattened on {001} and elongated on [010]. It is pale yellow with a vitreous luster, and shows bright yellow-green fluorescence under both SW and LWUV light.

Figure 273. Glassy yellowish acicular bluelizardite forming minute spherical aggregates on sandstone, with other sulfates, from the TL at the Blue Lizard mine, Utah. FOV is ~5 mm. RJL4609

Bluelizardite is monoclinic with a=21.151, b=5.347, c=34.671 Å, and β=104.913°. The structure is topologically unique among known structures of uranyl minerals and inorganic compounds. It is based on clusters of uranyl pentagonal bipyramids and sulfate tetrahedra. Two uranyl pentagonal bipyramids are linked through the two vertices of SO_4 groups. The remaining three vertices of each UO_7 bipyramid are occupied by SO_4 groups, linked through one corner. The eight independent Na^+ cations are linked through the Na-O bonds along with hydrogen bonds (involving H...O and H...Cl bonds) into a 3D framework. The Raman spectrum is dominated by symmetric stretching modes of the $(UO_2)^{2+}$ groups and SO_4 tetrahedra, and by O-H stretching and bending modes in the water molecules (Plášil et al. 2014c).

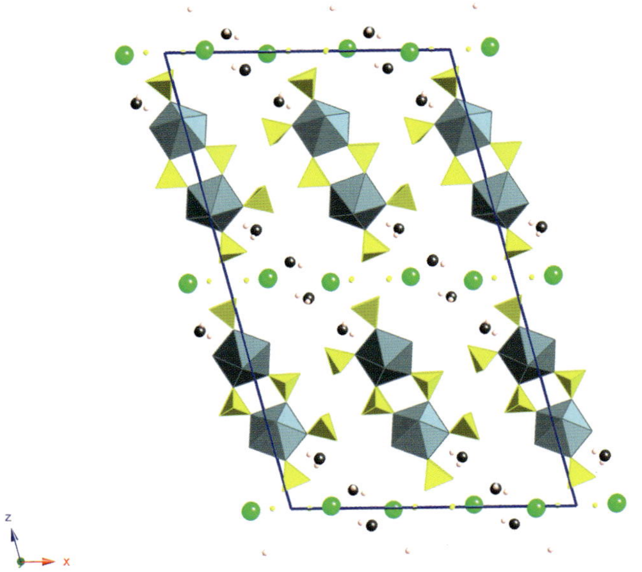

Figure 274. The uranyl sulfate topology of bluelizardite, viewed along [010], showing pairs of uranyl bipyramids linked by SO_4 tetrahedra. Water molecules are shown as gray O atoms and pink H atoms, with Cl^- shown in bright green.

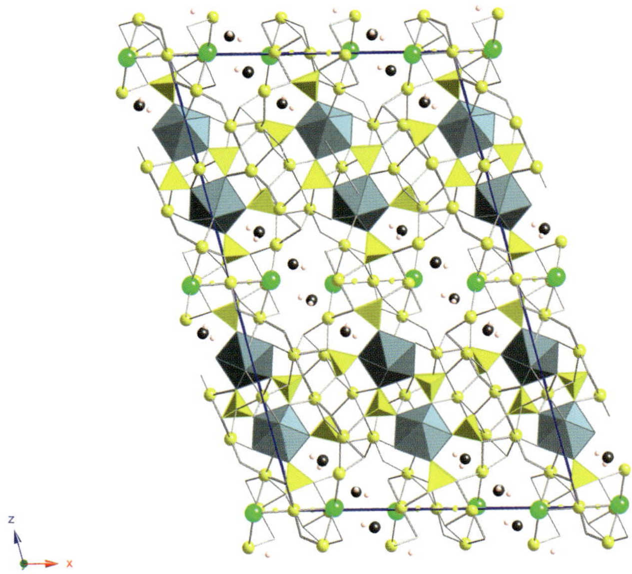

Figure 275. The full structure with Na added (yellow atoms with stick bonds) to show how the clusters are linked into a framework. The framework is further strengthened via hydrogen bonds to O and Cl ions.

Bobcookite $NaAl(UO_2)_2(SO_4)_4 \cdot 18H_2O$
Bobcookite was described from the Blue Lizard mine, San Juan Co., Utah, where it occurs as a secondary alteration phase in association with boyleite, chalcanthite, dietrichite, gypsum, hexahydrite, johannite, pickeringite, rozenite, and wetherillite. It forms lime green to greenish-yellow massive veins and irregular columnar to crude prismatic crystals, which are often somewhat curved and sometimes composite. Crystals are transparent with vitreous luster, and fluoresce bright greenish white in both LW and SWUV. Bobcookite is moderately hygroscopic and easily soluble in water (Kampf et al. 2015b).

Bobcookite is triclinic with a=7.784, b=10.567, c=11.230 Å, α=68.885, β=70.910, and γ=87.057°. The structure contains $[(UO_2)(SO_4)_2(H_2O)]$ chains linked by $NaO_4(H_2O)_2$ octahedra to form layers. Hydrogen bonds to insular $Al(H_2O)_6$ octahedra and isolated H_2O groups hold the structure together. Raman spectroscopy showed a number of features, which were interpreted as O-H stretching and H-O-H bending modes of free or weakly hydrogen-bonded H_2O molecules, SO_4 stretching and bending modes, U-O stretching modes, and UO_2 bending vibrations.

Figure 276. Lime green bobcookite with pink boyleite from the Blue Lizard mine, Utah; FOV is 6 mm wide. Reproduced with the kind permission of the Mineralogical Society of Great Britain and Ireland from a paper by Kampf et al. (2015b)

Figure 277. Glassy, bright green bobcookite, with massive pale pink to cream boyleite, from the Blue Lizard mine; sample of material studied by the authors of the original description. RJL4613

Figure 278. The sheet structure of bobcookite, showing uranyl suflate chains cross-linked by Na (yellow octahedra).

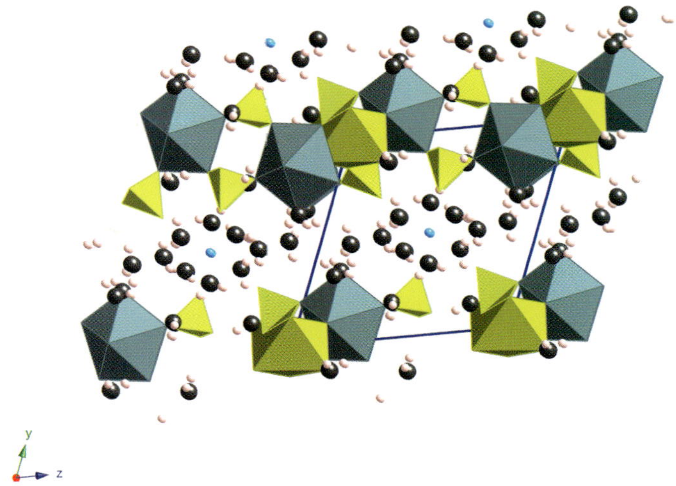

Figure 279. The structure viewed parallel to the sheet, showing Al (light blue atoms) in the interlayer, surrounded by water molecules (gray O and pink H atoms).

Deliensite $Fe(UO_2)_2(SO_4)_2(OH)_2 \cdot 7H_2O$
Deliensite was described from Mas d'Alary, Lodeve, France, where it formed as an alteration product of uraninite and primary sulfides. It forms microcrystalline crusts and spherical aggregates (to 4 mm) of pale yellow to grayish-white tabular crystals. Individual crystals are less than about 0.35 mm long and 0.06–0.15 mm wide, and have one perfect cleavage on {100}. It is transparent to translucent with vitreous luster, and is not fluorescent. It was found in the oxidation zone, associated with uraninite, granular pyrite, and gypsum (Vochten, Blaton, and Peeters 1997).

Deliensite was the second uranyl sulfate discovered at the Mas d'Alary uranium deposit (after rabejacite). It can be distinguished from rabejacite by the grayish-white color of individual flakes and by its vitreous luster. Elsewhere in France, it occurs at the L'Ecarpiere mine, Getigne, Loir Atlantique; and at Mas Lavayre, Lodeve. It has also been noted at several localities in the Jachymov district, Czech

Figure 280. Earthy, pale yellow deliensite on friable matrix, from Mas Lavayre, Lodeve, France. FOV is ~ 1 cm wide. RJL4055

Figure 281. Elongated prismatic deliensite with jarosite (yellowish) melanterite (colorless with a glassy lustre) on strongly altered uraninite from Jeroným mine, Abertamy deposit, Jachymov ore district, Western Bohemia, Czech Republic. FOV is 2.5 mm; photo by P. Škácha. Reproduced with the kind permission of the Mineralogical Society of Great Britain and Ireland from a paper by Plášil et al. 2012b)

Figure 282. Aggregate of prismatic [001] deliensite crystals from Jeroným mine, Abertamy deposit, Jachymov ore district, Western Bohemia, Czech Republic (sample MGL 92633). Reproduced with the kind permission of the Mineralogical Society of Great Britain and Ireland from a paper by Plášil et al. 2012b)

Republic, including: with uranopilite on massive altered uraninite on the dumps of the Schweitzer mine; with johannite, zippeite, gypsum, and disseminated uraninite in highly altered gangue at the Geister mine; and with rozenite, melanterite, and jarosite on a strongly altered vein of uraninite with granular pyrite or marcasite (Plášil et al. 2012b).

Deliensite is orthorhombic, with a=15.851, b=16.248, and c=6.894 Å. The structure contains uranyl-sulfate sheets with a phosphuranylite topology, consisting of dimers of edge-sharing uranyl pentagonal bipyramids linked by corner-sharing with sulfate tetrahedra. The sheets lie in the (100) plane and are decorated by $[Fe^{2+}O(H_2O)_5]$ octahedra; two weakly bonded H_2O molecules are present in the interlayer. The $[Fe^{2+}O(H_2O)_5]$ octahedron is linked directly to one sheet via the uranyl oxygen atom, but doesn't bridge from one sheet to the next; adjacent sheets are linked only via hydrogen bonds (Plášil et al. 2012b).

Figure 283. The uranyl sulfate sheet in deliensite, viewed along [100] to show the linkage between the uranyl dimers and the sulfate tetrahedra.

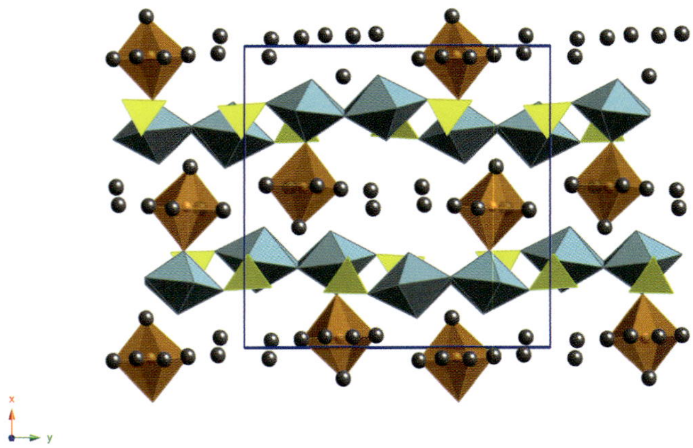

Figure 284. The structure viewed along [001], parallel to the sheets, showing how the Fe (brown octahedra) are coordinated to five H_2O groups (gray) in the interlayer and one uranyl oxygen in the sheet. Because adjacent sheets are only linked by hydrogen bonding, there is an easy cleavage parallel to the sheets.

Fermiite $Na_4(UO_2)(SO_4)_3{\cdot}3H_2O$

Fermiite was described from the Blue Lizard mine, San Juan Co., Utah, where it occurs as a secondary alteration product in association with blödite, bluelizardite, chalcanthite, epsomite, gypsum, hexahydrite, kröhnkite, manganoblödite, oppenheimerite, sideronatrite, tamarugite, and wetherillite. It forms pale greenish-yellow, prismatic crystals, as isolated individuals, sub-parallel intergrowths, and irregular aggregates. Crystals are transparent with vitreous luster and bright greenish-white fluorescence in both LW and SWUV (Kampf et al. 2015c).

Fermiite is orthorhombic with a=11.887, b=7.878, and c=15.386 Å. In the structure, adjacent uranyl pentagonal bipyramids are linked to one another via two bridging SO_4 groups forming infinite chains. The $[(UO_2)(SO_4)_3]^{4-}$ chain in fermiite is parallel to [100]. The chains are linked by bonds involving Na-O polyhedra to form a framework. Raman spectral features were assigned to O-H stretching modes of structurally non-equivalent H_2O, SO_4 stretching and bending modes, and UO_2 stretching modes.

Figure 285. Fermiite crystals from the Blue Lizard mine, Utah; field of view is 2 mm wide. Reproduced with the kind permission of the Mineralogical Society of Great Britain and Ireland from a paper by Kampf et al. (2015c)

Figure 286. SEM image of fermiite crystals on hexahydrite from the Blue Lizard mine, Utah; in the backscatter mode, the fermiite appears brighter because of the greater scattering power of heavy elements such as U. Reproduced with the kind permission of the Mineralogical Society of Great Britain and Ireland from a paper by Kampf et al. (2015c)

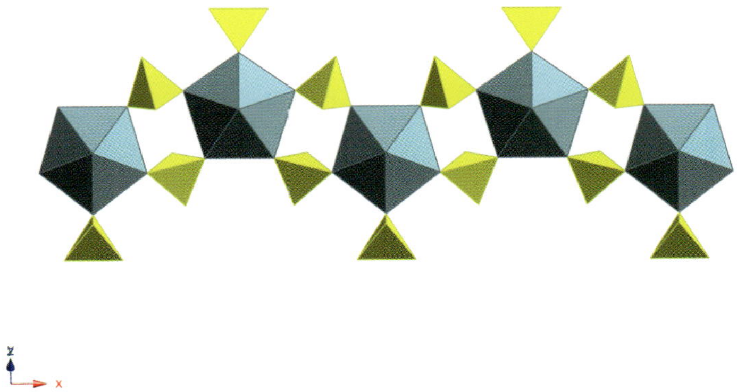

Figure 287. The uranyl sulfate chain in fermiite, viewed normal to the chain.

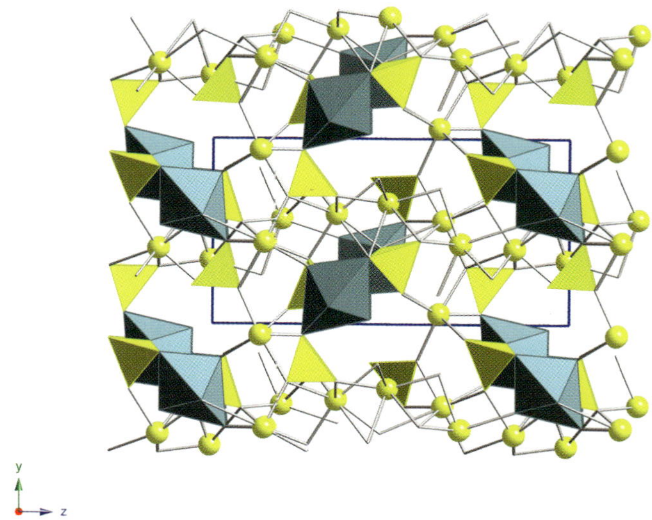

Figure 288. The full structure of fermiite, viewed along [100], i.e., parallel to the chains, showing linkage by Na (yellow atoms with stick bonds).

Jachymovite $(UO_2)_8(SO_4)(OH)_{14} \cdot 13H_2O$
Jachymovite was described from the Jachymov district, Krusny Hory Mountains, Czech Republic. It was found as an oxidation product on dolomite-uraninite veins associated with gypsum, uranopilite, and uraninite. Small patches ("some cm^2") of acicular crystals up to 0.1 mm long were found; individual crystals are translucent yellow with vitreous luster and are fluorescent yellow. The species is chemically similar to uranopilite and metauranopilite (Čejka et al. 1996). It has also been reported from the La Creusaz uranium deposit, Valais, Switzerland.

As a historical note, the name "jachymovite" was originally given to a hydrous copper-uranium silicate in 1935. That material was later discredited as being identical to cuprosklodowskite. Because the (discredited) name had been out of use for more than fifty years, the CNMMN-IMA allowed it to be used for this new species when the proposal was submitted in 1994.

Indexing the powder XRD data indicates the structure is monoclinic, with a=18.554, b=9.276, c=13.533 Å, and β=125.56°. The complete structure has not been determined.

Johannite $Cu(UO_2)_2(SO_4)_2(OH)_2 \cdot 8H_2O$
Johannite was one of the earliest-known uranium sulfates, studied by J. F. John in 1821 and formally described by Haidinger (1830) based on specimens from Jachymov, Czech Republic. Although less common than some of the other sulfates, johannite is one of the few that occasionally form good crystals It is found in a variety of habits from thick tabular to prismatic crystals to lath-like fibrous crystals and microcrystalline crusts in various shades of green. Crystals are transparent to translucent with vitreous luster. Particularly fine crystals have been found at the TL, where sharp prismatic green crystals form divergent clusters and sprays associated

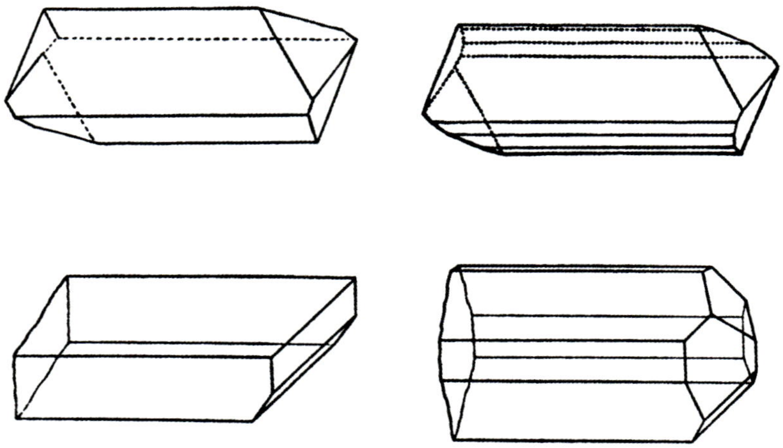

Figure 289. Drawings of johannite crystals from Jachymov, adapted from Goldschmidt (1918).

Figure 290. Green johannite forming rosettes ~5 mm wide, associated with yellow zippeite (or possibly uranopilite) from the TL at Jachymov, Czech Republic. RJL379

Figure 291. Green johannite, yellow zippeite, and other secondary phases incrusting the rock surface, from Roughton Consols, Bodmin Moor, Cornwall, England. RJL1014

Figure 292. Detail of the sample in the previous photo, showing glassy, pale green johannite and earthy yellow zippeite. RJL1014

with zippeite, gypsum, and other secondary sulfates. Haidinger remarked, "This species is as rare as it is beautiful. The only specimens hitherto known were found in opening some old works near the mine of Elias at Joachimsthal, in Bohemia, in the year 1819, as a coating on fragments of uranium-ore."

Johannite has been reported from numerous localities, including: Ronneburg, Johanngeorgenstadt, and Wölsendorf, Germany; the Mas d'Alary uranium deposit and the Limouzat mine, France; the South Wheal Bassett, Roughton Consols, and Geevor mines, Cornwall, England; the Taboshar uranium deposit, Tadzhikistan; and the Mounana mine, Gabon. It has been identified at a number of mines in the Colorado Plateau region including: the Kirk, Wood, and Schwartzwalder mines, Colorado; the Oyler, Mi Vida, and Happy Jack mines in Utah; and the Hillside mine in Arizona.

Material described as *gilpinite* by Larsen and Brown (1917) from Gilpin Co., Colorado, and Cornwall, England, was later shown by Larsen and Berman (1926) to be identical with johannite.

Figure 293. Apple-green microcrystalline johannite with yellow zippeite, from the Mi Vida mine, San Juan Co., Utah. RJL3048

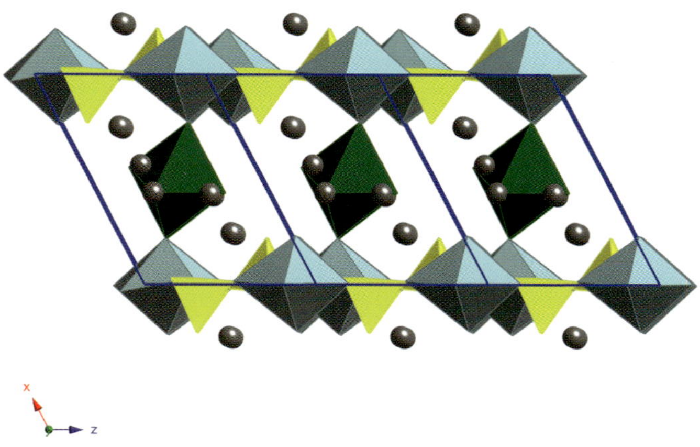

Figure 294. Crystal structure of johannite, showing octahedrally coordinated Cu ions (dark green) bridging between the uranyl sulfate sheets.

Johannite is triclinic, with a=8.903, b=9.499, c=6.812 Å, α=109.87, β=112.01, and γ=100.40°. The structure contains uranyl sulfate sheets of composition $(UO_2)_2(SO_4)_2(OH)_2$, in which pairs of edge-sharing $UO_5(OH)_2$ pentagonal bipyramids are linked by SO_4 groups. The sheets lie parallel to (100). The interlayer contains elongated $[Cu(H_2O)_4O_2]$ octahedra and water molecules. Each copper octahedron shares uranyl oxygens with both adjacent sheets, thereby directly linking the sheets together (Mereiter 1982).

Leydetite $Fe(UO_2)(SO_4)_2·11H_2O$
Leydetite was described from the supergene zone at Mas d'Alary, Lodeve, Herault, France, associated with abundant pyrite, uraninite, calcite, quartz, clay minerals, gypsum, and deliensite. It forms pale yellow to greenish, transparent to translucent, tabular crystals up to ~2 mm. Crystals have a strong vitreous luster and are not fluorescent.

Leydetite is monoclinic, with a=11.302, b=7.729, c=21.815 Å, and β=102.402°. The crystal structure contains uranyl sulfate sheets based on the protasite anion topology. The sheet consists of UO_7 bipyramids, which share four of their equatorial vertices with SO_4 tetrahedra. Each SO_4 tetrahedron, in turn, shares two of its vertices with UO_7 bipyramids. The remaining unshared equatorial vertex of the bipyramid is occupied by H_2O, which extends hydrogen bonds within the sheet to one of a free vertex of the SO_4 tetrahedron. Sheets are stacked perpendicular to the c direction. In the interlayer, Fe^{2+} ions and H_2O groups link to the sheets on either side via a network of hydrogen bonds. Leydetite is isostructural with the synthetic compound $Mg(UO_2)(SO_4)_2(H_2O)_{11}$ (Plášil et al. 2013a).

Figure 295. Aggregate of leydetite crystals from Mas d'Alary, Lodéve, France. The dominating faces are {001}, then {010} and {100}. Faces {101} and {011} (and their symmetrically related equivalents) also occur. FOV is 2.5 mm; photo by V. Bourgoin. Reproduced with the kind permission of the Mineralogical Society of Great Britain and Ireland from a paper by Plášil et al. (2013a)

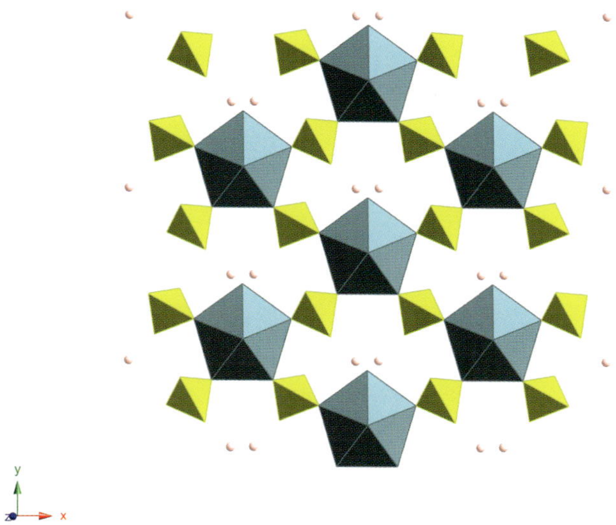

Figure 296. The uranyl sulfate sheet of leydetite, viewed along [001]. The two H atoms associated with the water molecule on the equatorial vertex of the uranyl polyhedron are shown in pink. Hydrogen bonds extend between each of these H atoms and the nearest oxygens on the adjacent SO_4 groups.

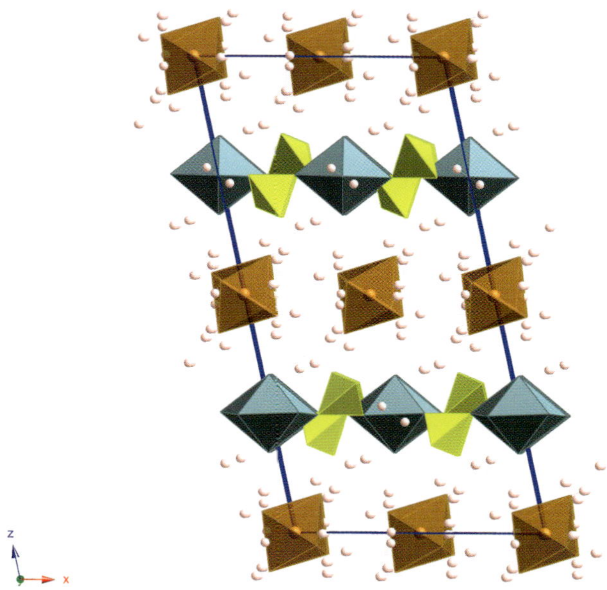

Figure 297. The leydetite structure viewed parallel to the sheets, showing $Fe(H_2O)_6$ (brown octahedra) and additional H_2O groups (shown as pink H atoms with wire bonds) in the interlayer. Linkage between layers involves only hydrogen bonds.

Mathesiusite $K_5(UO_2)_4(SO_4)_4(VO_5)\cdot 4H_2O$

Mathesiusite was discovered along with adolfpateraite in an old mine adit on the Geschieber vein of the Svornost mine, Jachymov district, Czech Republic. It appears to be very rare, occurring on only a few specimens. It forms greenish yellow radiating aggregates of acicular crystals on gangue or fragments of country rock that are devoid of primary U mineralization. It is strongly fluorescent yellowish green in both LW and SWUV light. Associated minerals include gypsum, schoepite, čejkaite, and adolfpateraite. The assemblage is interpreted to be of very recent formation, connected to acid mine drainage in the old mine workings (Plášil et al. 2014).

Mathesiusite is tetragonal, with a=14.970 and c=6.817 Å. The crystal structure contains topologically unique heteropolyhedral sheets based on $[(UO_2)_4(SO_4)_4(VO_5)]^{5-}$ clusters. These clusters arise from linkages between corner-sharing quartets of uranyl pentagonal bipyramids, which define a square-shaped void at the center that is occupied by V^{5+} cations. Each pair of uranyl pentagonal bipyramids shares two vertices of SO_4 tetrahedra. Each SO_4 shares a third vertex with another cluster to form the sheets. The K^+ ions lie between the sheets, together with a single H_2O group. The corrugated sheets are stacked perpendicular to c. These heteropolyhedral sheets are similar to those in the structures of synthetic uranyl chromates (Plášil et al. 2014).

Figure 298. Greenish-yellow radiating aggregate of mathesiusite composed of fine, needle-like crystals in association with bright-yellow adolfpateraite on reddish gangue. FOV 3.0 mm; photo by P. Škácha. Reproduced with the kind permission of the Mineralogical Society of America from a paper by Plášil et al. (2014)

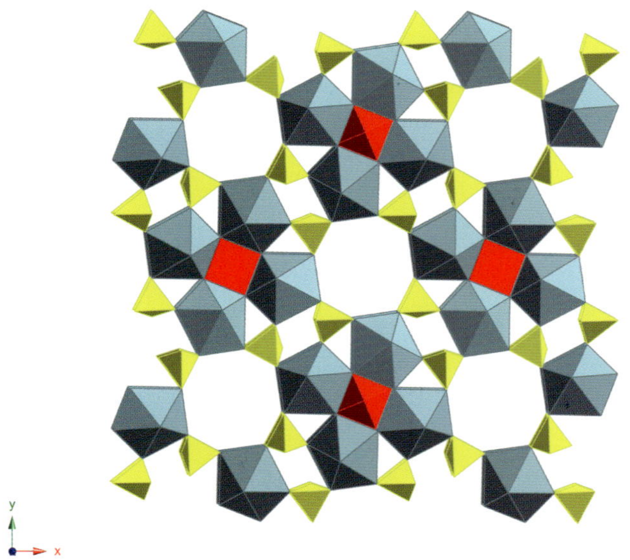

Figure 299. The heteropolyhedral sheet structure of mathesiusite, viewed along [001], showing clusters made of four UO_7 bipyramids (blue-gray), four SO_4 tetrahedra (yellow), and one VO_5 square pyramid (red).

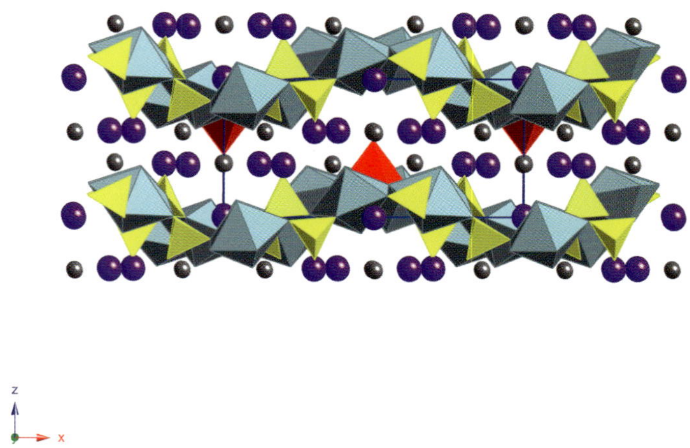

Figure 300. The structure viewed parallel to the sheets, showing K^+ (purple) and water (dark gray) in the interlayer. Blue lines indicate the unit cell.

Meisserite $Na_5(UO_2)(SO_4)_3(SO_3OH)(H_2O)$

Meisserite was described from the Blue Lizard mine, White Canyon district, Utah, where it occurs in sandstone, associated with chalcanthite, copiapite, ferrinatrite, gypsum, johannite, belakovskiite, and bluelizardite. The assemblage is attributed to post-mining alteration of uraninite and sulfides. Meisserite crystals are typically intergrown with other uranyl sulfates and are only rarely found as isolated crystal aggregates. The prismatic crystals range up to 0.3 mm long, and are elongated on [100]. They are pale green to yellowish green with a vitreous luster. Fluorescence is bright yellow-green under both SW and LWUV. Meisserite is somewhat hygroscopic.

Meisserite is triclinic with a=5.3232, b=11.5105, c=13.5562 Å, α=102.865, β=97.415, and γ=91.462°. The structure contains uranyl pentagonal bipyramids linked by SO_4 groups to form chains. Na^+ cations bond to O atoms in the chains and to an SO_3OH group and an H_2O group between the chains, thereby forming a heteropolyhedral framework. The Raman spectrum is dominated by the symmetric stretching modes of $(UO_2)^{2+}$, $(SO_4)^{2-}$ and also weaker O-H stretching modes (Plášil et al. 2013e).

Figure 301. Sample of meisserite, ~4 cm across, forming a bright greenish-yellow crust on matrix from the TL at the Blue Lizard mine, Utah. RJL4608

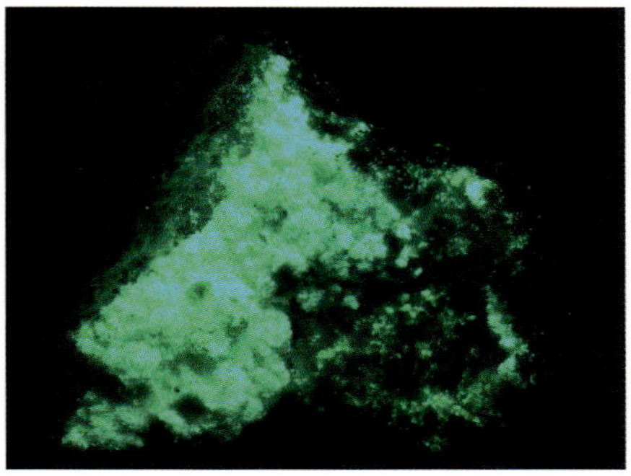

Figure 302. The sample in the previous figure, viewed in SWUV, showing the prominent fluorescence. RJL4608

Figure 303. Meisserite (greenish-yellow) in typical association with belakovskiite (IMA2013-075) (whitish fibrous) and blödite, from the Blue Lizard mine, Utah. FOV is 2 mm. Reproduced with the kind permission of the Mineralogical Society of Great Britain and Ireland from a paper by Plášil et al. (2013e)

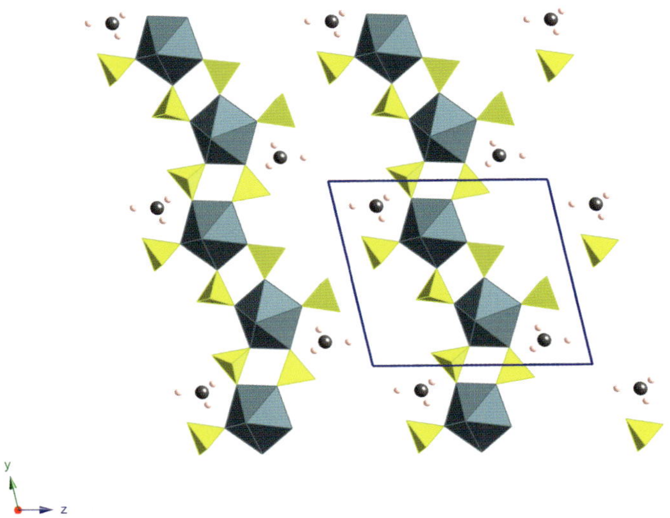

Figure 304. The uranyl sulfate topology of meisserite, viewed along [001], showing uranyl bipyramids linked by SO_4 tetrahedra to form chains extending along [010]. Water molecules are shown as gray O atoms and pink H atoms.

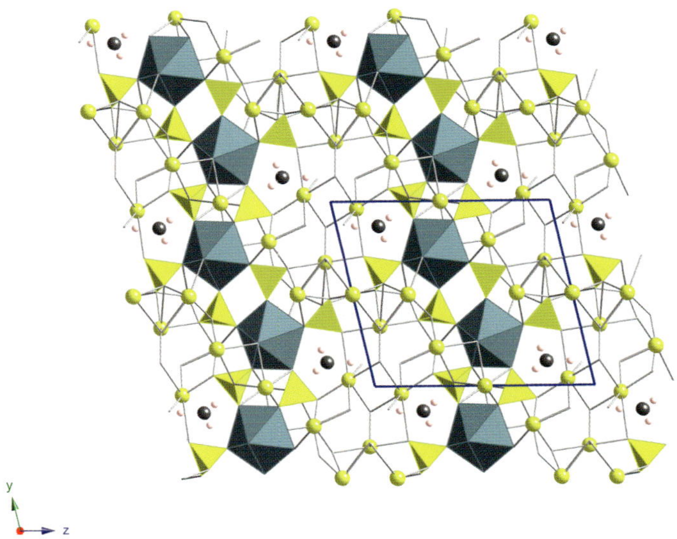

Figure 305. The full structure with Na added (yellow atoms with stick bonds) to show how the chains are linked into a framework.

Oppenheimerite $Na_2(UO_2)(SO_4)_2 \cdot 3H_2O$
Oppenheimerite was described from the Blue Lizard mine, San Juan Co., Utah, where it occurs as an alteration product in association with blödite, bluelizardite, chalcanthite, epsomite, fermiite, gypsum, hexahydrite, kröhnkite, manganoblödite, sideronatrite, tamarugite, and wetherillite. It forms pale greenish-yellow prismatic crystals in subparallel intergrowths up to several milimeters across, with surfaces made up of small stepped faces. Crystals are transparent with vitreous luster and bright yellowish green fluorescence in UV. It is easily soluble in water (Kampf et al. 2015c).

Oppenheimerite is triclinic with a=7.958, b=8.195, c=9.805 Å, α=65.967, β=70.281, and γ=84.517°. In the structure, adjacent pentagonal bipyramids are linked to one another via two bridging SO_4 groups forming infinite chains. The $[(UO_2)(SO_4)_2(H_2O)]^{2-}$ chain in oppenheimerite is parallel to [11-1]. The chains are linked by bonds involving Na-O polyhedra to form a framework. Raman spectral features were assigned to O-H stretching modes of structurally non-equivalent H_2O, SO_4 stretching and bending modes, and UO_2 stretching modes.

Figure 306. Oppenheimerite with hexahydrite (white) on blodite; FOV is 5 mm wide. Reproduced with the kind permission of the Mineralogical Society of Great Britain and Ireland from a paper by Kampf et al. (2015c)

Figure 307. SEM photo showing subparallel intergrowth of oppenheimerite crystals. Reproduced with the kind permission of the Mineralogical Society of Great Britain and Ireland from a paper by Kampf et al. (2015c)

Figure 308. The uranyl sulfate chain in oppenheimerite, viewed normal to the chain.

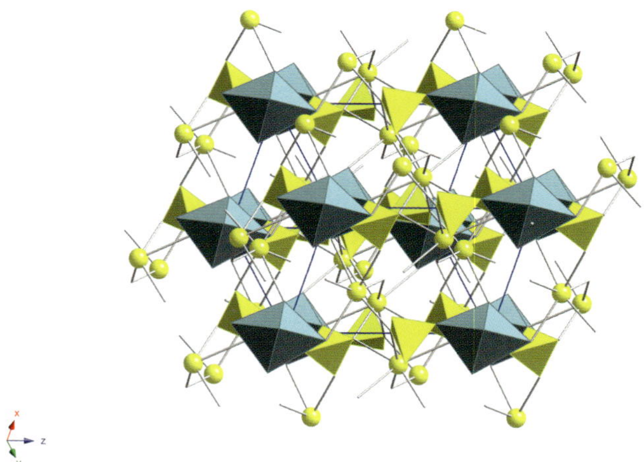

Figure 309. The full structure of oppenheimerite, viewed diagonally to the unit cell, i.e., parallel to the chains, showing linkage by Na (yellow atoms with stick bonds).

Plášilite $Na(UO_2)(SO_4)(OH) \cdot 2H_2O$
Plášilite was described from the Blue Lizard mine, San Juan Co., Utah, where it occurs in and on sandstone as a low-temperature secondary mineral formed by post-mining weathering of uraninite. It forms greenish yellow, prismatic crystals, generally as long, thin blades elongated on [001] and flattened on {100}. Crystals are commonly twinned. It is transparent with vitreous luster, and is fluorescent bluish-white in both LW and SWUV. Associated minerals include atacamite, blödite, brochantite, calcite, chalcanthite, dickite, gerhardtite, gypsum, hexahydrite, johannite, manganöblodite, natrozippeite, and tamarugaite (Kampf et al. 2015a).

Figure 310. Thin bladed plášilite crystals with chalcanthite on quartz, from the Blue Lizard mine, Utah; FOV is 1.8 mm. Reproduced with the kind permission of the Czech Geological Society from a paper by Kampf et al. (2015a)

Figure 311. Sprays of plášilite with tamarugaite (white) on blödite and quartz, from the Blue Lizard mine, Utah; FOV is 3 mm wide. Reproduced with the kind permission of the Czech Geological Society from a paper by Kampf et al. (2015a)

Plášilite is monoclinic with a=8.712, b=13.837, c=7.047 Å, and β=112.13°. The structure contains uranyl sulfate sheets of composition $[(UO_2)_2(SO_4)_2(OH)_2]^{2-}$ lying parallel to (010) and having the phosphuranylite topology. The sheets are linked together by chains of edge-sharing $NaO_2(H_2O)_4$ octahedra running parallel to [001]. Prominent features of the Raman spectrum are interpreted as symmetric stretching modes of the $(UO_2)^{2+}$ and (SO_4) groups, and O-H stretching modes of the H_2O groups.

Figure 312. Acicular yellow plášilite with blue chalcanthite and other phases, from the Blue Lizard mine, Utah. RJL4614

Figure 313. The uranyl sulfate sheet in plášilite, viewed normal to the sheet. The topology is the same as that of the phosphuranylite sheet.

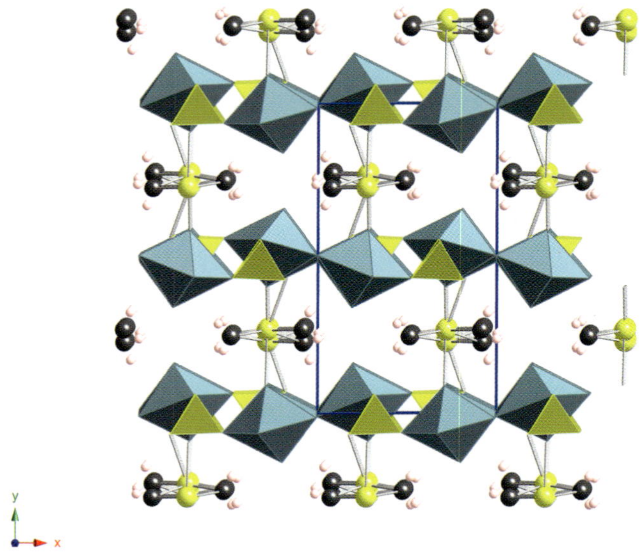

Figure 314. The plášilite structure viewed along [001]; the $NaO_2(H_2O)_4$ octahedral chains, viewed end-on, are presented as Na atoms (yellow with stick bonds) and water molecules (gray oxygen and pink hydrogen atoms).

Rabejacite $Ca(UO_2)_4(SO_4)_2(OH)_6 \cdot 6H_2O$

Rabejacite was described from the Rabejac uranium deposit, near Lodeve, France, and was also found nearby at Mas d'Alary. It forms minute (0.1 mm) tabular or acicular crystals and rounded nodules up to 0.3 mm. The crystals are bright yellow to amber, transparent to translucent with vitreous luster; UV fluorescence is light yellow. The mineral was derived from the alteration of pitchblende, and at the type locale it is associated with gypsum, fontanite, seelite, and various other secondary uranium minerals (Deliens and Piret 1993). It has also been reported from the La Creusaz uranium prospect, Valais, Switzerland; the Rio Giulis uranium deposit, Brione, Italy; the Jachymov district, Czech Republic; and the Ranger No. 1 deposit, Northern Territory, Australia.

Rabejacite is orthorhombic, with a=8.73, b=17.09, and c=15.72 Å. Note that Plášil et al. (2011a) argue that rabejacite "should be considered as an additional member of the zippeite group, based on many similar characteristics."

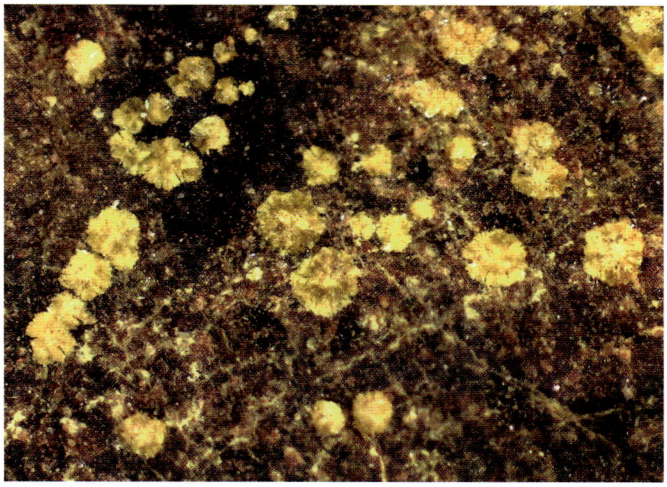

Figure 315. Tiny spherical aggregates of bright yellow rabejacite crystals on matrix, from the TL at Rabejac, Lodeve, France. FOV is ~3.5 mm wide. RJL3907

Uranopilite $(UO_2)_6(SO_4)(OH)_{10} \cdot 12H_2O$
Metauranopilite $(UO_2)_6(SO_4)(OH)_{10} \cdot 6H_2O$ (?)

Uranopilite was one of the earliest known uranyl sulfates; analyses were reported by Dauber in 1854 and by Schultz in 1882, based on material from Jachymov, Czech Republic. The species as it is presently defined was based on more rigorous analyses published by Nováček in 1935 [see Frondel (1952) for a discussion of the history of these studies]. Jachymov is generally recognized as the TL.

The mineral forms fragile yellow crusts and coatings composed of felted masses of microscopic acicular to lath-like crystals. The color ranges from bright lemon yellow to straw yellow; UV fluorescence is bright yellow-green. The

Figure 316. Earthy, golden yellow uranopilite forming a crust on matrix from the TL at Jachymov, Czech Republic. Sample is ~5 cm wide. RJL388

Figure 317. The uranopilite specimen in the previous photo, viewed in SWUV. RJL388

fine, acicular crystals sometimes give the crusts a silky appearance, in contrast to the generally dull, earthy habit of zippeite, with which it often occurs. However, it must be emphasized that these minerals cannot be confidently distinguished by appearance or fluorescence, and microanalysis or X-ray diffraction is needed to make a definitive identification.

Important historical localities for uranopilite include: Jachymov, Pribram, and Dlazov, Czech Republic; Johanngeorgenstadt, Germany; and Wheal Owles, St. Just, Cornwall. It is also known from: Great Bear Lake and elsewhere in Northwest Territories, Canada; La Crouzille, Haute-Vienne, and Grury, Saone-et-Loire, France; Urgeirica mine, Viseu district, Portugal; and Shinkolobwe, Congo. Notable occurrences in the US include: the Happy Jack mine, Utah, where it forms thick coatings on the walls, associated with johannite and zippeite minerals; and the Snowball claim, Fremont Co., Wyoming.

Uranopilite is triclinic, with a=8.896, b=14.029, c=14.339 Å, α=96.610, β=98.472, and γ= 99.802°. Burns (2001a) reported a structure refinement using a single crystal from Jachymov: The structure contains six symmetrically distinct U^{6+} cations, each of which is coordinated by seven ligands forming $U[O,OH,H_2O]_7$ pentagonal bipyramids. The S^{6+} cation is coordinated by four O atoms in a tetrahedral arrangement. The uranyl bipyramids share vertices and edges, resulting in a cluster of composition $[(UO_2)_6O_6(OH)_6(H_2O)_6]^{6-}$. These clusters are linked through SO_4 tetrahedra, which share each of their vertices with different uranyl bipyramids, creating charge-neutral uranyl sulfate chains of composition $[(UO_2)_6(SO_4)O_2(OH)_6(H_2O)_6]$ that extend along [100]. The uranyl sulfate chains are linked to form the extended structure by hydrogen bonds bridging directly between the chains and to interstitial H_2O groups. The uranyl sulfate chain in uranopilite is novel in minerals and synthetic compounds. (Most other uranyl sulfate minerals contain uranyl sulfate *sheets*.)

The original concept of metauranopilite (the "beta-uranopilite" of Nováček) was based on a single specimen from Jachymov, Czech Republic. The material had grayish, dirty green, or brownish, acicular to lath-like crystals and chemical analysis suggested a composition similar to that of uranopilite but with about half as much water. This led Nováček to infer that the mineral was a natural dehydration

Figure 318. Yellow, microscopically fibrous uranopilite forming tiny, rounded aggregates, from Les Bois Noirs, Loire, France. FOV is ~5 mm wide. RJL794

Figure 319. Structure of uranopilite viewed along [001] showing uranyl sulfate chains of composition $[(UO_2)_6(SO_4)O_2(OH)_6(H_2O)_6]$ extending along [100], which are held to one another via hydrogen bonding.

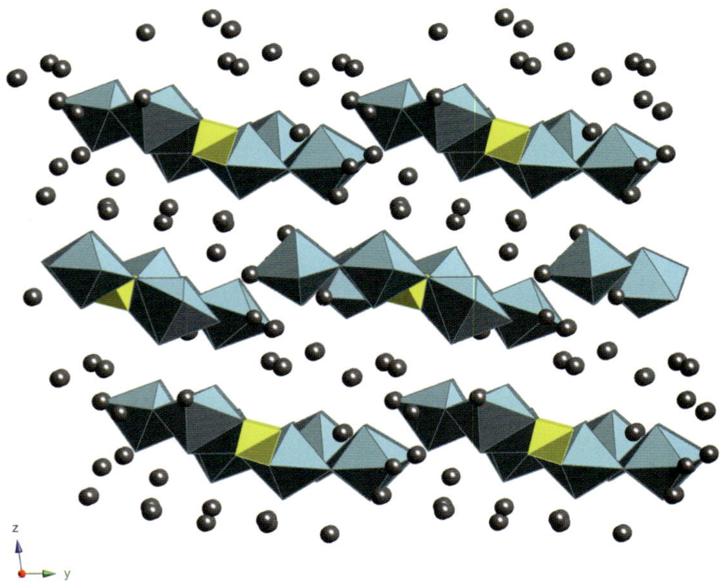

Figure 320. The uranopilite structure viewed along [100], i.e., parallel to the chains, showing the interstitial water molecules (gray). The structure is linked together via hydrogen bonding.

product of uranopilite. Later work on thermal dehydration and rehydration of uranopilite made it doubtful that the material described by Nováček could be such a dehydration product (Frondel 1952). In addition to the TL, metauranopilite has been reported from the Mas Lavayre mine, Lodeve, France; and Menzenschwand, Germany.

Dehydration of uranopilite from the South Alligator River, Northern Territory, Australia, was studied by Frost et al. (2006). The material consisted of elongated crystals up to 50 μm long and 5 μm wide. The authors noted, "Thermogravimetry combined with mass spectrometry shows that dehydration occurs at ~31°C resulting in the formation of metauranopilite. The first dehydration step over 20–71°C corresponds to a decrease of 5.4 wt. %, equivalent to 6.076 H_2O." No structural data are given for the dehydrated, or "meta-" form, but the weight loss suggests that it would have 6 H_2O remaining.

It is appropriate to note that over sixty years ago Frondel commented, "The status of metauranopilite is uncertain, and further work on the mineral is desirable." That is, regrettably, still the case at this writing, particularly with regard to the fine structural details of the dehydrated phase.

Wetherillite $Na_2(Mg,Zn)(UO_2)_2(SO_4)_4 \cdot 18H_2O$

Wetherillite was described from the Blue Lizard mine, San Juan Co., Utah, where it occurs as a secondary alteration phase in association with bobcookite, boylite, chalcanthite, dietrichite, gypsum, hexahydrite, johannite, pickeringite, and rozenite. It forms pale greenish-yellow prisms or blades with irregular terminations, up to ~1 mm long, in subparallel intergrowths, divergent sprays, and jackstraw aggregates. Crystals are transparent, with vitreous luster; no UV fluorescence was reported. Wetherillite is easily soluble in water (Kampf et al. 2015b).

Wetherillite is monoclinic with a=20.367, b=6.833, c=12.903 Å, and β=107.879. The structure contains $[(UO_2)(SO_4)_2(H_2O)]$ sheets parallel to {100}. Edge-sharing chains of $Na(H_2O)_5O$ polyhedra link adjacent uranyl sulfate sheets forming a weakly bonded three-layer sandwich of composition $[Na_2(UO_2)_2(SO_4)_4(H_2O)_8]^{2-}$. The sandwich layers are linked to one another by hydrogen bonds through insular $Mg(H_2O)_6$ octahedra and isolated H_2O groups. Features in the Raman spectrum were assigned to O-H stretching and H-O-H bending modes of free or weakly bonded H_2O molecules, SO_4 stretching modes, UO_2 stretching and bending modes, and Na-O stretching modes.

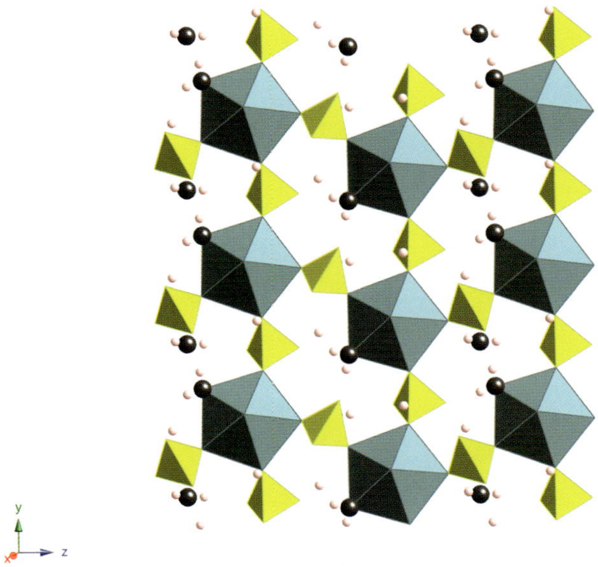

Figure 322. The uranyl sulfate sheet in wetherillite, viewed normal to the sheet.

Figure 321. Wetherillite prisms and blades from the Blue Lizard mine, Utah; FOV is 3 mm wide. Reproduced with the kind permission of the Mineralogical Society of Great Britain and Ireland from a paper by Kampf et al. (2015b)

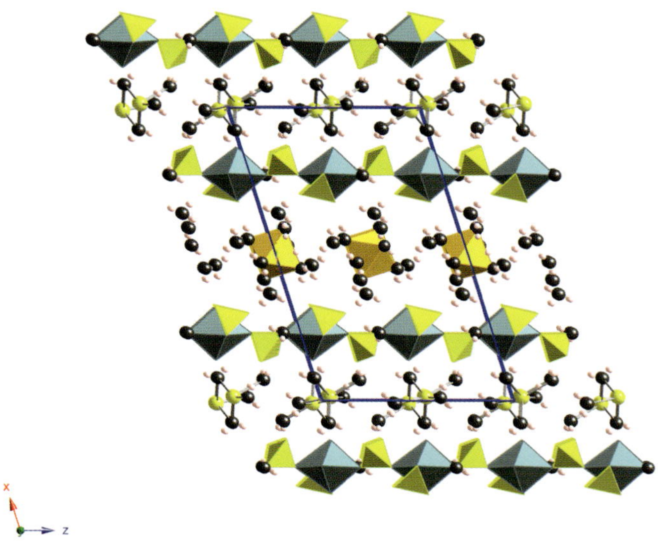

Figure 323. The complete wetherillite structure, viewed along [010], showing two $[Na_2(UO_2)_2(SO_4)_4(H_2O)_8]^{2-}$ sandwich layers, with Na shown as translucent bright yellow polyhedra. These layers are bonded to each other via $Mg(H_2O)_6$ octahedra (gold).

SELENITES, TELLURITES, AND ARSENITES

These minerals are treated together in this section because of the similar behavior of selenium, tellurium, and arsenic in many geochemical processes. The individual minerals in this section show a number of structural motifs and have formed in a range of geological settings.

Table 47. Uranyl selenites, tellurites, and arsenites

Species	Formula	Symmetry
Selenites		
Demesmaekerite	$Cu_5Pb_2(UO_2)_2(SeO_3)_6(OH)_6 \cdot 2H_2O$	tric
Derriksite	$Cu_4UO_2(SeO_3)_2(OH)_6 \cdot H_2O$	orth
Guilleminite	$Ba(UO_2)_3(SeO_3)_2O_2 \cdot 3H_2O$	orth
Haynesite	$(UO_2)_3(OH)_2(SeO_3)_2 \cdot 5H_2O$	orth
Larisaite	$Na(H_3O)(UO_2)_3(SeO_3)_2O_2 \cdot 4H_2O$	mon
Marthozite	$Cu(UO_2)_3(SeO_3)_2O_2 \cdot 8H_2O$	orth
Piretite	$Ca(UO_2)_3(SeO_3)_2(OH)_2 \cdot 4H_2O$	orth
Tellurites		
Cliffordite	UTe_3O_9	cub
Moctezumite	$Pb(UO_2)(TeO_3)_2$	mon
Schmitterite	$UO_2(TeO_3)$	orth
Markcooperite	$Pb_2(UO_2)Te^{6+}O_6$	mon
Arsenite		
Chadwickite	$H(UO_2)(AsO_3)$	tet

Selenites

Uranyl selenite minerals form where selenium-bearing sulfide minerals are undergoing oxidation and dissolution; in all these minerals, selenium is in the tetravalent state as the selenite ion, $(SeO_3)^{-2}$. Most uranyl selenites are found at the Musonoi deposit, Congo, where the source of selenium is the oxidation of Se-rich digenite (Deliens et al. 1981).

Demesmaekerite $Cu_5Pb_2(UO_2)_2(SeO_3)_6(OH)_6 \cdot 2H_2O$
Demesmaekerite was described from the Musonoi mine, Kolwezi, Congo (Cesbron, Bachet, and Oosterbosch 1965). It forms triclinic crystals that are elongated on [001] and flattened on (100). The crystals are bottle green when fresh; with dehydration the color darkens to olive green or brownish and the material becomes more opaque. It is not fluorescent in UV.

Figure 325. SEM photo showing the typical morphology of demesmaekerite crystals. RJL815

Figure 324. Radiating group of dark green demesmaekerite crystals from Musonoi, Kolwezi, Congo. RJL815

Figure 326. Partially dehydrated demesmaekerite forming a cluster of flattened, olive-green crystals in a vug in massive ore from Musonoi, Congo. FOV is ~1 cm wide. RJL4653

Figure 327. Yellow schoepite pseudomorph after demesmaekerite crystals, from Musonoi, Kolwezi, Congo. The crystal group is about 6 mm tall. RJL1699

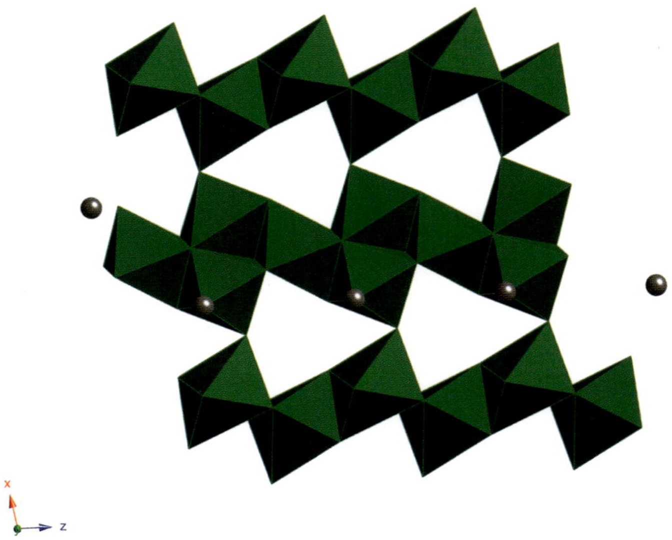

Figure 328. The copper octahedral sheet structure of demesmaekerite, viewed along [101]. Gray atoms represent H_2O.

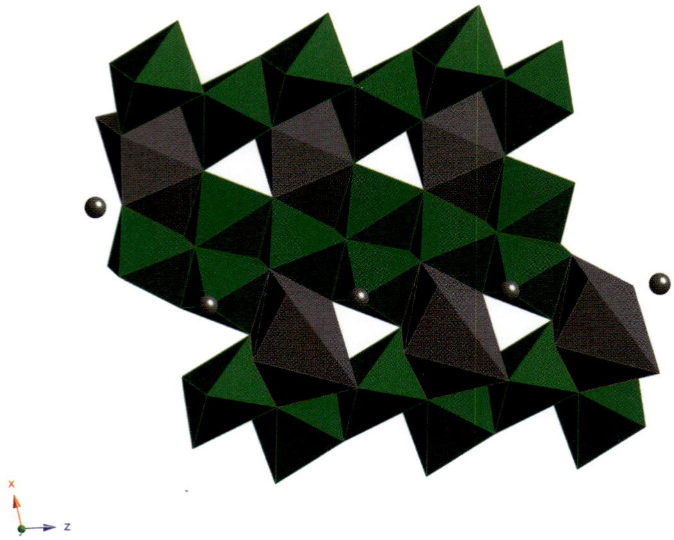

Figure 329. The octahedral sheet with Pb polyhedra (gray) added.

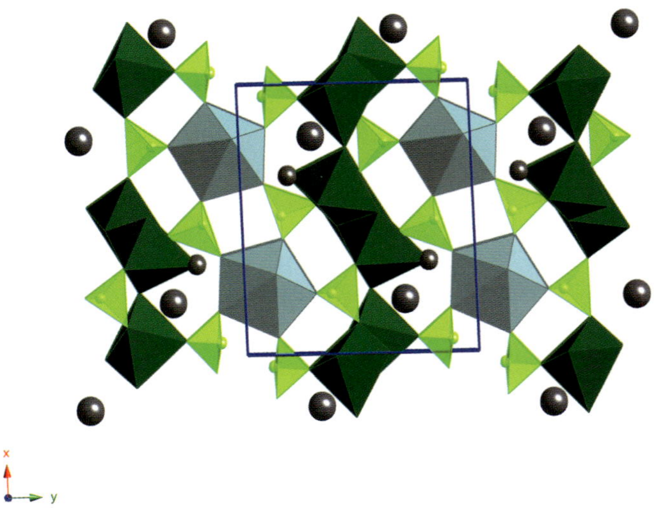

Figure 330. Crystal structure of demesmaekerite, viewed along [001]. The SeO_3 groups (bright green) consist of triangular pyramids with the Se atom at the apex. These share oxygens with the uranyl polyhedra and with the Cu sheets. In this view, Pb are shown as large gray atoms and H_2O are smaller gray oxygen atoms.

Demesmaekerite forms in the lower part of the oxidation zone at the Musonoi mine, in association with cuprosklodowskite, kasolite, malachite, derriksite, guilleminite, and chalcomenite in Se-rich digenite. Pseudomorphs of schoepite after demesmaekerite have also been found at Musonoi.

Demesmaekerite is triclinic with a=11.955, b=10.039, c=5.639 Å, α=89.78, β=100.36, and γ=91.34°. The structure consists of layers of $Cu(O,OH,H_2O)_6$ octahedra lying parallel to (010) that are linked together by oblique chains formed by oxygen bridges that link uranyl and SeO_3 groups (Ginderow and Cesbron 1983a).

Derriksite $Cu_4UO_2(SeO_3)_2(OH)_6 \cdot H_2O$

Derriksite is one of the rarer of the uranyl selenites in the Musonoi deposit, Congo (Cesbron, Pierrot, and Verbeek 1971). It forms minute euhedral crystals less than about 0.7 mm across, which are elongated on [001] and flattened to slightly rounded on (100), with good cleavage on (010). The color ranges from clear green to malachite green and rarely bottle green. It is not fluorescent in UV.

Derriksite forms in the lower part of the oxidation zone in association with dark green demesmaekerite, acicular yellow guilleminite, green to yellow marthozite, and blue chalcomenite. As are the associated species, derriksite is an alteration product of primary uraninite and selenian digenite.

Derriksite is orthorhombic with a=5.570, b=19.088, and c=5.965 Å. The structure consists of $Cu(O,OH)_6$ octahedra in layers parallel to (010). The octahedral sheets are linked together by uranyl selenite chains (Ginderow Cesbron 1983b).

Figure 331. Very dark green derriksite microcrystals, forming a patch ~2 mm across, on a thin crust of apple-green cuprosklodowskite lining a small vug in massive malachite ore, from Musonoi, Congo. RJL4635

Figure 332. Green micro crystals of derriksite on acicular yellow guilleminite, from Musonoi, Congo. RJL688

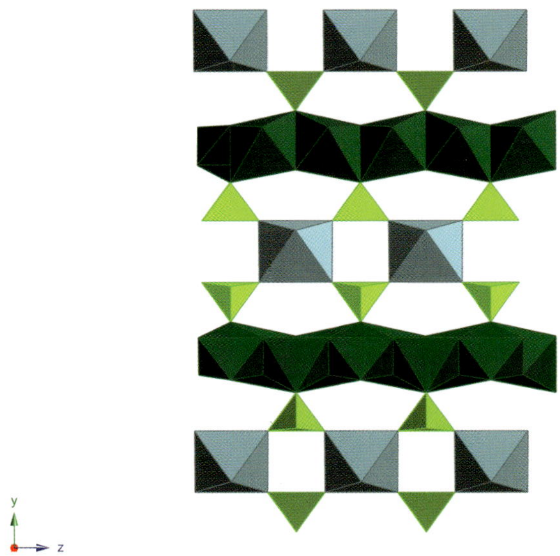

Figure 333. The derriksite structure, viewed along [100], showing corrugated sheets of octahedrally coordinated Cu (dark green). Adjacent sheets are bonded via chains of unit composition $UO_2(SeO_3)_2$ that run parallel to [001].

Guilleminite $Ba(UO_2)_3(SeO_3)_2O_2{\cdot}3H_2O$

Guilleminite was described from the Musonoi deposit, Congo, where it forms canary-yellow powdery to microcrystalline coatings, silky to fibrous masses, and small, spherical nodules in cavities in the host rock. The crystals are typically opaque to translucent, rarely transparent in very small crystals. The tabular crystals are elongated parallel to the *z*-axis and flattened on (010). Guilleminite is not fluorescent in UV (Pierrot, Toussaint, and Verbeek 1965).

Guilleminite occurs in the oxidation zone at Musonoi, in association with malachite, wulfenite, and other secondary uranium minerals, including: uranophane, kasolite, vandenbrandeite, cuprosklodowskite, and meta-torbernite. It is also frequently associated with the other uranyl selenites.

Guilleminite has also been found occasionally at Shinkolobwe, Congo, where it is associated with hydrated uranium oxides, rutherfordine, and uranophane.

Guilleminite is orthorhombic with *a*=7.084, *b*=7.293, and *c*=16.881 Å. The structure comprises U atoms on two distinct sites, one with (UO_8) coordination and the other with (UO_7) coordination. The UO_7 groups are joined into edge-sharing dimers, which in turn share edges with UO_8 groups to form (U_3O_{14}) chains. These chains run parallel to [100] and are cross-linked in the [001] direction by SeO_3 groups, forming $[(UO_2)_3(SeO_3)_2O_2]$ sheets that lie parallel to (010). These sheets are then linked in the *y*-direction by the Ba atoms as well as by hydrogen bonding involving H_2O that is also located in the interlayer region. Interestingly, the sheets have an identical topology to the sheets in phosphuranylite, with SeO_3 groups substituting for PO_4 (Cooper and Hawthorne 1995).

Figure 334. Bright yellow guilleminite with platy, olive-green marthozite, on Se-rich digenite, from Musonoi, Congo. RJL1140

Figure 336. The $[(UO_2)_3(SeO_3)_2O_2]$ sheet structure of guilleminite viewed along [010].

Figure 335. Yellow, powdery guilleminite with malachite from Musonoi, Congo. RJL341

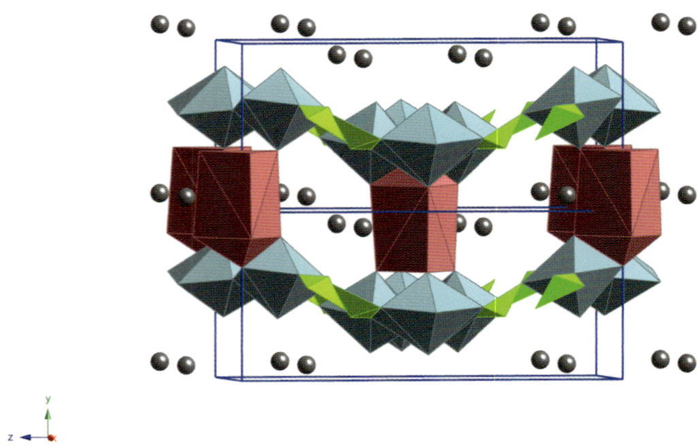

Figure 337. The guilleminite structure viewed just off [100] showing the uranyl selenite sheets, which are in turn bonded together via Ba (magenta polyhedra) and some hydrogen bonding from the water (gray) in the interlayer. Two unit cells are shown.

Haynesite $(UO_2)_3(OH)_2(SeO_3)_2 \cdot 5H_2O$
Haynesite was described based on materials collected by Patrick Haynes at the Repete Mine, San Juan Co., Utah (Deliens and Piret 1991). It forms transparent amber-yellow vitreous crystals, as tablets elongated on [001] or as acicular prismatic rosettes to several milimeters in diameter. The mineral forms thin crusts on mudstones and sandstones of the Morrison formation. When the crystals are very small, the material forms yellowish smears and tiny felted masses; microscopic examination at 20× reveals minute, pale yellow fibrous crystals. At the larger sizes, the crystals are deep golden yellow to amber and are clearly visible to the naked eye. Associated minerals at the TL include andersonite, boltwoodite, larisaite, ferroselite, uraninite, marcasite, and gypsum.

The Repete mine is described in detail by Haynes (1993), who notes that haynesite was found at only three places in the mine. At one, gypsum and ferroselite are the only visible associated minerals. At the other two places, haynesite is associated with andersonite, boltwoodite, ferroselite, gypsum, marcasite, and selenium.

Haynesite is orthorhombic with a=8.025, b=17.43, and c=6.935 Å. A complete structure refinement has not been reported to date.

Figure 338. A typical example of haynesite from the TL at the Repete mine, Utah. To the eye it appears to form earthy to slightly felted masses on the surface, but magnification reveals the minute golden fibrous crystals. RJL1715

Figure 339. Flattened rosettes of haynesite, ~1 mm in diameter, on mudstone from the Repete mine, Utah. This habit is characteristic of crystallization within very thin seams in the host rock. RJL3149

Figure 340. Detail of the specimen shown in the previous photo. RJL3149

Larisaite $Na(H_3O)(UO_2)_3(SeO_3)_2O_2 \cdot 4H_2O$

Larisaite is a newly described mineral discovered by reexamining materials found in a collection from the Repete Mine, San Juan Co., Utah (Chukanov et al. 2004). The mineral forms yellow to yellow-green tabular or lath-like plates to ~1 mm long and rosettes to ~2 mm wide, in association with montmorillonite, quartz, haynesite, wolsendorfite, andersonite, uranophane, gypsum, and calcite. The crystals are fluorescent green under SWUV.

Larisaite is monoclinic with a=6.981, b=7.646, c=17.249 Å, and β=90.039°. The structure, like that of guilleminite, consists of $[(UO_2)_3(SeO_3)_2O_2]$ layers arranged parallel to (010). Within the interlayer are Na^+, K^+, and H_3O^+ ions as well as water molecules. Both the Na^+ and the K^+ sites are partially occupied. The K^+ ions occupy the same sites as does Ba^+ in guilleminite, whereas the Na ions occupy a new site with shorter interatomic distances. Water and H_3O^+ ions are distributed over six sites in the interlayer, arranged so that contacts between H_3O^+, K^+, and Na^+ are avoided (Chukanov et al. 2004).

Figure 341. Radiating sprays of amber larisaite crystals on matrix from the TL at the Repete mine, Utah; material studied by the authors of the original description. RJL2561

Figure 342. Another larisaite from the Repete mine, Utah, here forming a yellow crust composed of spherical aggregates of microscopic crystals on matrix; FOV is ~5 mm. RJL3677

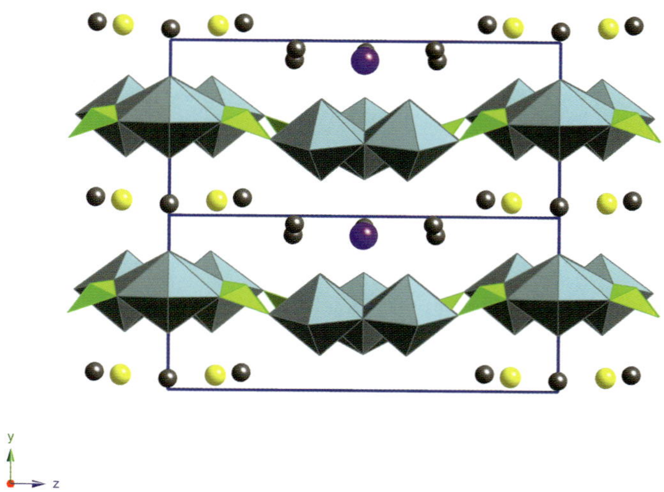

Figure 343. Crystal structure of larisaite, showing $[(UO_2)_3(SeO_3)_2O_2]$ layers lying parallel to (010). The interlayer contains Na^+ (yellow), K^+ (purple), and water and hydronium ions (gray). Two unit cells are shown.

Marthozite $Cu(UO_2)_3(SeO_3)_2O_2 \cdot 8H_2O$

Marthozite was described from the Musonoi mine, Kolwezi, Congo (Cesbron, Oosterbosch, and Pierrot 1969). It forms distinctive triangular or truncated pyramidal crystals that are flattened on (100); the (100) faces are typically striated parallel to [001]. The color ranges from yellowish green to greenish brown. Marthozite is transparent when fresh, becoming opaque upon dehydration; at room temperature it spontaneously dehydrates to a meta-phase. It is not fluorescent in UV.

Marthozite forms in the lower part of the oxidation zone at the Musonoi mine, in association with cuprosklodowskite, kasolite, malachite, derriksite, demesmaekerite guilleminite, sengierite, and chalcomenite.

Marthozite is orthorhombic with a=6.988, b=16.454, and c=17.223 Å. The structure comprises sheets of $[(UO_2)_3(SeO_3)_2O_2]$ like those in guilleminite. Neighboring sheets are linked by Cu^{2+} ions in the interlayer and also by

Figure 344. Marthozite crystals to ~3 mm, scattered on massive digenite ore from Musonoi, Congo. Here, the pearly luster parallel to the sheet structure can be seen. RJL4536

hydrogen bonding that involves both the H_2O groups that are coordinated to the Cu and interstitial H_2O groups that are not bonded to any cations. The hydrogen bonding arrangement in marthozite is very different from that in guilleminite. In marthozite, there are eight unique interlayer H_2O groups; four of these H_2O groups bond directly to Cu^{2+} and four are held in the structure solely by H bonds. In guilleminite, there are two unique interlayer H_2O groups, both of which bond to the interlayer Ba atoms. Despite the fact that the Cu^{2+} ions are much smaller than the Ba^+ ions in guilleminite, the sheets in marthozite are somewhat farther apart (8.227 vs. 7.293 Å) because of the different positioning of the interlayer ions (Cooper and Hawthorn 2001).

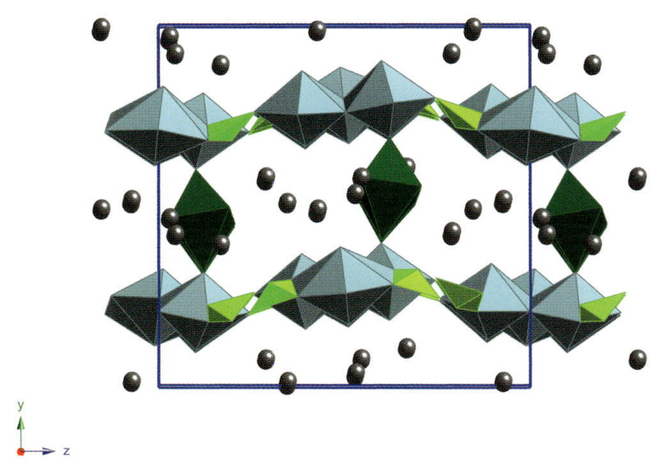

Figure 345. Crystal structure of marthozite viewed along [100], showing the $[(UO_2)_3(SeO_3)_2O_2]$ sheets, bonded together by Cu (dark green polyhedra) and by hydrogen bonding from the water groups (gray).

Piretite $Ca(UO_2)_3(SeO_3)_2(OH)_4 \cdot 4H_2O$

Piretite was described at the Shinkolobwe mine, Congo. It forms lemon yellow, elongated tabular crystals, generally

2–3 mm long and 1 mm wide. It is transparent to translucent, with a pearly luster, and is not fluorescent. Crystals are flattened on (001) with good cleavage on {001}; they are typically of poor quality and occur as curved sheets.

It typically forms crusts on the surface of uraninite crystals in association with masuyite or other Pb-uranyl oxyhydroxide species, and is the product of alteration of primary uraninite and primary Se-rich sulfides such as penroseite, with Ca derived from the host rocks.

Piretite is orthorhombic with a=7.010, b=17.135, and c=17.606 Å (Vochten et al. 1996).

Figure 346. Cliffordite (yellow earthy blebs) with mackayite (olive green), on matrix from Moctezuma, Sonora, Mexico. Mackayite is $Fe^{3+}Te_2O_5(OH)$. RJL338

Tellurites

All three of the known uranyl tellurite minerals are found at the San Miguel and Moctezuma mines in Moctezuma, Sonora, Mexico. One of them, cliffordite, also occurs at Shinkolobwe in the Congo. These minerals form where sulfide minerals are undergoing oxidation and the Te is presumably an impurity in the sulfides. In contrast to the uranyl selenites, all of the uranyl tellurites are anhydrous. A uranyl tellurate, markcooperite, was described from two sites on Otto Mountain, California. It has a unique structure and no close affinities to the uranyl tellurites, nor is it the Te analog of any known uranyl sulfates or arsenates. It is treated in this section because it shares a somewhat similar paragenesis with the tellurites.

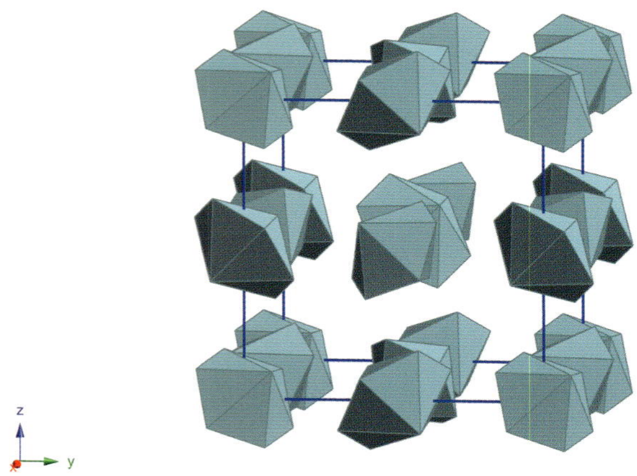

Figure 347 The locations of uranyl polyhedra in the unit cell of cliffordite. As is often the case, the polyhedra are flattened somewhat because the U-O bond distances are about 2 Å in the axial direction (the uranyl ion itself) whereas the U-O bond distances to the six "equatorial" oxygens are 2.3 to 2.5 Å.

Cliffordite UTe_3O_9

Cliffordite is one of two minerals named in honor of Clifford Frondel (1907–2002), in recognition of his profound contributions to uranium mineralogy. It forms minute (0.2 mm) bright sulfur-yellow octahedral crystals along joint surfaces in the oxidized zone of a tellurium-gold-silver deposit at the San Miguel mine, near Moctezuma, Sonora, Mexico, in association with mackayite, barite, quartz, and limonite. Gaines (1969) describes the geological setting. "The San Miguel Mine is about 10 km northwest of the Moctezuma Mine in an environment of tertiary andesite porphyrys and altered tuffs. Quartz veins cutting these rocks sometimes contain substantial amounts of barite and pyrite, and in local shoots may be mineralized with native tellurium and other tellurides including gold and silver minerals. Oxidation of these veins has given rise to a number of oxidized tellurium minerals, especially mackayite, besides abundant jarosite and limonite and other iron oxides coating fractures. Minerals recognized in very minor amounts include tetrahedrite, rickardite, emmonsite, tellurite, schmitterite (a new uranyl tellurite, the description of which is in preparation), cliffordite, and cerargyrite."

Cliffordite also occurs in small amounts at the nearby Moctezuma mine in association with native tellurium and paratellurite.

Cliffordite is cubic with a=11.370 Å and Z=8. The structure contains 8-coordinated uranium, in which the uranyl ion and six additional oxygens form a flattened polyhedron. The TeO_3 groups form distorted tetrahedra (Galy and Meunier 1971).

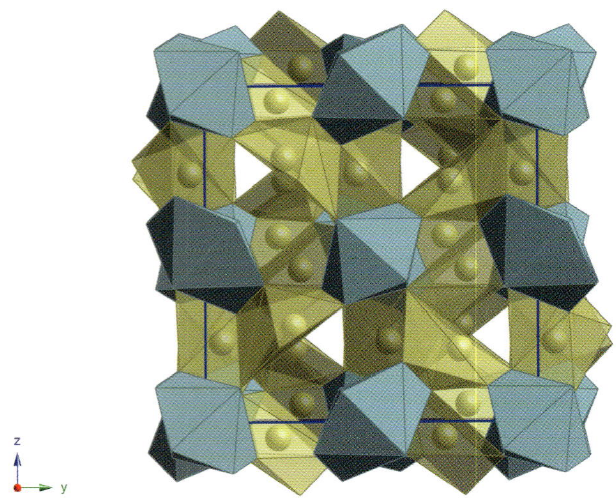

Figure 348. The cliffordite structure viewed along [100] showing the connectivity of the UO_8 and TeO_3 (pale gold) polyhedra.

Moctezumite $Pb(UO_2)(TeO_3)_2$

Moctezumite was described from the Moctezuma mine, Moctezuma, Sonora, Mexico, in association with zemannite, burkhardtite, schmitterite, other tellurides, and pyrite. This mine is also sometimes referred to as "La Bambolla." Moctezumite forms small, thin-bladed crystals up to 3 mm long and 1 mm wide, but generally smaller. The crystals are elongated parallel to the [010] and flattened parallel to (201); most crystal faces are curved. There is a perfect cleavage on (100). Fresh material is bright orange, turning dark to brownish-orange with alteration. It is reported that moctezumite may alter pseudomorphously to schmitterite (Gaines 1965).

Moctezumite is monoclinic with a=7.813, b=7.061, c=13.775 Å, and β=93.71°. The dominant structural theme is a layer structure parallel to (001). Within each layer the uranyl bipyramids form zigzag chains parallel to the y-axis by sharing edges. These chains alternate with chains of edge-sharing Pb coordination polyhedra. The Pb-U layers, stacked along [001], are bonded by Te_2O_6 units composed of TeO_3 and TeO_4, which share an oxygen atom (Swihart et al. 1993).

Figure 349. Orange moctezumite crystals in a cavity on drusy quartz, from the TL at Moctezuma, Mexico. FOV is ~3 mm wide. RJL2802

Figure 350. A spray of moctezumite ~1 mm wide, that has altered to a pale yellow phase, possibly schmitterite. RJL2802

Figure 351. The layer structure of moctezumite viewed along [001], showing chains of Pb (translucent gray) and U (blue gray) polyhedra lying parallel to [010].

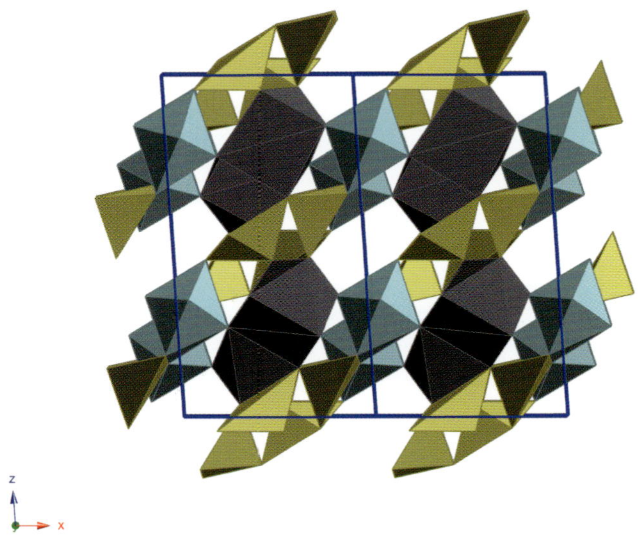

Figure 352. The moctezumite structure viewed along [010] showing the TeO_3 and TeO_4 groups (gold polyhedra) connecting adjacent layers. Two unit cells are shown.

Schmitterite $UO_2(TeO_3)$

Schmitterite was described from the Moctezuma mine ("La Bambolla"), Moctezuma, Sonora, Mexico, where it occurs sparsely in association with emmonsite, moctezumite, and other tellurites; it is also known from the San Miguel mine in the Moctezuma district. It forms minute bladed crystals, typically elongated on [100] and flattened on {001} or as small micaceous rosettes up to about 1 mm diameter. The crystals are very pale straw yellow, with a pearly luster. As noted above, it occasionally forms pseudomorphs after moctezumite (Gaines 1971).

Schmitterite has also been found at Shinkolobwe, Congo, dispersed on material containing uraninite, siegenite, covellite, chalcosite, and pyrite; the schmitterite rosettes are associated with orange-yellow coatings of meta-saleeite (Deliens et al., 1981).

Schmitterite is orthorhombic with a=10.161, b=5.563 and c=7.862 Å. The structure is based on chains of edge-sharing uranyl pentagonal bipyramids runing parallel to [001]. The chains are linked by TeO_4 polyhedra (Meunier and Galy 1973).

Figure 353. Pale yellow to cream schmitterite forming small, earthy rounded aggregates to ~0.5 mm, on porous quartz, from the TL at Moctezuma, Mexico. RJL4625

Figure 354. Crystal structure of schmitterite, viewed along [100], showing the uranyl chains lying parallel to [001].

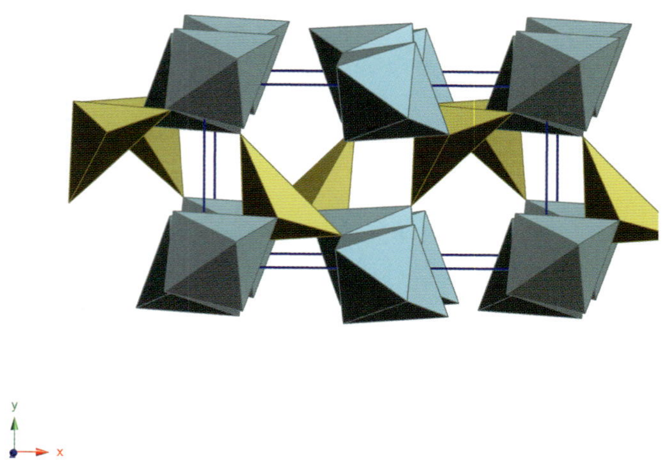

Figure 355. The complete structure of schmitterite, viewed just off [001], showing the connectivity between the TeO_4 groups and the uranyl chains.

Markcooperite $Pb_2(UO_2)Te^{6+}O_6$

Markcooperite was described from the Aga mine on Otto Mountain, ~2 km northwest of Baker, California; it was also found nearby in the Bird Nest drift on the south flank of the mountain. It forms orange, pseudotetragonal prisms to 0.2 mm, and botryoidal intergrowths to 0.3 mm. It has an adamantine luster and is not fluorescent.

It occurs in small vugs and fracture surfaces in brecciated quartz veins in association with Br-rich chlorargyrite, iodargyrite, khinite-*4O*, wulfenite, and several new tellurites, including housleyite, thorneite, ottoite, and timroseite. The complex assemblage is interpreted to have formed via partial oxidation of primary sulfides and arsenides (e.g., galena and hessite) during or after brecciation of the quartz veins. The source of the U hasn't been determined (Kampf et al. 2010).

Markcooperite is monoclinic with a=5.722, b=7.748, c=7.889 Å, and β=90.833°. The structure is based on sheets of corner-sharing uranyl square bipyramids and tellurate octahedra, with Pb atoms between the sheets. Markcooperite is the first compound to show Te^{6+} substitution for U^{6+} within the same crystallographic site. It is structurally related to synthetic $Pb(UO_2)O_2$.

Figure 356. Minute orange films and aggregates of markcooperite on colorless quartz, from the TL at the Aga mine, California. RJL4027

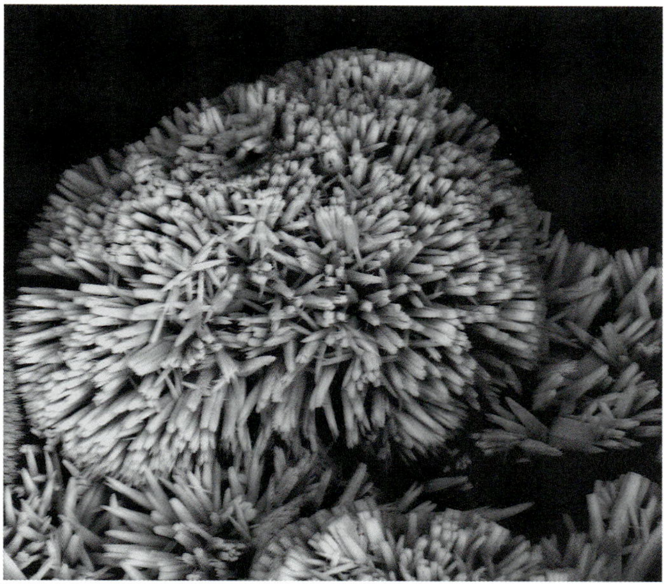

Figure 357. SEM image (backscatter mode) of a cluster of markcooperite crystals. FOV is 30 µm wide.

Figure 358. SEM image of the surface of a botryoidal aggregate of markcooperite crystals. FOV is 75 µm wide. [SEM photos reproduced with the kind permission of the Mineralogical Society of America from a paper by Kampf et al. (2010)].

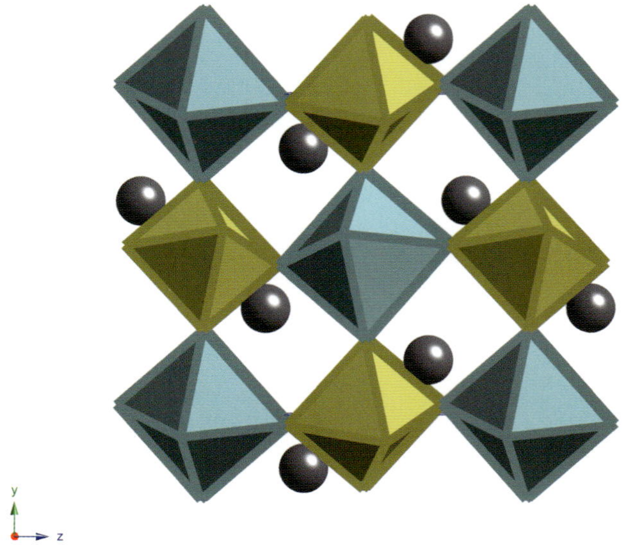

Figure 359 The structure of markcooperite viewed along [100], to emphasize the uranyl tellurite sheet formed by corner-sharing UO_6 square bipyramids (blue-gray) and TeO_6 octahedra (gold). Note, however, that the center polyhedron actually contains some Te (~$U_{.75}Te_{.25}$), which closely matches the empirical formula $Pb_{2.05}U_{.80}Te_{1.18}O_{7.99}Cl_{.01}$.

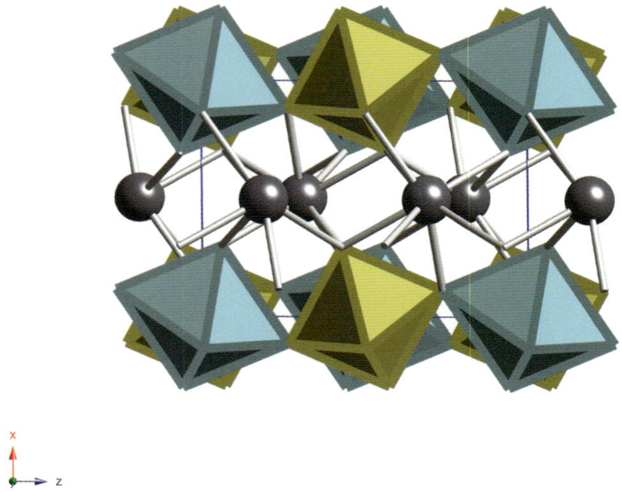

Figure 360. The structure viewed along [010] to show how the sheets are bonded together by sharing oxygens with the Pb (gray atoms with stick bonds).

Arsenite

An anhydrous uranyl arsenite mineral, chadwickite, was described from a mine dump in the central Black Forest, Germany. It does not appear to have any close relationship to other species.

Chadwickite $H(UO_2)(AsO_3)$

Chadwickite was described from the Sophia Mine near Wittichen, central Black Forest, Germany. It forms yellow, earthy to scaly crusts and minute single grains up to about 20 µm. Single grains are partly transparent. The grains are tetragonal with a tabular habit on {001}, occasionally showing a rectangular or square form and a "more or less perfect" cleavage on {001}. It is not fluorescent. It was found on the mine dump on uraninite, in association with metakahlerite, abernathyite, erythrite, pitticite, and another uranyl arsenite.

Chadwickite is tetragonal with a=11.00 and c=15.96 Å, based on powder XRD data and comparison with the same composition synthesized at room temperature (Walenta 1998).

MOLYBDATES, TUNGSTATES, NIOBATES, AND TITANATES

Uranyl molybdates are important ore minerals at some localities and are also common accessory minerals in roll-front deposits and other settings where both uraninite and Mo-bearing species are being weathered. In these minerals the Mo is present as Mo^{+6} and the U may be of mixed valence, U^{+4} and U^{+6}. It is likely that these minerals are more abundant than commonly believed, because of the fact that they are often difficult to distinguish in the field and are often intergrown with one another in fine-grained masses (Finch and Murakami 1999). The uranyl molybdates include calcurmolite, cousinite, deloryite, iriginite, moluranite, tengchongite, and umohoite.

There is only one uranyl tungstate, uranotungstite, which has been reported from several localities in Germany.

Carlosbarbosaite is the only naturally occurring uranyl niobate described to date, although several other compounds with related structures have been synthesized.

Holfertite is, at present, the only naturally occurring uranyl titanate. Its structure is related to that of carlosbarbosaite.

Table 48. Uranyl molybdates, tungstates, niobates, and titanates

Species	Formula	Symmetry
Molybdates		
Calcurmolite	$Ca(UO_2)_3(MoO_4)_3(OH)_2 \cdot 11H_2O$	mon
Cousinite	$Mg(UO_2)_2(MoO_4)_2(OH)_2 \cdot 5H_2O$	?
Deloryite	$Cu_4(UO_2)(MoO_4)_2(OH)_6$	mon
Iriginite	$[(UO_2)Mo_2O_7(H_2O)_2](H_2O)$	orth
Moluranite	$H_4U^{4+}(UO_2)_3(MoO_4)_7 \cdot 18H_2O$	amorph
Mourite	$U^{4+}Mo^{6+}{}_5O_{12}(OH)_{10}$	mon
Sedovite	$U(MoO_4)_2$	orth
Tengchongite	$Ca(UO_2)_6(MoO_4)_2O_5 \cdot 12H_2O$	orth
Umohoite	$[(UO_2)MoO_4(H_2O)](H_2O)$	tric
Tungstate		
Uranotungstite	$(Fe^{2+},Ba,Pb)(UO_2)_2(WO_4)(OH)_4 \cdot 12H_2O$	orth
Niobate		
Carlosbarbosaite	$(UO_2)_2Nb_2O_6(OH)_2 \cdot 2H_2O$	orth
Titanate		
Holfertite	$U^{6+}{}_{1.75}TiO_{7.5}Ca_{0.25}(H_2O)_3$	hex

Calcurmolite $Ca(UO_2)_3(MoO_4)_3(OH)_2 \cdot 11H_2O$
Calcurmolite was discovered in the early 1950s at a U-Mo deposit that Pekov (1998) identifies as the Sokh-Karasu area of the Kadzharan molybdenum deposit, upper Kokch River, Kafan district, Armenia. A preliminary description was given by L. S. Rudnitskaya, who tentatively called it "kadzharanite." Rudnitskaya (1959) published somewhat more details, including XRD data and chemical analysis, referring to the mineral as "calcium uranium molybdate," with the formula $Ca(UO_2)_3(MoO_4)_3(OH)_2 \cdot 8H_2O$. The name calcurmolite was given by A. S. Povarennykh in 1962 (perhaps to avoid an overt identification of the locale). Fedorov (1963) refined the formula using material discovered at a second locale, the Kyzylsai Mo-U deposit, Kazakhstan. Calcurmolite forms honey-yellow prismatic crystals to about 1.5 mm and massive to platy aggregates. UV fluorescence is strong yellowish-green. The mineral occurred in chalcedony veins, associated with uranophane, uranospinite, halloysite, betpakdalite, and ferrimolybdite in the oxidation zone, sometimes forming pseudomorphs after uraninite or uranophane.

It has also been reported from Tongbiguan village, Yunnan Province, China, and from the Rabejac and Mas d'Alary deposits, Herault, France. At Rabejac, it forms bright yellow spherules to ~0.5 mm on sandstones and pelites, associated with uranophane and brick-red nodules of umohoite (Deliens 1992).

Recent microanalytical studies of calcurmolites from Kadzharan, Armenia; and Kyzylsai, Kazakhstan (Sidorenko et al. 2005); suggest that calcurmolite may contain significant Na and somewhat more water than previously reported, leading to the proposed formula $(Ca,Na)_2(UO_2)_3Mo_2(O,OH)_{11} \cdot nH_2O$, where n = 10–15 and Na ≤ 0.6 apfu. A pure Ca end member composition was not found among the specimens examined. Microscopic examination also suggested that the mineral might have formed by the pseudomorphic replacement of uranophane by molybdate.

The microscopic fibrous texture of calcurmolite crystals, intergrown with uranophane, has thus far precluded a full structure refinement. Sidorenko et al. (2005), based on powder XRD patterns, report a monoclinic unit cell with a=16.30, b=25.49, c=19.50 Å, and β=90°7'.

Figure 361. Bright yellow crusts and small rounded masses of calcurmolite from Mas d'Alary, Herault, France. RJL4424

Cousinite $Mg(UO_2)_2(MoO_4)_2(OH)_2 \cdot 5H_2O$

Cousinite is an inadequately described mineral noted at Shinkolobwe, Congo, where it was found as an alteration product of ore that contained molybdenite and uraninite. It formed thin, black, bladed crystals with highly vitreous luster. Neither the structure nor the physical properties were determined, and the formula was based on chemical analysis of 270 mg of material. The derivation of the name was not given (Vaes 1958). Cousinite is regarded by IMA as a questionable or doubtful species.

Figure 362. An old sample from the US National Museum, Division of Mineralogy No. 140639, labeled "cousinite w. uraninite" from Shinkolobwe. The cabinet-sized mass of uraninite is extremely radioactive, and has various surface alteration products, along with areas of what might be described as "black bladed crystals." Is this possibly an example of that poorly described mineral? Further research is clearly needed! RJL4636

Deloryite $Cu_4(UO_2)(MoO_4)_2(OH)_6$

Deloryite was described from the "Zeunerite zone" in the southern part of the Cap Garonne mine, near Le Pradet, Var, France, where it occurred on a dendritic quartz gangue in association with metazeunerite, atacamite, paratacamite, malachite, tourmaline, and barite. It forms dark green to black crystals up to 3 × 1 × 0.3 mm and rosettes to about 6 mm. The crystals tend to be tabular with a rounded termination; they are transparent to nearly opaque with a vitreous to greasy luster, and are not fluorescent (Sarp and Chiappero 1992).

Deloryite is monoclinic, with a=19.94, b=6.116, c=5.520 Å, and β=104.18°. It is structurally related to derricksite, $Cu_4(UO_2)(SeO_3)_2(OH)_6$. The structure consists of sheets of distorted edge-sharing $Cu(O,OH)_6$ octahedra parallel to (100). The octahedral layers are linked by columns, built up of alternating pseudotrigonal UO_6 bipyramids and pseudotrigonal MoO_5 bipyramids that share edges to form Mo_2O_8 dimers (Pushcharovsky, Rastsvetaeva, and Sarp 1996).

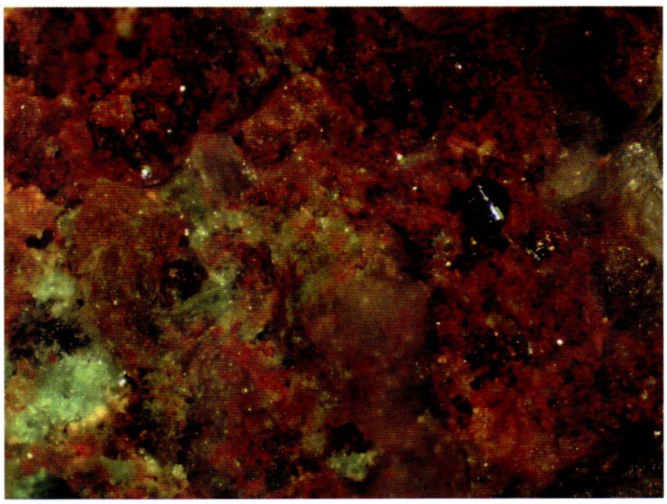

Figure 363 Tiny black crystal of deloryite, on iron-stained matrix with green corroded zeunerite, from the TL at the Cap Garonne mine, near Le Pradet, France. RJL3813

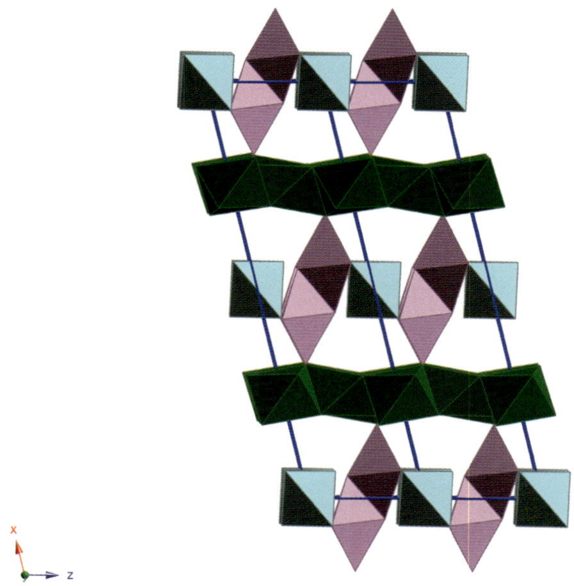

Figure 364. Crystal structure of deloryite showing sheets of octahedrally coordinated Cu (dark green), joined via UO_6 bipyramids (blue gray) and Mo_2O_8 dimers (magenta).

Iriginite $[(UO_2)Mo_2O_7(H_2O)_2](H_2O)$
Iriginite was found at the Aleksandrovskii Golets Mo-U deposit, Udokan Range, Northern Transbaikal region, Russia, by G. Yu. Epshtein in 1951 (Pekov 1998). At the TL, iriginite forms fine-grained, dense yellow aggregates, sometimes as pseudomorphs after brannerite, in granulated albitite in association with other U-Mo species (Epshtein 1959; Soboleva and Pudkovkina 1957).

It is one of the most abundant secondary minerals in the oxidation zones of many hydrothermal U-Mo deposits. Reported localities include: the Kyzylsai Mo-U deposit, Kazakhstan; the Rabejac deposit, Herault, France; the Hervey's Range deposit, Queensland, Australia; the White River area, Pennington Co., South Dakota (Stephenson 1964); in San Juan Co., Colorado; Cerro Picacho, Sonora, Mexico; and Shinkolobwe, Congo.

Iriginite is orthorhombic, with a=6.705, b=12.731, and c=11.524 Å. The crystal structure, determined using synthetic material, consists of individual uranyl ions coordinated by five additional O atoms to form a pentagonal bipyramid. Individual Mo^{6+} ions are coordinated by five O atoms and one H_2O group to form a distorted octahedron. The U and Mo coordination polyhedra are linked by edge sharing to form charge-balanced sheets of composition $[(UO_2)Mo_2O_7(H_2O)_2]$. The sheets are linked together by hydrogen bonds involving H_2O groups in the interlayer (Krivovichev and Burns 2000a).

Figure 365. Waxy yellow film of iriginite from the type locale, Aleksandrovskiy Golets Mo-U deposit, Transbaikal, Russia. RJL3145

Figure 366. Yellow iriginite on matrix from Cerro Picacho, Bacoachi, Sonora, Mexico. RJL536

Figure 367. Dull yellow earthy iriginite from Ben Lomand, Queensland, Australia. RJL1349

Figure 368. The sheet structure of iriginite, viewed along [010], showing the coordination polyhedra of U (blue gray) and Mo (magenta), forming sheets of composition $[(UO_2)Mo_2O_7(H_2O)_2]$ with zero net charge. Gray balls represent water molecules that form one corner of the Mo coordination polyhedron.

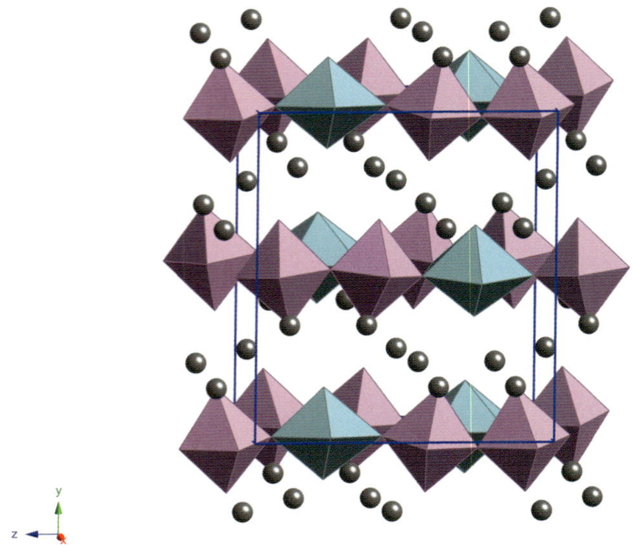

Figure 369. The structure viewed just off [100], showing how the charge-neutral sheets are linked together solely by hydrogen bonding via the interlayer H_2O.

Moluranite $H_4U^{4+}(UO_2)_3(MoO_4)_7 \cdot 18H_2O$
Moluranite, along with iriginite, was originally found by G. Yu. Epshtein in 1951 at the Aleksandrovskii Golets Mo-U deposit (Pekov 1998). Moluranite is black (translucent brown in thin sections) and has a resinous luster. X-ray diffraction revealed no crystal structure. It occurs in fine fissures in granulated albitite, in association with molybdenite, chalcopyrite, and galena. The moluranite forms colloform masses with brannerite and other U-Mo phases (Epshtein 1959; Soboleva and Pudovkina 1957).

Moluranite has also been reported from Cerro Picacho, Bacoachi, Sonora, Mexico; and from the Dajishan U-Mo-Hg deposit, Xianyi Co., Guizhou, China.

Figure 370. Small black pods of moluranite in matrix from Cerro Picacho, Bacoachi, Sonora, Mexico. RJL538

Mourite $U^{4+}Mo^{6+}_5O_{12}(OH)_{10}$

Mourite was described from the oxidation zone of the Kyzylsai Mo-U deposit, southwestern Balkhash region, Kazakhstan. It forms violet nodules and crusts in the supergene zone; the hypogene ore at this locale contains pitchblende, molybdenite, jordisite, pyrite, and other sulfides. Umohoite is found directly below the supergene zone, which contains "U-Mo blacks," iriginite, U-rich powellite, natrouranospinite, tyuyamunite, and uranophane, along with goethite, kaolinite, and jarosite. Water at the mine contains acid sulfate. Mourite is found growing on corroded umohoite in places, and at the mine mourite decomposes to ilsemannite (an amorphous hydrous Mo oxide). Individual crystals are minute, rectangular plates a few μm across, arranged in fan-shaped, vermiform, or radiating-fibrous aggregates (Kopchenova et al. 1962).

Material collected in 1955-56 by A. D. Weeks at the Boso-Hackney prospect, Tordillo Hill, Karnes Co., Texas, was later shown to be mourite. There, the mineral forms spherulitic masses of purple bladed fibers elongated along [010] and flattened on {100}, with a pronounced cleavage on {100}. Mourite is monoclinic, with a=24.44, b=7.18, c=9.90 Å, and β=102°12' (Smith, Marinenko, and Weeks 1971).

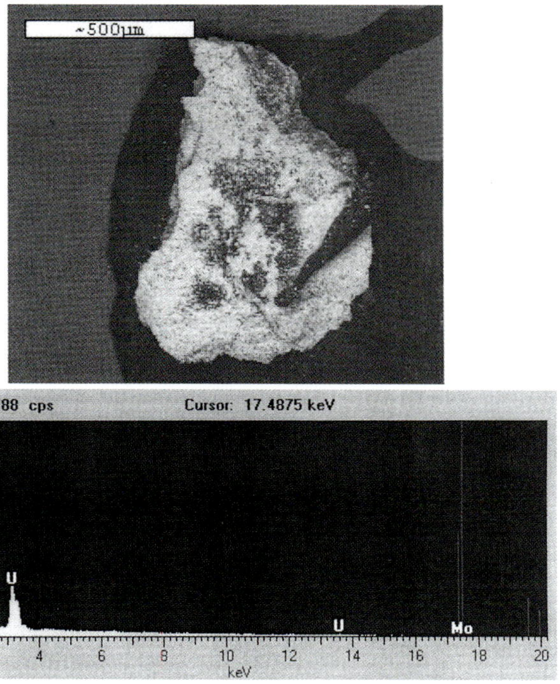

Figure 371. SEM photo of a small crust of mourite from the type locale, and the corresponding analytical data. Courtesy of A. J. Nikischer. RJL3671

Sedovite $U(MoO_4)_2$

Sedovite was described from the Kyzylsai Mo-U deposit, Chu-Ili Mountains, Kazakhstan, where it forms thin reddish-brown crusts and radial aggregates of nearly opaque prismatic crystals to ~0.2 mm long. It occurs in the supergene zone, associated with wulfenite, mourite, calcurmolite, pitchblende, molybdenite, and gypsum. Mourite forms on the sedovite. Based on indexing the powder XRD data, sedovite is reported to be orthorhombic (but possibly monoclinic) with a=3.36, b=11.08, and c=6.42 Å (Skvortsova and Sidorenko 1965).

It has also been reported from Cerro Picacho, Bacoachi, Sonora, Mexico; and from Pinkney Point, Yarmouth Co., Nova Scotia, Canada.

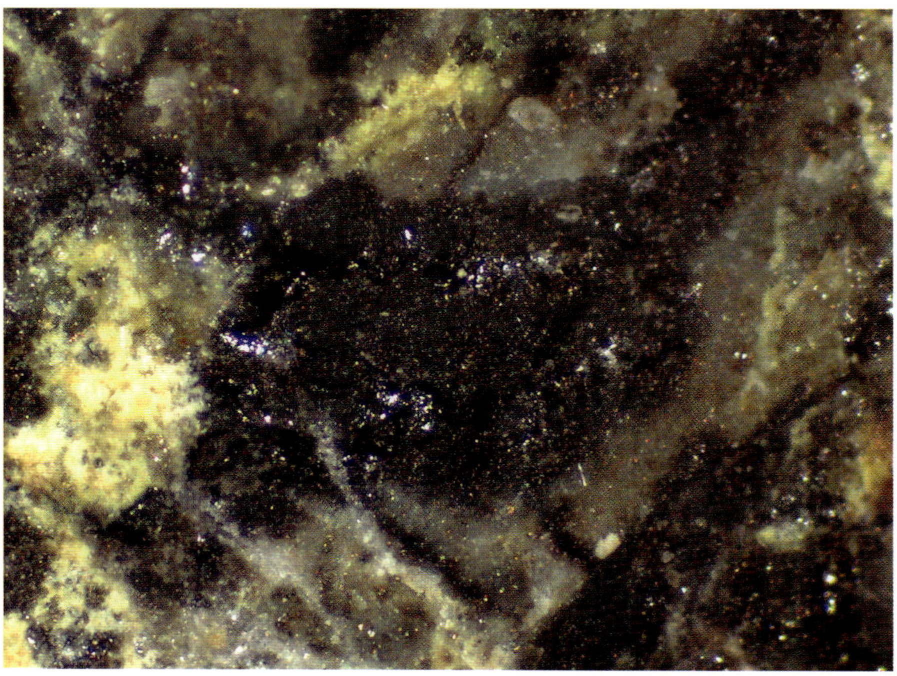

Figure 372. An ore sample from Cerro Picacho, Bacoachi, Sonora, Mexico. Black masses are likely a mixture of sedovite and moluranite; pale yellow earthy crust is iriginite. RJL537

Tengchongite $Ca(UO_2)_6(MoO_4)_2O_5 \cdot 12H_2O$
Tengchongite is a calcium uranyl molybdate that differs from calcurmolite in that calcurmolite has UO_3:MoO_3 = 1:1, whereas for tengchongite the ratio is 3:1. It was described from the oxidation zone of a uranium mineralization at the contact of an ocular migmatite and a migmatitic gneiss at Tongbiguan village, Tengchong Co., Yunnan Province, China. It forms small yellow, tabular to micaceous grains, in association with studtite, calcurmolite, and kivuite. It is not fluorescent. The structure is orthorhombic with a=15.616, b=13.043, and c=17.716 Å (Chen et al. 1986).

Umohoite $[(UO_2)MoO_4(H_2O)](H_2O)$
Umohoite was the first U-Mo species to be described, having been identified by Brophy and Kerr (1953) in ore samples from the Freedom No. 2 mine, Marysvale, Utah. At the TL it forms black to bluish-black specks and thin veinlets in the alteration zone of clay minerals along a U-bearing vein. Associated species include fluorite, pyrite, quartz, uraninite, ilsemannite, and jordisite.

At the Lucky Mc mine, Fremont Co., Wyoming, umohoite was found intergrown with gypsum in association with iron sulfides and U oxides in a zone where oxidation to uranyl minerals had not yet occurred. This was interpreted as a transient stage, because schoepite and uranophane were found within ~10 cm of the umohoite. Fresh material is blue black or dark green, but in places where secondary uranyl species are forming, the umohoite alters to an unidentified yellow uranium molybdate (Coleman and Appleman 1957).

It is also reported from the Alyce Tolino mine, Cameron, Arizona. There, radioactive blue-black sooty masses and carbonaceous trash replacements were found. It was noted that, "After a few days' exposure to the atmosphere, a freshly broken surface of this material develops a bright blue water soluble molybdenum efflorescence which is mainly ilsemannite. Such areas have been found to contain umohoite, although the aggregate differs in megascopic appearance from previously described crystals. The three US localities, as well as an occurrence in the USSR, indicate that this unusual hydrous uranium-molybdate may be more widespread than was considered likely at the time of the original discovery. Knowledge of the existence of several other localities on the Colorado Plateau where secondary molybdenum is associated with uranium suggests the possibility that further study may disclose an even more widespread distribution of umohoite," (Hamilton and Kerr 1959).

Figure 374. The sheet structure of umohoite, built of UO_7 bipyramids (blue-gray) and $MoO_5(H_2O)$ octahedra (magenta with the attached water molecule shown in gray).

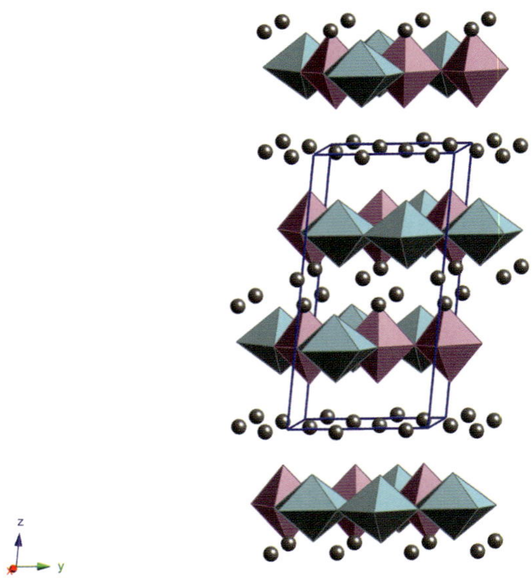

Figure 375. The structure viewed parallel to the sheet, showing that there are two different interlayer structures at z≈0.5 (middle of the unit cell) and z≈0 (top and bottom of the unit cell) with different hydrogen bonding schemes as discussed in the text.

Figure 373. Orange-red crust of microcrystalline umohoite, associated with pale yellow uranophane, from Rabejac, France. RJL4306

Other locales include: the Rabejac and Mas d'Alary deposits, Herault, France; Shinkolobwe, Congo; and the Kyzylsai Mo-U deposit, Kazakhstan. Interestingly, at Rabejac, France, the mineral typically forms tiny, brick-red spherical aggregates.

Umohoite is often partially replaced by iriginite. Krivovichev and Burns (2000) discuss the structural changes needed to obtain the iriginite anion topology from that of umohoite, either by adding Mo^{6+} ions or removing U^{6+} ions. The latter mechanism seems most likely, given the high mobility of U^{6+} in ground water.

Using single-crystal precession photographs, Kamhi (1959) suggested a primitive orthorhombic unit cell, but noted that c=12.4 and 14.5-17.0 Å depending on ambient humidity and temperature. The reported instability of umohoite led various workers to suggest that the structure might contain U^{4+}. However, Krivovichev and Burns (2000) refined the structure and found no evidence of U^{4+}. Their structure refinement showed that umohoite is triclinic with a=6.375, b=7.529, c=14.628 Å, α=82.64, β=85.95, and γ=89.91°.

The structure contains U coordination polyhedra (pentagonal bipyramids) sharing edges with octahedrally coordinated Mo^{6+} to form sheets of composition $[(UO_2)MoO_4(H_2O)]$ lying parallel to (001). (Each Mo cation is bonded to five atoms of O and one H_2O group.) The sheets are bonded to one another via hydrogen bonds. Interestingly, the structure contains two different interlayer structures with different kinds of hydrogen bonding. The interlayer at z≈0.5 contains only the water molecules that are part of the Mo coordination polyhedra. Those H_2O are weakly bonded to the Mo^{6+} cations in the sheets, and each donates H bonds that are accepted by O(2) and O(5) anions in the adjacent sheet. The interlayer at z≈0 contains unattached H_2O groups, each of which donates H bonds that are accepted by adjacent sheets. The structural formula shown above may also be expressed as $UMoO_6 \cdot 2H_2O$.

Sidorenko et al. (2003) discuss the possible role of interlayer Na ions in stabilizing the structure of Na-rich umohoites, and suggest the following general formula for U-deficient compositions: $(UO_2)_{1-x}[MoO_4H_2O]Na_{2x}(H_2O)_{1-2x}$.

Uranotungstite $(Fe^{2+},Ba,Pb)(UO_2)_2(WO_4)(OH)_4 \cdot 12H_2O$
Uranotungstite was described based on material found at two different localities in Germany: the Menzenschwand deposit in the southern Black Forest, and the Clara mine near Oberwolfach in the central Black Forest. Microprobe analysis of the Menzenschwand sample indicated the following composition: $(Fe_{.38}Ba_{.37}Pb_{.36})U_{1.82}W_{.99}H_{28.94}O_{24}$, corresponding to the ideal formula given above. Semi-quantitative analysis of Clara mine material showed only a trace of Ba and no Pb. The mineral forms crusts on quartz, metauranocircite, and metaheinrichite; other associates include bergenite, metatorbernite, metazeunerite, and schoepite. The orthorhombic, lath-like crystals are yellow, orange, or brownish, forming spherules to ~0.3 mm diameter. It is not fluorescent (Walenta 1985).

Carlosbarbosaite $(UO_2)_2Nb_2O_6(OH)_2 \cdot 2H_2O$
Carlosbarbosaite was described from the Jaguaraçu pegmatite, Jaguaraçu municipality, Minas Gerais, Brazil. It forms long, flattened, lath-like crystals with a very simple orthorhombic habit, elongated along [001] and flattened on (100). Crystals range up to 120 μm long and 2–5 μm thick,

Figure 376. Pale yellow aggregates of carlosbarbosaite as a late-stage cavity filling in albite, from the TL at the Jaguaraçu pegmatite, Minas Gerais, Brazil. RJL4532

Figure 377. The UO_5 (blue-gray) and $Nb_2\varphi_8$ (pink) double-chain linkage in carlosbarbosaite, viewed just off [100].

Figure 378. The U-Nb framework viewed along [001] showing $Ca(H_2O)_4$ groups (blue and gray balls) occupying the tunnels. Wire bonds are shown between Ca and the uranyl oxygens.

and are cream to yellow with vitreous luster. Individual crystals are transparent and massive aggregates are translucent; no fluorescence was observed in either SW or LWUV. The mineral occurs as a late-stage cavity filling in albite, closely associated with zircon, muscovite, kaolinite, and columbite-Fe (Atencio et al. 2012). The mineralogy and geology of the Jaguaraçu pegmatite is given by Cassedanne and Alves (1994).

The structure is orthorhombic with a=14.151, b=10.394, and c=7.529 Å. It contains a single U site, which is populated by U atoms and vacancies. The U site is surrounded by seven O atoms in a pentagonal bipyramidal arrangement. The Nb site is coordinated by four O atoms and two OH groups forming an Nbφ6 octahedral arrangement. Octahedrally coordinated Nb polyhedra share edges and corners to form $Nb_2O_6(OH)_2$ double chains, and edge-sharing U polyhedra form UO_5 chains. The $Nb_2O_6(OH)_2$ and UO_5 chains share edges to form an open U-Nb-φ framework with tunnels along [001] that contain $Ca(H_2O)_4$ clusters. Carlosbarbosaite is closely related to a family of synthetic U-Nb-O framework tunnel structures. The structure of carlosbarbosaite resembles that of holfertite (Atencio et al. 2012).

Holfertite $U^{6+}_{1.75}TiO_{7.5}Ca_{0.25}(H_2O)_3$

The occurrence in Topaz Valley, Utah, of a presumed new mineral containing U, Ti, and Ca was first reported by Foord et al. (1995) based on material collected in 1988. Analytical work was hampered by the small size of the crystals, the limited amount of material, and the fact that the tiny crystals were hollow and inhomogeneous. Later collecting during 2000–02 produced enough material to fully characterize and describe the new mineral as holfertite (Belakovskiy et al. 2006). At the type locale, Starvation Canyon (formerly Searle Canyon), Thomas Range, UT, holfertite forms minute yellow needles up to 1 × 5 mm with a hexagonal outline and frequently a hexagonal internal cavity. Individual needles are usually perched on or piercing hematite crystals or associated with topaz or red beryl. The deposit is a pneumatolytic assemblage in cavities and fractures in rhyolite. Holfertite is also found at several nearby workings including the Maynard Topaz mine and at a pseudobrookite locality 1 mile south of Pismir Wash.

Figure 379. Minute yellow holfertite crystals on a topaz crystal from the TL at Searle Canyon, Utah, associated with quartz, hematite, and other phases. RJL2896

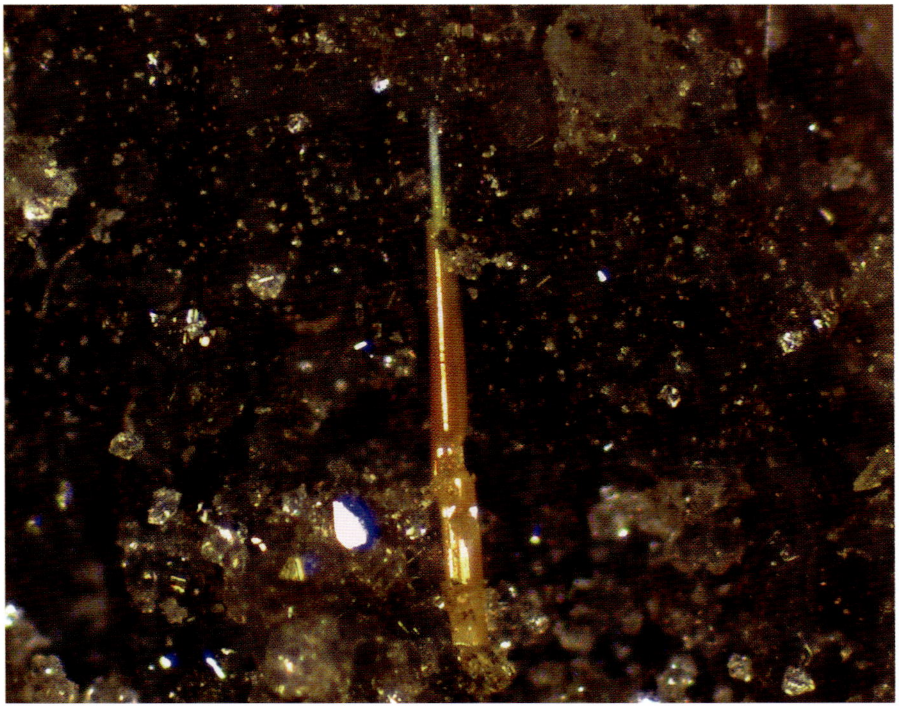

Figure 380. Acicular, orange-yellow holfertite crystal, ~1 mm long, with topaz and drusy quartz, from the TL at Searle Canyon, Utah. RJL3679

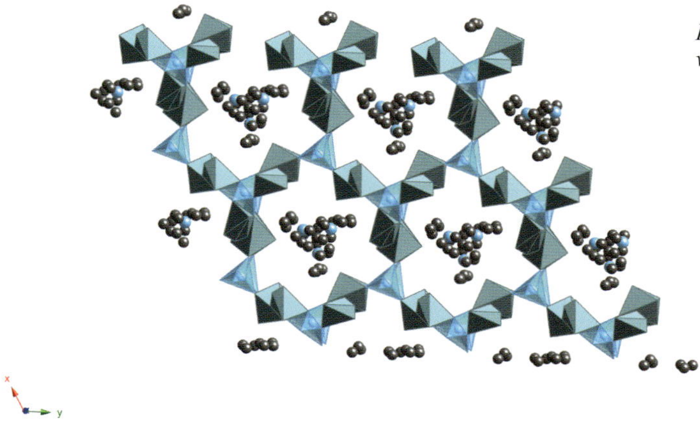

Figure 381 The structure of holfertite showing the U-Ti framework with channels containing Ca and H_2O.

Holfertite is hexagonal with *a*=10.824, *b*=10.824, and *c*=7.549 Å. Taking into account the U(1A) and U(2) sites with 75 and 77% occupancy, U^{6+} polyhedra share common edges to form a $[UO_4]^{2-}$ chain extending along [001]. There are four [5]-coordinated Ti sites. The Ti(1) and Ti(2) sites are fully occupied, and theTi(3) and Ti(4) sites are half-occupied. The Ti(1) and Ti(2) trigonal bipyramids share common vertices to form $[TiO_4]^{4-}$ chains along [001]. Although some structural details remain unclear because of partial occupancies, in the generalized model of the structure, U chains and Ti chains link through common vertices to form a framework with channels along [001]. The channels have a diameter of ~8 Å and occlude eight sites that are partly occupied by Ca atoms and (H_2O) groups, giving the total channel constituents as $\{Ca_{0.25}(H_2O)_{3.00}\}$ apfu. In the crystal structure of holfertite, only five of the twenty-eight sites are fully occupied. Because of the order-disorder nature of holfertite's crystal structure it has not been possible to definitively state whether (OH) is present or absent. Thus, a second formula that differs slightly from the one given above may be equally plausible based on the analyses presented thus far (Sokolova et al. 2005).

Thorium molybdates

Two thorium molybdates have recently been discovered in a Mo-Bi mineralization at Sardinia, Italy.

Table 49. Thorium molybdates

Species	Formula	Symmetry
Ichnusaite	$Th(MoO_4)_2 \cdot 3H_2O$	mon
Nuragheite	$Th(MoO_4)_2 \cdot 2H_2O$	mon

Ichnusaite $Th(MoO_4)_2 \cdot 3H_2O$

Ichnusaite was described from the Mo-Bi mineralization of Su Seinargiu, Sarroch, Cagliari, Sardinia, Italy. It forms colorless, thin tabular crystals up to 200 µm long, associated with muscovite, xenotime-(Y), and nuragheite. It has a pearly to adamantine luster and perfect {100} cleavage. Ichnusaite is typically intergrown with nuragheite on a very fine scale, occurring in vugs in quartz veins. Its formation is attributed to alteration of the molybdenite ore.

Ichnusaite is monoclinic, with a=9.6797, b=10.3771, c=9.3782 Å, and β=90.00°. The structure consists of $ThO_7(H_2O)_2$ and MoO_4 polyhedra forming electroneutral sheets of composition $[Th(MoO_4)_2(H_2O)_2]^0$ lying parallel to (100). Adjacent sheets are linked via hydrogen bonding to H_2O groups in the interlayer (Orlandi et al. 2014).

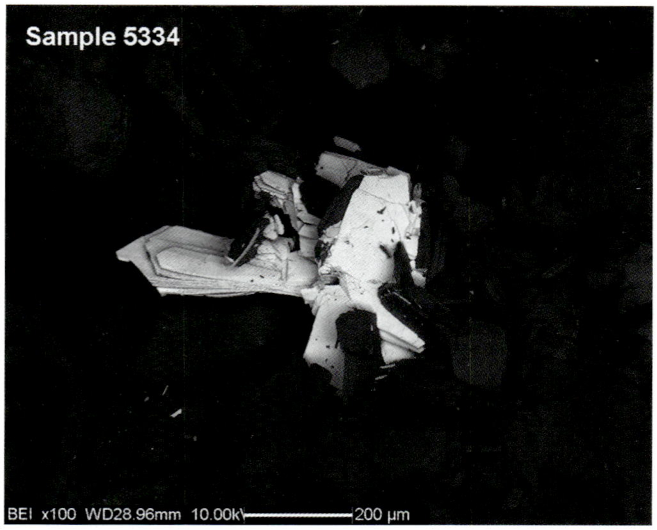

Figure 382. SEM image of tabular ichnusaite crystals on quartz, from the TL at Su Seinargiu, Sardinia. Reproduced with the kind permission of the Mineralogical Society of America from a paper by Orlandi et al. (2014)

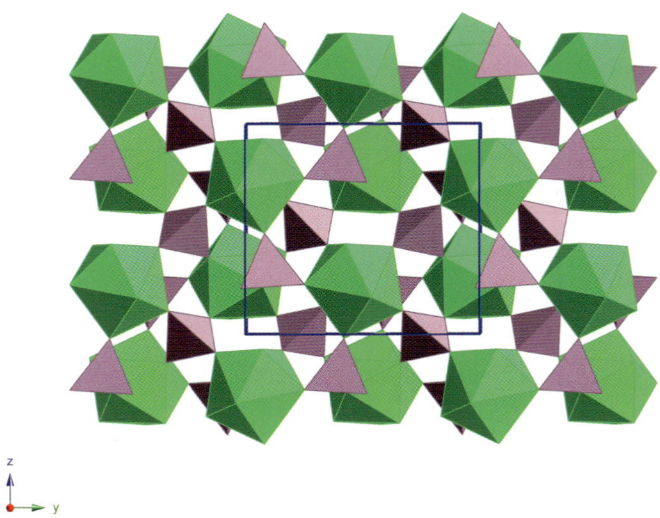

Figure 383. Topology of the thorium molybdate sheet in ichnusaite, showing the linkages between Th (green polyhedra) and Mo (lavender tetrahedra).

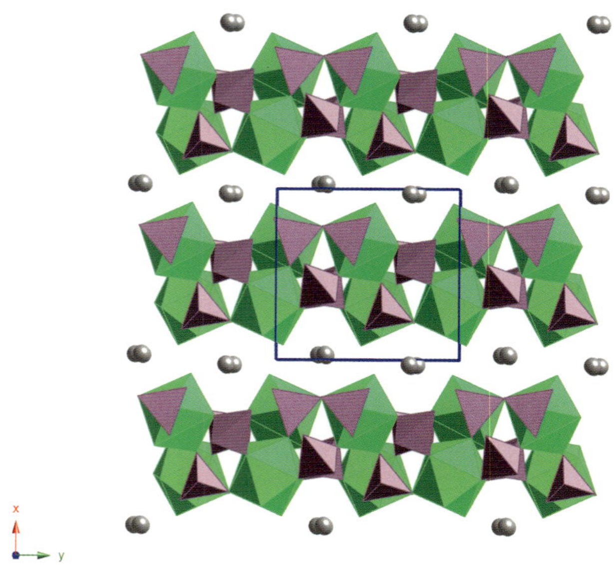

Figure 384. The structure viewed along [001], parallel to the sheets, showing interlayer H_2O groups (gray). Adjacent sheets are linked via hydrogen bonding.

Nuragheite $Th(MoO_4)_2 \cdot 2H_2O$

Nuragheite was described from the Mo-Bi mineralization of Su Seinargiu, Sarroch, Cagliari, Sardinia, Italy. It forms colorless, thin tabular crystals up to 200 μm long, with a pearly to subadamantine luster. Associated minerals include ichnusaite, muscovite, and xenotime-(Y). The assemblage is interpreted to be related to hydrothermal alteration of the Mo-Bi ore (Orlandi et al. 2015).

Nuragheite is monoclinic, with a=7.358, b=10.544, c=9.489 Å, and β=91.88°. The structure consists of (100) layers formed by 9-coordinated $ThO_8(H_2O)$ polyhedra and MoO_4 tetrahedra. Adjacent sheets are bonded via sharing of oxygens at the O2 site between Mo2 and Th polyhedra. Some short O-O bond lengths suggest the presence of some hydrogen bonding as well.

Figure 385. SEM image of nuragheite crystals (light) on drusy quartz (dark gray) from the original description. Reproduced with the kind permission of the Mineralogical Society of America from a paper by Orlandi et al. (2015)

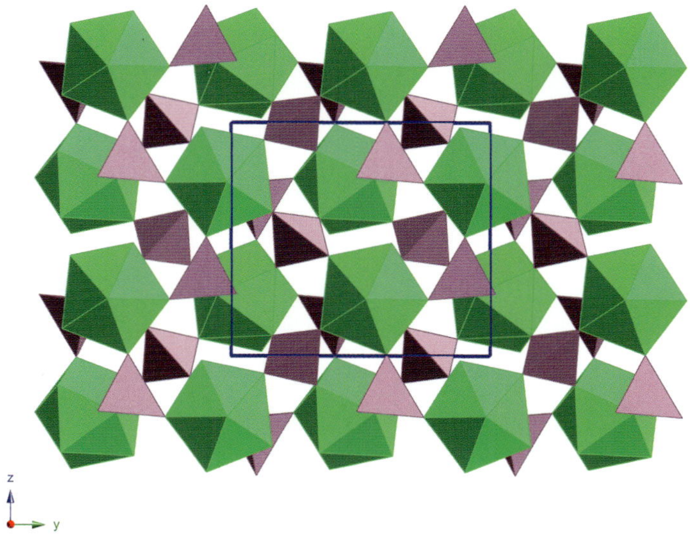

Figure 386. The thorium molybdate sheet of nuragheite, viewed along [100].

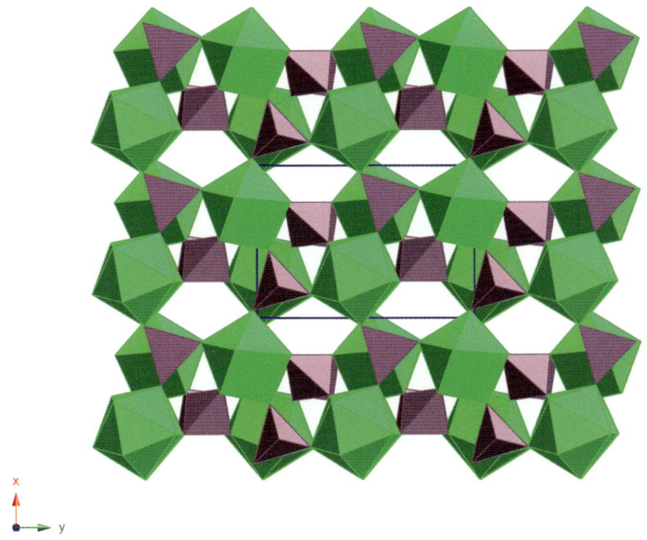

Figure 387. The structure viewed along [001], showing that adjacent sheets are linked directly through MoO_4 tetrahedra to form a continuous network.

PHOSPHATES AND ARSENATES

Autunite Group

The autunite group is divided into two subgroups based on crystal structure differences related to the degree of hydration. For each cation (Ca, Cu, Mg, Ba, etc.) and anion (PO_4 or AsO_4) species there are usually several hydration states, with the lower hydration being generally more stable at room temperature and ordinary humidity. When there is a single lower hydration state it is designated by the prefix "meta-", (e.g., torbernite and metatorbernite). When many stable hydration states are observed, they may be designated by Roman numerals (e.g., "metauranocircite I" and "metauranocircite II"). For a collector devoted to accurate labeling, this presents a dilemma. Should a partially dehydrated sample be labeled "autunite," "autunite with meta-autunite," or "meta-autunite pseudomorph after autunite?" It is the author's view that the real value of a mineral specimen lies in its information content. What can the rock tell us about its origin? The original mineral assemblage can provide valuable insights into the thermodynamic conditions that were present at the time of formation, and labeling practices should recognize that this information is more important in most cases than artifacts that might have occurred later during storage. The label should therefore reflect (to the extent it is known) the species that was originally formed.

The autunite subgroup minerals are tetragonal, orthorhombic, or monoclinic uranyl phosphates and arsenates with the general formula: $\mathbf{A}(UO_2)_2(\mathbf{X}O_4)_2{\cdot}8\text{-}12H_2O$, where $\mathbf{A}$ = Ba, Ca, Cu, Fe^{+2}, ½(HAl), Mg, Mn^{+2}, or $Na_3(UO_2)$, and $\mathbf{X}$ = As^{+5} or P^{+5}. The subgroup includes arsenuranospathite, autunite, heinrichite, kahlerite, nováčekite, rauchite, sabugalite, saléeite, torbernite, trögerite, uranocircite, uranospathite, uranospinite, and zeunerite.

The meta-autunite subgroup minerals are tetragonal or orthorhombic uranyl phosphates and arsenates with the general formula: $\mathbf{A}(UO_2)_3(\mathbf{X}O_4)_2{\cdot}4\text{-}8H_2O$, where $\mathbf{A}$ = Al, Ba, Ca, Co, Cu, Fe^{+2}, $(H_3O)_2$, K_2, Mg, $(NH_4)_2$, or Zn, and $\mathbf{X}$ = As^{+5} or P^{+5}. The subgroup includes abernathyite, bassetite, chernikovite, chistyakovaite, lehnerite, meta-ankoleite, meta-autunite, metaheinrichite, metakahlerite, metakirchheimerite, metalodevite, metanatroautunite, metanováčekite, metarauchite, metatorbernite, metauranocircite, metauranospinite, metazeunerite, natrouranospinite, threadgoldite, trögerite, uramarsite, and uramphite.

The crystallography of the group has been studied extensively by X-ray methods (e.g., Donnay and Donnay 1955; Ross, Evans, and Appleman 1964; Locock and Burns 2003a, 2003c; Locock et al. 2004; Locock, Burns, and Flynn 2004, 2005). De Abeledo, Galloni, and Benyacar (1968) used selected-area electron diffraction on minute {001} platelets to determine cell parameters of metazeunerite, metaheinrichite, and metanováčekite, and compared their results to data reported from various earlier studies.

Table 49. Autunite group

Mineral	Formula	Symmetry
Autunite subgroup		
Arsenuranospathite	$Al(UO_2)_2(AsO_4)_2F{\cdot}20H_2O$	orth
Autunite	$Ca(UO_2)_2(PO_4)_2{\cdot}10\text{-}12H_2O$	orth
Fritzscheite[a]	$Mn^{2+}(UO_2)_2[(PO_4),(VO_4)]_2{\cdot}10H_2O$ (?)	tet
Heinrichite	$Ba(UO_2)_2(AsO_4)_2{\cdot}10\text{-}12H_2O$	mon
Kahlerite	$Fe^{2+}(UO_2)_2(AsO_4)_2{\cdot}10\text{-}12H_2O$	tet
Nováčekite	$Mg(UO_2)_2(AsO_4)_2{\cdot}12H_2O$	tet
Rauchite	$Ni(UO_2)_2(AsO_4)_2{\cdot}10H_2O$	tric
Sabugalite	$H_{0.5}Al_{0.5}(UO_2)_2(PO_4)_2{\cdot}8H_2O$	mon
Saléeite	$Mg(UO_2)_2(PO_4)_2{\cdot}10H_2O$	mon
Torbernite	$Cu(UO_2)_2(PO_4)_2{\cdot}8\text{-}12H_2O$	tet
Uranocircite	$Ba(UO_2)_2(PO_4)_2{\cdot}10H_2O$	mon
Uranospathite	$Al_{1-x}\square_x[(UO_2)(PO_4)]_2(H_2O)_{20+3x}F_{1-3x}$	orth
Uranospinite	$Ca(UO_2)_2(AsO_4)_2{\cdot}10H_2O$	orth
Zeunerite	$Cu^{2+}(UO_2)_2(AsO_4)_2{\cdot}10\text{-}16H_2O$	tet
Meta-autunite subgroup		
Abernathyite	$K_2(UO_2)_2(AsO_4)_2{\cdot}8H_2O$	tet
Bassetite	$Fe^{2+}(UO_2)_2(PO_4)_2{\cdot}8H_2O$	mon
Chernikovite	$(H_3O)_2(UO_2)_2(PO_4)_2{\cdot}6H_2O$	tet
Chistyakovaite	$Al(UO_2)_2(AsO_4)_2(F,OH){\cdot}6.5H_2O$	mon
Lehnerite	$Mn^{2+}(UO_2)_2(PO_4)_2{\cdot}8H_2O$	mon
Meta-ankoleite	$K_2(UO_2)_2(PO_4)_2{\cdot}6H_2O$	tet
Meta-autunite	$Ca(UO_2)_2(PO_4)_2{\cdot}2\text{-}6H_2O$	
Metaheinrichite	$Ba(UO_2)_2(AsO_4)_2{\cdot}8H_2O$	
Metakahlerite	$Fe^{2+}(UO_2)_2(AsO_4)_2{\cdot}8H_2O$	tet
Metakirchheimerite	$Co(UO_2)_2(AsO_4)_2{\cdot}8H_2O$	tet
Metalodevite	$Zn(UO_2)_2(AsO_4)_2{\cdot}10H_2O$	tet
Metanatroautunite	$Na_2(UO_2)_2(PO_4)_2{\cdot}8H_2O$	tet
Metanováčekite	$Mg(UO_2)_2(AsO_4)_2{\cdot}4\text{-}8H_2O$	
Metarauchite	$Ni(UO_2)_2(AsO_4)_2{\cdot}8H_2O$	tric
Metatorbernite	$Cu(UO_2)_2(PO_4)_2{\cdot}8H_2O$	tet
Metauranocircite	$Ba(UO_2)_2(PO_4)_2{\cdot}6\text{-}8H_2O$	mon
Metauranospinite	$Ca(UO_2)_2(AsO_4)_2{\cdot}8H_2O$	
Metazeunerite	$Cu^{2+}(UO_2)_2(AsO_4)_2{\cdot}8H_2O$	tet
Natrouranospinite	$(Na_2,Ca)(UO_2)_2(AsO_4)_2{\cdot}5H_2O$	tet
Przhevalskite[b]	$Pb(UO_2)_2(PO_4)_2{\cdot}2H_2O$	tet
Threadgoldite	$Al[(UO_2)(PO_4)]_2(OH){\cdot}8H_2O$	mon
Trögerite	$(H_3O)_2(UO_2)_2(AsO_4)_2{\cdot}6H_2O$	tet
Uramarsite	$(NH_4,H_3O)_2(UO_2)_2(AsO_4,PO_4)_2{\cdot}6H_2O$	tet
Uramphite	$(NH_4)_2(UO_2)_2(PO_4)_2{\cdot}6H_2O$	tet

[a]Fritzscheite is a possibly questionable species.

[b]Przhevalskite has some structural similarities to the meta-autunite group.

Figure 388. The uranyl phosphate sheet in autunite, viewed along [010], showing corner-sharing UO_6 bipyramids (blue-gray) and PO_4 tetrahedra (dark gray).

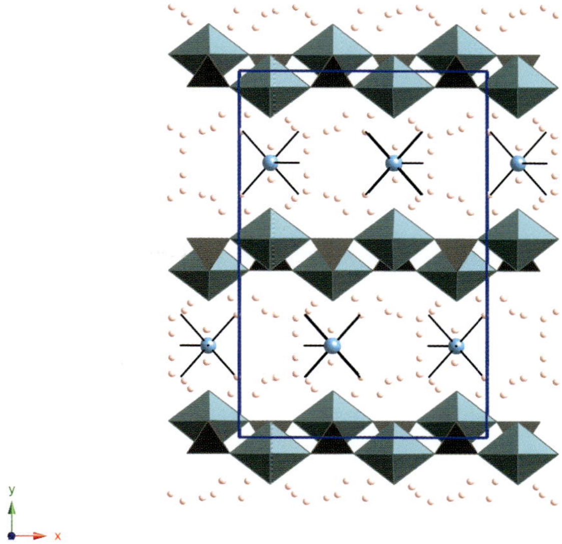

Figure 389. The structure of autunite, viewed along [001]. Ca ions are shown as blue atoms with stick bonds to nearby H_2O groups (shown as pairs of pink H atoms).

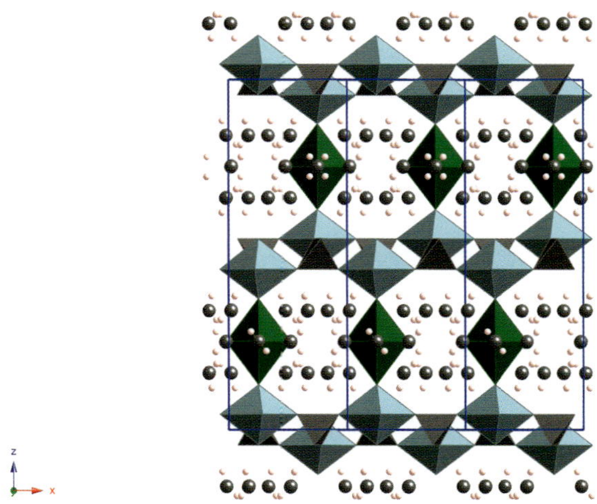

Figure 390. The structure of torbernite, viewed parallel to the sheet, showing distorted Cu octahedra (dark green) directly bridging between the sheets, along with H_2O groups (pink and gray).

The basic structural unit in all of these minerals is the autunite-type sheet, whose composition is $[(UO_2)(PO_4)]^-$ or $[(UO_2)(AsO_4)]^-$, and whose structure consists of uranyl square bipyramids that share equatorial vertices with phosphate or arsenate tetrahedra. These sheets are separated, and bonded together, by an interlayer containing cations and water molecules. The pronounced layer structure led miners and early mineralogists to refer to the group as "uranium micas," and German authors continue to use "Uranglimmer" as a generic term for minerals of the group.

The structures of about a dozen of the various species have been fully determined, and each is influenced by the size and charge of the interlayer cations. Hydrogen bonding also plays an important role, particularly in those cases where there is no direct linkage between the sheets and the interlayer cations. In general, the AsO_4 analogs have slightly greater cell dimensions than their PO_4 counterparts (~0.15 Å in the sheet and ~0.1 Å normal to the sheet). Furthermore, the true symmetry depends on the arrangement of the interlayer water molecules, so many early reports considered the structures to be tetragonal simply because X-ray methods at the time couldn't accurately locate the H_2O groups. Modern structure refinements have improved our understanding of the local structural details and shown that "anomalous" optical properties are not anomalous after all if the true structures have lower symmetry than previously thought.

The structure of autunite was determined by Locock and Burns (2003a) using synthetic material. The interlayer contains Ca ions coordinated by seven H_2O groups. The Ca is centered between uranyl oxygens in the adjacent sheets on either side, but at a fairly large distance from them (3.275 Å); this possibly furnishes a direct, but weak, bond. However, the sheets are mostly bonded together via hydrogen bonding involving the H_2O groups.

In torbernite, each interlayer Cu^{2+} ion lies at the center of a distorted octahedron, with short bonds to four H_2O groups and longer bonds to uranyl oxygens on adjacent phosphate sheets (Locock and Burns 2003).

In heinrichite, the large divalent interlayer cation (Ba^{2+}, ionic radius 1.34 Å) is coordinated to eight interlayer H_2O groups and to the uranyl oxygens in adjacent sheets. The overall symmetry is monoclinic; by analogy, uranocircite is also assumed to be monoclinic (Locock, Burns, and Flynn 2005).

Figure 391. The structure of heinrichite, viewed along [010], showing the monoclinic symmetry. AsO_4 groups in the sheets are shown as dark maroon tetrahedra. Ba ions (magenta) are coordinated to H_2O groups and to uranyl oxygens.

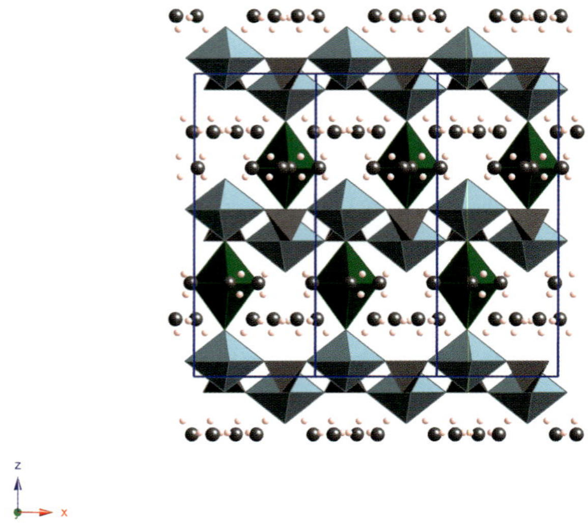

Figure 392. The structure of metatorbernite viewed along [010]. Comparing this to the structure of torbernite shown in Figure 390, one can see the significant reduction in the spacing between adjacent uranyl phosphate sheets as interlayer water is lost.

Well over a century ago it was well established that members of the group can spontaneously dehydrate in air from a higher to a lower hydration state; they clearly retain the attributes of a single crystal while certain optical properties shift (Gaubert 1904; Hallimond 1916, 1920). Before the advent of X-ray diffraction, mineralogists could only make structural inferences based on morphological crystallography and careful optical measurements. In this context, the conclusions of Hallimond (1920) are impressive:

> "The occurrence of natural crystalline material removes all doubt that metatorbernite I has a perfect crystalline structure. In the transition from torbernite we have, therefore, a morphotropic change that can be brought about, not only by a re-growth in the new form, but by the direct removal of water from the crystal network, which at the same time undergoes a homogeneous deformation and assumes new constants proper to the lower hydrate. It has already been remarked . . . that the water is removed by way of the edges of the crystals and is apparently unable to escape through the basal planes, and also that parallel growths of barium chloride on the basal plane have been recorded for both substances. There is thus every reason to think that the horizontal dimensions of the two crystals are similar, and that the additional water is held in layers that alternate with basal layers having the constitution of metatorbernite. If this is the case, the morphotropic change should consist in a decrease in the vertical axis of the crystal to an extent proportional to the change in molecular volume due to the transition. From the data already given, the molecular volume of torbernite is 313, that of metatorbernite is 254; the ratio c:a for torbernite is 2.97, that for metatorbernite is 2.28. Thus the ratio of volumes is 0.81, that of the *c*-axes is 0.76. In view of the inaccuracy of the crystal measurement for metatorbernite this may be regarded as a satisfactory confirmation of the above conclusions."

Modern crystal structure refinements have clarified the structural changes that accompany dehydration (Takano 1961; Ross 1963; Ross, Evans, and Appleman 1964; Locock et al. 2004; Suzuki et al. 2005). As an example, Locock and Burns (2003c) note that, "Although the topologies of the uranyl phosphate sheets in torbernite and metatorbernite are identical, the stacking of the sheets is quite different. Considering only the uranyl phosphate sheets, from a geometrical point of view, the transition from torbernite to metatorbernite involves the relative displacement of every second uranyl phosphate sheet by half the *a* unit-cell dimension along both [100] and [010]."

The minerals of this group generally form fairly thin tabular crystals in various shades of yellow to green. Color and transparency can vary with the degree of hydration. The crystals may be individual tablets, foliated masses, or divergent stacks or "books." Crystals occasionally form truncated bipyramids.

Most of the species in this group are strongly fluorescent. The Cu-containing species such as torbernite and zeunerite are normally not fluorescent because Cu quenches the fluorescence. However, anomalously fluorescent torbernites have been observed, and this is attributed to a very thin overgrowth of fluorescent autunite on the surface of the

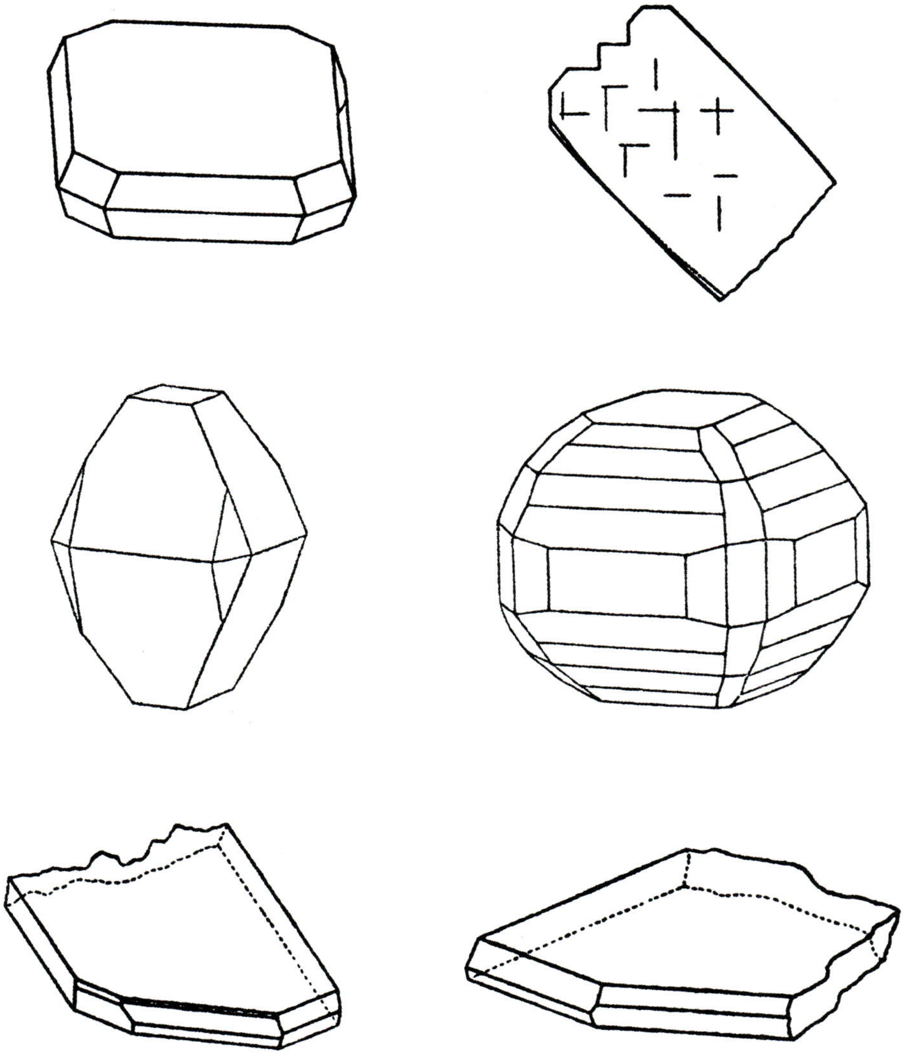

Figure 393. Drawings of some natural crystals, modified from Goldschmidt (1918, 1923). Top: autunite from Cornwall, England (left), and Johanngeorgenstadt, Germany (right). Middle: two torbernites from Cornwall. Bottom: trogerite from Schneeberg, Germany (left), and zeunerite from Schneeberg (right).

Figure 394. A pegmatite sample ~2 cm across, with crusts of tabular greenish autunite crystals and bright yellow microcrystalline uranophane, from Chalk Mountain, Spruce Pine district, North Carolina. RJL1027

Figure 395. The specimen in the previous figure, viewed in SWUV, showing the intense fluorescence of the autunite compared to the relatively non-fluorescent uranophane. RJL1027

torbernite. Intergrowths of bassetite and uranospathite can also show anomalous fluorescence.

In the entries that follow, corresponding members of the two subgroups will be treated together in order to avoid unnecessary repetition.

Abernathyite $K_2(UO_2)_2(AsO_4)_2 \cdot 8H_2O$
Abernathyite was described from ore samples at the Fuemrol No. 2 mine, Emery Co., Utah, and named for Jess Abernathy, operator of the mine. (Although the name of the mine appears to be a misspelling of "fumarole," this is the name shown in the original papers; readers may encounter other spellings in various references.) The host rock lies in vanadiferous and asphaltic sandstones in the Shinarump conglomerate. Abernathyite forms crystalline coatings lining fractures in the sandstone, associated with earthy scorodite. Other minerals noted at the mine include native arsenic, realgar, orpiment, jarosite, and metazeunerite. The tiny (0.5 mm) yellow crystals are tabular with a perfect basal cleavage. It was noted that the crystals are "unusually clear and transparent for a mineral of the metatorbernite group." Fluorescence is moderate yellow-green in SW and LWUV (Thompson, Ingram, and Gross 1956).

In addition to the TL, abernathyite has been reported from several places on the Colorado Plateau; also the Rivieral mine, Herault, France; and Sailauf, Bavaria, Germany.

Abernathyite is tetragonal, with a=7.18 and c=18.13 Å. The crystal structure was described in detail by Ross and Evans (1964) as part of a broader study of the role of H_2O in this group of minerals.

Arsenuranospathite $Al(UO_2)_2(AsO_4)_2F \cdot 20H_2O$
Arsenuranospathite was described from Menzenschwand, Germany, based on two specimens collected in 1959. Minute amounts were also found at Wittichen in the Black Forest (Walenta 1978). At the TL, it occurs in crusts on barite, associated with limonite and various secondary U species, including zeunerite, uranocircite, heinrichite, uranophane, and studtite. The crusts are made up of minute, elongated, tabular to needle-like crystals up to ~0.3 mm long. Crystals are transparent to translucent, pale yellow to nearly white, with weak greenish fluorescence in UV. In the fully hydrated state, the unit cells of arsenuranospathite and uranospathite contain significantly more water than other members of the group. Walenta attributed this to the small ionic radius of Al^{3+} and a general trend of increasing H_2O with decreasing ionic radius and increasing charge of the additional cation.

Arsenuranospathite has also been reported from: the Rabejac deposit, Herault, France; Menzenschwand, Germany; Jachymov, Czech Republic; the La Creusaz prospect, Switzerland; and the Bota-Burum deposit, Kazakhstan.

Arsenuranospathite is orthorhombic, with a=30.070, b=7.147, and c=7.193 Å. The structure of uranospathite was refined by Locock, Kinman, and Burns (2005); although they didn't determine the structure of arsenuranospathite, it was presumed to be isostructural with uranospathite. If so, the true structural formula would differ slightly from earlier formulas. (See the entry for **uranospathite** and the cited paper for further details and discussion.) More recently, Chukanov et al. (2009) examined a number of samples of uranospathite and arsenuranospathite from Menzenschwand, using a variety of analytical techniques. They conclude that the name arsenuranospathite has been applied to at least two different species—arsenuranospathite *sensu stricto* with n = 20 H_2O, and its partially dehydrated counterpart with n = 8-10 H_2O, which is tetragonal or pseudotetragonal.

Autunite $Ca(UO_2)_2(PO_4)_2 \cdot 10\text{-}12H_2O$
Meta-autunite $Ca(UO_2)_2(PO_4)_2 \cdot 2\text{-}6H_2O$
Autunite is named for the locality near Autun, France, that produced the material first analyzed by Berzelius in 1824 (Phillips, Brooke, and Miller 1852). Before that time, autunite was confused with torbernite and both minerals were lumped under the term "uranite" and variations thereof. As in other members of the group, the water content can vary reversibly over some range. Under very dry conditions autunite slowly dehydrates to meta-autunite. The tabular crystals may be thin to thick tabular with a rectangular or, less commonly, octagonal outline. At several localities, the crystals form thickly stacked groups or "books" that stand on edge and produce very distinctive, attractive specimens. Both species are bright yellow to green, usually becoming paler as dehydration proceeds, and both strongly fluoresce yellow-green in SW and LWUV.

Autunite is formed by the oxidation and weathering of hydrothermal veins or other primary uraninite deposits at ordinary conditions of temperature and pressure. It is one of the most abundant of the secondary uranium minerals and has been mined as ore in several places, notably Portugal

Figure 396. Autunite crystals to ~3 mm, thickly scattered on granite from Margnac, France. RJL2971

and Madagascar (Frondel, 1958). Minerals often associated with autunite include phosphuranylite, torbernite, saléeite, uranophane, and uranophane-β. Autunite is occasionally altered to phosphuranylite.

Many localities around the world have produced excellent specimens of autunite. A few of the more notable deposits are: the Daybreak Mine, Opportunity, Spokane Co., Washington; Mount Painter in the Flinders Range, South Australia; in the Rum Jungle area, Northern Territory, Australia; in Germany it occurs at several mines in the Johanngeorgenstadt-Eibenstock district, Saxony; at Sabugal and several other localities in Portugal; at the São Pedro mine and elsewhere in Brazil. In the Ruggles pegmatite near Grafton Center, New Hampshire, autunite forms as an alteration product of dendritic uraninite. In Cornwall, England, it is found at St. Austell and at Redruth; it occurs at Shinkolobwe and other localities in the Congo,

Figure 397. Autunite crystals and rosettes to ~1 cm completely covering the surface of fine-grained granite from Luzy, France. RJL663

Figure 398. Thin tabular autunite crystals to 3 mm on dense, iron-rich matrix from the Apex mine, Nevada. RJL2644

Figure 399. Autunite crystal ~6 mm wide, on stalactitic goethite, from the São Pedro claim, Malacacheta, Minas Gerais, Brazil. RJL4404

but there it is generally much less abundant than torbernite. It is also a minor constituent of the sandstone-type uranium deposits of the Colorado Plateau.

The structure of autunite and its dehydration products has been studied by a number of workers. Brichard and Brasseaur (1958) compared the powder XRD patterns and dehydration curves of natural and synthetic autunite. Takano (1961) presented the structures and cell dimensions of autunite, meta-autunite I, and meta-autunite II. Most studies suggested that autunite is tetragonal; however, it is optically biaxial in the plane of the sheets, which would be anomalous in a tetragonal crystal. Locock and Burns (2003a) determined that the structure of synthetic autunite is orthorhombic, with a=14.014, b=20.712, and c=6.996 Å.

Figure 400. A small bassetite crystal from the type locale at Basset mines, Cornwall, associated with minor yellow phosphuranylite. RJL480

Figure 401. Micaceous scales of brown to greenish bassetite from the Dzhentuar deposit, Kazakhstan. FOV is ~6 mm. RJL3680

Bassetite $Fe^{2+}(UO_2)_2(PO_4)_2 \cdot 8H_2O$

Bassetite was described by Hallimond (1915) from samples found at the Basset group of mines, Redruth, Cornwall, England. The mineral forms small, square or rectangular, tabular crystals in stacked or fan-like aggregates typical of the group. Color ranges from yellow or bronzy yellow to brown or olive green; it is not fluorescent. Hallimond reported optical and morphological properties in detail and showed various forms of zoning and twinning. He also described cavities in gossan deposits in which honey-yellow bassetite had formed epitaxially on stout torbernite crystals. He remarked that, "the crystal measurements are so remarkably close to those of torbernite that it can be regarded with almost complete certainty as the isomorphous calcium salt . . ." Such a composition would, of course, correspond to autunite, but at the time autunite from the TL hadn't been thoroughly characterized, and the properties of bassetite clearly differed from those of autunite. Frondel (1954) confirmed earlier reports that bassetite is actually the ferrous iron analog of autunite, and provided additional data on its optical and crystallographic properties and dehydration behavior.

Bassetite is rarer than autunite or torbernite. It has been reported from a number of locales in Cornwall, England; some other occurrences include: the Los Ratones and Pedro Alvaro mines, Spain; Arcu su Linnarbu, Sardinia, Italy; the Dzhentuar deposit, Kazakhstan; and at a number of places in the western US.

Bassetite is monoclinic with a=6.98, b=17.07, c=7.02 Å, and β=90°32' (Vochten, de Grave, and Pelsmaekers 1984).

Chernikovite $(H_3O)_2(UO_2)_2(PO_4)_2 \cdot 6H_2O$

The hydronium analog of meta-autunite was prepared synthetically as early as 1898 and various authors used the term "hydrogen autunite" to describe this phase as well as naturally occurring material reported from several unspecified localities in the Soviet Union. Because the term "hydrogen autunite" is potentially misleading, Atencio (1988) proposed the name chernikovite in honor of A. A. Chernikov, who described the original occurrence. The TL is identified by Pekov (1998) as the Karakat U deposit, Karamazar Mountains, Tajikistan. There, it forms thin micaceous plates associated with autunite, meta-autunite, natroautunite, and other species in cracks of volcanic rock near an orebody confined to a fault. It is pale yellow with intense yellow-green fluorescence in UV.

At Perus, São Paulo, Brazil, it forms minute (10μm) oriented inclusions in autunite or meta-autunite crystals (Atencio and Hypolito 1994).

Chernikovite from the TL is tetragonal with a=7.031 and c=9.035 Å. Ross (1955) described material synthesized by Frondel and reported the cell dimensions as a=7.020 and c=9.043 Å; she stated that the mineral is "closely similar to or isostructural with meta-autunite." Atencio (1988) reported slightly different values for the material from Perus, Brazil, but also noted that the reported cell parameters might actually represent a pseudo-cell so that the true space group could be any one of several possibilities.

This compound has received considerable study over the years because it is an ionic conductor with high proton conductivity at room temperature. Accurate structure determinations have been reported in the physics literature for the synthetic analog.

Chistyakovaite $Al(UO_2)_2(AsO_4)_2(F,OH)\cdot 6.5H_2O$
Chistyakovaite, the As analog of threadgoldite, was described in 2006, based on studies of museum samples that had been collected at the Bota-Burum deposit, southern Kazakhstan, by E. V. Kopchenova in the 1950s. It forms imperfect yellow crystals up to 2 mm across and intergrown clusters. Associated minerals include natrouranospinite, scorodite, mansfieldite, metazeunerite, trögerite, and arseniosiderite. The assemblage is the result of supergene alteration of hydrothermal pitchblende and arsenopyrite

Chistyakovaite is monoclinic, with a=19.99, b=9.79, c=19.62 Å, and β=110.7°, based on powder XRD. The poor quality of the crystals precluded a full structure determination, but the mineral is presumably isostructural or homeotypical with threadgoldite (Chukanov et al. 2006).

Fritzscheite $Mn^{2+}(UO_2)_2[(PO_4),(VO_4)]_2\cdot 10H_2O$ (?)
Fritzscheite is a questionable species described from a hematite deposit at Nove Hamry, Czech Republic (Breithaupt 1865). It was reportedly reddish brown to hyacinth-red, with vitreous to pearly luster, often as an overgrowth on autunite or torbernite. Other reported locales include Johanngeorgenstadt and Schneeberg, Germany.

Heinrichite $Ba(UO_2)_2(AsO_4)_2\cdot 10\text{-}12H_2O$
Metaheinrichite $Ba(UO_2)_2(AsO_4)_2\cdot 8H_2O$
Heinrichite and metaheinrichite were first studied in 1956 from the White King mine, near Lakeview, Oregon; around the same time, similar material was identified from several localities in the central Black Forest, Germany, and given the names *sandbergite* and *metasandbergite*. In a joint paper describing both occurrences, the authors established heinrichite and metaheinrichite as the accepted names (Gross et al. 1958). Both minerals form small (~1 mm) tabular, transparent to translucent, yellow to green crystals with vitreous to pearly luster. Fluorescence in both SW and LWUV is bright green to greenish yellow.

At the White King mine, the minerals occur in a light gray, altered, silicified rhyolite tuff as fracture coatings and lining small vugs. They occur at or near the surface; at greater depth the main U-containing phase is coffinite, which is associated with orpiment and realgar.

In the Black Forest, the minerals form from the alteration of pitchblende in Co-Ni-Bi-Ag-U veins, and are typically

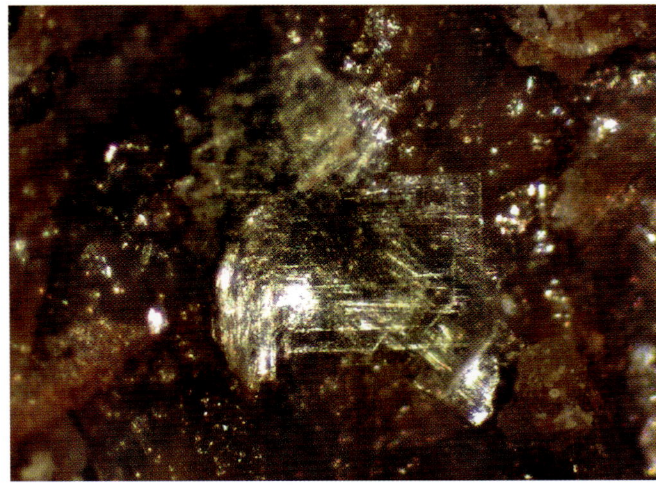

Figure 402. Pale yellow, thin tabular heinrichite crystal, ~1 mm long, on altered rhyolite from the co-TL at the White King mine, Lakeview, Oregon. RJL569

Figure 403. Tiny, light green heinrichite crystals from the co-TL at Wittichen, Germany. RJL510

Figure 404. Thin tabular, greenish-yellow metaheinrichite on granite from Schmidestollenhalde, Wittichen, Germany. RJL1183

found on granite and gangue minerals, associated with zeunerite, nováčekite, and other secondary species. Reported occurrences there include the Sophia mine, Wittichen; the Anton mine, near Schiltach; and at Reinerzau.

Gross et al. (1958) note that, "Both hydrates are common in the Black Forest material. Only traces of the higher hydrate were found during the study of the minerals from Oregon because heinrichite dehydrates readily to metaheinrichite and the laboratory investigations took place under conditions of low relative humidity. Only freshly broken crystals showed the lower refractive indices and these fragments very shortly dehydrated."

Synthetic heinrichite is monoclinic, with a=7.155, b=7.134, c=21.290 Å, and β=104.17°; the structure was refined by Locock, Burns, and Flynn (2005).

Kahlerite $Fe^{2+}(UO_2)_2(AsO_4)_2 \cdot 10\text{-}12H_2O$
Metakahlerite $Fe^{2+}(UO_2)_2(AsO_4)_2 \cdot 8H_2O$
Kahlerite was first noted as a non-fluorescent "autunite" from Huttenberg, Austria, in 1939. Tests at the time showed U, As, and H_2O, so the material was believed to be trögerite. Later chemical analysis showed the presence of Fe, indicating that it is actually the As analog of bassetite. Kahlerite is rare at the TL, forming tabular, lemon-yellow crystals to ~2 mm, associated with scorodite, symplesite, and pitticite, on altered loellingite (Meixner 1953).

Kahlerite has also been reported from: the Sophia and St. Joseph mine dumps, near Wittichen, and at Menzenschwand, Germany; at Mas d'Alary, France; and from the Southwick Cliffs, a coastal area south of Dalbeattie, Dumfires and Galway, Scotland (Braithwaite and Knight 1990).

Metakahlerite was described from the Sophia mine, near Wittichen, Baden, Germany, where it forms scaly, sulfur-yellow aggregates that are not fluorescent (Walenta 1958a). It is likely to be present at locales reported for kahlerite.

Kahlerite and metakahlerite are tetragonal. Based on powder XRD, the cell parameters for metakahlerite are a=20.25 and c=17.20 Å (Vochten, De Grave, and Pelsmaekers 1986).

Lehnerite $Mn^{2+}(UO_2)_2(PO_4)_2 \cdot 8H_2O$
Lehnerite was described from the Hagendorf-Sud pegmatite, Hagendorf, Germany (Mücke 1988), where it forms thin tabular crystals to ~1 mm in decomposed zwieselite. The crystals are bronze-yellow to ochre-yellow, with vitreous to pearly luster, and are not fluorescent. Associated minerals include fluellite, carlhintzeite, morinite, and pachnolite.

Lehnerite is monoclinic, with a=7.04, b=17.16, c=6.95 Å, and β=90.18°, based on powder XRD refinement. The formula suggests that lehnerite is the Mn^{2+} analog of bassetite; its relationship to fritzscheite is not clear.

Meta-ankoleite $K_2(UO_2)_2(PO_4)_2 \cdot 6H_2O$
Meta-ankoleite, the potassium analog of meta-autunite, was described from the Mungenyi pegmatite, Ankole district, southwest Uganda, where it was found intergrown with a phosphuranylite-like mineral in association with muscovite, quartz, and albite. It was also found in the Sebungwe district, Zimbabwe, in a poorly cemented feldspathic quartzite in association with quartz, microcline, monazite, spinel, zircon, and barite in a specimen of Karroo sandstone. It forms yellow plates up to ~1 mm across, with a micaceous cleavage; fluorescence is yellow-green in both SW and LWUV (Gallagher and Atkin 1966).

Meta-ankoleite has also been reported from: St. Yrieux, Haute-Vienne, France; Tirpersdorf, Saxony, Germany; Arcu su Linnarbu, Sardinia, Italy; the Virgin Valley district, Humboldt Co., Nevada; and Sterling Hill, New Jersey. The fully hydrated analog has thus far not been reported in nature.

The powder XRD patterns of material from both Uganda and Zimbabwe are "almost identical with one another and are also generally similar to that of metatorbernite, with which meta-ankoleite is homostructural if not isostructural." The structure is tetragonal with a=6.99 and c=8.89 Å. There was no evidence that any material existed in a higher hydration state. The type material contains a small amount of Ba, expressed by the empirical formula $(K_{1.7}Ba_{0.2})(UO_2)_2(PO_4)_2 \cdot 6H_2O$ (Gallagher and Atkin 1966).

Metakirchheimerite $Co(UO_2)_2(AsO_4)_2 \cdot 8H_2O$
Metakirchheimerite was described from the Sophia shaft at Wittichen, Germany, where it had formed pale rose crusts on pitchblende, in association with metakahlerite, nováčekite, metaheinrichite, and erythrite. Individual crystals are minute, with pearly luster on the cleavage (Walenta 1958). The name was assigned by analogy to metatorbernite; fully hydrated material has not been found in nature.

It was also identified as fissure coatings in the Svornost shaft, Adit level, Evangelista vein, Jachymov, Czech Republic. There it forms small, orange-beige, tetragonal tabular crystals up to ~0.5 mm. The crystals aren't associated with any other secondary minerals. Analysis showed the presence of minor Fe, presumably substituting for some of the Co^{2+} (Ondruš et al. 2003a).

Metakirchheimerite is tetragonal with a=7.15 and c=8.62 Å.

Metalodevite $Zn(UO_2)_2(AsO_4)_2 \cdot 10H_2O$
Metalodevite was described from the oxidation zone of the Rivieral deposit near St. Martin du Bosc, Lodeve, France. It occurs in very small amounts through the alteration of primary ore containing pitchblende, arsenopyrite, sphalerite, pyrite, and coffinite. The tiny (<0.2 mm) platy crystals are pale pearly yellow to olive and are weakly fluorescent yellow-green (Agrinier et al. 1972).

Metalodevite has also been reported from: Mas d'Alary, Herault, France; and Sterling Hill, New Jersey.

Crystals are tetragonal, flattened on (001), with a=7.16 and c=17.20 Å.

Metanatroautunite $Na(UO_2)(PO_4)·3H_2O$
The terms "sodium autunite" and "natroautunite" were first applied to synthetic materials as early as 1929. Chernikov later applied the name sodium autunite to materials discovered at the Western Area of the Kuruk U deposit, near Khodzhent, northern Tajikistan (Pekov 1998). The name "sodium meta-autunite" has also been used by several authors. Based on a 1987 decision by the CNMNC it appears that metanatroautunite is the correct term for this mineral and that fully hydrated material with 10-16H_2O (which would properly be called *natroautunite*) has yet to be conclusively reported in nature (Mills 2004). Chernikov and Organova (1994) stated that fully hydrated crystals are stable only under conditions existing in the underground mine, and they decompose immediately to the *meta*- form when brought to the surface. However, a proposal was never submitted to resurrect the name for the higher hydrate and thus *natroautunite* isn't currently a recognized species.

At the TL in northern Tajikistan, platy, lemon-yellow to greenish-yellow crystals and masses are associated with schoepite, gypsum, kaolinite, and limonite (Chernikov, Krutetskaya, and Organova 1957). At the Lake Boga granite quarry, Victoria, Australia, it forms bright yellow, tabular crystals and clusters to ~4 mm on limonite-coated joint planes and in miarolytic cavities. It occasionally forms intergrowths with saléeite or torbernite (Mills 2004). It has also been reported from the Slim Buttes area, Harding Co., South Dakota.

A potassium-rich variety was described from a playa deposit 105 km NNE of Kalgoorlie, Western Australia, where it forms aggregates of 20-50 µm crystals as void and fracture infillings in near surface gypsiferous muds. The empirical formula indicates the interlayer cation content ~$[Na_{.56}K_{.29}(H_3O)_{.16}]$. Butt and Graham (1981) note that it "is the first reported occurrence of an intermediate member of a predicted solid solution series with potassium, sodium and hydroxonium end members. It differs in origin from other naturally occurring autunite and meta-autunite minerals in being a direct precipitate from groundwater, rather than an alteration product."

The crystal structure is tetragonal; structure refinement by Mills, Kampf, and Birch (2012) clarified the H_2O content and noted a doubling of the unit cell, with a=6.994 and c=17.510 Å. Material hydrated in water for several days by Chernikov and Organova (1994) had a=6.976 and c=8.99 [or 17.98] Å.

Natrouranospinite $(Na_2,Ca)(UO_2)_2(AsO_4)_2·5H_2O$
Natrouranospinite was first described (as *sodium uranospinite*) from the Bota-Burum U deposit, Chu-Ili Mountains, southwestern Balkhash region, Kazakhstan. It occurs in the oxidation zone of a hydrothermal deposit that contained pitchblende, arsenopyrite, galena, and pyrite, the oxidation of which produced realgar, orpiment, scorodite, mansfieldite, metazeunerite, trögerite, and arseniosiderite. Natrouranospinite is the most abundant uranyl mineral, and is sometimes replaced by uranophane. It forms fine, yellow-green to straw-yellow, tabular to elongated crystals up to ~2 cm long, as radial aggregates, or as square pseudomorphs after metazeunerite (sometimes retaining a metazeunerite core); UV fluorescence is strong yellow-green (Kopchenova and Skvortsova 1957).

It should be noted that the low water content actually makes this mineral the Na analog of metauranospinite; if a fully hydrated form were found, it isn't clear what it would be called.

The pure end member had been synthesized by Mrose (1951, 1953), who reported that the structure is tetragonal with a=7.12 and c=8.70 Å, based on powder XRD. Sidorenko, Chukanov, and Naumova (2004) reported powder XRD analyses of holotype material from Bota Burum, which is near the end-member composition, as well as of material from Djideli, Kazakhstan, which contains some Ca and P [~ $Na_{.86}Ca_{.40}(H_3O)_{.26}$ and ~ $(AsO_4)_{1.40}(PO_4)_{.58}$]. The type specimen has a=7.113 and c=8.64 Å; the sample from Djideli has a=7.059-7.063 and c=8.61-8.64 Å, showing that substitution of some PO_4 for AsO_4 decreases the a-dimension in the sheet, as one would expect.

Nováčekite $Mg(UO_2)_2(AsO_4)_2·12H_2O$
Metanováčekite $Mg(UO_2)_2(AsO_4)_2·4\text{-}8H_2O$
Nováčekite was described from Schneeberg, Saxony, Germany, by Frondel (1951a), who recognized it to be the arsenate analog of saléeite; saléeite had been noted at Schneeberg the previous year (Mrose 1950). At the TL, nováčekite forms straw-yellow to lemon-yellow, platy to tabular crystals, often in subparallel aggregates, on deeply altered quartz and limonite vein material. The secondary mineral suite is attributed to oxidation of Co, Ni, and Fe arsenides in the primary veins, and includes erythrite, annabergite, roselite, pharmacosiderite, and pharmacolite. Fluorescence in both SW and LWUV is dull green to bright lemon-yellow, depending on the amount of (PO_4) present.

In the Woodrow area, Laguna Reservation, Valencia Co., New Mexico, straw-yellow nováčekite coats iron-stained sandstone of the Jurassic Morrison formation (Stern and Annell 1954).

Members of the saléeite-nováčekite series have been reported from Cornwall, England (Ryback and Tandy 1992), mainly from Wheal Edward, St. Just. End-member nováčekite was found *in situ* on the footwall of a back-filled stope in abandoned workings at West Wheal Owles, St. Just. It forms botryoidal crusts of interlocking scales with an open "egg crate" structure. Individual crystals are typically 10–20 µm across, but can be as large as 0.1 mm (Elton, Hooper, and Jeal 1994).

Nováčekite has been reported from many locales, including: Rabejac and Mas d'Alary, France; Needle's Eye, Dalbeattie, Scotland; the Cherkasar deposit, Uzbekistan; Las

Figure 405. Sharp nováčekite crystals forming a thumbnail-sized group, from Brumado, Brazil. RJL1378

Animas mine, Chihuahua, Mexico; and at the White King mine, Oregon. Particularly large, sharp crystals are found at the Pedra Preta mine, Brumado, Bahia, Brazil. Metanováčekite is likely to be found at many of the same locales.

Frondel (1951a) also described intermediate compositions that suggested a complete series exists between saléeite and nováčekite, and at least partial series extend between nováčekite and both zeunerite and uranospinite. Nováčekite is tetragonal; relatively pure material from the TL has a=7.16 and c=20.19 Å. Analyzed material of intermediate composition in the nováčekite-saléeite series, with As:P ~ 1:1, has a=7.12 and c=20.14 Å. Comparing these values to those of saléeite (7.02 and 19.81 Å) illustrates the general trend that the arsenates have somewhat larger cell parameters than their phosphate analogs.

Przhevalskite $Pb(UO_2)_2(PO_4)_2{\cdot}2H_2O$
Przhevalskite was discovered in the upper part of the contact zone of the Dzherkamer U deposit, near Adrasman, Karamazar Mountains, Tajikistan, by V. G. Kruglova in 1946 (Pekov 1998). The mineral forms foliated aggregates of tiny (0.1–1 mm) bright yellow crystals with prominent cleavage on (001). Associated species include torbernite, autunite, dumontite, renardite, uranophane, halloysite, and wulfenite. The X-ray diffraction pattern of przhevalskite differs from those of other Pb uranyl phosphates and has some similarities to those of the meta-autunite group (Soboleva and Pudovkina 1957). The species is regarded by IMA as questionable or doubtful.

Synthetic material is reportedly tetragonal, with a=7.24 and c=18.22 Å.

Rauchite $Ni(UO_2)_2(AsO_4)_2{\cdot}10H_2O$
Metarauchite $Ni(UO_2)_2(AsO_4)_2{\cdot}8H_2O$
Rauchite was described from the Belorechenskoye deposit, Adygea Republic, Northern Caucasus, Russia, where it forms transparent to translucent, light yellowish-green lamellar crystals to 0.5 mm, in fans or clusters to ~2 mm across. Crystals have a vitreous luster and are not fluorescent. Rauchite is a supergene mineral, associated with dymkovite, annabergite, and goethite in cavities in a dolomite vein with primary uraninite, nickeline, and gersdorfite (Pekov et al. 2012a).

Metarauchite was described from the Schweitzer vein on the second level of the Eduard mine, Jachymov district, Czech Republic. It forms yellow to light greenish-yellow, thick tabular crystals to ~0.8 mm across. Crystals have a vitreous to pearly luster and are not fluorescent. At the TL, it forms on strongly altered aggregates of primary ore with relics of uraninite, arsenopyrite, and nickelskutterudite, and traces of native bismuth. Associated species include metazeunerite, erythrite, and gypsum. Metarauchite was also found in the Adam Heber mine, Neustädtel, Schneeberg district, Germany, where it occurs in the "Gang Adam Heber Flache" in strongly oxidized vein material at a depth of 10-12 m (Plášil et al. 2010a).

Rauchite is triclinic, with a=7.100, b=7.125, c=19.99 Å, α=92.41, β=94.92, and γ=90.42° (Pekov et al. 2012a). Metarauchite is triclinic, with a=7.194, b=9.713, c=13.202 Å, α=75.79, β=83.92, and γ=81.59° (Plášil et al. 2010a).

Sabugalite $H_{0.5}Al_{0.5}(UO_2)_2(PO_4)_2{\cdot}8H_2O$
Sabugalite was originally described from three pegmatites in Portugal: Mina da Quarta Seira, Sabugal, Beira province; Kariz, Minho province; and Bendada, Beira province. At Mina da Quarta Seira, it formed a thick crust on a brecciated mass of milky quartz and feldspar; other minerals noted from that locale include meta-autunite, saleeite, and phosphuranylite. At Kariz, it formed thick coatings admixed with and underlain by meta-autunite. At Bendada, the sabugalite was intergrown with meta-autunite on quartz/limonite vein material. Analysis showed that sabugalite is isostructural with fully hydrated autunite (Frondel 1951). It typically forms small, bright yellow, tabular or lath-like plates to ~1 mm and aggregated crusts; fluorescence is bright lemon-yellow in both SW and LWUV.

Other reported locales include: the Pedro Alvara and Ciudad Rodrigo vanadium mines, Salamanca, Spain; at Margnac and La Crouzille, France; the Sapucaia pegmatite mine, near Governador Valadares, Brazil; and the El Sharana mine, South Alligator Valley, Australia. It is found in several deposits on the Colorado Plateau, including: the Lucky Mc No. 20 mine, Fremont Co., Wyoming; the Happy Jack mine,

White Canyon, Utah; and the Huskon and Arrowhead claims, near Cameron Arizona.

Synthetic sabugalite is monoclinic, with a=19.426, b=9.843, c=9.850 Å, and β=96.161°. Vochten and Pelsmaekers (1983) used ion exchange to transform hydrogen autunite into meta-autunite and sabugalite. They note that, "It is evident that hydrogen autunite rarely occurs in a natural environment. The preferential formation of meta-autunite with respect to sabugalite can be explained by the fact that there are more Ca^{2+} ions than Al^{3+} ions present in circulating water. Meta-autunite can only transform into sabugalite if the local concentration of Al^{3+} ions is increased, as the result of some alteration of the surrounding rocks. As a consequence, sabugalite occurs locally and is frequently associated with meta-autunite."

Figure 406. Small yellow sabugalite crystals on brownish, granitic matrix, from Oliva de la Frontera, Badajoz, Spain. RJL2648

Saléeite $Mg(UO_2)_2(PO_4)_2{\cdot}10H_2O$

Saléeite was described from Shinkolobwe, Congo, where it formed yellow to yellow-green flattened tabular crystals to ~3 mm, typically in aggregates associated with torbernite and dewindtite in siliceous rock. The crystals show bright yellow-green fluorescence. It is fairly rare at the type locale (Thoreau and Vaes 1932).

Saléeite from the Weisser Hirsch mine, Schneeberg, Germany, containing some AsO_4 (P:As ~84.4:18.8) was described in detail by Mrose (1950), who showed it to be isostructural with fully hydrated autunite. It formed crystals to ~1 mm across, in limonite rock associated with uranophane and zeunerite.

A variety containing some Fe^{2+} in substitution for Mg from Arcu su Linnabaru, Sardinia, Italy, yielded the empirical formula $(Mg_{.83}Fe^{2+}_{.16}K_{.02})(UO_2)_2(PO_4)_2{\cdot}3.84H_2O$. It formed opaque, light yellow aggregates of small (0.5 mm) tabular crystals. The locality was explored for uranium in 1950, and mined by open pit and underground workings until around 1980–83. Uranium minerals reported there include meta-autunite, bassetite parsonsite, phosphuranylite, sabugalite, saleeite, and metatorbernite (Vochten and Van Springel 1996).

Saléeite has been reported from many locales worldwide; a few notable examples are: the Cunha Baixa and Quarta Seira mines, Sabugal, Portugal (Frondel 1951a); the South

Figure 407. Yellow platy saléeite, with green torbernite and crusts of dewindtite, from Shinkolobwe, Congo. RJL334

Figure 408. Crust of minute yellow saléetite aggregates and green metazeunerite, from Wheal Edward, St. Just, Cornwall, England. RJL778

Figure 409. Yellow-tan saléeite crystals to ~2 mm on matrix from Rum Jungle, Northern Territory, Australia. RJL571

Figure 410. The specimen in the previous photo, shown in LWUV light. RJL571

Figure 411. A thumbnail-sized cluster of saléeite crystals from the Ranger #3 mine, Northern Territory, Australia. RJL2588

Figure 412. The specimen in the previous photo, shown in SWUV light, illustrating the spectacular fluorescence, which partially contributes to the color seen in daylight. RJL2588

Terras, Bassett, and Roughton Consols mines, Cornwall, England; in Australia, excellent specimens occur at the Ranger mine, and at Rum Jungle; at the Corrego do Urucum pegmatite, Galileia, Brazil; and the Sue mine, Gila Co., Arizona.

The crystal structure, refined using a crystal from the Ranger 1 mine, Northern Territory, Australia, is monoclinic (pseudotetragonal), with a=6.951, b= 19.948, c= 9.896 Å, and β=135.17°. The Mg^{2+} ions are coordinated to six of the ten interlayer H_2O groups, but the last four are not bound to the Mg. The structure becomes pseudotetragonal because the positions of the last few water molecules distort it from true tetragonal symmetry (Miller and Taylor 1986).

Threadgoldite $Al[(UO_2)(PO_4)]_2(OH)\cdot 8H_2O$
Threadgoldite was originally described from the Kobokobo beryl-columbite pegmatite, Kivu province, Congo (Deliens and Piret 1979). At the TL, it occurs with other Al uranyl phosphates (upalite and phuralumite), meta-autunite, and phosphuranylite; it forms greenish-yellow micaceous crystals up to ~1 mm long. Fluorescence is green in LW and pale green in SWUV.

The species was named in honor of Ian Threadgold, who had previously described similar material as unknown "Mineral H" from the Rockhole mine, Northern Territory, Australia (Threadgold 1960). Although Threadgold's original samples were no longer available for study, recent work has confirmed the existence of threadgoldite in archived samples from several localities in the South Alligator Valley district (Henry, Pogson, and Williams 2005). Those mines and prospects, dating from the late 1950s, have been reclaimed and are therefore no longer available to collectors or researchers.

Threadgoldite has also been reported from: Arcu su Linnabaru, Cagliari Province, Sardinia, Italy; and the Cactus Rat mine, Yellow Cat district, Utah.

Threadgoldite is monoclinic with a=20.169, b=9.847, c=19.719 Å, and β=110.71°. The structure (Piret, Declerq, and Wauters-Stoop 1979; Kosrawan-Sazedj 1982) differs from that of uranospathite in that the interlayer Al is present as a dimer with composition $[Al_2(OH)_2(H_2O)_8]$. There are four more H_2O positions held in the structure by hydrogen bonding. Threadgoldite also contains OH, which is not present in uranospathite.

Figure 413. Yellow lath-like threadgoldite forming a thin crust on granite, from the TL at Kobokobo, Kivu, Congo. FOV is ~3 mm. RJL4615

Figure 414. Bright yellow, powdery threadgoldite on muscovite, also from Kobokobo, Congo. FOV is ~5 mm wide. RJL784

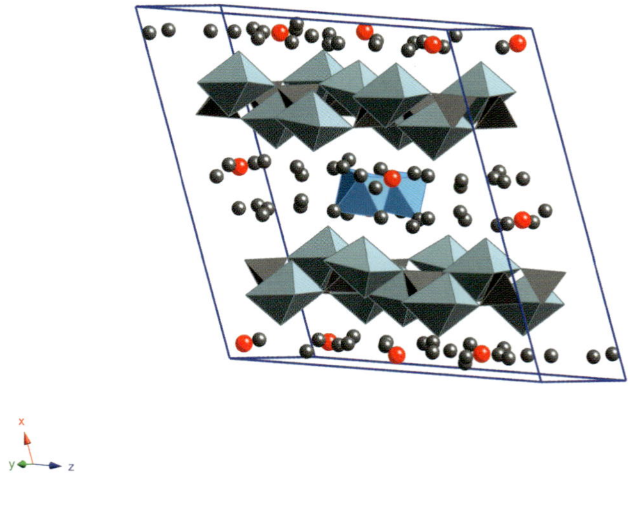

Figure 415. The unit cell of threadgoldite, viewed just off [010], showing the dimer consisting of two edge-sharing distorted Al coordination octahedra (light blue), water oxygens (gray), and OH oxygens (red) in the interlayer between uranyl phosphate sheets.

Torbernite $Cu(UO_2)_2(PO_4)_2 \cdot 8\text{-}12H_2O$
Metatorbernite $Cu(UO_2)_2(PO_4)_2 \cdot 8H_2O$
The name torbernite was first used by Werner around 1790, making this the first uranium mineral named for a person, the Swedish chemist Torber Bergmann (1735-84). At the time there was some confusion regarding the true chemical identities of the autunite group minerals, in particular whether or not copper was an essential constituent or an impurity. Berzelius showed that autunite and torbernite were distinct species around 1819-26; zeunerite/metazeunerite were distinguished from torbernite in 1872 and metatorbernite was separated in 1916 (Frondel, 1958). The vein deposits at Jachymov, Czech Republic, are regarded as the TL. Crystals are commonly thin to thick tabular on (001) with a rectangular or octagonal outline; less commonly the crystals form pyramids terminated by small (001) faces. The crystals often form subparallel or slightly divergent groups or crusts; less frequently seen are scaly or lamellar aggregates and granular or earthy masses. The species has been found as parallel growths with autunite,

Figure 416. Torbernite crystal group, ~2 cm wide, on matrix from the Gunnislake mine, Cornwall. RJL769

Figure 417. Spectacular torbernite specimen, ~6 × 8 cm, from the noteworthy find at Margabal, France. RJL2417

Figure 418. Small but lustrous torbernite crystals from the classic uranium locality at Menzenschwand, Germany, where many members of the group occur (Markl and Wolfsried 2011). Sample is ~1 cm wide. RJL4584

Figure 419. Transparent, emerald-green torbernite crystals lining a cavity ~ 2 cm wide, from Katanga, Congo. RJL274

uranospinite, bassetite, and metazeunerite; when viewed on (001) one mineral will be seen forming a rim around the other. Torbernite crystals are various shades of green and are not fluorescent. Lyon (1956) described a torbernite specimen from Rum Jungle, Northern Territory, Australia, which showed a vivid green fluorescence. The crystals were composites of torbernite and autunite in parallel growth on a very fine scale (0.05–0.5 mm across and 0.005–0.015 mm thick), and the fluorescence was confined to the autunite laths.

As in other members of the group, there is some zeolitic water, although samples kept in ambient conditions typically contain close to $12H_2O$. Desiccation or slight heating will cause the crystals to dehydrate to metatorbernite (Hallimond 1916, 1920).

Figure 420. Torbernite crystals to ~3 mm from Mina Los Azuls, Copiapo, Chile. RJL2970

Figure 422. Torbernite crystals to ~2 mm, thickly covering matrix from La Luz, Sonora, Mexico. RJL3076

Figure 421. Opaque green metatorbernite crystals on matrix from Shinkolobwe, Congo. RJL2570

Figure 423. Metatorbernite crystals forming a solid mass ~2 × 3 cm, from Trancoso, Portugal. RJL768

Figure 424. Bipyramidal metatorbernite crystals to ~3 mm, showing conspicuous color zoning on cleavage sections, from Cunha Baixa, Viseu, Portugal. RJL347

Torbernite is usually found in the gossan of veins containing primary uraninite and chalcopyrite or other copper minerals. It ranks with autunite as one of the commonest secondary uranium minerals. Commonly associated species include autunite, metazeunerite, uranophane, parsonsite, and dewindtite. Some notable localities include: at the Gunnislake mine and elsewhere in Cornwall, England; several of the mines in Katanga, Congo; Margabal, France; Menzenschwand, Germany; La Luz, Sonora, Mexico; and several localities in Portugal.

Single crystal structure refinement using synthetic material showed that torbernite is tetragonal, with a=7.027 and c=20.807 Å. Metatorbernite is also tetragonal with a=6.976 and c=17.349 Å (Locock and Burns 2003c).

Trögerite $(H_3O)_2(UO_2)_2(AsO_4)_2 \cdot 6H_2O$

Trögerite was described by Weisbach (1871, 1873) from the Walpurgis vein in the Weisser Hirsch mine, Germany. It was thought to have the composition $(UO_2)_3(AsO_4)_2.12H_2O$, but Frondel (1958) reviewed some of the problems with reported analyses, and by analogy to "hydrogen uranospinite" suggested that $H_2(UO_2)_2(AsO_4)_2 \cdot 8H_2O$ was a more likely formula. Locock et al. (2004) further point out that compounds with uranyl:arsenate = 3:2 aren't consistent with the stoichiometry and bond valence requirements of the autunite structure. By this reasoning, trögerite likely has the normal 2:2 ratio and is isostructural with chernikovite. The formula given above is based on these arguments. Readers will, however, encounter various conflicting formulas in the literature.

At the TL, trögerite forms thin, lemon-yellow, tabular crystals to ~1 mm, associated with walpurgite, uranospinite, uranospathite, asselbornite, zeunerite, uranosphaerite, erythrite, and cobaltian wad. It has also been reported from: the Rabejac, Riveiral, and Mas d'Alary deposits, Herault, France; Marbruie, Scotland; the Bota-Burum deposit, Kazakhstan; the Bald Mountain district, Lawrence Co., South Dakota; and the Green River area, Emery Co., Utah.

The structural relationships of trögerite to other members of the group were studied by Mrose (1953) and Ross and Evans (1965), primarily using synthetic materials. Ross and Evans (1965) suggest that "complete solid solution series probably exist" between abernathyite and trögerite and between trögerite and uramphite. Assuming that trögerite corresponds to the synthetic "hydrogen autunite" of Mrose (1953), the structure is tetragonal, with a=7.16 and c=8.80 Å.

Uramarsite $(NH_4,H_3O)_2(UO_2)_2(AsO_4,PO_4)_2 \cdot 6H_2O$

Uramarsite was recently described from the supergene zone of the Bota-Burum U-Mo deposit, southwestern Balkhash area, southern Kazakhstan, where it was associated with calcite, arsenopyrite, pyrite, galena, chistyakovaite, natrouranospinite, metazeunerite, trögerite, and several other secondary minerals. It is a hypergene mineral interpreted to have formed by oxidation of the primary pitchblende-sulfide minerals in fracture zones of hydrothermally altered porphyritic felsites. Uramarsite and chistyakovaite form blocky pseudocrystals with sharp boundaries between the two mineral phases. Uramarsite tends to form imperfect, platy crystals up to about 2 mm, along with films and small flattened square crystals to about 0.1 mm. The mineral is pale green; crystals have a vitreous luster, whereas films tend to be dull. It is fluorescent green in UV light (Sidorenko et al. 2007).

Analysis of the type material gave the empirical formula $[(NH_4)_{1.5}(H_3O)_{.72}Na_{.09}](UO_2)_{2.02}[(As_{.61}P_{.39})O_4]_2 \cdot 6.1H_2O$. Crystal quality was insufficient for single crystal studies; refinement using powder XRD indicated that uramarsite is tetragonal, with a=7.19 and c=9.15 Å, but the true c cell edge might be twice the indicated value.

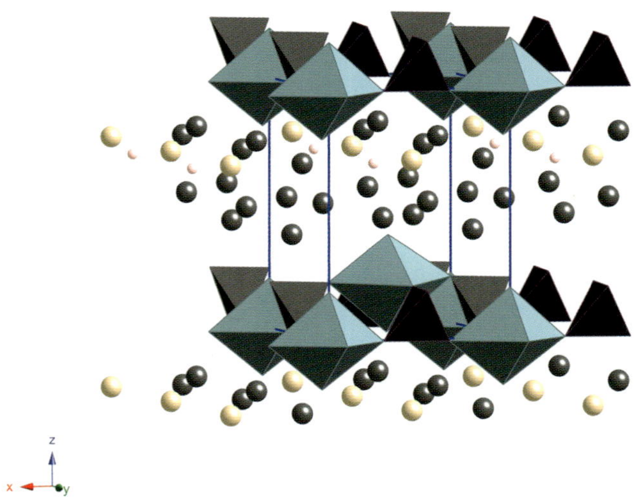

Figure 425. Structure of uramarsite, showing uranyl phosphate-arsenate sheets, with water (gray), H_3O (pink) and NH_4 groups (tan) in the interlayer.

Uramphite $(NH_4)_2(UO_2)_2(PO_4)_2 \cdot 6H_2O$

Uramphite was discovered in 1950 in the oxidized zone of the uranium-coal deposit at Tura-Kavak, Kyrgyzstan (Pekov 1998). It was found in fractures in the coal 20–50 m below the surface. The mineral forms bottle-green to pale green, square tabular crystals (~0.2 mm), small rosettes, and lichen-like crusts. UV fluorescence is medium yellow-green. DTA indicates endothermic breaks at 150°C (loss of H_2O) and 500°C (loss of NH_3 and destruction of the crystal lattice), and an exothermic break at 630°C (crystallization of uranium pyrophosphate) (Nekrasova 1957).

The original report presented powder XRD data for both natural and synthetic material. By analogy to uramarsite, the structure is tetragonal with a=7.02 and c=18.08 Å.

Figure 426. Minute specks of uramphite on coal from the TL at the Tura-Kavak deposit, Kyrgyzstan. RJL3676

Uranocircite $Ba(UO_2)_2(PO_4)_2 \cdot 10H_2O$

Metauranocircite $Ba(UO_2)_2(PO_4)_2 \cdot 6\text{-}8H_2O$

Barium uranyl phosphates have been found in nature in three different states of hydration: uranocircite, with $10H_2O$; metauranocircite I, with $8H_2O$; and metauranocircite II, with $6H_2O$. The original material from Bergen, Falkenstein, Germany, described under the name uranocircite by Weisbach (1877), was later shown to be the *meta-* I form. At Bergen, it was found in quartz veins in granite. Uranocircite is visually indistinguishable from autunite. Intermediate compositions are known, including a Ca-rich metauranocircite from Madagascar, and it is likely that a complete series extends between the Ca and Ba end members (Nuffield and Milne 1953).

Because of the ease with which the hydration state can vary with ambient humidity, it is likely that the species may coexist at any particular locale, and also that the hydration will change during storage. A sample that has been out of the ground for any significant time should be assumed to be in one of the *meta-* states.

Some reported localities are: in Germany at Bergen, Falkenstein (TL), at Schwartzenberg, Saxony, at Wölsendorf, Bavaria, and at Menzenschwand, Black Forest; near Antsirabe, Madagascar; the Mounana mine, Gabon; at the São Pedro claim, Minas Gerais, Brazil; and in the US in the Wind River Basin, Fremont Co., Wyoming, in Lawrence Co., South Dakota, and near Cameron, Arizona.

Pulou (1957) described near end-member metauranocircite from the Entraygues region, Aveyron, France. Pulou

Figure 427. "Uranocircite" (probably metauranocircite I) from Streuberg, Bergen, Saxony, Germany: a very rich specimen covered with stacks of crystals to ~1 cm on a side. RJL3111

Figure 428. Tabular metauranocircite covering massive, iron-stained quartzite, from Bergen, Saxony. RJL777

noted that although the mineral is considered fairly rare, it was so abundant in this area as to form an exploitable ore deposit.

The structures of the *meta*- I and *meta*- II forms were refined by Locock, Burns, and Flynn (2005), who suggested that metauranocircite I is probably isostructural with its arsenate analog, metaheinrichite. Synthetic uranocircite is monoclinic, with a=7.01, b=6.99, c=21.2 Å, and β=103.9°. Metauranocircite I is monoclinic, with a=6.943, b=17.634, c=6.952 Å and β=90.024°.

Figure 429. Thick, tabular uranocircite crystal ~15 mm wide, from the São Pedro claim, Malacacheta, Minas Gerais, Brazil. RJL4277

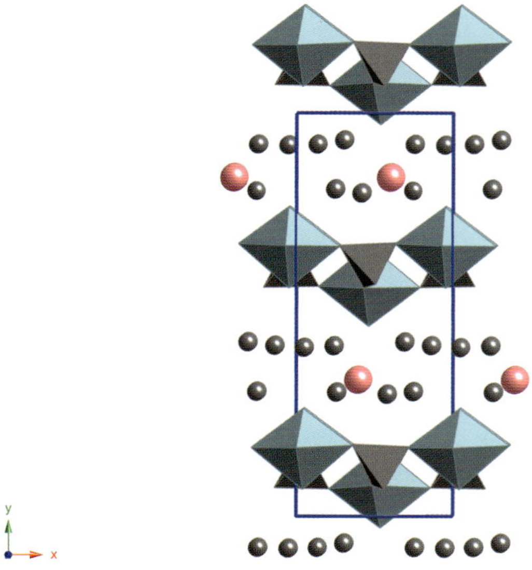

Figure 430. The structure of metauranocircite-I.

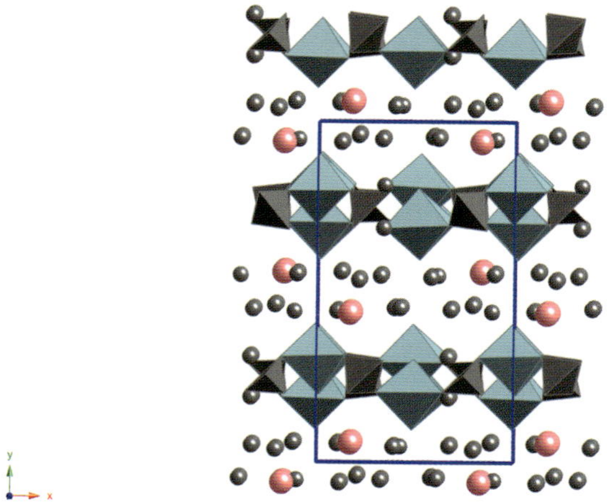

Figure 431 The structure of metauranocircite-II, showing the contraction of the c dimension compared to metauranocircite-I.

Uranospathite $Al_{1-x}\square_x[(UO_2)(PO_4)]_2(H_2O)_{20+3x}F_{1-3x}$, 0<x<0.33
Uranospathite was described by Hallimond (1915) based on material from Redruth, Cornwall, England. The type material consisted of fan-like groups of thin yellow rectangular plates in black gossan. Hallimond recognized it as a member of the autunite group (generically called *uranite* at the time) and noted that its low index of refraction suggested it was a new species even though there wasn't enough material for chemical analysis. Frondel (1954), possibly working with impure material, confirmed some of the physical properties. Reinvestigation by Walenta (1978) established the species as a member of the autunite group containing Al and showed that it is more highly hydrated than other members of the group. Both Hallimond and Walenta noted the parallel growth of bassetite and uranospathite in the same crystal.

Uranospathite has also been reported from: La Crouzille and Sagnes mines, Haut-Vienne, France; the Pedro Alvaro mine in Salamanca and at El Padregal, Badajoz, Spain; the Krunkelbach Valley, Menzenschwand, Germany; and at Radium Hill, South Australia.

Uranospathite is orthorhombic, with *a*=30.020, *b*=7.008, and c=7.049 Å. The structure contains the autunite-type uranyl phosphate sheet, with isolated $[Al(H_2O)_6]$ octahedra held in the interlayer by hydrogen bonding to water molecules above and below the octahedra. The Al sites are not fully occupied (~86%); however, there is a net positive charge, which is compensated by the presence of F^-. The interlayer spacing is unusually large at 15.01 Å, compared to ~10 Å for most members of the group (Locock, Kinman, and Burns 2005).

Figure 432. Yellow, lath-like uranospathite microcrystals from Pedro Alvaro, Salamanca Province, Spain. RJL804

Figure 433. Uranospathite from the Krunkelbach deposit, near Menzenschwand, Black Forest, Germany. FOV is ~4 mm wide. RJL2705

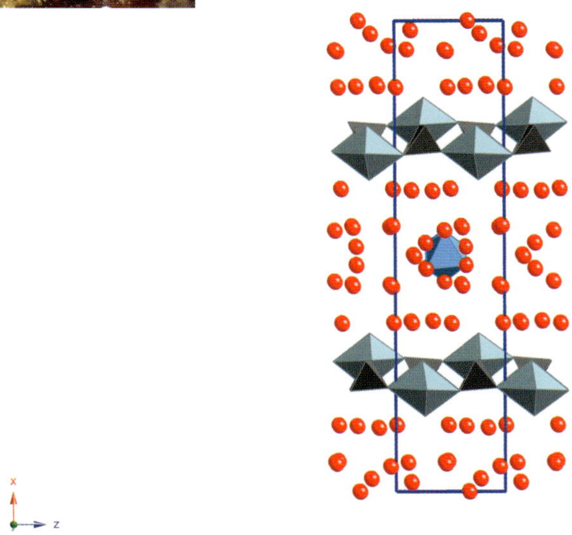

Figure 434. Crystal structure of uranospathite, showing octahedrally coordinated Al (light blue), held in the interlayer via hydrogen bonding to H_2O molecules (red).

Uranospinite $Ca(UO_2)_2(AsO_4)_2 \cdot 10H_2O$
Metauranospinite $Ca(UO_2)_2(AsO_4)_2 \cdot 8H_2O$
Uranospinite, the arsenate analog of autunite, was discovered at the Weisser Hirsch mine, Neustädtel, near Schneeberg, Germany, in 1871; the name and description were published by Weisbach (1873, 1877). The mineral was formed through alteration of uraninite and primary arsenides (e.g., smaltite) in a hydrothermal vein (Frondel 1958). Crystals are thin tabular, generally <1 mm, in aggregates and crusts; color is lemon-yellow to siskin-green. Associated species include zeunerite, walpurgite, uranosphaerite, and trögerite. Synthetic crystals fluoresce bright lemon-yellow in LWUV, less brightly in SWUV (Mrose 1953).

Reported localities include: the Clara mine, the Sophia mine, and the Menzenschwand deposit, Germany; Rabejac, France; Cherkasar, Uzbekistan; the Talmessi mine, Iran; and Radium Hill, South Australia. In the US, it has been reported from: the White King mine, Lakeview, Oregon (a locale more notable for heinrichite); the Orphan mine, Arizona; and several sites in Utah.

Mrose (1953) synthesized uranospinite and several of its analogs and characterized their properties, suggesting that uranospinite was tetragonal. Locock, Burns, and Flynn (2005) refined the structures of several minerals with the autunite structure and predicted that uranospinite should be orthorhombic, with *a*=14.35, *b*=20.66, and *c*=7.17 Å.

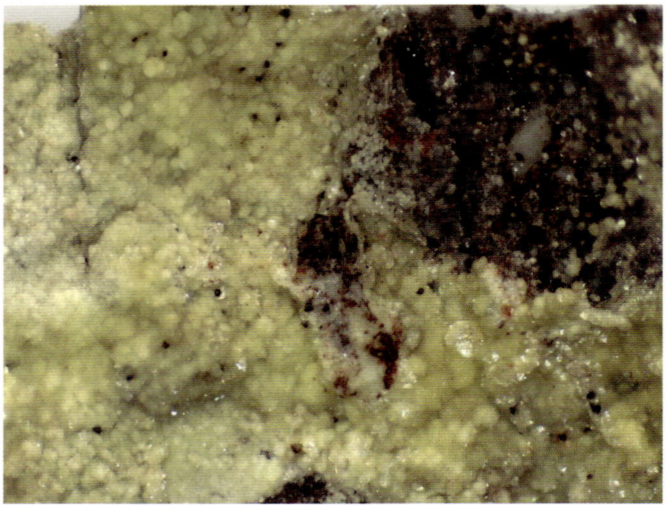

Figure 435. Minute, somewhat pearly, pale yellowish-green scales of uranospinite forming tiny globular aggregates from the No. 16 mine, Pribram, Czech Republic. FOV is ~4 mm. RJL4616

Zeunerite $Cu^{2+}(UO_2)_2(AsO_4)_2 \cdot 10\text{-}16H_2O$
Metazeunerite $Cu^{2+}(UO_2)_2(AsO_4)_2 \cdot 8H_2O$
Zeunerite was described from the Weisser Hirsch mine, near Schneeberg, Germany, by Weisbach (1873, 1877), who recognized it to be the arsenate analog of torbernite. Zeunerite crystals typically resemble those of torbernite, being transparent emerald-green when fully hydrated, and paler and more opaque when dehydrated. Zeunerite is not fluorescent.

The original "zeunerite" of Weisbach appears to have been metazeunerite, but fully hydrated zeunerite has also been identified at Schneeberg. Thus the material originally analyzed was likely a dehydration product and not a directly formed metahydrate (Frondel 1958). The water content of zeunerite varies between about 10 and 16 H_2O; under ordinary conditions it is usually around 10. Equilibra-

Figure 436. Bright green metazeunerite crystals to ~5 mm, scattered on altered silicic matrix, from Cinovec, Bohemia, Czech Republic. RJL2381

Figure 437. Platy metazeunerite crystals on matrix from Adelaide River, Northern Territory, Australia. RJL572

Figure 438. Tabular metazeunerite crystals lining a seam ~1 mm thick, from the Copper stope, Majuba Hill, Pershing Co., Nevada. RJL782

Figure 439. Equant metazeunerite crystals, ~5 mm wide, from Erongo, Namibia. RJL2783

tion with ambient humidity is fairly rapid: fully hydrated material ($16H_2O$) will dehydrate to ~$10H_2O$ upon standing in air overnight (Frondel 1951). J. W. Frondel (1951) compared natural zeunerite from Tintic, Utah, with synthetic material and noted, "Study of synthetic zeunerite indicated the existence of two and possibly three hydrates, one with 16 to $10H_2O$, one with 8 to $5H_2O$, and possibly a still lower hydrate. In this respect the synthetic material is similar to autunite and torbernite, both of which have been found in the fully hydrated form and as lower hydrates, meta I and meta II. Natural zeunerite corresponds in water content and X-ray structure to meta-autunite I and to synthetic zeunerite with $8H_2O$."

Zeunerite and metazeunerite are found in the oxidized zone of deposits containing both uraninite and primary arsenide minerals such as skutterudite or enargite. Typical associated species include trögerite, walpurgite, uranospinite, erythrite, mimetite, pharmacosiderite, olivenite, and chalcophyllite.

Many localities have been reported: as with other members of the group, both zeunerite and metazeunerite will often be found at a particular occurrence even if only one of the two has been formally documented. Reported locales include: in England at Wheal Gorland, Wheal Owles, and Wheal Edwards; Cap Garonne, Rabejac, and Margnac, France; Cinovec, Czech Republic; in Australia in the Bali Lo copper prospect, Western Australia, in the Dome Rock copper mine, South Australia, and in the South Alligator Valley, Northern Territory; in the Erongo Mountains, Namibia; the Pedra Preta mine, Bahia, Brazil; in the US, at Majuba Hill, Nevada, at the Centennial Eureka mine, Utah, and at the Grandview mine, Arizona.

The structures of Cu-containing members of the autunite group were refined by Locock and Burns (2003c) using synthetic materials. Zeunerite is isostructural with torbernite; it is tetragonal with a=7.1797 and c=20.857 Å. Metazeunerite is isostructural with metatorbernite; it is tetragonal with a=7.1094 and c=17.416 Å.

Phosphuranylite Group

The phosphuranylite group is based on a structural sheet with U:P = 3:2 suggesting the general formula $(M^{p+})_m[(UO_2)_3(O,OH)_2(PO_4)_2]_{p\cdot m/2}(H_2O)_n$. This concept of the group includes: althupite, arsenovanmeersscheite, arsenuranylite, bergenite, dewindtite, dumontite, françoisite-(Ce), françoisite-(Nd), hügelite, kivuite, mundite, phosphuranylite, phuralumite, phurcalite, renardite, upalite, vanmeersscheite, metavanmeersscheite, and yingjiangite. Many authors consider renardite to be identical to dewindtite; the IMA officially lists it as "questionable." Kivuite is also regarded by some as a doubtful species, perhaps a mixture of several minerals, and it has been discredited by the IMA. Neither mineral will be treated in detail here.

Table 50. Phosphuranylite group

Mineral	Formula	Symmetry
Althupite	$AlTh(UO_2)[(UO_2)_3O(OH)(PO_4)_2]_2(OH)_3{\cdot}15H_2O$	tric
Arsenuranylite	$Ca(UO_2)_4(AsO_4)_2(OH)_4{\cdot}6H_2O$	orth
Arsenovanmeersscheite	$U^{4+}(UO_2)_3(AsO_4)_2(OH)_2{\cdot}4H_2O$	orth
Bergenite	$Ca_2Ba_4[(UO_2)_3O_2(PO_4)_2]_3{\cdot}16H_2O$	mon
Dewindtite	$Pb_3[H(UO_2)_3O_2(PO_4)_2]_2{\cdot}12H_2O$	orth
Dumontite	$Pb_2[(UO_2)_3O_2(PO_4)_2]{\cdot}5H_2O$	mon
Françoisite-(Ce)	$Ce[(UO_2)_3O(OH)(PO_4)_2]{\cdot}6H_2O$	mon
Françoisite-(Nd)	$Nd[(UO_2)_3O(OH)(PO_4)_2]{\cdot}6H_2O$	mon
Hügelite	$Pb_2[(UO_2)_3O_2(AsO_4)_2]{\cdot}5H_2O$	mon
Metavanmeersscheite	$U^{6+}[(UO_2)_3(OH)_6(PO_4)_2]{\cdot}2H_2O$	orth
Mundite	$Al(UO_2)_3(OH)_3(PO_4)_2{\cdot}5.5\ H_2O$	orth
Nielsbohrite[a]	$K(UO_2)_3(AsO_4)(OH)_4{\cdot}H_2O$	orth
Phosphuranylite	$Ca(UO_2)[(UO_2)_3(OH)_2(PO_4)_2]_2{\cdot}12H_2O$	orth
Phuralumite	$Al_2[(UO_2)_3(OH)_2(PO_4)_2](OH)_4{\cdot}10H_2O$	mon
Phurcalite	$Ca_2[(UO_2)_3O_2(PO_4)_2]{\cdot}7H_2O$	orth
Upalite	$Al[(UO_2)_3O(OH)(PO_4)]{\cdot}7H_2O$	mon
Vanmeersscheite	$U^{6+}[(UO_2)_3(OH)_6(PO_4)_2]{\cdot}4H_2O$	orth
Yingjiangite	$(K_2,Ca)(UO_2)_7(PO_4)_4(OH)_6{\cdot}6H_2O$	orth?

[a]Actually a framework structure, but with uranyl arsenate chains like those in the phosphuranylite group.

The structures are based on sheets consisting of dimers of edge-sharing uranyl pentagonal bipyramids that in turn share edges with uranyl hexagonal bipyramids to form chains running parallel to [001]. Phosphate or arsenate tetrahedra share edges with a given chain and vertices with the next chain to form sheets of general composition $[(UO_2)_3(\mathbf{T}O_4)_2\mathbf{X}_2]$ where $\mathbf{T}$ = P or As and $\mathbf{X}$ = O or OH. The interlayer region between adjacent sheets is occupied by monovalent to tetravalent cations and/or complex ions, along with OH or H_2O

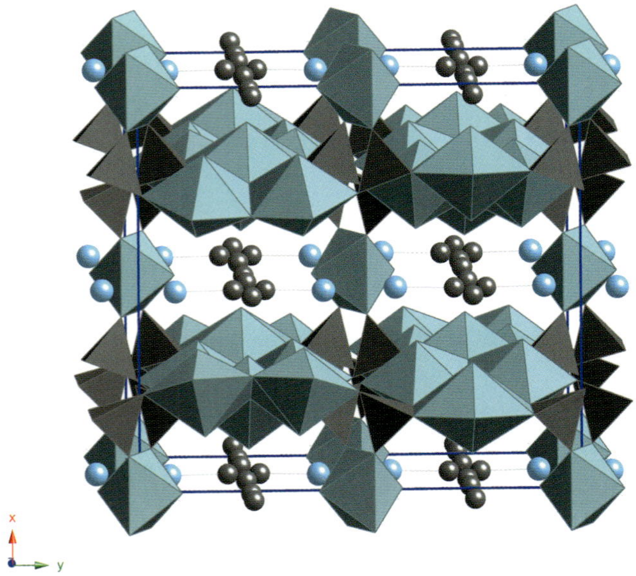

Figure 441. Crystal structure of phosphuranylite, viewed just off [001], showing two uranyl phosphate sheets, with H_2O (gray), Ca (blue atoms), and UO_6 groups (blue-gray octahedra) linking the sheets via oxygens on the phosphate groups.

Figure 440. The sheet in phosphuranylite, viewed along [100], showing the chains of edge-sharing uranyl polyhedra, connected by PO_4 tetrahedra (dark gray).

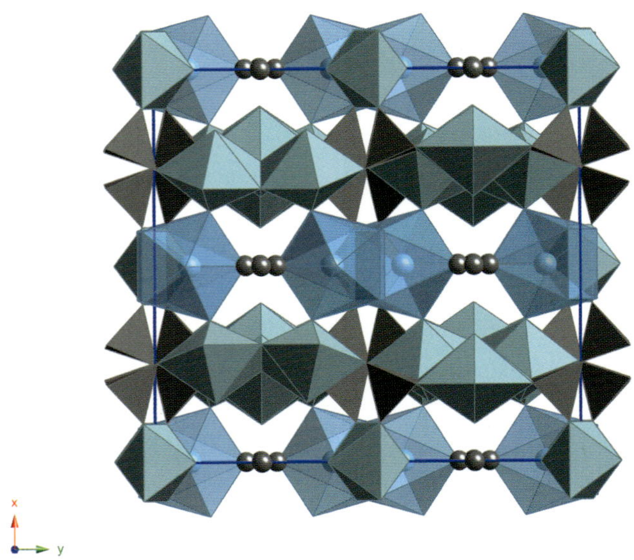

Figure 442. In this view, the coordination polyhedra of Ca (translucent blue) have been added to show the additional linkages to adjacent sheets via bonds to the Ca.

groups. The phosphuranylite sheet derives some geometrical flexibility through variations in the orientation of the PO_4 or AsO_4 tetrahedra, giving rise to different geometrical isomers of the sheet (Krivovichev and Plašil 2013).

Compared to the autunite group, phosphuranylite-group minerals form under more alkaline conditions at the end of the acidic stage in the development of oxidation zones. They tend to replace autunite at a pH of ~7.5–8.5, and contain either O^{2-} or OH^-, which reflects crystallization under alkaline conditions (Krivovichev and Plašil 2013).

Althupite $AlTh(UO_2)[(UO_2)_3O(OH)(PO_4)_2]_2(OH)_3{\cdot}15H_2O$
Althupite was described from a uraniferous pegmatite at Kobokobo, Kivu, Congo, where it occurred as thin, transparent amber-yellow, tabular crystals to ~0.1 mm long, and flattened on (100). It is not fluorescent. The tiny, triclinic crystals are found as isolated individuals and parallel groups, occupying small cavities in white, kaolinized feldspar. Associated species include beryl and columbite.

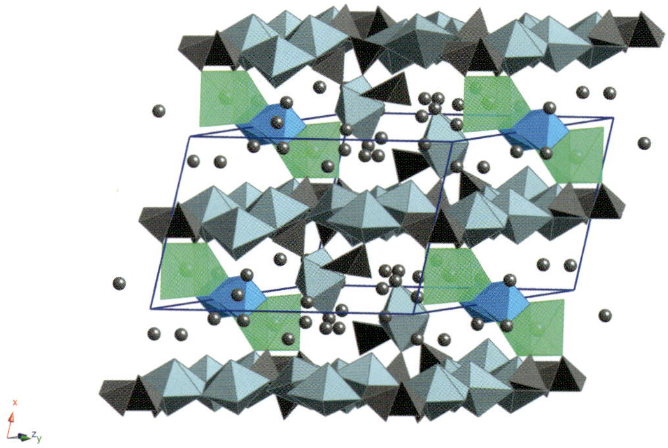

Figure 443. The crystal structure of althupite, showing $[(UO_2)_3O(OH)(PO_4)_2]^{3-}$ sheets, connected by Al (small joined pairs of blue octahedra), Th (translucent green polyhedra), and UO_7 bipyramids. Gray atoms represent the oxygen in H_2O groups.

Althupite was the first mineral of the group found to have a triclinic structure, with a=10.935, b=18.567, c=13.504 Å, α=72.64, β=68.20, and γ=84.21°. Typical of the group, the dominant structural theme features sheets of overall composition $[(UO_2)_3O(OH)(PO_4)_2]^{3-}$. The layers are connected by joined pairs of octahedrally coordinated Al, one Th polyhedron (a tri-capped trigonal prism), and a UO_7 bipyramid (Piret and Deliens 1987).

Arsenovanmeersscheite $U^{4+}(UO_2)_3(AsO_4)_2(OH)_2{\cdot}4H_2O$
Arsenovanmeersscheite was described from the uranium deposit at Menzenschwand, Germany (Walenta and Theye 2007). Markl and Wolfsried (2011) note that vanmeersscheite and arsenovanmeersscheite form intense yellow, elongated, very thin tabular crystals. Individual crystals are very small (~0.1 mm) and form cm-sized coatings. Associated minerals include schoepite, studtite, and uranocircite.

Arsenuranylite $Ca(UO_2)_4(AsO_4)_2(OH)_4{\cdot}6H_2O$
Arsenuranylite, the As analog of phosphuranylite, was discovered in 1954 in the oxidation zone of the Cherkasar U deposit, Uzbekistan (Pekov 1998). It formed via the oxidation of As-bearing sulfides, in association with metazeunerite, uranospinite, and nováčekite, which it replaces. Arsenuranylite may in turn be replaced by schoepite and paraschoepite. It forms orange-yellow, lichen-like masses of minute, scaly crystals (Belova 1958).

Figure 444. A small but historical specimen: holotype material from the Cherkasar deposit, Uzbekistan, originally studied by Larisa Belova, showing powdery orange-yellow arsenuranylite, with green metazeunerite and bright yellow schoepite. RJL1674

Arsenuranylite is orthorhombic; the powder XRD pattern, indexed by analogy to phosphuranylite, gives a=15.40, b=17.40, and c=13.768 Å. The Raman spectrum was presented by Frost, Čejka, and Dickfos (2009) and interpreted by assigning the observed Raman bands to OH stretching, ν_3 $(UO_2)^{2+}$ antisymmetric stretching, ν_1 $(UO_2)^{2+}$ and $(AsO_4)^{3-}$ symmetrical stretching modes.

Bergenite $Ca_2Ba_4[(UO_2)_3O_2(PO_4)_2]_3 \cdot 16H_2O$

Bergenite was originally found on a mine dump from the SDAG Wismut 254 shaft at Streuberg, near Bergen on the Trieb, Vogtland, Germany (Bültemann and Moh 1959) where it was associated with uranocircite, torbernite, renardite, autunite, and other secondary uranium minerals. Bergenite forms clusters and coatings of bright yellow thin tabular crystals. It is weakly fluorescent pale green under UV light according to several recent handbooks, although the original description noted "weak orange-brown" fluorescence. Material from this author's research collection shows weak green fluorescence under SWUV.

In the original description, bergenite was assumed to be the (end member) Ba analog of phosphuranylite, with the proposed formula $Ba(UO_2)_4(PO_4)_2(OH)_4 \cdot 8H_2O$. That compound had been synthesized earlier by Ross (1956), who called it barium-phosphuranylite. The authors rejected that name as potentially confusing and named the species for the TL. Later studies showed that Ca is an essential constituent.

Piret and Deliens (1981) re-examined bergenite, using material from the holotype specimen, and concluded that although bergenite is a member of the phosphuranylite group, it is not completely equivalent to synthetic "barium-phosphuranylite." Based on microprobe analysis, they suggested the formula $Ba_{4/3}Ca_{2/3}(UO_2)_3(PO_4)_2(OH)_4 \sim 5.5(H_2O)$.

Figure 445. A historical specimen of bergenite from Mechelgrun, near Bergen, Vogtland, Germany, dating from the late 1960s. A yellow mass about 1 cm long, of thin, platy bergenite crystals rests in a quartz-lined vug. RJL2248

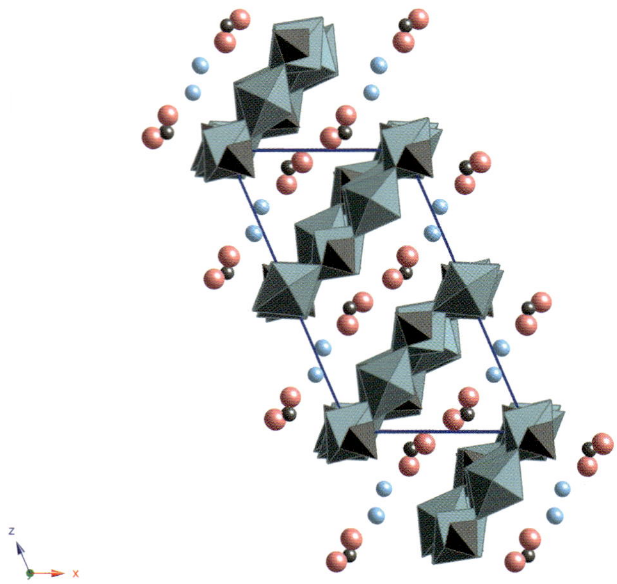

Figure 446. The crystal structure of bergenite, viewed along [010], showing uranyl phosphate sheets running diagonally through the unit cell, with Ca (pale blue), Ba (pink), and water (gray) in the interlayer.

The TL has since been reclaimed; bergenite is also found at Menzenschwand, Germany, and in graphitic uranium ore of the Black Slate Formation, South Korea.

Bergenite was originally thought to have orthorhombic symmetry; however, a sample from the TL was determined to be monoclinic with a=10.092, b=17.245, c=17.355 Å, and β=113.678°. Based on this refinement, Locock and Burns (2003b) presented the ideal formula given above. The structure contains chains of UO_7 and UO_8 bipyramids. The chains are linked together by phosphate tetrahedra to form sheets with an unusual geometry. The interlayer contains one Ca atom, two Ba atoms, and eight H_2O groups.

Dewindtite $Pb_3[H(UO_2)_3O_2(PO_4)_2]_2 \cdot 12H_2O$

Dewindtite was described from Kasolo, Congo, where it was first isolated as minute crystals impregnating torbernite. It forms canary-yellow, powdery to very rarely compact masses. In the microscope, individual crystals are very small (0.02–0.03 mm), translucent, pale-yellow tablets with generally square or rectangular outline. The edges are often somewhat ragged, suggesting incipient corrosion or dissolution (Schoep 1922a). In the commentary accompanying the English abstract, the abstractor (E. T. W.) suggested that the species might belong to the autunite group, based on its apparently tetragonal crystals. Later, Schoep (1928) reported another new mineral, *renardite*, found in a similar association at Kasolo, and resembling dewindtite and dumontite. Frondel and Cuttita (1954) determined that *renard-*

ite is isostructural with dewindtite and phospuranylite and that it is the Pb analog of phosphuranylite. *Renardite* has since been questioned as a species and is now generally regarded as identical to dewindtite.

In addition to the type locale, dewindtite has been reported from: Shinkolobwe and Swambo, Congo; in Germany at Wölsendorf, Streuberg, and Schneeberg; La Faye, France; Urgeriça, Portugal; the Ranger mine, Australia; and several places in Canada.

Dewindtite is orthorhombic with a=16.032, b=17.265, and c=13.605 Å. The structure comprises uranyl phosphate sheets parallel to (100), linked by Pb in the interlayer; the formula given above is based on the structural analysis (Piret, Piret-Meunier, and Deliens 1990).

Figure 447. Golden yellow, somewhat micaceous dewindtite forming a crust with green torbernite, from Mine La Faye, Grury, Saone et Loire, France. The author obtained the specimen in 1983, at which time it was labeled "renardite." RJL575

Figure 448. Thin golden-brown tabular dewindtite crystals on green torbernite, also from La Faye, France. RJL1030

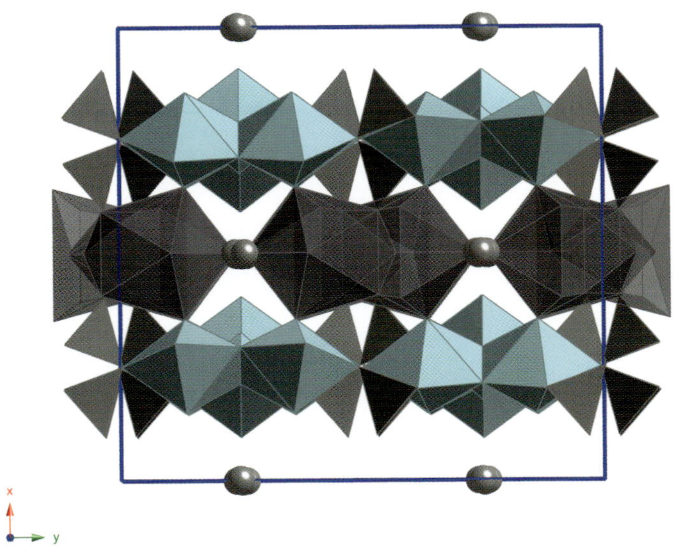

Figure 449. The dewindtite crystal structure, showing uranyl phosphate sheets with Pb (large translucent gray polyhedra) and water (small gray atoms) in the interlayer.

Dumontite $Pb_2[(UO_2)_3O_2(PO_4)_2]·5H_2O$
Dumontite was originally found in pockets of compact torbernite at Shinkolobwe, Congo, where it formed flattened, elongated, prismatic crystals as isolated individuals and stellate groups. Individual crystals are prismatic, up to 2 mm long × 0.15 mm wide, elongated parallel to [001], flattened on (100), and often striated parallel to the elongation. The crystals are pale yellow to golden yellow, and are not fluorescent (Schoep 1924).

Dumontite is fairly rare at Shinkolobwe. Other reported locales include: Kobokobo, Kivu, Congo; the White Oak property, near Nogales, Arizona; the Green Monster mine, Clark Co., Nevada; and the Ranger mine, Australia.

Dumontite is monoclinic with a=8.119, b=16.820, c=6.983 Å, and β=109.03°. The structure contains $[(UO_2)_3O_2(PO_4)_2]$ sheets, parallel to (100), connected by Pb^{2+}, which are bonded to eight oxygens forming a distorted square antiprism. Each antiprism is, in turn, bonded to two others, forming chains parallel to [010] (Piret and Piret-Meunier 1988).

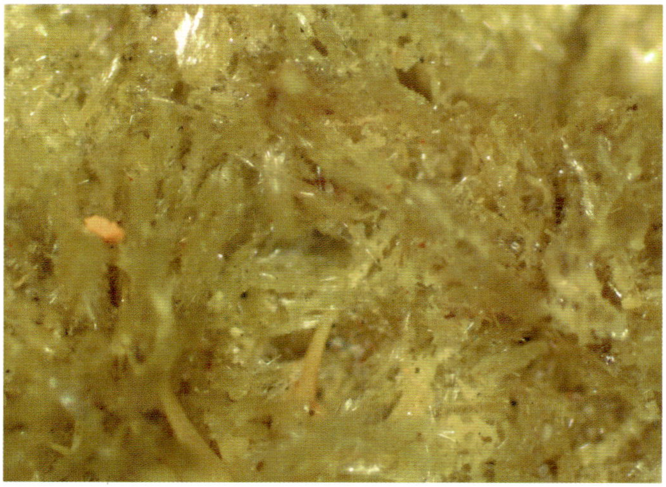

Figure 450. Pale yellow, flattened needles of dumontite from the type locale at Shinkolobwe, Congo. RJL340

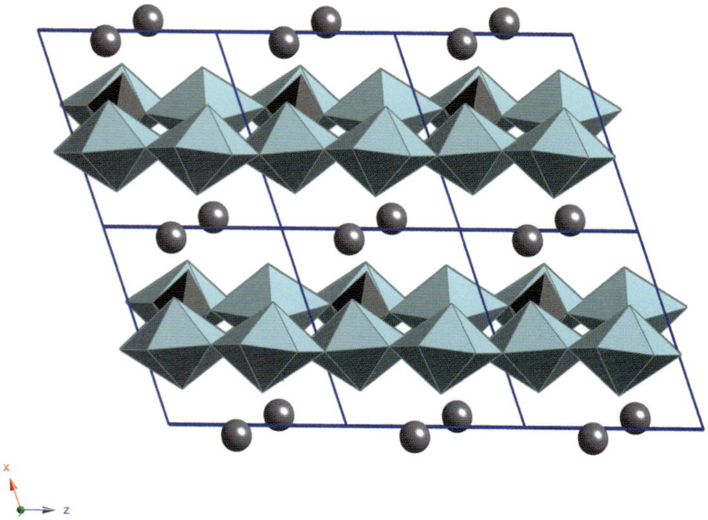

Figure 452. The structure of dumontite, showing the $[(UO_2)_3O_2(PO_4)_2]$ sheets, one per unit cell, with Pb (gray atoms) in the interlayer. Water groups are not shown.

Figure 451. Pale yellow mass of dumontite, ~2 cm wide, on malachite from Shinkolobwe, Congo. RJL4340

Figure 453. The dumontite structure viewed normal to the uranyl phosphate sheet, showing how the Pb coordination polyhedra (translucent gray) form chains running parallel to [010].

Françoisite-(Nd) $Nd[(UO_2)_3O(OH)(PO_4)_2]·6H_2O$
Françoisite-(Nd) was one of several rare-earth minerals discovered during expansion of the Kamoto East Co-Cu mine, Congo, in the late 1980s (Deliens, Piret, and Van Der Meerssche 1990). At the TL, françoisite-(Nd) forms aggregates of elongated, yellow, tabular crystals to ~0.3 mm long, flattened on (010) and elongated on [001], which are not fluorescent. Associated species include uraninite, schoepite, uranophane, curite, kamotoite-(Y), and schuilingite (Piret, Deliens, and Piret-Meunier 1988). In the type material, Nd is the dominant REE, followed by Sm, Y, Ce, Pr, Dy, and La, in order of decreasing abundance. The Nd content is about four times that of Ce.

A variety containing significant amounts of Ce (Ce:Nd ~ 1:1) was later found in tiny amounts at the natural fission reactor site at Bangombé, Gabon (Janeczek and Ewing 1996). Ce-rich françoisite-(Nd) is also found at the Number 2 workings on Radium Ridge, near Arkaroola, Northern Flinders Range (Brugger et al. 2004).

Françoisite-(Nd) is monoclinic with a=9.298, b=15.605, c=13.668 Å, and β=112.77°. The structure contains uranyl phosphate sheets lying parallel to (010), bonded together by the rare-earth elements in the interlayer (Piret, Deliens, and Piret-Meunier 1988).

Pure françoisite-(Nd), i.e., containing no other REE, was synthesized from solutions of neodymium acetate, uranyl acetate, and phosphoric acid. The product was characterized by XRD, TGA, and both IR and Raman spectroscopy in order to gain insights into the fundamental properties of this mineral phase as they might relate to radioactive waste behavior and remediation strategies (Armstrong et al. 2011).

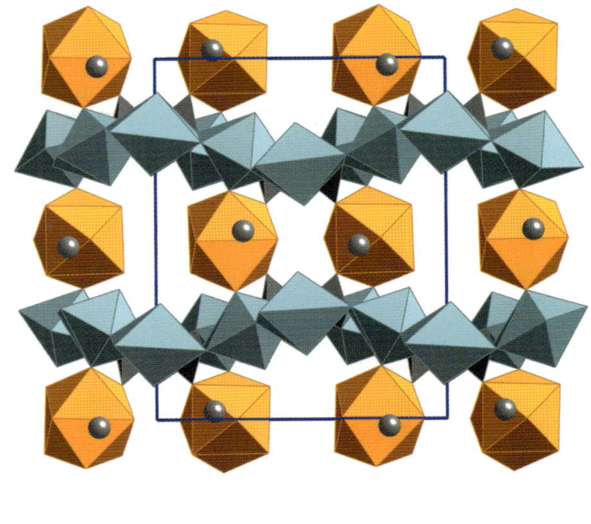

Figure 454. Crystal structure of françoisite-(Nd), viewed along [100], showing two $[(UO_2)_3O(OH)(PO_4)_2]$ sheets, bonded by rare-earth atoms (orange polyhedra); water groups are gray. The unit cell is shown in blue.

Françoisite-(Ce) $Ce[(UO_2)_3O(OH)(PO_4)_2]·6H_2O$
The Ce-dominant analogue of françoisite-(Nd) was based on a description of materials from two co-type locales: the La Creusaz uranium prospect, Switzerland, and the Number 2 workings, Radium Ridge, Australia (Meisser et al. 2010). At Le Creusaz, it forms radial aggregates of lemon-yellow prismatic crystals, elongated on [001] and up to ~100 µm long, growing directly on gneiss adjacent to the uraniferous hydrothermal breccia. Crystals are transparent with vitreous luster, and are not fluorescent. At Radium Ridge, it forms scale-shaped crystals to 1 mm long, mixed with microscopic inclusions of a kaolinite mineral.

At both of the co-type locales, françoisite-(Ce) is part of the secondary mineral assemblage formed through alteration of a primary uranium (±REE) mineralization, although the occurrences are significantly different.

At La Creusaz, it was found in the "Gisinger" surface scratching (~630 m^2). The mineralization comprises hydrothermal breccia veins containing uraninite and pyrite, which were subject to several later episodes of mineralization, partial remobilization involving intense fluid circulation, and strong mechanical and chemical weathering. The françoisite-(Ce) is associated with abundant uranophane, novačekite-metanovačekite, jarosite, and minor françoisite-(Nd), metatorbernite, metazeunerite, arsenuranospathite, and hyalite opal. Primary allanite-(Ce), monazite-(Ce), and fluorapatite are locally abundant accessory minerals in the gneisses on which françoisite-(Ce) is found.

The Radium Ridge province contains numerous small hematite-U-Cu-Nb-REE deposits, along with a large epithermal system of complex quartz/fluorite veins and breccias. At the Number 2 workings, a small lens of coarse hematite-quartz ore contained monazite-(Ce), xenotime-(Y),

Figure 455. Yellow microcrystals of françoisite-(Ce) forming minute tufted aggregates on gneiss, from the TL at the La Creusaz U prospect, Les Marecottes, Switzerland. FOV is ~5 mm wide. RJL3597

an ishikawaite-samarskite solid solution, Ca-Fe phosphate, and an unidentified U(Pb,Mn) oxide that was always fully replaced by a secondary mineral assemblage. Secondary uranium minerals mostly occur in cavities in the epithermal quartz and in cavities formed by the dissolution of the U(Pb,Mn) oxide mineral; françoisite-(Ce) was found in the latter setting. It is directly associated with metatorbernite, barite, an abundant kaolinite mineral, and cesarolite(?). Other secondary minerals reported include curite, boltwoodite, billietite, hyalite opal, kasolite, rutherfordine, schoepite, metaschoepite, soddyite, spriggite, metatorbernite, uranophane-β, and weeksite.

Material with a composition very near Ce:Nd=50:50 was noted in sandstone beneath the natural fission reactor at Bangombé, Gabon (Janaczek and Ewing 1996).

Françoisite-(Ce) is monoclinic; based on the refinement of powder XRD data, and assuming it is isostructural with françoisite-(Nd), the unit cell parameters are *a*=9.296, *b*=15.53, *c*=13.719 Å, and β=112.39°.

Hügelite $Pb_2[(UO_2)_3O_2(AsO_4)_2]·5H_2O$
Hügelite, one of the few arsenate members of the group, was described from the Michael mine, near Lahr, in the Black Forest, Germany, by Dürrfield (1913) as a lead zinc vanadate hydrate. Walenta and Wimmenauer (1961) determined that the mineral is actually a hydrated lead uranyl arsenate structurally related to dumontite. It forms monoclinic, somewhat tabular, brown to orange-yellow crystals up to ~3 mm long, associated with hallimondite, widenmannite, zeunerite, mimetite, and cerussite, in cavities in hornstone breccia.

Hügelite is monoclinic with *a*=31.066, *b*=17.303, *c*=7.043 Å, and β=96.492°. The structure contains sheets of uranyl hexagonal and pentagonal bipyramids and AsO_4 tetrahedra, oriented parallel to (100). The interlayer contains Pb atoms, each of which is coordinated to two oxygens from uranyl ions, two oxygens from AsO_4 groups, and three H_2O groups. Although hügelite is *chemically* the As analogue of dumontite, the two species are not isostructural; the uranyl arsenate sheet in hügelite is geometrically the same as the uranyl phosphate sheet in dumontite and vanmeersscheite, but the unit cell is significantly larger. The size increase is probably necessary to accommodate the As atoms (which are larger than P) and also the resulting small variations in the coordination around the U and Pb atoms (Locock and Burns 2003c).

Mundite $Al(UO_2)_3(OH)_3(PO_4)_2·5.5\ H_2O$
Mundite is a very rare species described from Kobokobo, Congo, where it had formed in the oxidation zone of a uraniferous granite pegmatite. It forms golden-yellow, transparent crystals, tabular on {010} with rectangular cross-section; the crystals are weakly fluorescent pale green in UV. Associated minerals include tabular meta-autunite, canary-yellow nodules of ranunculite, and other Al-U phosphates (Deliens and Piret 1981a).

Mundite is orthorhombic with *a*=17.08, *b*=30.98, and *c*=13.76 Å, based on powder XRD data. Although a complete structure refinement hasn't been reported, the powder pattern and cell parameters indicate that mundite belongs to the phosphuranylite structural group.

Figure 456. The crystal structure of hügelite, showing $[(UO_2)_3O_2(AsO_4)_2]$ sheets, parallel to (100), with Pb atoms (large gray) and water groups (small gray) in the interlayer. Note the significantly larger unit cell compared to dumontite.

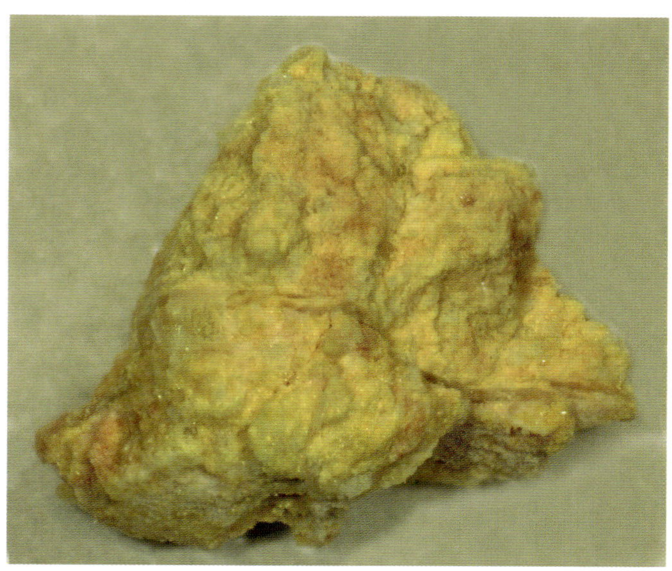

Figure 457. A tiny sample, ~2 mm across, consisting of coatings of microscopic yellow mundite from the type locale at Kobokobo, Congo. RJL712

Nielsbohrite $K(UO_2)_3(AsO_4)(OH)_4 \cdot H_2O$
Nielsbohrite was described from the Menzenschwand U deposit in the Krunkelbach Valley, southern Black Forest, Germany, and is the sixth new radioactive species described from there. As an interesting historical note, mining at Menzenschwand ceased around 1990, the mine is flooded, and the dumps have been reclaimed. The type specimens of nielsbohrite, first received in 1983 while the mine was still operating, had come from a private collection and were presumably collected from the dumps. Although X-ray diffraction and microprobe work suggested that a new species was present, the mineral wasn't fully characterized until 2002 and the full description was published in 2009. This demonstrates the value of careful collection *and preservation* of interesting specimens, and also shows the slow and painstaking scientific work that must be done when only minute amounts of material are available for study.

In the type material, nielsbohrite forms minute (up to 150 μm), euhedral, rhombohedron-like single crystals and clusters on quartz or barite in cavities, associated with hematite, pyrite, schoepite, and metazeunerite. The crystals are various shades of yellow and are not fluorescent; fresh crystals are transparent with a vitreous luster, whereas altered crystals are dull and nontransparent. The mineral is interpreted to have formed in the oxidized zone of the deposit, with U derived from uraninite, As from arsenopyrite, löllingite, or safflorite, and K from orthoclase in the country rock (Walenta et al. 2009).

Nielsbohrite is orthorhombic with a=8.193, b=11.430, and c=13.500 Å. The structure comprises uranyl arsenate chains, similar to chains in the phosphuranylite group, running parallel to [001]. But in contrast to the sheets formed in phosphuranylite, the chains in nielsbohrite are cross-linked to form a framework. Potassium ions and H_2O lie in channels running parallel to [001].

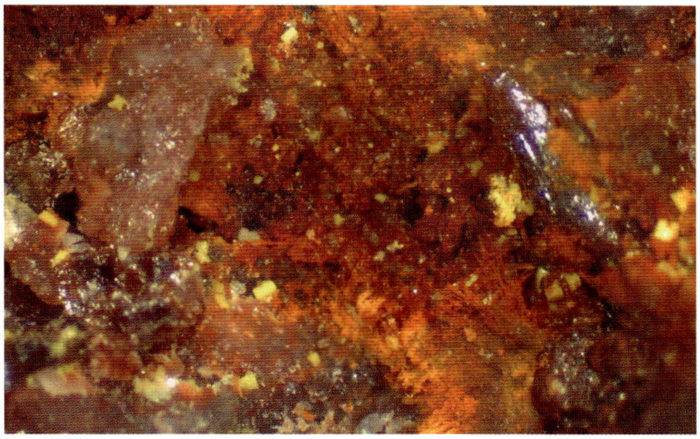

Figure 458. Minute yellow nielsbohrite crystals on matrix from the TL at Menzenschwand, Black Forest, Germany. FOV is ~3 mm. RJL4166

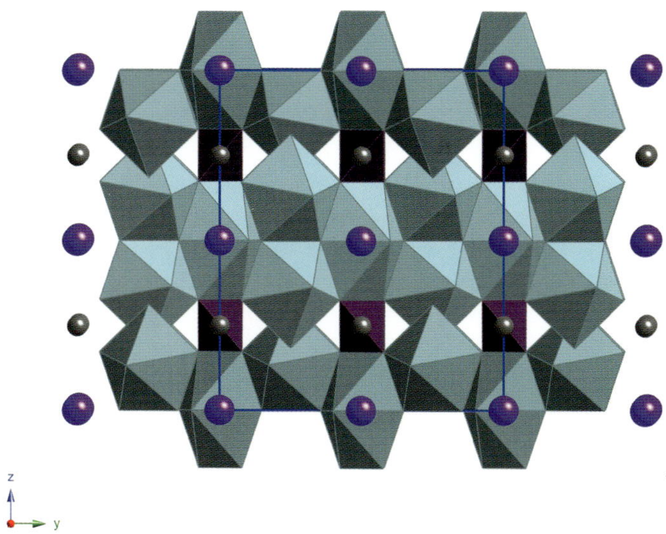

Figure 459. Crystal structure of nielsbohrite viewed along [100], showing uranyl arsenate chains arranged parallel to [001]. K (purple atoms) and H_2O groups (gray) lie in channels in the structure.

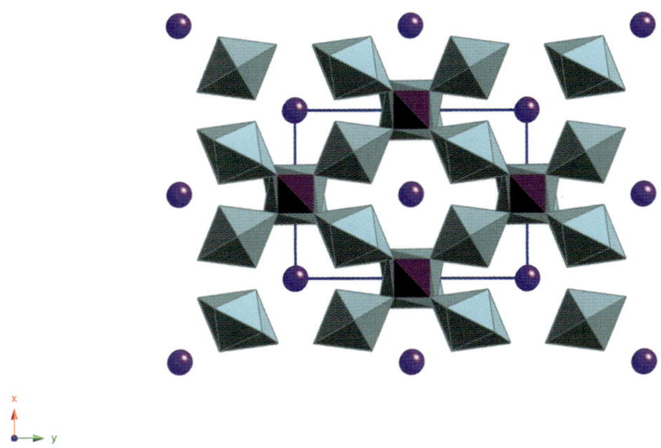

Figure 460. The structure viewed along [001], showing how the chains are cross-linked to form a framework with K in the channels.

Phosphuranylite $Ca(UO_2)[(UO_2)_3(OH)_2(PO_4)_2]_2 \cdot 12H_2O$
Phosphuranylite was (inadequately) described from the Flat Rock pegmatite, Mitchell Co., North Carolina, by F. A. Genth (1879), who considered the mineral to be a uranyl phosphate (Dana 1879). Frondel (1950a) re-examined the type material, determined that it contained Ca, and showed it to be isostructural with dewindtite. At the TL, it forms yellow films and coatings on quartz, feldspar, muscovite, and garnet; cubic molds of uraninite crystals were also found. The coatings contain microscopic, thin scales with a rectangular outline and a distinctive deep golden yellow color; the crystals are not fluorescent.

Figure 461. Rich golden-yellow scaly phosphuranylite crystals filling a seam in pegmatite from Margnac, France. RJL573

Phosphuranylite is widely distributed (well over 100 locales) but usually found in relatively small amounts. The most common mode of occurrence, of which the type material is typical, is an alteration product of uraninite in radioactive pegmatites, e.g., at the Ruggles and Palermo pegmatites, Grafton Co., New Hampshire; Bedford, New York; and Keystone, South Dakota. The phosphate and calcium content are usually derived from the weathering of apatite by meteoric water in the weathering zone; phosphuranylite is especially abundant at the Ruggles pegmatite, which contains large amounts of primary apatite in close association with uraninite (Frondel 1950). Commonly seen associated species include members of the autunite group, uranophane, becquerelite, curite, parsonsite, haiweeite, and fluorescent hyalite opal. It has been noted at several important uranium ore deposits, including: Bergen on the Trieb and Wölsendorf, Germany; La Crouzille and Margnac, France; Sabugal, Portugal; and numerous locales in the Colorado Plateau region.

Phosphuranylite is orthorhombic with a=15.836, b=17.325, and c=13.724 Å. Early structural work on material from Urgeirica, Portugal (Hogarth and Nuffield 1954), suggested that phosphuranylite and dewindtite are isostructural, but that their chemical formulas cannot be written in a way that makes them strictly analogous. A modern structural refinement (Piret and Piret-Meunier 1991) showed that the structure comprises $[(UO_2)_3(OH)_2(PO_4)_2]$ sheets, joined by Ca and (UO_6) in the interlayer, as discussed earlier in the introductory section.

Phuralumite $Al_2[(UO_2)_3(OH)_2(PO_4)_2](OH)_4 \cdot 10H_2O$
Phuralumite is one of several very rare Al-U phosphates described from the beryl-columbite pegmatite at Kobokobo, Congo. At the TL, phuralumite forms crusts and veinlets of tiny (0.5 mm) lemon-yellow prismatic crystals that are not fluorescent. Associated species include meta-autunite, phosphuranylite, threadgoldite, and other Al uranyl phosphates (Deliens and Piret 1979b).

Figure 462. Bright yellow globular coating of minute, lath-like phosphuranylite crystals from Marysvale, Utah. FOV is ~1 cm. RJL2746

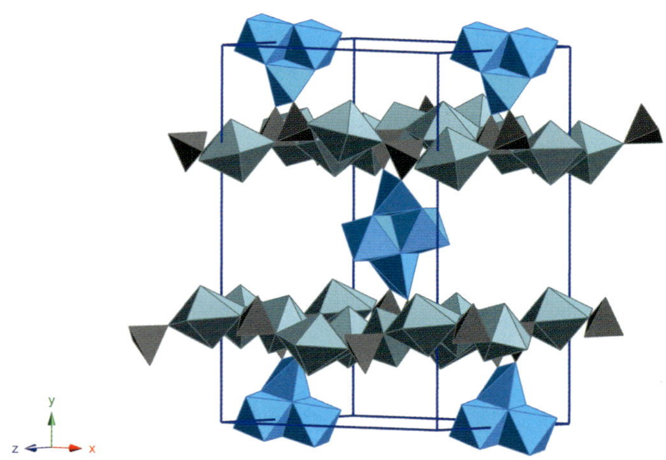

Figure 463. Crystal structure of phuralumite, showing uranyl phosphate sheets lying parallel to (010), connected via Al_4O_{14} groups (blue polyhedra forming clusters).

Phuralumite is monoclinic with a=13.87, b=20.79, c=9.38 Å, and β=112°. The structure comprises sheets of composition $[(UO_2)_3(OH)_2(PO_4)_2]^{2n-}{}_n$ connected via Al_4O_{14} groups consisting of two octahedra and two trigonal bipyramids sharing edges (Piret, Piret-Meunier, and Declerq 1979).

Phurcalite $Ca_2[(UO_2)_3O_2(PO_4)_2]\cdot 7H_2O$
Phurcalite was described from Streuberg, Bergen on the Trieb, Saxony, Germany, where it formed flattened, tabular crystals to ~1 mm long. The crystals are yellow with a vitreous to adamantine luster, and are not fluorescent. Phurcalite is typically found in uraniferous granites and granite pegmatites, lining cracks and fissures in the weathering zone. Associated species include autunite, uranophane, and other secondary minerals (Deliens and Piret 1978).

Phurcalite has been reported from a number of locales, including: the Merrivale quarry, Devon, England; Nisa, Portugal; Davignac, France; Shinkolobwe, Congo; and in tourmaline-bearing granitic pegmatite at Perus, São Paulo, Brazil. Notable US locales include the Posey mine, White Canyon district, Utah; and the Shale Hills, Kern Co., California.

Phurcalite is orthorhombic with a=17.415, b=16.035, and c=13.598 Å. The structure was refined using unusually large (~5 mm) crystals collected from the pegmatite at Perus, Brazil, yielding the structural formula given above. The structure consists of $[(UO_2)_3O_2(PO_4)_2]^{4n-}{}_4$ sheets, connected via the coordination polyhedra of Ca in the interlayer. Linkage also involves hydrogen bonding via the interlayer H_2O groups (Atencio et al. 1991).

Figure 464. A colorful association: bright yellow phurcalite and green malachite, from the Posey mine, White Canyon district, San Juan Co., Utah. RJL2643

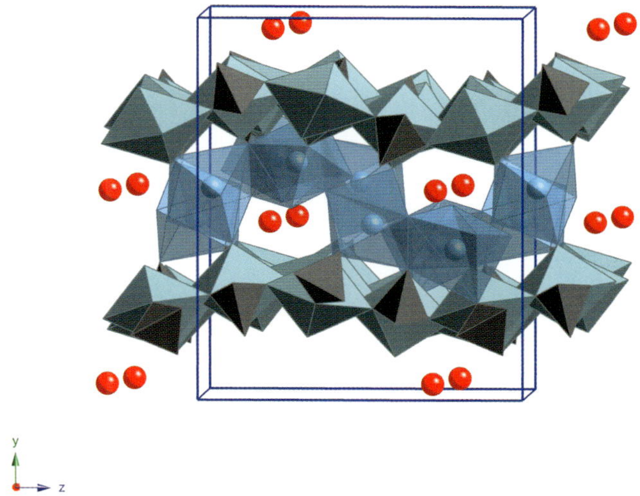

Figure 466. The structure of phurcalite, showing the uranyl phosphate sheets lying parallel to (010). The coordination polyhedra of the interlayer Ca (rendered translucent for clarity) are of two types: capped trigonal prisms, and dodecahedra. Water groups are shown in red.

Figure 465. Minute phurcalite crystals on shale from the Shale Hills, Kern Co., California. RJL350

Upalite $Al[(UO_2)_3O(OH)(PO_4)]·7H_2O$

Upalite is a very rare mineral, one of several described from the uraniferous pegmatite at Kobokobo, Congo. It forms amber-yellow needles to ~0.33 mm long, in association with meta-autunite, phosphuranylite, threadgoldite, and other uranyl species. Upalite is not fluorescent (Deliens and Piret 1979b).

Upalite was originally thought to be orthorhombic, but later studies showed it is actually monoclinic, with the structural formula given above. Cell dimensions are a=13.704, b=16.82, c=9.332 Å, and β=111.5°. The crystals always have a very fine-scale lamellar twinning by reflection on {001} giving them pseudo-orthorhombic symmetry (Piret and Declerq 1983). The structure comprises uranyl phosphate sheets lying parallel to (010). Octahedrally coordinated Al lies in the interlayer, along with H_2O groups. Some $Al(O,OH)_6$ octahedra connect the sheets by sharing oxygens with PO_4 groups, whereas others are held by hydrogen bonding only.

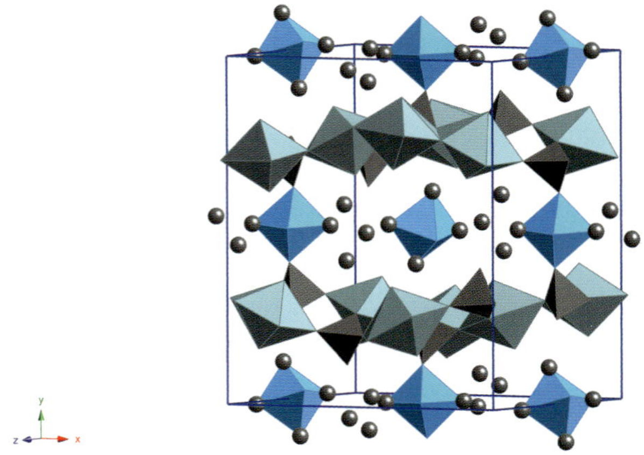

Figure 467. Crystal structure of upalite, showing uranyl phosphate sheets arranged parallel to (010). Note the two distinct types of Al sites: $AlO_2(H_2O)_4$ octahedra (center) are not directly connected to the sheets, whereas $AlO_4(H_2O)_2$ octahedra (center left and center right) share oxygens with phosphate groups in adjacent sheets. Gray atoms represent water.

Vanmeersscheite $U^{6+}[(UO_2)_3(OH)_6(PO_4)_2]·4H_2O$
Metavanmeersscheite $U^{6+}[(UO_2)_3(OH)_6(PO_4)_2]·2H_2O$

Vanmeersscheite and metavanmeersscheite were described from the pegmatite at Kobokobo, Congo, where both species form elongated, platy crystals, which are translucent canary-yellow and exhibit strong green fluorescence in both SW and LWUV. Associated species include meta-autunite, upalite, threadgoldite, and studtite (Piret and Deliens 1982).

Vanmeersscheite also occurs at Menzenschwand, Germany (Markl and Wolfsried 2011).

Both species are orthorhombic, with a=17.06, b=16.76, and c=7.023 Å (vanmeersscheite) and a=34.18, b=33.88, and c=14.074 Å (metavanmeersscheite). The crystal structure comprises sheets of composition $[(UO_2)_3(OH)_6(PO_4)_2]^{2-}$ with U^{6+} and water in the interlayer. The Raman spectrum of vanmeersscheite was presented by Frost, Čejka, and Dickfos (2009) and interpreted by assigning the observed Raman bands to $(PO_4)^{3-}$ and $(UO_2)^{2+}$ stretching vibrations, OH stretching vibrations, and $(PO_4)^{3-}$ bending vibrations. The average U-O bond length calculated from Raman data (1.8 Å) is shorter than that originally reported by Piret and Deliens (1982) from crystal structure data (1.88 Å), suggesting that the structure of vanmeersscheite should be revised somewhat.

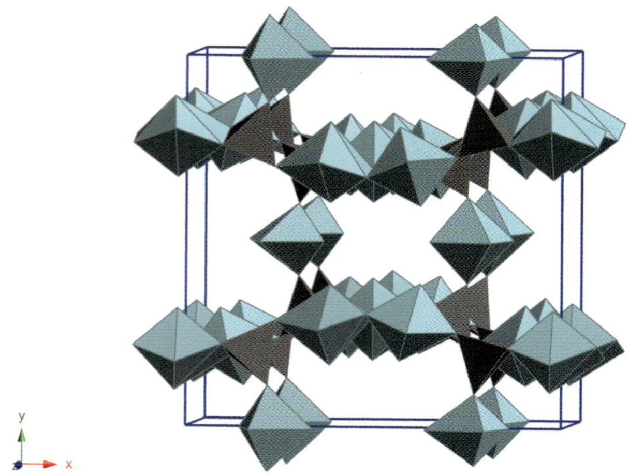

Figure 468. Crystal structure of vanmeersscheite, showing uranyl phosphate sheets parallel to (010), with distorted UO_6 octahedra bridging between adjacent sheets by sharing oxygens with PO_4 groups (gray tetrahedra). Water groups (not shown) also occupy the interlayer.

Yingjiangite $(K_2,Ca)(UO_2)_7(PO_4)_4(OH)_6·6H_2O$

Yingjiangite was described from the oxidized zone of a deposit containing uraninite and uranothorite at Tongbiguan Village, Yingjiang Co., Yunnan Province, China. It occurs as golden-yellow to yellow, compact, microcrystalline aggregates of individual grains ~10–15 μm, with weak yellowish-green fluorescence. Associated species include studtite, calcurmolite, tengchongite, and autunite. Although the complete crystal structure wasn't determined, the X-ray powder pattern showed similarities to that of phosphuranylite (Chen, Huang, and Gu 1990).

Yingjiangite has also been reported from the Xiazhuang U deposit, Guangdong Province, China.

Figure 469. Thin yellow film on pegmatite from the Ruggles mine, New Hampshire, in which microprobe analysis suggested the presence of yingjiangite, fourmarierite, rutherfordine, and uranophane (J. Vaydak analysis). RJL3837

Walpurgite Group

All of the minerals treated in this section share a common 1:2 ratio of uranyl:phosphate or uranyl:arsenate. Beyond that, some of the species are clearly more closely related than others. Parsonsite is isostructural with its arsenate analog, hallimondite. Walpurgite and phosphowalpurgite are likewise arsenate and phosphate analogs, and orthowalpurgite is the orthorhombic polymorph of walpurgite. Lakebogaite and ulrichite have unusual structures; interestingly, however, they share the same type locale.

Table 51. Walpurgite group

Mineral	Formula	Symmetry
Hallimondite	$Pb_2[(UO_2)(AsO_4)_2](H_2O)_n$	tric
Lakebogaite	$CaNa_2Fe^{3+}{}_2H(UO_2)_2(PO_4)_4(OH)_2(H_2O)_8$	mon
Orthowalpurgite	$(BiO_4)_4(UO_2)(AsO_4)_2{\cdot}2H_2O$	orth
Parsonsite	$Pb_2[(UO_2)(PO_4)_2](H_2O)_n$	tric
Phosphowalpurgite	$(UO_2)Bi_4O_4(PO_4)_2{\cdot}2H_2O$	tric
Ulrichite	$CaCu(UO_2)(PO_4)_2{\cdot}4H_2O$	mon
Walpurgite	$(UO_2)Bi_4O_4(AsO_4)_2{\cdot}2H_2O$	tric

Hallimondite $Pb_2[(UO_2)(AsO_4)_2](H_2O)_n$ $0 \leq n \leq 0.5$

Hallimondite is a rare mineral described from the Michael mine, a historic lead-zinc vein near Reichenbach in the Black Forest, Germany. The species was actually discovered during re-examination of old specimens of hügelite that had been collected before 1914; later efforts to find more samples in the area were not successful (Walenta 1965). The mineral occurs as small crystals and coatings in cavities and fractures in quartz, associated with barite, galena, hügelite, and mimetite. The arsenate is derived from native arsenic, which is abundant in the deposit. The crystals are small (<0.4 mm) but well-developed, flattened, and somewhat elongated in the *z*-direction. The yellow color distinguishes it from the orange to brownish hügelite with which it is intergrown.

Hallimondite has also been reported from the Bühlskopf, near Ellweiler, Rhineland-Palatinate, Germany.

Walenta (1965) showed that hallimondite is triclinic and isostructural with its phosphate analogue, parsonsite, and assumed that both were anhydrous. More recently, the structures of both minerals were refined using synthetic material, and the resulting formula given above reflects a better understanding of the role of water. The structure contains UO_7 bipyramids that share a side to form dimers that are linked by edge- and vertex-sharing AsO_4 tetrahedra to form chains parallel to [001]. The uranyl arsenate chains are joined together by the Pb coordination polyhedra. Water groups lie in cavities and are possibly zeolitic in nature; no network of hydrogen bonding exists, and the water doesn't seem to be critical to maintaining stability of the structure. The cell parameters are a=7.1154, b=10.4780, c=6.8572 Å, α=101.178, β=95.711, and γ=86.651° (Locock, Burns, and Flynn 2005a).

Lakebogaite $CaNa_2Fe^{3+}{}_2H(UO_2)_2(PO_4)_4(OH)_2(H_2O)_8$

Lakebogaite was found at the Lake Boga quarry in northwestern Victoria, Australia, in 2000. At the time of its formal description (Mills et al. 2008) half a dozen specimens were known, containing about 35 crystals. This species, along with other secondary U minerals, tends to be concentrated along joint planes in the miarolytic granite. Crystals are

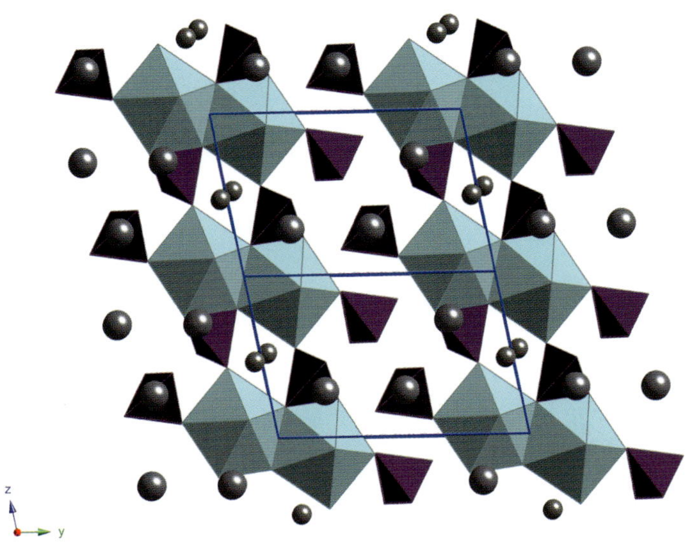

Figure 470. The structure of hallimondite, viewed along [100], showing the uranyl arsenate chains running parallel to [001]. Dark purple tetrahedra represent AsO_4; large gray atoms are Pb; small gray atoms are H_2O oxygens.

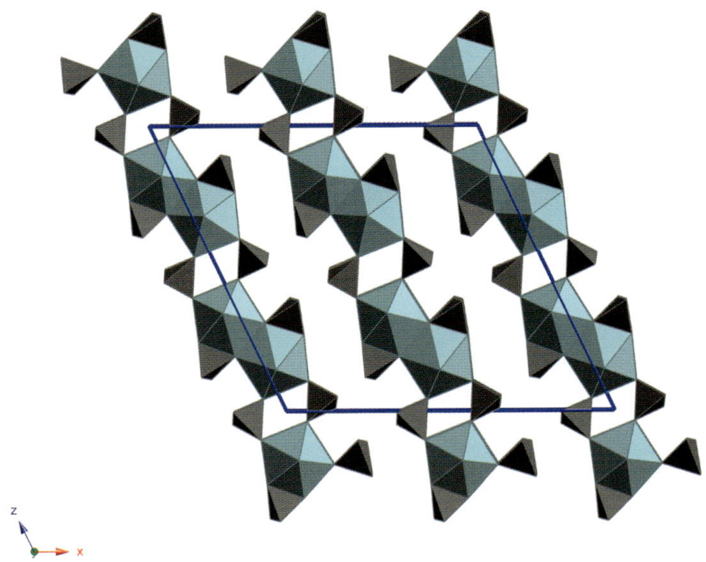

Figure 471. Structure of lakebogaite viewed along [010] showing $[(UO_2)_2(PO_4)_4]$ chains running parallel to [001].

found as small (up to ~0.4 mm) blocky to prismatic individuals and clusters. They are transparent, lemon yellow, with a vitreous luster. The authors note that, "Because U is rapidly fixed in uranyl phosphates . . . it is likely that crystallization has taken place only at short distances from the sites of oxidation. It is likely . . . that lakebogaite and other uranyl phosphates have acted to sequester uranium released during weathering of the granite over time scales of several hundred thousand years."

Lakebogaite is monoclinic, with *a*=19.644, *b*=7.096, *c*=18.703 Å, and β=115.692°. The structure is different from those of other uranyl minerals and synthetic compounds, although it does have some structural features in common with parsonsite and hallimondite. It is built from pairs of edge-sharing UO_7 pentagonal bipyramids linked by corner-sharing PO_4 tetrahedra to form chains parallel to [001]. The chains, with composition $[(UO_2)_2(PO_4)_4]$, are cross-linked through corner-sharing between PO_4 tetrahedra and $FeO_4(OH)_2$ octahedra. These octahedra are joined by corner-sharing through OH groups, forming chains parallel to [010], creating a three-dimensional network. Sodium and calcium ions lie in channels formed by this network, coordinated to water molecules that also lie in those channels. Additional H_2O groups occupy other channels and aren't coordinated to any cations; they are likely held in place by hydrogen bonding (Mills et al. 2008).

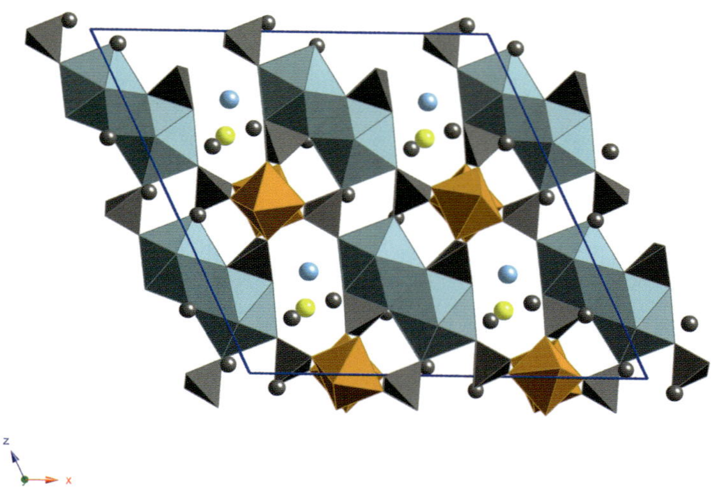

Figure 472. The structure viewed in the same direction but showing how the chains are cross-linked by octahedrally coordinated Fe^{3+} (brown). Ca^{2+} ions (blue), Na^+ ions (yellow), and H_2O (gray) occupy channels in the structure.

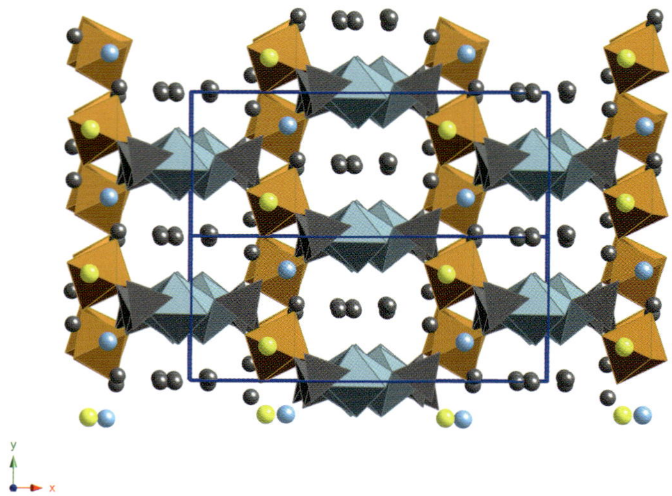

Figure 473. The same structure viewed along [001] showing the chains formed by Fe^{3+} octahedra and the channels containing Na, Ca, and water.

Orthowalpurgite $(BiO_4)_4(UO_2)(AsO_4)_2{\cdot}2H_2O$
Orthowalpurgite was described from a single specimen found in 1980 on the mine dump at Schmiedestollen, Wittichen, Germany, where it forms elongated, transparent yellow, tabular crystals to ~0.3 mm long. As the name indicates, it is the orthorhombic polymorph of walpurgite. In the type specimen, orthowalpurgite is associated with quartz, anatase, and preisingerite. The mineral assemblage is attributed to the weathering of emplectite, wittichenite, or other Bi-bearing ore minerals (Krause, Effenberger, and Brandstätter 1995).

Orthowalpurgite is orthorhombic, with *a*=5.592, *b*=13.324, and *c*=20.685 Å. The structure is similar to that of walpurgite, differing in the configuration of the $Bi_4O_4(AsO_4)_2{\cdot}2H_2O$ layers, the twisting of AsO_4 groups in the $UO_2(AsO_4)_2$ chains, some details in the coordination of oxygen atoms, and the acceptors for hydrogen bonding to the water groups.

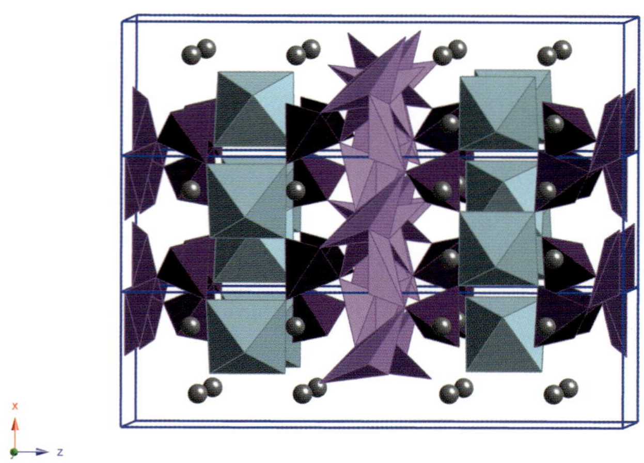

Figure 474. Crystal structure of orthowalpurgite, viewed nearly parallel to the uranyl arsenate chains, oriented for direct comparison to that of walpurgite (shown later). Bi ions are shown in magenta and H_2O groups are gray. Three unit cells are shown.

Parsonsite $Pb_2[(UO_2)(PO_4)_2](H_2O)_n$ $0 \leq n \leq 0.5$
Parsonsite was described from Kasolo, Congo, where it was found in earthy masses on torbernite. The minute tabular crystals are pale brown, due mainly to microscopic inclusions, with a greasy luster (Schoep 1923). Frondel (1950) provided a detailed description of parsonsite from the Ruggles pegmatite, New Hampshire, where it forms crusts of lath-like crystals along fracture surfaces in massive quartz and feldspar near masses of uraninite and gummite, and from Wölsendorf, Germany, where it forms tufted aggregates of microscopic crystals with uranocircite on dark purple fluorite. Parsonsite is not fluorescent.

It has been reported from several locales in France, and it is the main U ore mineral at Lachaux, Puy-de-Dome, France. Some other locales are: Tragos, Portugal; Arcu su Linnabaru, Sardinia, Italy; the Ranger mine, Northern Territory, Australia; and at several places in Gunnison Co., Colorado.

The crystal structure of a sample from La Faye, Grury, France, was shown to be triclinic with a=6.842, b=10.384,

Figure 475. Yellow, glassy to opaque microcrystals of parsonsite richly coating tufaceous matrix, from Grury, Saone et Loire, France. RJL2647

Figure 476. Detail of a similar specimen from Grury, France, showing the amber, acicular parsonsite crystals filling the surfaces of a small cavity in the ore. RJL1029

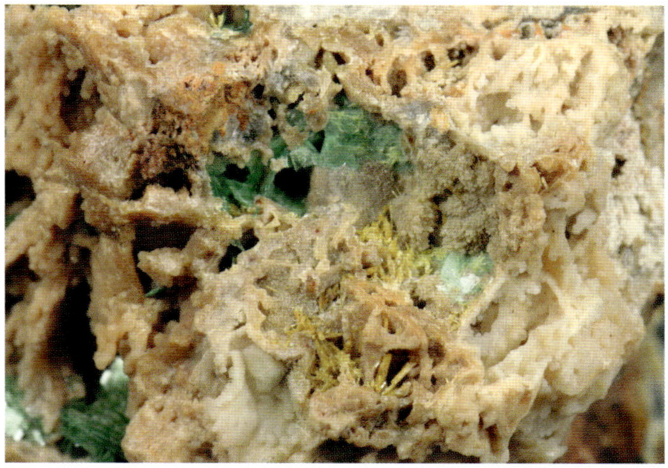

Figure 477. Tiny amber parsonsite crystals associated with green tabular autunite, from Lachaux, France. RJL2649

Figure 478. Detail of the parsonsite crystals in the previous photo. RJL2649

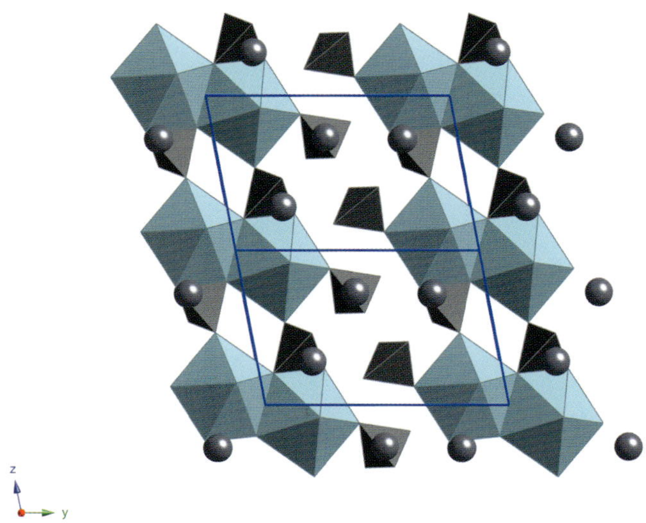

Figure 479. The structure of parsonsite, viewed along [100], showing the uranyl phosphate chains running parallel to [001]. Dark gray tetrahedra represent PO_4; large gray atoms are Pb. Water isn't shown (see the structure diagram for hallimondite).

c=6.670 Å, α=101.27, β=98.19, and γ=86.39°. It was the first uranyl phosphate mineral whose structure is based on chains of polyhedra rather than sheets (Burns 2000). Further studies using synthetic material (Locock, Burns, and Flynn 2005a) clarified the role of H_2O in the structures of parsonsite and its AsO_4 analogue, hallimondite. (For further discussion, see the entry for **hallimondite**.)

Phosphowalpurgite $(UO_2)Bi_4O_4(PO_4)_2 \cdot 2H_2O$
The partial substitution of phosphate for arsenate in walpurgite had been documented as early as the 1940s; one analysis presented by Evans (1950) implies a PO_4:AsO_4 ratio around 1:1.3. The presence of phosphate, although not explicitly mentioned, may be inferred from the reported As content in much older analyses of walpurgite as well (Frondel 1958). The true phosphate-dominant analogue was recently described as phosphowalpurgite from old mine dumps at Smrkovec, western Bohemia, Czech Republic (Sejkora et al. 2004). The authors note that materials at the TL "reveal a broad variation in $(AsO_4)^{3-}$ and $(PO_4)^{3-}$ content. Electron-microprobe analyses show that both As- and P dominant phases occur at this locality and that *walpurgite with some (PO_4) content is the more abundant*" (emphasis added), so true phosphowalpurgite is somewhat of a rarity even at the TL. This historic deposit was mined for silver in the sixteenth century, then around 1917–18 bismuth was recovered from the mine dumps, and later the area was prospected for uranium in the early 1950s. From samples on the dump it is inferred that the ore minerals (native Bi, galena, Ag-bearing sulfides, Ni and Co arsenides, and uraninite) are present as disseminated aggregates in quartz gangue. The diverse secondary mineral assemblage includes autunite, metatorbernite, zippeite, minerals of the atelestite and preisingerite groups, and numerous others.

Phosphowalpurgite forms clusters, up to ~1 cm, of somewhat irregular tabular crystals. Individual crystals average 0.1–0.3 mm; the crystals are translucent brownish gray and are not fluorescent.

Phosphowalpurgite is triclinic, with a=7.060, b=10.238, c=5.464 Å, α=101.22, β=109.93, and γ=87.93°. The IR spectra of phosphowalpurgite, PO_4-rich walpurgite, and walpurgite are similar, further indicating that the structures are closely related and a complete solid-solution series likely exists. At Smrkovec, compositions range from about $(PO_4)_{.8}(AsO_4)_{.2}$ to $(PO_4)_{.25}(AsO_4)_{.75}$.

Ulrichite $CaCu(UO_2)(PO_4)_2 \cdot 4H_2O$
Ulrichite was described from the Lake Boga granite, Victoria, Australia (Birch, Mumme, and Segnit 1988), and to date it is known only from there. It forms apple-green to lime-green acicular crystals up to 1 mm long and 0.05 mm thick, or flat prisms. The crystals are transparent to translucent and are not fluorescent. Ulrichite occurs in miarolytic cavities in the granite, with turquoise and chalcosiderite; other associated minerals include fluorapatite, libethenite, torbernite, saléeite, and cyrilovite.

Ulrichite is monoclinic with a=12.784, b=6.996, c=13.007 Å, and β=91.92°. The structure consists of a "heteropolyhedral" sheet made up of UO_7 bipyramids, PO_4 tetrahedra, and irregular CaO_8 polyhedra. The UO_7 and CaO_8 polyhedra form rows parallel to [010]. The Cu^{2+} plays the role of an interlayer cation, and is associated with four close and two very distant oxygens, forming an elongated octahedron. The Cu polyhedra link the sheets together by sharing corners with the PO_4 tetrahedra; further interlayer connection involves hydrogen bonds (Kolitsch and Giester 2001).

Figure 480. Bright green spray of ulrichite crystals on matrix from the type locale, Lake Boga, Australia. RJL2890

Figure 481. The structure of ulrichite, viewed parallel to [001], showing the sheet formed by UO_7 (blue-gray), CaO_8 (blue) and PO_4 (dark gray) polyhedra arranged in rows running parallel to [010].

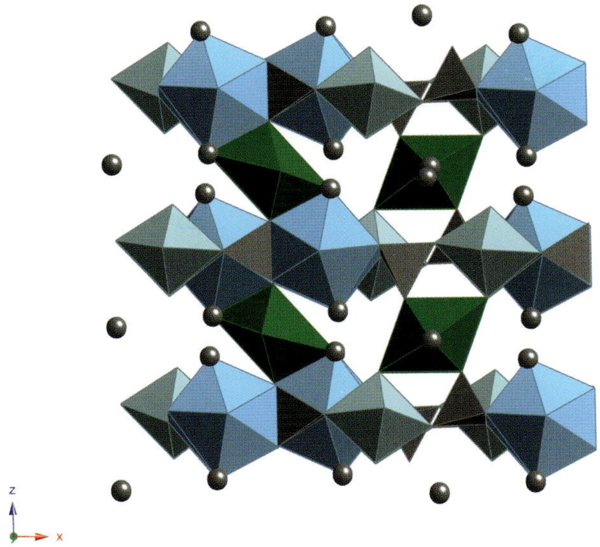

Figure 482. The ulrichite structure viewed along [010] (i.e., parallel to the sheets) to show that the sheets are connected directly via CuO_6 octahedra (dark green) and hydrogen bonding involving the water groups (gray).

Walpurgite $(UO_2)Bi_4O_4(AsO_4)_2{\cdot}2H_2O$
Walpurgite was described by Weisbach (1873, 1877) from the Walpurgis vein at the Weisser Hirsch mine, Schneeberg, Germany, where it had formed in the oxidation zone of a vein that contained uraninite, Co and Ni arsenides, and native Bi. The small crystals are waxy yellow to straw yellow, and are often twinned, giving a pseudomonoclinic habit much like gypsum. At the type locale, it is associated with trögerite, zeunerite, uranosphaerite, torbernite, and uranospinite, all of which (except torbernite) were described by Weisbach as new species from this mine. Walpurgite crystals tend to be fairly opaque but more transparent at the terminations; the opacity is attributed to finely disseminated impurities such as Bi carbonate, as well as air bubbles (Evans 1950).

Walpurgite is fairly rare, but has been reported from: Jachymov, Czech Republic; and at Schmiedestollen and Nussbach, Black Forest, Germany.

At a locality near Dalbeattie, Scotland, a vein filling of granular crystalline native Bi is bordered by a greenish yellow alteration rim containing bright green mammillary growths and radiating needles of mixite, small (0.2 mm) brown spherules of bismutite, and bright yellow, transparent, long-prismatic walpurgite crystals with oblique terminations similar to the habit noted at the TL (Braithwaite and Knight 1990).

Walpurgite also occurs at Erongo, Namibia, where it forms small (<1 mm) tabular, amber-brown to yellow crystals, associated with purple fluorite, topaz, and quartz (Cairncross and Bahmann 2006).

Evans (1950) first determined the structure and noted that previous descriptions had emphasized that the crystals are often twinned to somewhat resemble gypsum. This work clarified the earlier morphological and analytical data of Weisbach (1877) and Fischer (1946).

Figure 483. An amber-yellow walpurgite crystal ~0.5 mm long, from the Schmiedestollen mine, Wittichen, Germany, showing the gypsum-like habit. RJL834

Figure 484. Amber walpurgite crystals richly covering drusy quartz, from the Schmiedestollen mine, Wittichen, Germany; sample is ~5 mm wide. RJL835

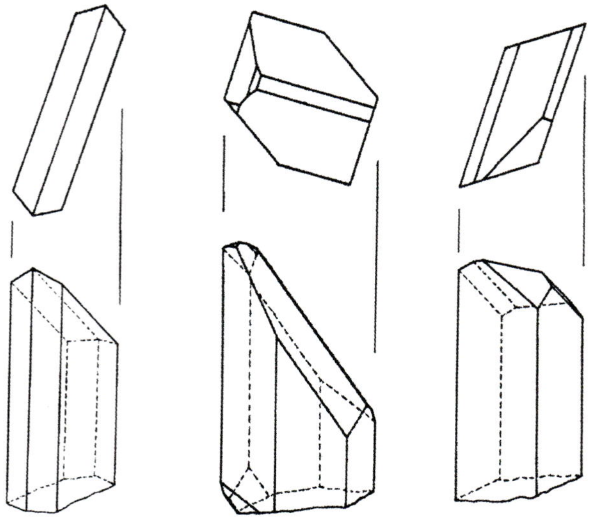

Figure 486. Drawings of walpurgite crystals, adapted from Evans (1950). Left to right: twin from Schneeberg, single crystal from Schneeberg; single crystal from Joachimsthal.

Figure 485. SEM photo of the specimen in the previous figure, showing the thin tabular habit of the walpurgite crystals, about 200 µm long, on drusy quartz, from the Schmiedestollen mine, Germany. RJL835

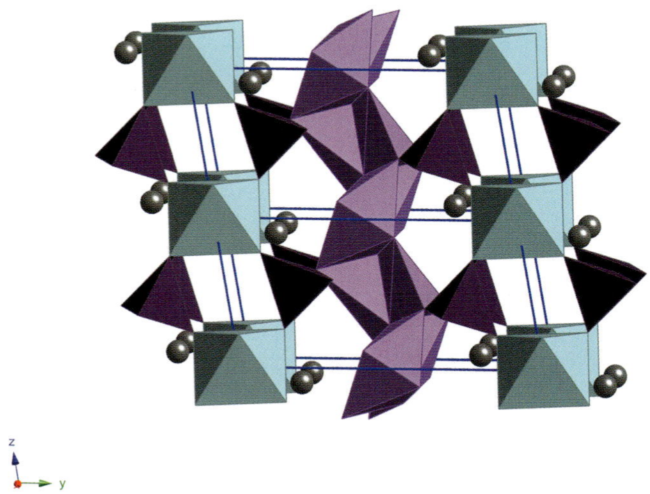

Figure 487. Crystal structure of walpurgite, showing the sheet oriented parallel to (010) formed by Bi (magenta polyhedra) and AsO_4 groups (dark purple). $UO_2(AsO_4)_2$ chains run parallel to [001]. Two unit cells are shown.

Walpurgite is triclinic, with *a*=7.135, *b*=10.426, *c*=5.494 Å, α=101.47, β=110.82, and γ=88.20°. The structure comprises complex layers of composition $Bi_4O_4(AsO_4)_2 \cdot 2H_2O$, that extend parallel to (010). Each layer is built up from a network of Bi^{2+} and oxygens, to both sides of which AsO_4 and H_2O groups are attached. UO_6 octahedra link these layers parallel to [010] via AsO_4 groups, simultaneously forming $UO_2(AsO_4)_2$ chains parallel to [001] (Mereiter 1982).

Other phosphates and arsenates

The species treated in this section don't have clear relationships to the groups of phosphates and arsenates discussed above, or to each other. The minerals have various ratios of uranyl:phosphate or uranyl:arsenate and in most cases their structures haven't been completely determined. Interestingly, Doynikova and Sidorenko (2006) proposed that coconinoite, furongite, and moreauite form a natural group with a presumed uranyl:(phosphate+sulfate) ratio of 2:5, based on similarities in their powder X-ray diffraction patterns and physical properties. However, this would require a significant revision in the formulas of furongite and moreauite, and Doynikova and Sidorenko provide no definitive new chemical analyses to support this.

Several of the species, štěpite, vyacheslavite, and vysokyite, actually contain U^{4+} rather than $(UO_2)^{2+}$, but are treated with the secondary minerals because they have formed as secondary products of post-mining alteration of various primary mineral phases.

One species, eylettersite, is a thorium phosphate.

Table 52. Other phosphates and arsenates

Mineral	Formula	Symmetry
Asselbornite	$(Pb,Ba)(UO_2)_4(BiO)_3[(As,P)O_4]_2(OH)_7{\cdot}4H_2O$	cub
Coconinoite	$Fe^{3+}{}_2Al_2(UO_2)_2(PO_4)_2(SO_4)(OH){\cdot}20H_2O$	mon
Dymkovite	$Ni(UO_2)_2(As^{3+}O_3)_2{\cdot}7H_2O$	mon
Eylettersite	$[Th,(H_3O)]Al_3[(PO_4),(H_4O_4)](OH,H_2O)_6$	hex
Furongite	$Al_2(UO_2)(PO_4)_2(OH)_2{\cdot}8H_2O$	tric
Kamitugaite	$PbAl(UO_2)_5[(P,As)O_4]_2(OH)_9{\cdot}9.5H_2O$	tric
Moreauite	$Al_3(UO_2)(PO_4)_3(OH)_2{\cdot}13H_2O$	mon
Ranunculite	$AlH(UO_2)(PO_4)(OH)_3{\cdot}4H_2O$	mon
Seelite	$Mg[(UO_2)(AsO_4)]_2{\cdot}4H_2O$	mon
Šreinite	$Pb(BiO)_3(UO_2)_4(PO_4)_2(OH)_7{\cdot}4H_2O$	cub
Štěpite	$U^{4+}(AsO_3OH)_2{\cdot}4H_2O$	tet
Triangulite	$Al_3(UO_2PO_4)_4(OH)_5{\cdot}5H_2O$	tric
Vochtenite	$(Fe^{2+},Mg)Fe^{3+}[UO_2/PO_4]_4(OH){\cdot}12\text{-}13H_2O$	mon
Vyacheslavite	$U^{4+}(PO_4)(OH){\cdot}nH_2O$	orth
Vysokýite	$U^{4+}[AsO_2(OH)_2]_4{\cdot}4H_2O$	tric
Xiangjiangite	$(Fe,Al)(UO_2)_4(PO_4)_2(SO_4)_2(OH){\cdot}22H_2O$	tet (?)

Asselbornite $(Pb,Ba)(UO_2)_4(BiO)_3[(As,P)O_4]_2(OH)_7{\cdot}4H_2O$
Asselbornite was described from Schneeberg, Saxony, Germany, where it occurs in quartz gangue in association with metauranospinite, uranophane, and uranosphaerite. It forms translucent brown to yellow idiomorphic cubic crystals to ~0.3 mm, with greasy to adamantine luster. The grains are optically isotropic and light yellow in thin section.

Microprobe analysis showed some chemical zoning in the grains, with Pb decreasing and Ba increasing slightly from core to rim. The empirical formula for the cores of the grains is: $Pb_{.77}Ba_{.35}U_{5.78}Bi_{4.18}As_{1.63}P_{.46}O_{30}{\cdot}9.16H_2O$, leading to the ideal formula shown above, with Pb>Ba and As>P. Asselbornite is isometric with a=15.66 Å (Sarp, Bertrand, and Deferne 1983).

Asselbornite has also been reported from Tirpersdorf, Vogtland, Germany.

Sejkora and Čejka (2007) reexamined material from the TL and found a few points where Ba>Pb and others where (Ca+Mg)>Pb. Applying the 33% rule to the ternary Ba-Pb-Ca system, such points would presumably represent new species if they were to be fully characterized. The authors also offered a corrected formula, based on comparison with šreinite, and that formula is the one given here.

Coconinoite $Fe^{3+}{}_2Al_2(UO_2)_2(PO_4)_2(SO_4)(OH){\cdot}20H_2O$
Coconinoite was described from the Sun Valley mine, Coconino Co., Arizona, where it forms creamy yellow, soft microcrystalline aggregates, with some lath-like grains up to ~6 × 20 μm long. It is weakly fluorescent green in SWUV. The fine grain size precluded accurate determination of many physical properties and made chemical analysis difficult; the currently accepted formula given above was considered "probable" in the original report, and the symmetry was given as "possibly monoclinic" (Young, Weeks, and Meyrowitz 1966).

The mineral had first been collected in 1953 in the White Canyon area, San Juan Co., Utah. The original study included samples from the Jomac mine, Utah; unspecified localities in New Hampshire and Wyoming; two mines in Arizona; and

Figure 488. Coconinoite forming earthy, cream-colored patches ~1 cm wide, on friable sandstone from the Jomac mine, Utah. RJL4568

Figure 489. Pale yellow, earthy blebs of coconinoite forming a mass ~1 cm across, on sandstone from the White Canyon mining district, San Juan Co., Utah. RJL2645

several from unknown locales submitted to the USGS by the public. It occurs in the oxidized zone of sandstone-type deposits in Triassic rocks at four mines on the Colorado Plateau. At the Jomac mine it is associated with gypsum, jarosite, limonite, quartz, clay, and coalified wood. At the Sun Valley mine it forms seams up to 1 mm thick, mainly along bedding planes in the sandstone; visible gypsum crystals occur with it. The mineral occurs in a similar geological setting at the Blackwater No. 4 mine, Apache Co., Arizona, and the Huskon No. 7 mine, near Cameron, Coconino Co., Arizona.

Young, Weeks, and Meyrowitz (1966) note, "The association of coconinoite with gypsum, jarosite, limonite and clay indicates an environment of oxidizing pyritic, low-valent uranium ore, such as found in various places in White Canyon. Coconinoite is more likely to occur in uranium deposits which contain little or no vanadium or copper."

Coconinoite from the oxidized zone at the Kyzylkum U-V deposit was studied by analytical TEM; electron diffraction indicated monoclinic symmetry (Belova et al. 1993). Further refinement, also using electron diffraction, gave the cell dimensions as a=12.46, b=12.96, c=17.22 Å, and β=105.7° (Doynikova and Sidorenko 2006). Raman spectroscopy suggests that significant hydrogen bonding exists in the structure (Frost, Palmer, and Čejka 2011).

Dymkovite $Ni(UO_2)_2(As^{3+}O_3)_2 \cdot 7H_2O$
Dymkovite was described from the Belorechenskoye deposit, Adygea Republic, Northern Caucasus, Russia. It forms long-prismatic, lath-like to acicular crystals up to 0.5 mm long and 0.05 mm thick, elongated along [010]. Crystals form radiating sprays and random groups to ~1.5 mm across; crusts up to 2 × 2 mm^2 and 0.05 mm thick are also found. Individual crystals are transparent bright yellow with vitreous luster, and are not fluorescent; crusts are translucent and light yellow to light greenish yellow. Dymkovite is a supergene mineral formed in cavities in a dolomite vein with primary uraninite, nickeline, and gersdorfite, associated with rauchite, annabergite, and goethite.

Dymkovite is monoclinic with a=17.99, b=7.034, c=6.634 Å, and β=99.62°. The structure is based on $[(UO_2)(As^{3+}O_3)]^-$ sheets formed by chains of edge-sharing $[UO_7]$ pentagonal bipyramids and $(As^{3+}O_3)$ triangular pyramids, which are linked through hydrogen bonds involving disordered $[Ni(H_2O)_6]^{2+}$ octahedra and additional H_2O molecules in the interlayer. Dymkovite is the Ni-dominant, nearly arsenate-free analogue of seelite, $Mg(UO_2)_2[(As^{3+}O_3)_{1.4}(As^{5+}O_4)_{0.6}] \cdot 7H_2O$ (Pekov et al. 2012b).

Eylettersite $[Th,(H_3O)]Al_3[(PO_4),(H_4O_4)](OH,H_2O)_6$
Eylettersite was described from the Kobokobo pegmatite, Kivu, Congo, where it forms small, white to creamy nodules in the pegmatite. Associated minerals include feldspar, zircon var. cyrtolite, columbite, apatite, and phosphuranylite. It is regarded as a member of the crandallite group, and some material at the TL is interpreted as an alteration product of Th-rich crandallite (Van Wambeke 1972).

There is a general lack of agreement in the literature on the correct formula. In the original description, Van Wambeke presented three empirical formulas for slightly different compositional variants. All are trigonal with a=6.98–6.99 and c=16.66–16.72Å. He attributes these chemical differences to alteration processes that led to **A** site deficiencies and to partial replacement of (PO_4,SiO_4) groups by H_4O_4. The simplified formula given above emphasizes this "hydrogarnet" substitution, although without a full structure refinement, there will always be some ambiguity on the assignment of hydrogen to OH, H_2O, H_3O, or H_4O_4 groups. In all of Van Wambeke's examples, $PO_4 > H_4O_4 >> SiO_4$. Small amounts of U, Ba, Pb, and other metals substitute for some of the Th on **A**.

Furongite $Al_2(UO_2)(PO_4)_2(OH)_2 \cdot 8H_2O$
Furongite was described from western Hunan, China, where it occurs in the oxidation zone of an illuvial-type uranium deposit in a Lower Cambrian carbonaceous shale. It forms tabular crystals and dense cryptocrystalline aggregates, associated with variscite, evansite, opal, halloysite, limonite, and autunite. Crystals are bright yellow to lemon yellow, translucent, with vitreous luster (Hunan team 1976).

The mineral was also reported from Kobokobo, Kivu, Congo, by Deliens and Piret (1985a), who considered it "relatively common" in the oxidation zone of the beryl-columbite pegmatite there. The powder XRD pattern is quite similar to that of the Chinese material, but the indicated composition for Kobokobo is quite different: $Al_2(UO_2)_2(PO_4)_3(OH) \cdot 13.5H_2O$. This discrepancy has never been satisfactorily resolved. To further complicate matters, Deliens and Piret (1985a) chose a new cell orientation with β and γ obtuse. Both teams agree that furongite is triclinic. Reported cell dimensions are: *a*=17.87, *b*=14.18, *c*=12.18 Å, α=67.8, β=77.5 and γ=79.9° (Hunan); and *a*=19.27, *b*=14.17, *c*=12.14 Å, α=67.6, β=115.5 and γ=94.6° (Kobokobo).

Kamitugaite $PbAl(UO_2)_5[(P,As)O_4]_2(OH)_9 \cdot 9.5H_2O$
Kamitugaite was described from the pegmatite at Kobokobo, Congo, where it is associated with minor triangulite, threadgoldite, dumontite, and studtite. It forms thin, elongated tabular yellow crystals up to 0.5 mm long. Observed forms are {100}, {010}, {001}, and {101}, with cleavages on (010) and (001).

The structure is triclinic with *a*=10.98, *b*=15.96, *c*=9.068 Å, α=95.1, β=96.1, and γ=89.0° (Deliens and Piret 1984).

Moreauite $Al_3(UO_2)(PO_4)_3(OH)_2 \cdot 13H_2O$
Moreauite was described from the pegmatite at Kobokobo, Congo, where it forms nodules or crusts of minute greenish yellow tabular crystals to ~0.2 mm. Individual crystals are flattened on (100), and when elongated along [010] may show a pointed termination forming an angle of 81o. It is strongly fluorescent green in SW and LWUV. Associated species include furongite, ranunculite, and phosphosiderite. The structure is monoclinic with *a*=23.41, *b*=21.44, *c*=18.34 Å, and β=92° (Deliens and Piret 1985).

Ranunculite $AlH(UO_2)(PO_4)(OH)_3 \cdot 4H_2O$
Ranunculite was originally described from Kobokobo, Congo, where it occurs in the uraniferous section of a pegmatite primarily worked for beryl and columbite. It forms gold-yellow nodules, typically 0.3 mm but occasionally up to 1 mm, which may also form a continuous botryoidal crust. The bright yellow color (*ranunculus* is the Latin word for buttercup) helps to distinguish the species from other Al-U phosphates, which tend to be paler shades of yellow. Ranunculite is not fluorescent under UV. Interestingly, the nodules are made up of very thin lamellae, rather than radiating fibers, and when crushed they disaggregate into thin plates with a rounded outline.

Based on XRD and optical measurements, the structure is apparently monoclinic, pseudo-orthorhombic with *a*=11.10, *b*=17.7, *c*=18.0 Å, and β near 90° (Deliens and Piret 1979c).

Seelite $Mg[(UO_2)(AsO_4)]_2 \cdot 4H_2O$
This mineral was first noted in samples collected in 1955 by Pierre Bariand at the Talmessi copper mine, Iran, and a preliminary description was published in 1963, without proposing a name. The structure was determined nearly thirty years later (Bachet, Brassy, and Cousson 1991) and a formal proposal was submitted to name the mineral seelite. At about the same time, Deliens and Piret submitted a proposal for a similar material from the uranium deposit at Rabejac, France. Ultimately, a joint description of the new species (Bariand et al. 1993), citing both localities, was approved under the name seelite, which had priority by a few weeks.

At Talmessi, seelite forms transparent, canary-yellow, tabular crystals, flattened on (100), to ~1 mm long, in tufted spherules and divergent sprays. It was found only in the 30-m level of the mine, in a uraniferous horizon. It occurs in cracks and vugs in arsenide ore, associated with annabergite, talmessite, zaratite, and uranospinite. At Rabejac, it forms radial aggregates of transparent, bright yellow crystals to ~0.5 mm long × 0.1 mm wide.

Analysis of the material from Rabejac indicated the empirical formula: $Mg[(UO_2)(AsO_3)_x(AsO_4)_{1-x}]_2 \cdot 7H_2O$ with x ~0.7. In the structure analysis of Talmessi material, all of the arsenic is pentavalent (i.e., AsO_4), corresponding to x=0 in the Rabejac formula.

Figure 490. Bright yellow ranunculite as microscopic yellow crystals forming tiny spherical aggregates richly covering matrix, from the TL at Kobokobo, Congo. RJL3548

Figure 491. Minute yellow acicular seelite crystals in flattened sprays on matrix, from Rabejac, France. RJL4561

Seelite is monoclinic with a=18.207, b=7.062, c=6.661 Å, and β=99.65° (Talmessi); and a=18.194, b=7.071, c=6.670 Å, and β=99.70° (Rabejac). The structure consists of uranyl arsenate layers lying parallel to (001), with Mg and water in the interlayer.

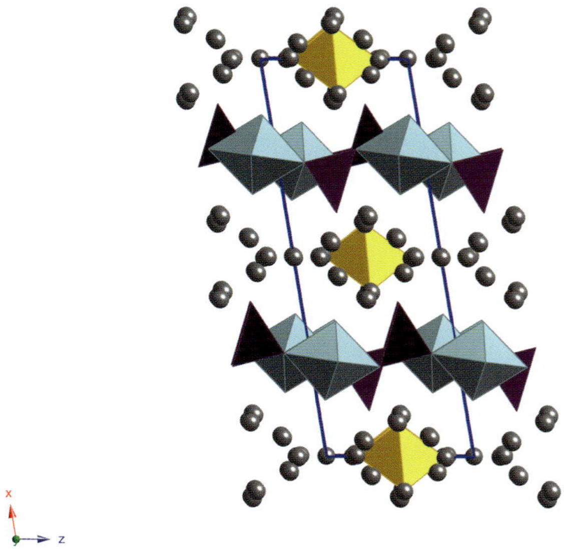

Figure 492. The structure of seelite, viewed along [010], showing the uranyl arsenate sheets with Mg (gold polyhedra) and H_2O groups (gray) in the interlayer.

Šreinite $Pb(BiO)_3(UO_2)_4(PO_4)_2(OH)_7 \cdot 4H_2O$
Šreinite, the phosphate analog of asselbornite, was described from a dump near Horní Halže, near Medĕnĕc, Krušné Hory Mountains, Czech Republic, where it occurs in quartz gangue. It forms irregular, yellow to orange-yellow coatings to ~0.2 mm thick, with a rugged, glossy surface. No crystals are visible; the surface patches have vitreous to subadamantine luster, and are not fluorescent. Associated minerals include uranosphaerite, goethite, Pb-bearing phosphuranylite, metatorbernite, bismutoferrite, kasolite, and uranophane; the assemblage is attributed to supergene weathering of uraninite, galena, and native bismuth (Sejkora and Čejka 2007).

Šreinite is cubic with a=15.5729 Å. Detailed reexamination of asselbornite and comparison with sreinite led to the following observations: 1. Samples of asselbornite range from ~$As_{.8}P_{.2}$ to $As_{.5}P_{.5}$, whereas samples of šreinite range from ~$As_{.48}P_{.52}$ to $As_{.32}P_{.68}$, so natural compositions close to either the pure As or pure P end members have not been found. 2. Asselbornite from the TL contains significant amounts of Ba, Ca, and Mg (representing, in principle, two potential species, the Ca and Ba analogs of asselbornite). 3. The ideal formula of asselbornite needs to be revised so that it is consistent with that of šreinite. Sejkora and Čejka (2007) also presented IR spectra for the two minerals and assigned the observed bands to various structural bending and stretching modes.

Štĕpite $U^{4+}(AsO_3OH)_2 \cdot 4H_2O$
Štĕpite was described from the Geschieber vein, Jachymov district, Czech Republic, where it forms emerald-green crystalline crusts on altered native arsenic. Associated species include arsenolite, bĕhounekite, claudetite, gypsum, kaatialaite, and vysokyite. Individual crystals are tabular, with vitreous luster, and are not fluorescent. The original material was found around 1980 on the tenth level of the Svornost mine, in a lens of arsenic (with traces of arsenopyrite, proustite, and pyrite) at the intersection of the Geschieber and Geyer veins. The orebody had been exposed to a moist environment for more than forty years, during which a variety of supergene alteration products had formed, including štĕpite and three other new U minerals. Štĕpite has also been found on the fifth level of Svornost, still in the Geschieber vein, where it is similarly associated with weathered massive arsenic.

Štĕpite occurs in three distinctive morphological habits, which normally crystallize in close association: Type I, tabular crystals with dominant {111} and {001} faces, up to 0.6 mm wide, often intergrown to form aggregates to 6 mm across; Type II, more blocky, intergrown crystals; and Type III, minute radiating aggregates, 20–50 μm in diameter, on long prismatic crystals of another uranyl arsenate mineral, vysokýite.

Štĕpite is unusual in that it has formed as a supergene alteration product, but contains tetravalent uranium. It is tetragonal with a=10.9894 and c=32.9110 Å. The structure consists of sheets perpendicular to [001], made up of eight-coordinate uranium atoms and hydroxyarsenate "tetrahedra."

Figure 493. Emerald-green tetragonal crystals of štĕpite coating altered arsenic covered by arsenolite. FOV is 5 mm; photo P. Škácha. Reproduced with the kind permission of the Mineralogical Society of Great Britain and Ireland from a paper by Plášil et al. (2013c)

Figure 494. Backscattered-electron image showing typical type I štěpite crystals with prominent {001} pinacoid faces and smaller {111} pyramid faces. Reproduced with the kind permission of the Mineralogical Society of Great Britain and Ireland from a paper by Plášil et al. (2013c)

Figure 495. Backscattered-electron image showing flat, blocky crystals of type II štěpite. The image is 120 µm across. Reproduced with the kind permission of the Mineralogical Society of Great Britain and Ireland from a paper by Plášil et al. (2013c)

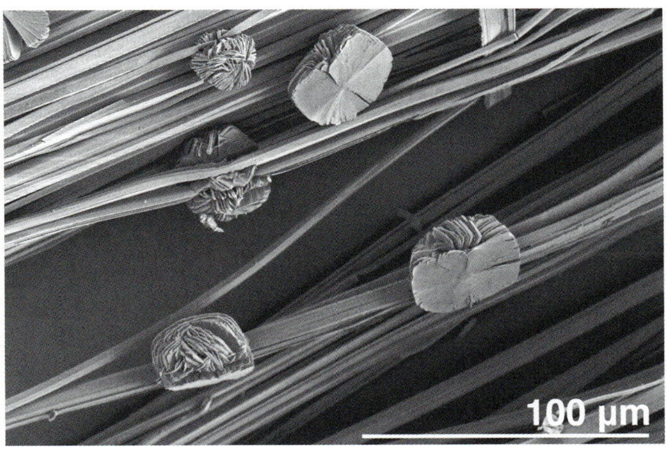

Figure 496. Backscattered-electron image showing spherical and hemispherical radiating platy type III štěpite crystal aggregates on long prismatic crystals of vysokýite, $U^{4+}[AsO_2(OH)_2]_4(H_2O)_4$. Reproduced with the kind permission of the Mineralogical Society of Great Britain and Ireland from a paper by Plášil et al. (2013c)

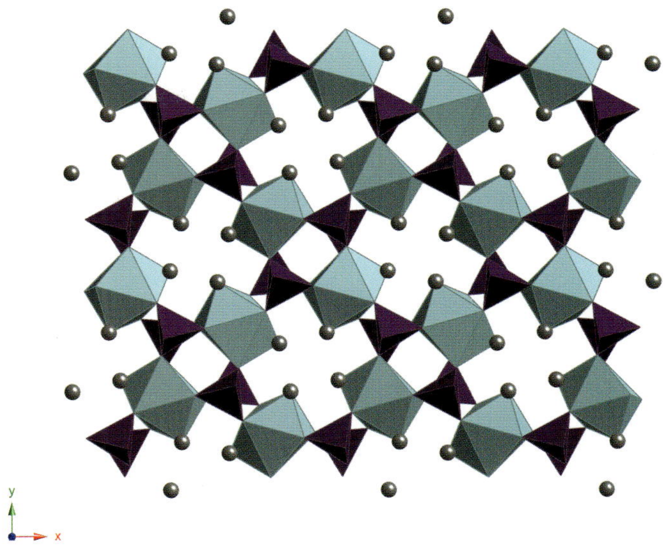

Figure 497. The sheet structure of štěpite, viewed along [001], showing the linkages of UO_8 (blue-gray) and $(AsO_3OH)^{2-}$ (purple) polyhedra.

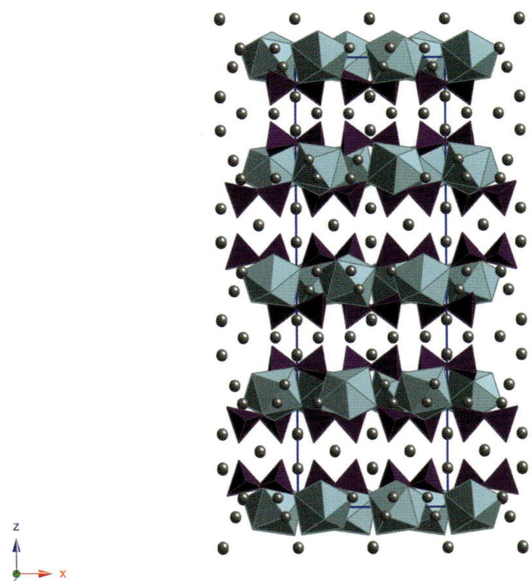

Figure 498. The structure viewed along [010]; the only link between adjacent sheets is through hydrogen bonding via interlayer H_2O (gray).

The ligands surrounding the uranium atom consist of six oxygen atoms that belong to the hydroxyarsenate groups and two oxygen atoms that belong to interlayer H_2O molecules. Each UO_8 polyhedron is connected to five other U polyhedra via six AsO_3OH groups. Adjacent electroneutral sheets, of composition $[U^{4+}(AsO_3OH)^{2-}{}_2]^0$, are linked by hydrogen bonds involving H_2O molecules in the interlayers and OH groups in the sheets (Plašil et al. 2013c).

Triangulite $Al_3(UO_2PO_4)_4(OH)_5 \cdot 5H_2O$
Triangulite was described from the quartz-K-feldspar-columbite zone of the pegmatite at Kobokobo, Congo, where it is associated with beryl, metamict zircon, and minor meta-autunite, phosphuranylite, and ranunculite. It forms bright yellow aggregates of flat, triangular, or rhombohedral crystals up to 0.2 mm long. Crystals have a vitreous luster and are not fluorescent.

Triangulite is triclinic with a=10.39, b=10.56, c=10.60 Å, α=116.4, β=107.8, and γ=113.4°. It is structurally related to ranunculite (Deliens and Piret 1982).

Vochtenite $(Fe^{2+},Mg)Fe^{3+}[UO_2/PO_4]_4(OH) \cdot 12\text{-}13H_2O$
Vochtenite was described from Wheal Basset, Redruth, Cornwall, England, where it occurs with transparent olive green bassetite and brown-yellow "oxidized bassetite", a phase studied by Vochten, de Grave, and Pelsmaekers (1984). Vochtenite forms crystal aggregates with a pseudo-quadratic outline, made up of small, subparallel crystals (0.5–1.0 mm) and having irregular fractures. It is brown, opaque in thick fragments, with a bronzy luster, and is not fluorescent. There is a prominent cleavage on {010}. Vochtenite is monoclinic with a=12.606, b=19.990, c=9.990 Å, and β=102.31°; the complete crystal structure has not been determined (Zwaan, Arps, and de Grave 1989).

Vyacheslavite $U^{4+}(PO_4)(OH) \cdot nH_2O$
Vyacheslavite was described from the Dzhantuar and Rudnoye uranium deposits, Auminzatau Mountains, Central Kyzykym Region, Uzbekistan [localities identified by Pekov (1998)]. It forms minute lamellar to isometric crystals to 8 μm long and 1.5 μm wide, as green powdery films on quartz and octahedral pyrite crystals in carbonaceous-siliceous schists, and small nests in phyllites. Associated species include nasturan, ningyoite, sphalerite, covellite, and chalcocite. Based on selected-area electron diffraction patterns, the structure is orthorhombic with a=6.96, b=9.10, and c=12.38 Å (Belova et al. 1984).

Vysokýite $U^{4+}[AsO_2(OH)_2]_4 \cdot 4H_2O$
Vysokýite was described from the Geschieber vein, Jachymov, Czech Republic, where it occurs on the altered surface of native arsenic, associated with běhounekite, štěpite, kaatialaite, arsenolite, claudetite, and gypsum. It forms long-fibrous, light green crystals to ~8 mm long, with perfect cleavage on (100) and (001). Crystals have an alabaster luster and are not fluorescent in UV.

Vysokýite is one of several novel minerals found on the tenth level of the Svornost mine in an altered lens of native As (see the entry for **štěpite**, above, for further details). The mass of native As contains no uraninite, so the U was apparently transported there by acidic solutions (as uranyl ions), and then reduction to U^{4+} occurred *in situ* on the arsenic lens (Plášil et al. 2013d).

Vysokýite is triclinic with a=10.749, b=5.044, c=19.1778 Å, α= 89.872, β=121.534, and γ=76.508°. Based on single-crystal structure refinement, the structure consists of UO_8 square antiprisms sharing all of their vertices with eight As-tetrahedra to form infinite chains parallel to [010]. These chains are linked by hydrogen bonds involving terminal (OH) groups of the double-protonated As-tetrahedra and water molecules located between the chains.

Figure 499. The uranium arsenate chain in vysokýite, running parallel to [010], with UO_8 polyhedra (blue-gray) connected by $AsO_2(OH)_2$ tetrahedra (purple); each arsenate group shares two vertices with neighboring U polyhedra.

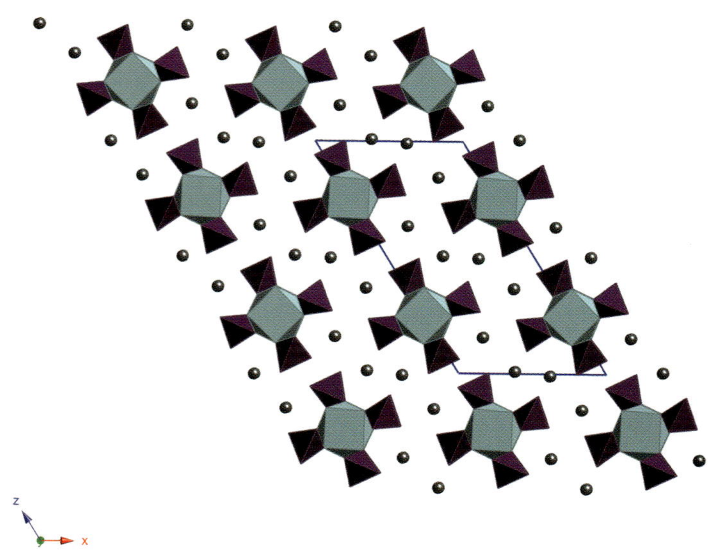

Figure 500. The structure viewed along [010]; adjacent chains are linked by a network of hydrogen bonds involving the H_2O groups (gray). One unit cell is outlined in blue. The observed cleavages lie parallel to the chains, as one would expect.

Xiangjiangite $(Fe,Al)(UO_2)_4(PO_4)_2(SO_4)_2(OH)\cdot 22H_2O$
Xiangjiangite was described from the oxidized zone of a Permian uranium deposit in Hunan, China (Xiangjiang = Hsiang River), where it is associated with sabugalite, variscite, pyrite, and quartz. It forms yellow to bright yellow, earthy microcrystalline aggregates; it is not fluorescent in SWUV light (Hunan team 1978).

It is possible that this mineral is related to the autunite group, based on the reported IR spectrum. There are some inconsistencies between the Chinese text of the original paper and its accompanying English abstract. The structure is reported to be tetragonal with a=7.17 and c=22.22 Å, although other references suggest that the true symmetry could be monoclinic or orthorhombic.

VANADATES

The uranyl vanadates are broadly divided into: 1. the carnotite group; and 2. other vanadates. Members of the carnotite group are, in general, fairly well characterized, whereas several of the other vanadates are poorly defined and further work is needed to elucidate their structures and their relationship to the carnotite group minerals. "Ferghanite," for example, might be closely related to or identical with metatyuyamunite; it isn't currently recognized as a valid species.

Table 53. Uranyl vanadates

Mineral	Formula	Symmetry
Carnotite group		
Carnotite	$K_2(UO_2)_2V_2O_8\cdot 3H_2O$	mon
Curienite	$Pb(UO_2)_2V_2O_8\cdot 5H_2O$	orth
Francevillite	$(Ba,Pb)(UO_2)_2V_2O_8\cdot 5H_2O$	orth
Margaritasite	$Cs_2(UO_2)_2V_2O_8\cdot H_2O$	mon
Metatyuyamunite	$Ca(UO_2)_2V_2O_8\cdot 3H_2O$	orth
Sengierite	$Cu(UO_2)_2(VO_4)_2(OH)_2\cdot 6H_2O$	mon
Strelkinite	$Na_2(UO_2)_2V_2O_8\cdot 6H_2O$	orth
Tyuyamunite	$Ca(UO_2)_2V_2O_8\cdot 5\text{-}8H_2O$	orth
Vanuranylite	$(H_3O,Ba,Ca,K)_2(UO_2)_2V_2O_8\cdot 4H_2O$	orth?
Other vanadates		
Metavanuralite	$Al(UO_2)_2(VO_4)_2(OH)\cdot 8H_2O$	tric
Rauvite	$Ca(UO_2)_2V^{5+}_{10}O_{28}\cdot 16H_2O$	?
Uvanite	$U^{6+}_2V^{5+}_6O_{21}\cdot 15H_2O$	orth?
Vanuralite	$Al(UO_2)_2(VO_4)_2(OH)\cdot 11H_2O$	mon

Uranyl vanadates are important ore minerals, e.g., in Colorado Plateau-type deposits and at the famous locality at Franceville, Gabon. Appleman and Evans (1965) note that, "In their highest oxidation states both U(+6) and V(+5) are rather soluble in dilute aqueous solutions. The greater part of these elements, therefore, would be lost into circulating ground waters, were the elements not fixed by the formation of very insoluble uranyl vanadate complexes of the carnotite type. Consequently, carnotite plays a crucial role in the geochemistry of oxidized uranium-vanadium ore deposits..."

Carnotite group

Minerals of the carnotite group are orthorhombic or monoclinic uranyl vanadates that are chemically related to the autunite group minerals but have a distinctive crystal structure. The minerals have the general formula: $\mathbf{A}(UO_2)_2V_2O_8\cdot nH_2O$, where **A** = Ca, Ba, Pb, K_2, Cs_2, Na_2, or H_3O, and n = 1 to 6. The group includes carnotite, curienite, francevillite, margaritasite, metatyuyamunite, sengierite, strelkinite, tyuyamunite, and vanuranylite.

The crystal chemistry of the carnotite group is dominated by the unique structure of the $[(UO_2)_2V_2O_8]_n^{2n-}$ sheets. First, the large uranyl ion is unlikely to be replaced by another divalent ion. Second, the asymmetric, five-fold coordination of the vanadium creates a square pyramid-shaped group rather than a regular tetrahedron; two VO_5 pyramids, in turn, share an edge to form a V_2O_8 dimer. Furthermore, elements such as P or As, which can sometimes substitute isomorphously for tetrahedrally coordinated vanadium compounds, cannot replace V in the square pyramid (carnotite) structure. (This could also explain the absence of V analogs of the autunite group minerals.) For these reasons, isomorphous substitutions are limited to the interlayer ions (K^+ in carnotite, Ca^{2+} in tyuyamunite, Ba^{2+} in francevillite, etc.) (Barton 1958).

Frost et al. (2005) used Raman spectroscopy to characterize several uranyl vanadates, including carnotite, curienite, francevillite, tyuyamunite, and metatyuyamunite. The minerals exhibit an intense band in the region 800-824 cm^{-1}, assigned

Figure 501. The $[(UO_2)_2V_2O_8]$ sheet in francevillite, consisting of UO_7 pentagonal bipyramids (blue-gray) and V_2O_8 dimers (red). Each V_2O_8 group consists of two VO_5 square pyramids, pointing in the +z and -z directions respectively and sharing one edge.

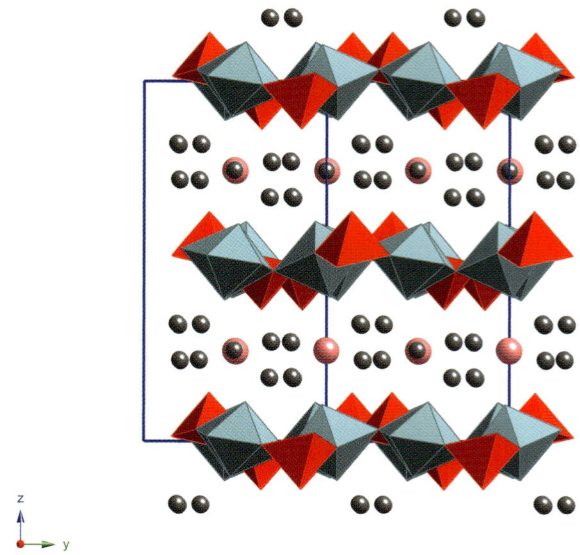

Figure 502. The structure of francevillite, viewed along [100], parallel to the puckered uranyl vanadate sheets, showing Ba (magenta) and H_2O (gray) in the interlayer. Two unit cells are indicated in blue.

to the ν_1 symmetric stretching vibrations of $(UO_2)^{2+}$ units. Another intense band lies in the 965-985 cm^{-1} region and is attributed to ν_1 (VO_3) symmetric stretching vibrations in the (V_2O_8) units. The authors note that, "The study of the vibrational spectroscopy of uranyl vanadates is complicated by the overlap of bands from the (VO_3) and $(UO_2)^{2+}$ units. Raman spectroscopy has proven most useful in assigning bands to these two units since Raman bands are sharp and well separated as compared with infrared bands. The uranyl vanadate minerals are often found as crystals on a host matrix and Raman spectroscopy enables their in-situ characterisation without sample preparation."

Carnotite $K_2(UO_2)_2V_2O_8{\cdot}3H_2O$

Carnotite was described from Montrose Co., Colorado, where it had been found by Charles Poulot. It formed weakly agglomerated yellow masses that were easily reduced to powder, associated with azurite and malachite. Analysis showed traces of Cu, Pb, and Ba, presumably substituting for K. Approximately ten tons of material was originally recovered (Friedel and Cumenge 1899).

Another interesting historical account, provided by Crook and Blake (1910), describes the occurrence of carnotite at a site near Olary railway station, South Australia. Studies of this deposit in 1906 by Mawson and his coworkers at Adelaide University included the first description of davidite. At that time, it was supposed that carnotite ". . . may be reasonably regarded as the vanadium analogue of autunite . . . ," reflecting the state of knowledge before the advent of X-ray crystallography. Crook and Blake obtained a sample of carnotite from Colorado, and based on a detailed comparison of their properties, suggested that the Colorado and Australian carnotites were probably identical.

Carnotite forms microscopic flattened rhomboidal crystals, as microcrystalline aggegates, crusts, and earthy masses. It is bright yellow to greenish yellow; when distinct crystals are present they have a pearly to waxy luster. Carnotite is an important component of sandstone-type ores and is widespread throughout the western US; it is also the one of the main ore minerals at the Tyuya Muyun deposit, Kyrgyzstan. Distinct crystals on malachite are found at Mashamba-West, Congo; at Radium Hill, near Olary, South Australia, it forms surface alterations on davidite-(La).

Barton (1958) studied carnotite in detail, and in particular explored several techniques for synthesizing the compound and many analogs. He synthesized the cesium analogue (later described as the mineral margaritasite) and also the thallium analog. Interestingly, a Tl-rich variety of carnotite has been reported from the Dzhentuar deposit, Kazakhstan.

The water content of natural carnotite varies from about 1-3 H_2O pfu. Barton (1958) synthesized completely anhydrous material, yet its powder XRD pattern was practically identical to that of natural carnotite. Weight loss on heating carnotite is smooth, with no sharp breaks, and DTA curves show no obvious exothermic or endothermic peaks. Barton concludes that, "none of the water in natural carnotite is essential to the structure. Under normal laboratory humidity the amount

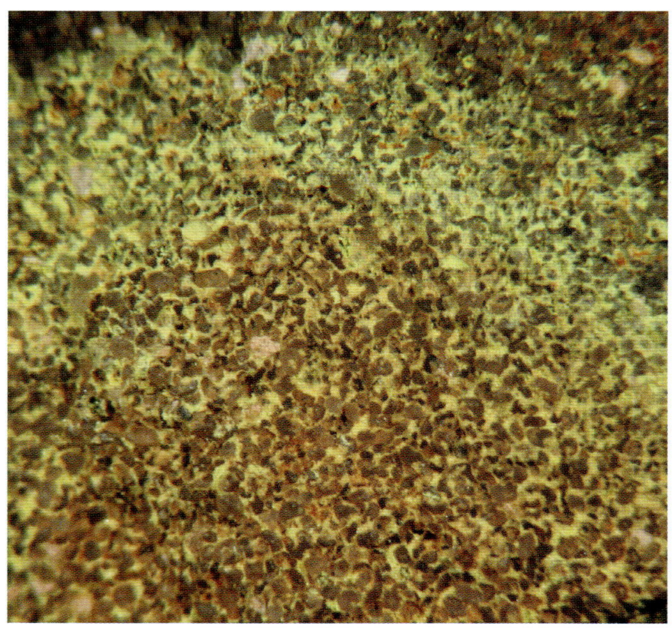

Figure 503. Optical micrograph of earthy yellow carnotite impregnating sandstone ore from Grants, New Mexico. One can see how this type of ore can be exploited by in-situ leaching, in which a chemical solution is pumped into the sandstone layer to simply dissolve the uranium minerals and yield a uranium-rich solution to be pumped out through an adjoining well. RJL69

Figure 504. Bright yellow powdery to microcrystalline carnotite/tyuyamunite on sandstone, from San Juan Co., Utah. RJL142

Figure 505. Butterscotch-colored microcrystals of carnotite from the Anderson mine, Yavapai Co., Arizona. RJL226

Figure 506. Earthy, bright yellow carnotite forming a crust on sandstone from the Harvey Black mine, Navajo Co., Arizona. Note that the mine is near Goulding, San Juan Co., Utah, but is actually in Arizona close to the Utah border. RJL4347

Figure 507. Bright yellow carnotite replacing wood, ~5 cm tall, from the McElroy mine, Henry Mountains, Utah. An accompanying Robert Roots label (ca. 1937) notes, "Part of a $7000 limb." RJL4460

Figure 508. Carnotite microcrystals forming a crust on gray sandstone from Comb Ridge, near Dinnehotso, Arizona. RJL3703

Figure 509. Pale yellow carnotite microcrystals on colorless quartz and gypsum from Chihuahua, Mexico. The material had been misidentified as nováčekite, possibly because of its bright green fluorescence, but the fluorescence appears to originate in nearby matrix areas (uranium opal?) and not in the yellow crystals, which were identified as carnotite by SEM/EDX. RJL389

Figure 510. Scaly yellow carnotite forming part of the alteration crust on davidite-(La) from Radium Hill, near Olary, South Australia. RJL2890

Figure 511. An ore fragment ~5 mm wide, from the Dzhentuar deposit, Kazakhstan, showing bands of greenish yellow thallium-rich carnotite and brown layers of goethite. RJL3681

of water in carnotite is sufficiently low so that there is little variation in the cell dimensions, although . . . some natural material may show a considerable range in the *c* dimension, presumably due to variation in water content. Carnotite does not seem to hydrate in a stepwise fashion as do autunite and tyuyamunite. X-ray diffraction examination shows that the natural carnotite crystals are noticeably disordered, especially in the *c* direction, as would be expected if there is disorder in stacking of the layers or variation in water content."

Synthetic carnotite is monoclinic with a=10.47, b=8.41, c=6.59 Å, and β=103°50' (Applemen and Evans 1965).

Curienite $Pb(UO_2)_2V_2O_8{\cdot}5H_2O$

Curienite was described from the Mounana mine, Franceville, Gabon, by Cesbron and Morin (1968) who recognized it to be the Pb analog of francevillite, with which it forms a series. At Mounana, curienite forms canary-yellow, powdery coatings on earlier-formed crusts of francevillite in mineralized sandstones. It is also found in association with chervetite and mottramite; in some cases the mottramite is pseudomorphically replaced by curienite. Microscopic examination shows that the curienite consists of stacks of lozenge-shaped tabular crystals (Cesbron and Bariand 1975).

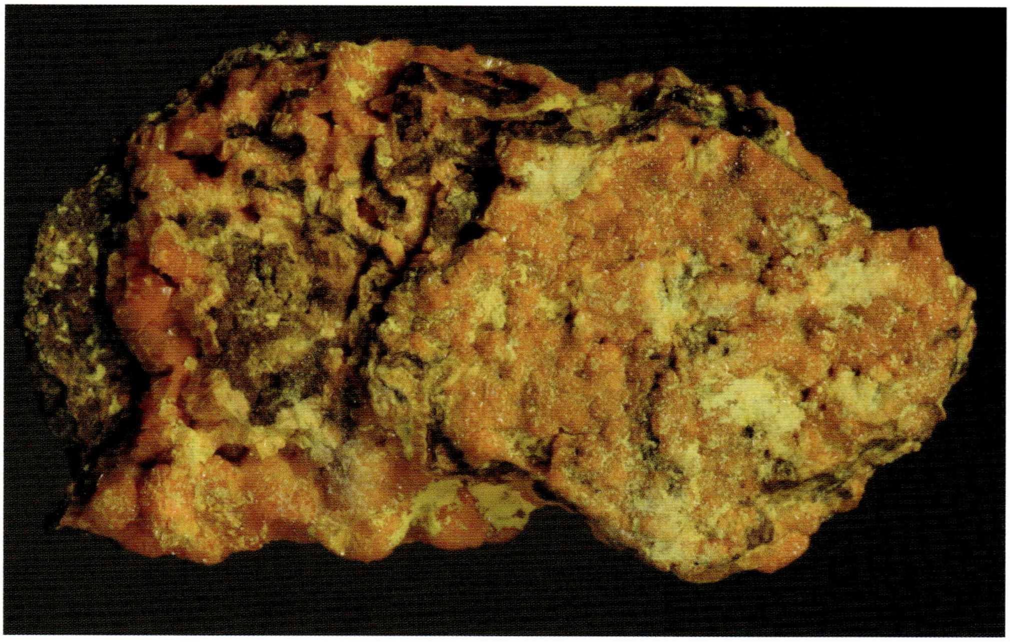

Figure 512. Yellow powdery curienite on crusts of orange francevillite from the type locale for both minerals, Mounana, Gabon. RJL903

Curienite is orthorhombic, with a=10.40, b=8.45, and c=16.34 Å. The crystal structure contains $[(UO_2)_2V_2O_8]$ sheets parallel to (001), which contain symmetry centers. Lead atoms, lying between these sheets, are surrounded by four water molecules, two oxygen atoms belonging to two VO_5 pyramids and two oxygen atoms belonging to two uranyl groups. The fifth water molecule appears to be interstitial (Borène and Cesbron 1971).

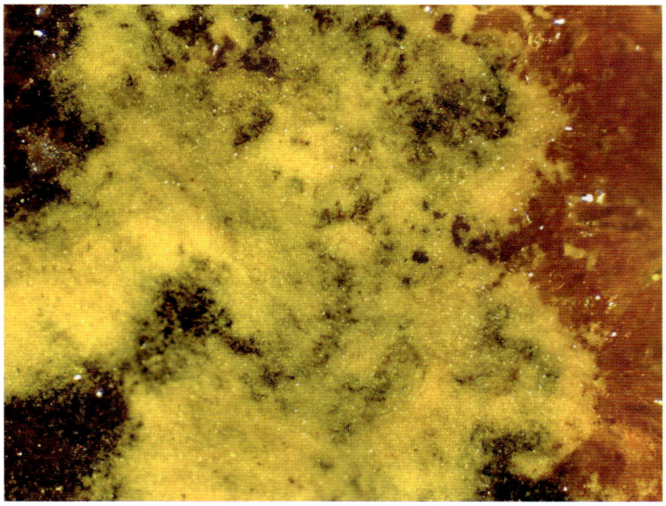

Figure 513. A closer view of the specimen shown in Figure 512, illustrating the radiating crust of francevillite (orange) with bright yellow curienite sprinkled about the surface. RJL903

Figure 514. Scanning electron micrograph showing minute tabular curienite crystals from a sample very similar to that shown in Figures 512–513, from Mounana, Gabon. Field of view is about 200 µm wide. RJL899

Francevillite $(Ba,Pb)(UO_2)_2V_2O_8{\cdot}5H_2O$
Francevillite was described from the Mounana mine, Franceville, Gabon, where it occurred as cryptocrystalline veinlets as well as tabular crystals several mm thick in sandstones. Crystals show perfect cleavage on (001), i.e., parallel to the sheet. The published analyses correspond to the composition $(Ba_{.67}Pb_{.33})(UO_2)_2(VO_4)_2{\cdot}5H_2O$, and the unindexed powder XRD pattern resembles that of metatyuyamunite, suggesting that francevillite is the Ba analog of metatyuyamunite (Branche et al. 1957). Francevillite typically forms tabular crystals, often in stacked or subparallel aggregates, in various shades of yellow-orange, greenish yellow, green, and brown. It is not fluorescent.

An excellent description of the geological setting at Mounana has been given by Cesbron and Bariand (1975). They note that francevillite is the most abundant uranyl

Figure 515. A closer view of francevillite on mounanaite from Mounana, Gabon, shown earlier in Figure 87, showing the fanlike aggregates of francevillite distributed on a layer of darker brown mounanaite. RJL1353

vanadate there, and crystals up to 8 mm across were not uncommon. Material with Ba:Pb = 2:1 is typical and those crystals are orange-yellow to yellow. However, crystals that are closer to the Ba end-member composition occur as well (13.9% BaO and 2.05% PbO, or ~$Ba_{.9}Pb_{.1}$) and these tend to be green.

From the type locale, brilliant orange crystals scattered on dark brown mounanaite make spectacularly colorful specimens. Small but excellent yellow-green crystals on black sandstone are found at the Pandora mine, San Juan Co., Utah. Other locales include the South Terras mine, Cornwall, England; Nussbach, Germany; and the Oktyabrskoe deposit, near Krasnokamensk, Siberia, Russia.

Francevillite is orthorhombic, with *a*=10.41, *b*=8.51, and *c*=16.76 Å (synthetic); and a=10.59, b=8.44, and c=16.06 Å (Mounana) (Cesbron and Morin 1968).

Figure 516. Scanning electron micrograph of another sample of francevillite from Mounana, Gabon, showing the crystal morphology in greater detail. Field of view is about 0.5 mm wide. RJL746

Figure 517. Tiny, olive-green clusters of francevillite crystals from the Pandora mine, San Juan Co., Utah. RJL2642

Figure 518. A closer view of the francevillite crystals in the previous photo. RJL2642

Margaritasite $Cs_2(UO_2)_2V_2O_8{\cdot}H_2O$
Margaritasite was first described from the Margaritas #2 uranium deposit, Sierra Pena Blanca, west of Chihuahua City, Mexico, where it forms part of the ore. It occurs in fine-grained aggregates of 1–3 μm tabular yellow crystals that are optically indistinguishable from carnotite. It is most easily distinguished from carnotite by XRD, as the d_{001} increases relative to carnotite because of the large size of the Cs^+ ion relative to K^+. The powdery aggregates typically fill tiny pores and relict phenocryst casts in a rhyodacite breccia of reworked tuff clasts in the lower Escuadra Formation. Associated minerals include uranophane, uranophane-β, autunite, and weeksite (Wenrich et al. 1982).

At the TL, margaritasite is probably of hydrothermal origin. Wenrich et al. (1982) note that geochemical data indicate that the necessary ratio of Cs:K can only be sustained in fairly high-temperature environments (200°C). This suggests that margaritasite is the product of local hydrothermal or pneumatolytic activity during or after uranium mineralization. It would thus be unlikely to find margaritasite in any Colorado Plateau type deposit; conversely, in uranium deposits of probable hydrothermal origin, any "carnotite" present should be examined as a potential new occurrence of margaritasite.

Chemical analysis of the type material yields the empirical formula $[Cs_{1.38}(H_3O)_{.35}K_{.29}]_{\Sigma 2.02}(UO_2)_{1.99}(V_2O_8)_{1.00}{\cdot}1.07H_2O$. It represents the natural material, with Cs:K~5:1, corresponding to the "Cs-carnotite" compound synthesized by Barton (1958). Wenrich et al. (1982) note that, "Unit cell parameters are a=10.514, b=8.425, and c=7.254, β=106.01° ($P2_1/a$, Z=2). Microprobe analyses of margaritasite and Cs-enriched carnotites synthesized by ion exchange of carnotite in aqueous CsCl solution at 200°C suggest a solid solution between carnotite and margaritasite, but X-ray powder patterns reveal that two discrete c dimensions exist with no intermediate value between them. The margaritasite has a (001) reflection at 12.7° (2θ) while that of carnotite lies at 13.8° (2θ); these peaks do not shift. It is likely that there are two distinct phases, perhaps as interlayered lamellae which are intergrown on a smaller scale than can be resolved by the electron microprobe beam."

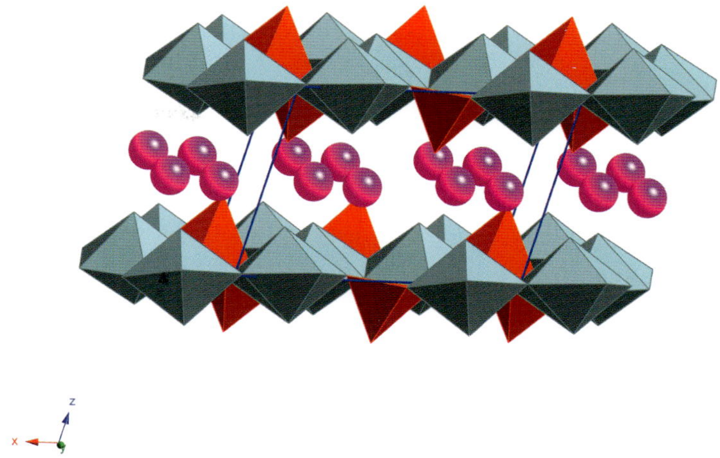

Figure 520. The structure of margaritasite, showing the large Cs⁺ ions (bright pink) in the interlayer region.

Figure 519. A small bleb of yellow earthy margaritasite, <1 mm long, lining a small pore in matrix from the type locale, Margaritas #2, Sierra Peña Blanca, Mexico. RJL418

Sengierite $Cu(UO_2)_2(VO_4)_2(OH)_2 \cdot 6H_2O$

Sengierite was described from the Luiswishi mine, Katanga, Congo, where it was found in fissures along a fault where chlorite/talc rock contacted dolomite containing minor sulfide mineralization, mostly as siegenite. The sengierite was associated with black oxides of Fe, Cu, Co, and Ni, as well as with yellow volborthite or with vandenbrandeite, malachite, and chrysocolla. Partially altered pitchblende was also found in pockets along the fault. Sengierite was interpreted to be "of supergene origin produced by deposition from solutions containing uranium, vanadium, and copper derived from earlier minerals," (Vaes and Kerr 1949). The species forms small green platy or flake-like crystals, generally less than 2 mm, with vitreous luster that coat fracture surfaces in the rock.

Sengierite was found as efflorescences on chalcocite at the Cole Shaft, Bisbee, Arizona (Hutton 1957). Other reported localities include: the Clara mine, Black Forest, Germany; Rabejac, Lodeve, France; and the Huemul mine, Mendoza, Argentina.

The initial report (Vaes and Kerr 1949) suggested that sengierite has orthorhombic symmetry; however, a later structure determination on material from Shinkolobwe showed it to be monoclinic, with a=10.62, b=8.10, c=10.11 Å, and β=103.67° (Piret, Declerq, and Wauters-Stoop 1980).

Figure 521. Emerald green, tabular sengierite crystals from the TL at Luiswishi, Congo. FOV is ~7 mm. RJL398

Figure 523. Detail of the sengierite in the previous photo, showing minute yellow-green platy sengierite crystals forming a spherical aggregate on massive black chalcocite. RJL1134

Figure 522. A classic example of sengierite on massive chalcocite from the 1,000-foot level, Cole Shaft, Bisbee, Arizona; documentation indicates that it was purchased in 1955. The sengierite forms minute masses of chartreuse-green tabular crystals on the dark metallic chalcocite, along with other secondary copper minerals. Field of view is about 3 cm wide. RJL1134

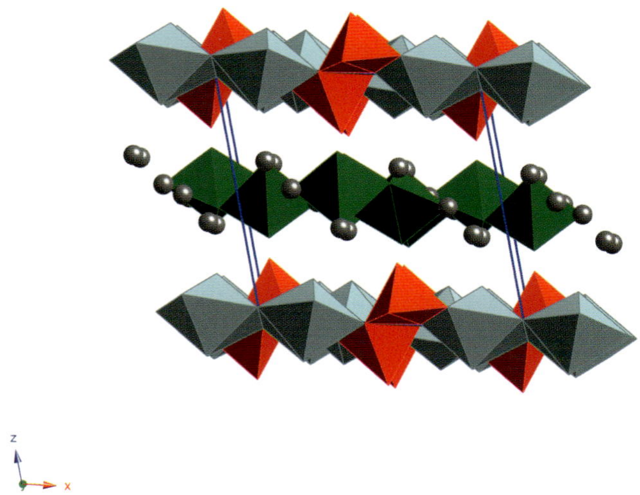

Figure 524. The structure of sengierite, viewed parallel to the sheets, showing Cu (dark green) and H_2O (gray) in the interlayer.

Strelkinite $Na_2(UO_2)_2V_2O_8{\cdot}6H_2O$
Strelkinite was originally described based on material from two separate occurrences, one in Uzbekistan and the other in Kazakhstan. The original report, following a common practice in the Soviet period, identified the locales rather vaguely as "Paleozoic carbonaceous-siliceous shales in one of the desert regions of the USSR" and "Devonian effusive rocks in other regions of the arid zone of the USSR" (Alekseeva et al. 1974). Pekov (1998) identifies these locales as follows: The first find, in 1965, was in cracks in coaly shales in the Kendyktas Mountains, Central Kyzylkum region, Uzbekistan. The later find was at Basaral, 10 km north of the Bota-Borum U deposit on the northeastern slope of the Chu-Ili Mountains, Southwestern Balkhash region, Kazakhstan; at this locale, strelkinite forms gold- to canary-yellow powdery aggregates and lamellar crystals to about 1.5 mm, associated with quartz, calcite, and hydrous iron oxides.

Strelkinite has also been reported from stone quarries along the Lisdan-Siwaga fault, 70 km southeast of Amman, Jordan.

It is interesting that Barton (1958) reported that synthesis of "Na-carnotite" from $NaVO_3$ melts was unpredictable and not easily repeated. Other workers had shown that the Na compound could easily be base exchanged in K^+ solutions to yield carnotite, and Barton's work showed that Na-carnotite dissolves in acid more rapidly and at a slightly higher pH than other members of the group. He suggests that, "These observations indicate that Na carnotite is relatively unstable compared to the carnotites which contain larger alkali cations. It seems unlikely that pure Na carnotite would occur in nature because sufficient amounts of K^+ are present in most natural solutions to base exchange the Na^+ by K^+. Apparently the decrease in stability of Na carnotite relative to K carnotite is related to the decrease in ionic size of Na^+ relative to K^+ [0.97 vs. 1.33 Å)." Nevertheless, analyses reported by Alekseeva et al. (1974) indicate compositions very near the Na end member, so at least under some circumstances the compound appears to be stable. Strelkhinite is orthorhombic with a=10.64, b=8.36, and c=32.27 Å.

Figure 525. Scaly, bright greenish yellow, strelkhinite crystals thickly coating the surface of carbonate rock from the Lisdan-Siwag fault, Jordan. RJL4537

Tyuyamunite $Ca(UO_2)_2V_2O_8{\cdot}5\text{-}8H_2O$
Metatyuymunite $Ca(UO_2)_2V_2O_8{\cdot}3H_2O$
Tyuyamunite was described from the Tyuya Muyun Cu-V-U deposit, Ferghana Valley, Kyrgyzstan (Nenadkevich 1912). Pekov (1998) provides a fascinating history of the find: The mineral "ferghanite" had been previously described by I. A. Antipov in 1908 as "hydrous uranium vanadate" based on what appears to have been faulty chemical analysis. Later work by Nenadkevich determined that all of the uranyl vanadate specimens contained calcium. He named this mineral tyuyamunite, and it is the predominant ore mineral at the locality. In the intervening years, a Ca-free uranyl vanadate has never been found at Tyuya Muyun despite ongoing studies of the ore minerals there, supporting the contention that the original analysis of "ferghanite" was mistaken. At the

Figure 526. Divergent sprays of golden yellow tyuyamunite crystals thickly lining a 2 cm cavity, associated with dark green fibrous malachite and colorless calcite rhombs from Musonoi, Congo. RJL797

Figure 527. Scanning electron micrograph of a small tyuyamunite crystal group removed from the sample in Figure 526, showing details of the crystal morphology. RJL797

Figure 528. Bright yellow crystals of tyuyamunite forming small rosettes on sandstone from the Marie mine, Pryor Mountains district, Montana. RJL2964

Figure 529. Another type of sample from the Pryor Mountains district: a plate of honey colored scalenohedral calcite crystals coated with a layer of microcrystalline tyuyamunite from the Dandy mine, Montana. RJL3055

Figure 530. Minute, thin tabular tyuyamunite crystals on matrix from the Ridenaur mine, Coconino Co., Arizona. RJL4351

Figure 531. Another example from the Ridenaur mine: a colorful association of tyuyamunite with dark green tangeite, $CaCu(VO_4)(OH)$, a member of the adelite-descloizite group. RJL4475

Figure 532. Detail of the specimen in the previous photo. RJL4475

TL, tyuyamunite forms powdery crack fillings and druses of bright yellow lamellar crystals to several mm in cavities in calcite or barite, in association with tangeite and malachite. This occurrence is discussed in greater detail by Chirvinsky (1925), who notes that, "Between the years 1904 and 1914 the mine yielded 1,000 tons of ore, from 700 tons of which 2.3 grams of metallic radium was extracted."

Tyuyamunite is the calcium analog of carnotite, and the two minerals have similar properties and modes of occurrence. Crystals are tiny, bright yellow, flattened scales and laths, usually in fanlike or radial aggregates or as an earthy component filling sandstone type deposits. Fluorescence under UV light is either very weak yellow-green or absent altogether. It is somewhat more likely than carnotite to form distinct crystals.

Figure 533. Earthy yellow metatyuyamunite forming a patch ~2.5 cm wide on friable sandstone from the Gould mine, South Dakota. RJL2803

Figure 534. Yellow earthy crust of metatyuyamunite from near Prewitt, New Mexico. RJL3134

Other notable locales include: numerous sites in the Colorado Plateau region; as fine specimens from the Marie and Dandy mines, Pryor Mountains district, Carbon Co., Montana; Rozmital, Czech Republic; Dortmorsbach, Germany; and Musonoi, Congo.

Metatyuyamunite was described based on material collected in 1948 at the Jo Dandy mine, Montrose Co., Colorado, by Stern et al. (1956), who reported that it had also been collected "from more than 35 localities on the Colorado Plateau and from several localities in Fall River County, S. Dak., and Campbell and Johnson Counties, Wyo." The properties of metatyuyamunite are similar to those of tyuyamunite, and the hydration-dehydration process is reversible over the range 3-8.5 H_2O. It is likely that metatyuyamunite exists at many or perhaps most reported tyuyamunite occurrences. Weeks and Thompson (1954) note that metatyuyamunite is especially abundant near Haystack Mountain and Laguna, Grants district.

Tyuyamunite is orthorhombic with a=10.63, b=8.36, and c=20.40 Å. Metatyuyamunite is orthorhombic with a=10.63, b=8.36, and c=16.96 Å (Stern et al. 1956).

Vanuranylite $(H_3O,Ba,Ca,K)_2(UO_2)_2V_2O_8{\cdot}4H_2O$

Vanuranylite was discovered in 1955 and somewhat incompletely described in 1965 from an unspecified locality in the USSR, where it forms bright yellow films and crusts of minute (0.03 mm) platy pseudohexagonal crystals along fractures in sandstones in the oxidation zone, associated with uranophane and soddyite (Buryanova, Strokova, and Shitov 1965). The locale has been identified as the Ust'-Yuk V-Se-U deposit, Tuva, Siberia (Pekov 1998).

The structure, determined by electron diffraction photography indicated the unit cell to be a=10.49, b=8.37, c=20.30 Å, and β=90°(?), close to the published values for tyuyamunite. However, quantitative analysis indicated a deficiency of cations, so the authors postulated the presence of hydronium ions in the interlayer to make up the balance of the charge, leading to the empirical formula $[(H_3O)_{1.2}Ba_{.19}Ca_{.09}K_{.08}Pb_{.03}](UO_2)_2V_2O_8{\cdot}4.2H_2O$, making it the hydronium analog of carnotite.

Other uranyl vanadates

Other uranyl vanadates include ferghanite, rauvite, uvanite, vanuralite, and metavanuralite. Several of these species are poorly defined and further work is needed to elucidate their structures and their relationship to the carnotite group minerals. Ferghanite, for example, might be closely related to or identical with metatyuyamunite; it is not recognized as a valid species.

Rauvite $Ca(UO_2)_2V^{5+}_{10}O_{28}{\cdot}16H_2O$

Rauvite was originally described from Temple Rock, and nearby Flat Top, San Rafael Swell, Emery Co., Utah, where it filled cracks and impregnated sandstone of the Shinarump formation, in association with carnotite, uvanite, metatorbernite,

Figure 535. Rauvite forming brownish bands in sandstone from the Monument #2 Mine, Apache Co., Arizona. RJL3670

and gypsum (Hess 1924, 1922). It is interesting to note that when the abstract of the Hess report was published in American Mineralogist, the editor commented that, "While this material is apparently new its metacolloidal nature makes it impossible to determine whether this is a single mineral or not. Its exact nature must be considered doubtful." Frondel (1958) describes it as a "hydrated vanadate of calcium and uranium of uncertain formula . . ." and further notes that its crystal system is not known and that the mineral seems to have a layer structure with extensive disorder. It tends to be very fine grained with variable physical and optical properties. Color ranges from purplish or bluish black to brownish red or dirty orange yellow. According to Frondel, "The mineral seems to be a hardened, gel-like precipitate, perhaps forming under more acid conditions than carnotite and tyuyamunite. It is probably derived by the alteration of uraninite and low-valence vanadium oxides, also possibly of tyuyamunite. In the oxidation of uraninite-montroseite ore, rauvite probably is the intermediate stage to carnotite and tyuyamunite, and it is generally associated with corvusite, vanadium hydromica [roscoelite], and some fully oxidized vanadium minerals."

Despite the fact that rauvite remains a poorly defined species, which IMA regards as questionable or doubtful, it has been reported from numerous localities on the Colorado Plateau and in the southern Black Hills. Some reported localities include: the Small Spot and Arrowhead mines, Mesa Co., and the Hummer mine, Montrose Co., Colorado; the Cactus Rat and Curvusite mines, Grand Co., Utah; the Monument No. 2 mine, Apache Co., Arizona; and the Road Hog No. 3A claim, Fall River Co., South Dakota.

Uvanite $U^{6+}_2V^{5+}_6O_{21}\cdot 15H_2O$

Uvanite was described from the vicinity of Temple Rock on the San Rafael Swell, Emery Co., Utah. Hess and Schaller (1914) described it as follows: "Uvanite is a brownish-yellow hydrous uranium vanadate. Its mode of occurrence is similar to that of carnotite which it resembles in general appearance, but it has not such a clear bright yellow color. . . . Some masses of the new mineral are a centimeter thick, but these are rather rare. The mineral occurs in sufficient quantity to be of economic importance. Examined microscopically, uvanite is seen to consist of very minute crystalline particles with a very high birefringence. The mineral is not soluble in water but dissolves very quickly in a solution of ammonium carbonate." From the chemical analysis they derived the formula $2UO_3\cdot 3V_2O_5\cdot 15H_2O$, but noted some uncertainty as to whether CaO (1.73%) and K_2O (0.3%) were essential to the composition or derived from admixed tyuyamunite and carnotite.

Uvanite, like rauvite, is a poorly described mineral and is considered questionable or doubtful by IMA. Frondel (1958) noted that the only reported analysis also contained Ca, which might be essential. Uvanite forms minutely crystalline masses and coatings that are not fluorescent. At the TL it is associated with carnotite, rauvite, hewettite, metatorbernite, and gypsum in sandstone. According to Frondel, "Only a few authentic specimens are known."

Vanuralite $Al(UO_2)_2(VO_4)_2(OH)\cdot 11H_2O$
Metavanuralite $Al(UO_2)_2(VO_4)_2(OH)\cdot 8H_2O$

Vanuralite was described from the supergene zone at the Mounana mine, Gabon, where it formed citron-yellow monoclinic plates, in association with Pb-rich francevillite and minor chervetite and brackebuschite (Branche et al. 1963). The crystals often form clusters, compact aggregates, and arborescent groups. Individual crystals are spear-shaped,

Figure 536. Pure arborescent mass of vanuralite about 1 cm tall from the type locale, Mounana, Gabon. RJL470

Figure 537. Golden yellow mass of vanuralite about 1 cm tall from Lavra Jabuti, Baixio, Minas Gerais, Brazil. RJL2889

elongated on [010] and flattened on (001), which is parallel to the perfect cleavage. At Mounana, vanuralite is often found on a fine-grained brown matrix consisting of mounanaite and limonite. The crystals are generally less than about 3 mm, but crusts or fracture fillings up to 4 cm thick were reported (Cesbron and Bariand 1975).

Metavanuralite was described by Cesbron (1970) based on material from Mounana. Cesbron noted that at 20°C the lower hydration state ($8H_2O$) is stable at relative humidity between 28 and 47%. The loss of $3H_2O$ lowers the symmetry from monoclinic to triclinic. The two species cannot be distinguished macroscopically, but because the loss of water is reversible, one can assume either or both may be present depending on ambient conditions.

Vanuralite has also been reported from Lavra Jabuti, Minas Gerais, Brazil; and the Carija quarry, Badajoz, Spain.

Vanuralite is monoclinic, with a=10.55, b=8.44, c=24.52 Å and β=103°. Metavanuralite is triclinic, with a=10.46, b=8.44, c=10.43 Å, α=75°53', β=102°50', and γ=90°. The transition from a monoclinic to a triclinic unit cell involves reduction of the distance between the planes of perfect cleavage (001) and slip parallel to [010] (Cesbron 1970).

SILICATES

Uranyl silicates are the most abundant uranyl minerals, owing to their relatively low solubility in most ground waters and the relative ubiquity of dissolved Si in those waters. For this reason, these minerals are also fairly stable and rarely alter to other species once formed. The minerals may be divided into several groups based on the atom ratio of uranium to silicon (Stohl and Smith 1981). The most U-rich species is soddyite with U:Si = 2:1. The 1:1 uranyl silicates comprise the uranophane group. More Si-rich species include the weeksite group, with a U:Si ratio of 2:5. The most Si-rich species is uranosilite with U:Si = 1:7.

Table 54. Uranyl silicates

Mineral	U:Si	Formula	Symmetry
Soddyite	**2:1**	$(UO_2)_2SiO_4 \cdot 2H_2O$	orth
Uranophane group	**1:1**		
Boltwoodite		$(K,Na)[(UO_2)SiO_3OH)] \cdot 1.5H_2O$	mon
Cuprosklodowskite		$Cu[(UO_2)_2(SiO_3OH)_2].6H_2O$	tric
Kasolite		$Pb(UO_2)SiO_4 \cdot H_2O$	mon
Natroboltwoodite[a]		$H(Na,K)(UO_2)SiO_4 \cdot 1.5H_2O$	mon
Oursinite		$(Co,Mg,Ni)(UO_2)_2Si_2O_7 \cdot 6H_2O$	orth
Sklodowskite		$(H_3O)_2Mg(UO_2)_2(SiO_4)_2 \cdot 4H_2O$	mon
Swamboite		$UH_6(UO_2)_6(SiO_4)_6 \cdot 30H_2O$	mon
Uranophane		$Ca[(UO_2)(SiO_3OH)]_2 \cdot 5H_2O$	mon
Uranophane-β		$Ca[(UO_2)(SiO_3OH)]_2 \cdot 5H_2O$	mon
Weeksite group	**2:5**		
Coutinhoite		$Th_xBa_{(1-2x)}(H_2O)_y(UO_2)_2Si_5O_{13} \cdot H_2O$	orth?
Haiweeite		$Ca[(UO_2)_2(SiO_3OH)_2(Si_3O_6)] \cdot 6H_2O$	orth
Metahaiweeite		$Ca(UO_2)_2Si_5O_{13} \cdot nH_2O$, n<5	mon
Weeksite		$K_2(UO_2)_2Si_5O_{13} \cdot 4H_2O$	mon
Uranosilite	**1:7**	$(UO_2)Si_7O_{15}$	orth

[a]Formula and symmetry by analogy to boltwoodite

Species with U:Si = 2:1 - Soddyite

Soddyite $(UO_2)_2SiO_4 \cdot 2H_2O$

Soddyite was described from Kasolo, Katanga Province, Congo (Schoep 1922) and is one of only three uranyl silicates that do not contain other metal cations (the others being uranosilite and swamboite). The crystals may form subparallel groups or divergent clusters; massive, fibrous, and earthy forms have also been reported. Opaque crystals are dull to bright yellow; transparent crystals are yellow to dark amber. Clear crystals have vitreous to subadamantine luster; cloudy crystals and massive aggregates can be subvitreous to dull or earthy. Crystals may be zoned, with a transparent rim and opaque core (Gorman 1952).

Structure and morphology

Soddyite is orthorhombic with *a*=8.334, *b*=11.212, and *c*=19.668Å. The structure consists of chains of edge-sharing UO_7 pentagonal bipyramids running parallel to [110], which are cross-linked by SiO_4 tetrahedra. The structure is further strengthened by hydrogen bonding between the water molecules and oxygen atoms on the uranyl polyhedra (Demartin, Gramaccioli and Pilati 1992).

Soddyite typically forms distinct bipyramidal crystals bounded by {111} and terminated either in a point or by the (001) faces; it may also form prisms elongated on [110]. Cleavage is perfect on {001} and good on {111}.

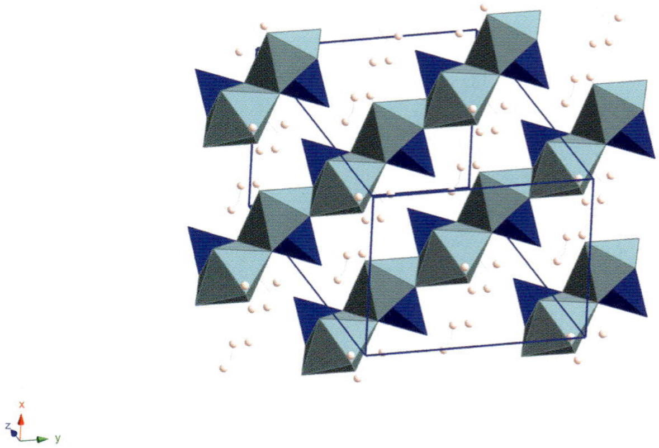

Figure 538. Crystal structure of soddyite, viewed at an oblique angle to highlight the chains of edge-sharing UO_7 polyhedra (light blue), SiO_4 tetrahedra (dark blue), and water (pink H atoms with wire bonds).

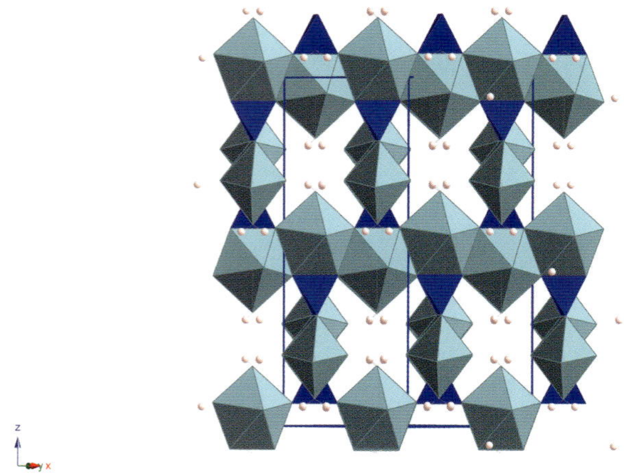

Figure 539. The soddyite structure viewed along [110] to show the cross-linking between chains.

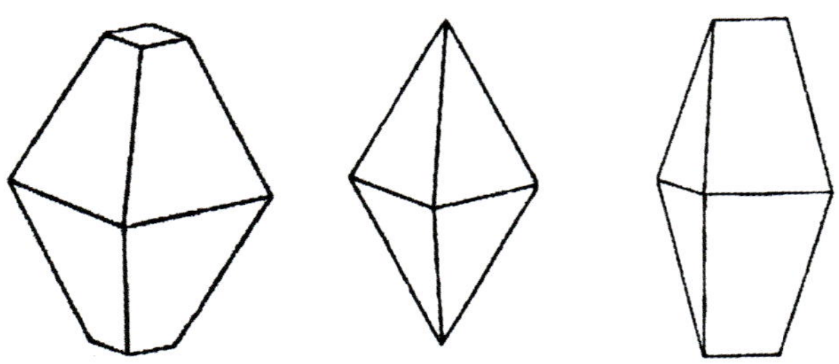

Figure 540. Crystal habits of soddyite, modified from Gorman (1952).

Figure 541. Opaque yellow soddyite crystals on matrix from Swambo, Congo, illustrating the typical blocky habit; some twinning can also be seen, for example in the large crystal at bottom. RJL3900

Formation and geochemistry

Soddyite is a secondary mineral formed by oxidation of uraninite. At the type locale at Kasolo it occurs in massive form, intimately admixed with earlier-formed curite, and as drusy crusts or fibrous veinlets. It is associated with later-formed sklodowskite, uranophane, and torbernite. It is common at Shinkolobwe as bipyramidal crystals up to 3 mm. At Musonoi and at Kolongwe it is associated with cuprosklodowskite and vandenbrandeite; it is particularly abundant at Swambo, associated with sklodowskite and black base-metal oxides (Deliens, Piret, and Comblain 1981).

Soddyite forms dense, straw-yellow pseudomorphs after uraninite at the Ruggles mine, Grafton Center, New Hampshire; it occurs at the Lucky Mc mine, Fremont Co., Wyoming; on sandstone at the Jackpile mine, Grants, New Mexico; with kasolite and uranothorite at the Steel City mine, Arizona; and in the Honeycomb Hills, Utah (Frondel 1958). It has also been reported from a pegmatite at Norabees, Namaqualand, South Africa, where it forms bladed encrustations on quartz, associated with malachite.

Species with U:Si = 1:1 - Uranophane Group

The uranophane minerals are hydrated uranyl silicates with the general formula: $\mathbf{A}[UO_2SiO_4]_y{\cdot}nH_2O$, where $\mathbf{A}$ = Pb^{+2}, Ca^{+2}, K, Na, Cu^{+2}, Mg^{+2}, Co^{+2}, U^{+6}. The group includes boltwoodite, cuprosklodowskite, kasolite, natroboltwoodite, oursinite, sklodowskite, swamboite, uranophane, and uranophane-β.

Figure 542. Soddyite crystals on sandstone from the Jackpile Mine, Grants, New Mexico. RJL343

Figure 543. Detail of the soddyite specimen in the previous photo, showing the prominent color zoning from an opaque yellow core to a transparent, dark amber rim. RJL343

Figure 544. Soddyite crystals to ~1 mm on matrix from Swambo, Congo. RJL901

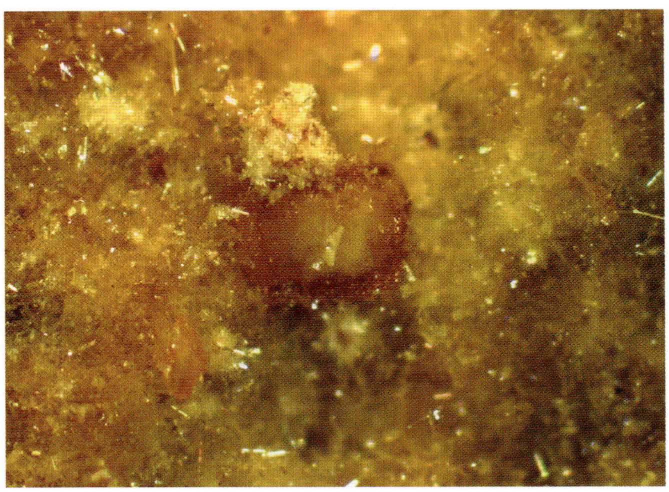

Figure 545. An interesting association from Swambo: a color-zoned soddyite crystal ~0.5 mm wide, surrounded by acicular uranophane (or possibly sklodowskite) crystals. RJL882

Structure and morphology

Minerals of the uranophane group contain sheets made up of uranyl bipyramids and silicate tetrahedra, with lower valence cations and H_2O groups situated in the interlayer region. The various members of the group are not truly isostructural because the structural details and overall symmetry depend on the number and charge of the lower valence cations. Furthermore, there are differences in the orientation of the SiO_4 tetrahedra: in uranophane the apical oxygens on the tetrahedra all point upward along one side of the chain and downward along the other side, whereas in uranphane-β the apical oxygens alternately point upward and downward along each side of the chain.

In these minerals, with the exception of kasolite, the cations alone don't provide the necessary formal charge balance. Therefore, some of the water must be treated as H^+, H_3O^+, or OH^-. When OH is part of the structure, it is typically considered to substitute for the apical oxygen on the silicate group, forming $SiO_3(OH)$ tetrahedra, as shown in several of the structural formulas. At least some portion of the water may be zeolitic.

Crystals are thin prismatic to acicular or fibrous, elongated parallel to the uranyl silicate chains. With the exception of cuprosklodowskite, which is green, all other species in the group are various shades of pale to bright yellow, amber, or yellowish brown.

Figure 546. The uranyl silicate sheet in uranophane, viewed along [100], showing chains of edge-sharing uranyl bipyramids (blue-gray) and SiO_4 tetrahedra (dark blue) linking the chains by sharing edges and vertices.

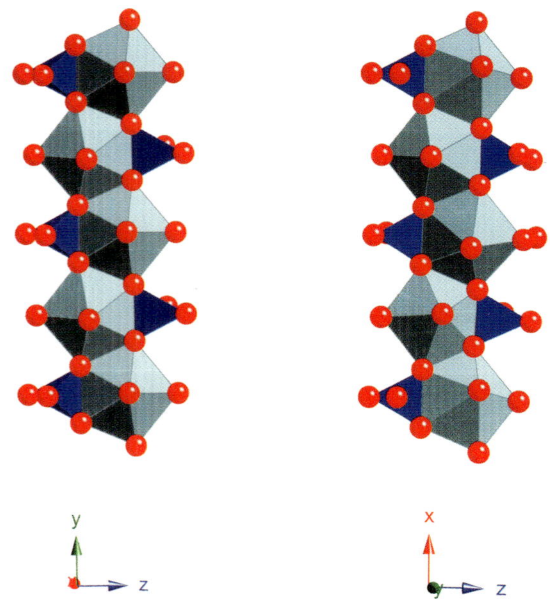

Figure 548. The uranyl silicate chains of uranophane (left) and uranophane-β (right) comparing the differing SiO_4 anion topologies. Oxygen atoms (red) are shown for added clarity.

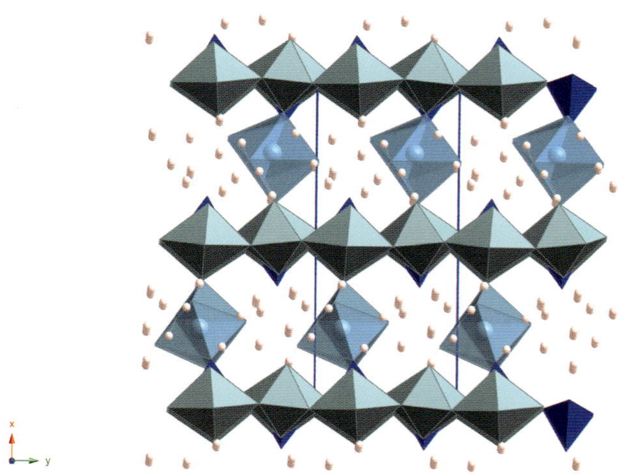

Figure 547. The uranophane structure viewed parallel to the sheets, showing the position of Ca (translucent blue octahedra) and H_2O (pink H atoms with wire bonds) in the interlayer. One unit cell is indicated in blue.

Figure 549. Felted masses of bright yellow uranophane on quartz from Grants, New Mexico, showing the typical acicular habit. RJL333

Formation and geochemistry

This group represents the most commonly occurring uranyl silicates. They form via the alteration of primary uraninite, often by meteoritic waters at near ambient temperatures. They can form directly on the uraninite, either as a crust or as a pseudomorphic replacement; alternatively the uranyl species can be carried by groundwater for some distance away from the primary minerals before precipitating the uranyl silicate phase.

The presence of extensive solid solution, e.g., between boltwoodite and natroboltwoodite, complicates the paragenetic relationships of uranyl minerals. Burns (1998c) points out that, "Mineral structures that are compatible with a range of compositions will tend to have larger stability fields than those that are compositionally inflexible. The determination or prediction of such fields of stability will require the study of crystals with a range of compositions, rather than thermodynamic measurements done for end-member compositions . . ."

The minerals

Boltwoodite $(K,Na)[(UO_2)(SiO_3OH)](H_2O)_{1.5}$

Boltwoodite was the first reported alkali uranyl silicate, based on type material from the Delta mine, Emery Co., Utah (Frondel and Ito, 1956). It forms pale yellow acicular crystals that may be mistaken for sklodowskite or uranophane. Individual crystals have a pearly luster on cleavage surfaces; radial aggregates may appear vitreous to silky depending on crystallite size. It typically forms coatings in fractures and cavities as well as compact to earthy pseudomorphs after uraninite. Depending on the space available for crystal growth, divergent tufts or sprays, flattened radial aggregates, or complete hemispherical masses may be found. Boltwoodite is fluorescent dull green under both LW and SWUV.

Boltwoodite is a fairly common secondary mineral and has been recognized at numerous localities in varied geological environments (Honea 1961). At the TL in Utah it occurs as yellow wartlike aggregates of fibers coating sandstone as well as in other specimens associated with becquerelite, fourmarierite, and uranyl sulfates. In the Williams quarry, Easton, Pennsylvania, it forms as an alteration product of thorian uraninite in serpentine. Pseudomorphs after uraninite from Alto Boqueirao, Brazil, and from the Little Indian mine, Colorado, are densely microcrystalline and pale straw yellow. Colorful micromount specimens from the New Method mine, near Amboy, California, feature pale yellow boltwoodite associated with quartz and minute clusters of very dark purple fluorite crystals.

Figure 551. Boltwoodite with black-purple fluorite from the New Method mine, near Amboy, San Bernardino Co., California. The crystal structure of boltwoodite was determined using material from this locality (Stohl and Smith 1981). RJL798

Figure 550. Hand-sized sample of boltwoodite on sandstone from the type locale, Pick's Delta mine, Utah. RJL893

Figure 552. Minute, golden-yellow spherical aggregates of boltwoodite, with dark green malachite, on sandstone from South Alligator District, Northern Territory, Australia. RJL837

Figure 553. Divergent sprays of boltwoodite on brown calcite crystals, from the Rössing area, Namibia. RJL1301

Figure 555. Two 5-mm radial aggregates of boltwoodite, characteristically growing on brown calcite, from the Goanikontes claim, Arandis, Namibia. RJL4323

Figure 554. Spherical aggregate of boltwoodite from the Rössing area, Namibia. RJL2589

Figure 556. Boltwoodite crystals ~3 mm long completely lining a cavity in carbonate rock from the Rössing area, Namibia. RJL3896

At the Little Indian No. 36 mine and the Lookout No. 22 mine, Gunnison Co., Colorado, it forms small, wartlike aggregates on becquerelite and as dense earthy pseudomorphs after uraninite. Associates reported at these localities include soddyite, metazeunerite, and other secondary minerals.

At the Sonia mine, near Guandacol, La Rioja Province, Argentina, alteration of zirconian uraninite produced boltwoodite as small botryoidal aggregates of fibrous crystals with cuprosklodowskite, and as interstitial fillings in sandstone surrounding lenticular pods of uraninite, associated with becquerelite.

In samples of highly argillized granitic gneiss from Myponga, South Australia, it forms pale yellow flattened radial aggregates between grains and cleavage surfaces of the friable host rock. In the South Alligator District, Northern Territory, Australia it forms minute bright yellow balls and needles on sandstone, associated with clear gypsum and some copper species.

The Rössing area, Namibia, is the source of what are at present arguably the best specimens of boltwoodite, in radial aggregates up to 2 cm across associated with clear quartz, calcite, gypsum, and other minerals. As discussed in the entry for **Namibia**, many fine specimens actually come from one or more nearby occurrences, such as Goanikontes; in the minerals trade, many specimens will simply be labeled "Rössing, Namibia" and without further details or information, the label should be read as the Rössing area.

Burns (1998c) determined the structure of sodium-rich boltwoodite, ~$(K_{.56}Na_{.43})$, from Rössing, Namibia, to be monoclinic with a=7.08, b=7.06, c=6.65 Å, and β=104.98°. He notes that, "...either K or Na can be accommodated in the interlayer between the sheets of silicate and uranyl polyhedra in the structure of boltwoodite, although this is achieved by the cations assuming different positions that result in different coordination polyhedra that are consistent with cation size."

Figure 557. Massive fibrous cuprosklodowskite in ore from Kasolo, Congo. FOV is 2 cm wide. RJL238

Cuprosklodowskite $Cu[(UO_2)_2(SiO_3OH)_2].6H_2O$

Cuprosklodowskite was described from Kalongwe, Congo (Vaes 1933). It forms acicular prismatic crystals elongated on the c-axis that occasionally show the forms {100}, {010}, and {110}. The mineral may form stellate fibrous aggregates of acicular crystals up to 1 cm long, silky to matted crusts, and radial-fibrous aggregates and nodules. It is pale green to emerald green, also olive green; fibrous samples can be nearly white. Like other uranyl minerals that contain copper, cuprosklodowskite is not fluorescent.

Cuprosklodowskite forms by the oxidation of deposits containing primary uraninite and copper minerals. It is found in most of the copper-cobalt deposits of southern Shaba, including Kolongwe, Kambove, Kasompi, Kolwezi, Luiswishi, Musonoi, and Shinkolobwe. The most common associates are vandenbrandeite, kasolite, malachite, and sklodowskite; less frequently seen are rutherfordine, schoepite, and other secondary minerals.

Figure 558. Nodule of cuprosklodowskite, ~2 × 5 cm, from Musonoi, Congo. The center is filled with malachite (dark green) and masses of yellowish rutherfordine and schoepite. RJL2571

Cuprosklodowskite is found at Grants, New Mexico, as well as at several localities in Utah, including Red Canyon and White Canyon in San Juan Co. and at Seven Mile Canyon near Moab in Grand Co.

In Canada it occurs at the Nicholson mine, Lake Athabaska, Saskatchewan, in association with kasolite, uranophane, and metazeunerite. Hogarth (1951) notes, "This assemblage is similar to that of Shinkolobwe, Katanga. Other analogies are apparent. In contrast to Bohemia where carbonates and sulphates of uranium predominate, silicates of uranium are the principal oxidation products at Nicholson and Katanga. Further, the Cu/Ni+Co ratios are high at both localities. As at Katanga the precious metals Au and Pt are present in significant amounts. Mineralogically therefore, the vein shows distinct similarities to the uranium occurrences in Katanga."

Other localities include: Johanngeorgstadt, Saxony, Germany; Jachymov, Czech Republic (Nováček 1935); and Amelal, Morocco.

Figure 559. Extremely delicate cuprosklodowskite needles forming spherical aggregates ~1 cm across, on matrix from Musonoi, Congo. RJL2942

Cuprosklodowskite is triclinic with a=7.052, b=9.268, c=6.655 Å, α=109.23, β=89.84, and γ=110.02°. The crystal structure was refined by Rosenzweig and Ryan (1975), who confirmed the general structural features and triclinic symmetry, and addressed the issue of formal charge balance. They noted, "The three oxygen Si-O distances in the SiO_4 tetrahedron which are shared with UO_2^{2+} ions are 1.59, 1.60, and 1.61 Å while the distance to the unshared oxygen atom is 1.67 Å, lending support to the hypothesis that the 2 protons per formula unit, which are necessary for charge balance, are attached to the unshared silicate oxygen. The structure was found to contain one water molecule more than previously reported; the formula $Cu[(UO_2)_2(SiO_3OH)_2].6H_2O$ is suggested."

Figure 560. Cuprosklodowskite needles to ~15 mm long filling a 3 × 4 cm cavity, from Kolwezi, Congo. Associated species are malachite (dark green), rutherfordine (pale yellow), and schoepite (bright yellow). RJL2617

Figure 561. Freestanding needles of cuprosklodowskite lining a vug ~1 cm wide, from Musonoi, Congo. RJL348

Kasolite $Pb(UO_2)SiO_4{\cdot}H_2O$

Kasolite was described from Kasolo (now Shinkolobwe), Congo (Schoep 1921a), where it is very common, most typically as divergent sprays or sheaf-like aggregates. It forms prismatic crystals (to 4 mm long); lath-like crystals elongated on [010] and flattened on (001), sometimes showing the forms {100}, {001}, {110} and {111}; also flattened radial clusters as well as massive aggregates. It is commonly one of the constituents of "gummite" as an alteration or pseudomorph after uraninite. The mineral may be various shades of yellow to orange with a subadamantine luster. It is not fluorescent.

At the TL, kasolite is often associated with uranophane, curite, and various uranyl phosphates. It is also commonly found at Menda, Swambo, Luiswishi, Kambove, Kalongwe, and Musonoi. Elsewhere in Africa, kasolite occurs at Franceville, Gabon.

Kasolite forms part of the "gummite" samples at Ruggles mine, New Hampshire, and has been reported from various localities in New York, Pennsylvania, Utah, and Arizona. In Nevada, kasolite is found in the East Walker River area in Lyon Co. and at Goodsprings in Clarke Co. It occurs at the

Figure 562. Radiating, flattened groups of orange kasolite from Shaba, Congo. RJL1371

Figure 563. Orange kasolite microcrystals with emerald-green torbernite, from Musonoi, Congo. RJL687

Figure 564. A cavernous mass of nearly pure orange kasolite, with drusy microcrystals lining cavities, from Musonoi, Congo. RJL332

Figure 565. Detail of the kasolite specimen in the previous figure. RJL332

Figure 566. Prismatic kasolite crystals to ~1 mm long with minor torbernite and other copper minerals, from Musonoi, Congo. RJL3902

Figure 567. Kasolite with green torbernite from Musonoi, Congo. RJL1026

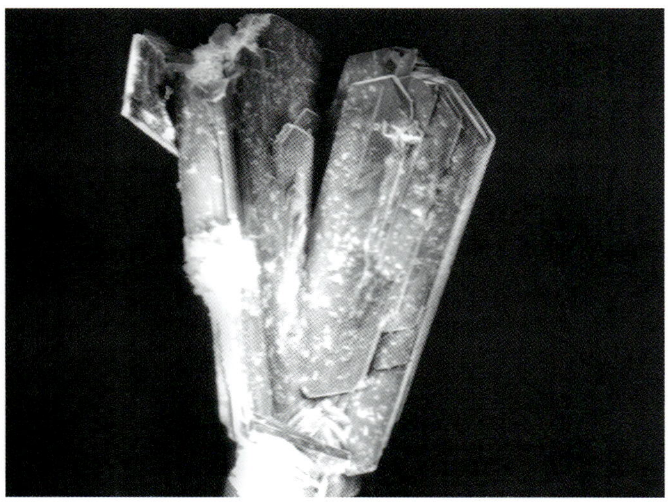

Figure 568. SEM photo of terminated kasolite crystals taken from the specimen in the previous photo, showing the crystal morphology. RJL1026

Nicholson mine in Saskatchewan, and at Great Bear Lake, Northwest Territories, Canada. In France, it has been found at a number of places, including: Kersagalec, Ligud, Morbihon; at Grury, Saone-et-Loire; and at Puy de Dome.

The crystal structure is monoclinic with a=6.705, b=6.926, c=13.286 Å, and β=105.06°. The uranyl silicate sheets are bonded together by the interlayer Pb^{2+} ions as well as by a network of hydrogen bonding involving the interlayer H_2O groups (Fejfarová et al. 2013).

Natroboltwoodite $H(Na,K)(UO_2)SiO_4 \cdot 1.5H_2O$
Natroboltwoodite (formerly *sodium boltwoodite*) forms pale yellow to white powdery aggregates in a near-surface oxidation zone, associated with calcite, limonite, manganese oxides, and clay minerals. It reportedly occurs sparingly in unspecified arid regions of the former USSR (Chernikov, Shashkin, and Gavrilova 1975). Pekov (1998) reports that the TL is an unnamed U occurrence within the Kyzylsai ore field, Chu-Ili Mountains, southwestern Balkhash region, Kazakhstan.

Burns (1998c) noted that natural boltwoodite can have a significant Na content, and the structures of the two minerals are closely related. Even though Na^+ is significantly smaller than K^+, the substitution only forces modifications of the interlayer H_2O positions rather than changes in the uranyl silicate sheets.

Oursinite $(Co,Mg,Ni)(UO_2)_2Si_2O_7 \cdot 6H_2O$
Oursinite was described from the oxidation zone at Shinkolobwe, Congo (Deliens and Piret 1983), where it is associated with other uranyl silicates including soddyite, sklodowskite, and kasolite, along with curite, schoepite, and metatorbernite. In another sample, needles of oursinite were intimately associated with nodules of lepersonnite-(Gd) and bijvoetite. In other samples, the various minerals were stained dark by hydrous cobalt oxides. Oursinite forms small tufts and radial-fibrous rosettes of pale yellow to off white transparent to translucent acicular crystals (the name is derived from the French word for "sea urchin" in allusion to its frequent appearance). The crystals are less than about 1 mm long and about 20 μm thick. Oursinite is formed from the oxidation of primary uraninite and nearby Co- and Ni-sulfides in the host rocks. It is not fluorescent.

Oursinite is orthorhombic with a=7.049, b=17.550, and c=12.734 Å. Based on powder XRD, Deliens and Piret (1983) noted that it was probably of the uranophane structural type. Later crystal structure refinement showed that oursinite contains a sheet based on the uranophane anion topology but with novel orientations of the silicate tetrahedra (Kubatko and Burns 2006).

Sklodowskite $(H_3O)_2Mg(UO_2)_2(SiO_4)_2 \cdot 4H_2O$
Sklodowskite was described from Shinkolobwe (Schoep 1924a, 1924b), and named in honor of Marie Sklodowska Curie. It forms acicular prismatic crystals elongated on [010] that occasionally show the forms {100}, {001}, {301}, and {011} but commonly have a tapering termination. The mineral may form rosettes, radial fibrous aggregates, mammillary crusts, and granular to earthy masses. It is pale to lemon yellow; fibrous samples can be nearly white. Sklodowskite is not fluorescent.

Interestingly, Schoep had briefly described a mineral he called *chinkolobwite* in 1923; when it was later shown that chinkolobwite and sklodowskite were identical, the name sklodowskite was retained and chinkolobwite was discarded, because the TL is actually at Kasolo Hill and not strictly in the village of Shinkolobwe. Schoep (1924a) noted that the original material formed pale yellow spherulites ~6–7 mm in diameter, but pale yellow, earthy material had also been found. He pointed out that sklodowskite resembles uranophane, but has a somewhat lower refractive index (1.63 vs. 1.67 in

Figure 569. Radiating groups of acicular sklodowskite on gypsum, from Charcas, San Luis Potosi, Mexico. RJL178

Figure 570. Yellow sklodowskite crystals lining cavities in a sample ~3 cm wide, from Musoni mine, Congo. RJL912

Figure 571. Optical photomicrograph of the sample in the previous figure, showing that the sklodowskite crystals have minute needles of uranophane growing from their tips. RJL912

Figure 572. SEM photo of the specimen in the previous figure. RJL912

Figure 573. Bright yellow sklodowskite microcrystals, richly distributed in friable matrix, from Mashamba, Congo. RJL2997

Figure 574. Detail of the crystals in the previous photo. The prismatic crystals contain an acicular phase, much like those in sample RJL912; close examination suggests that the acicular phase (uranophane?) formed first and was overgrown by the thicker sklodowskite crystals. Further study will be needed to determine if a true epitaxial relationship exists between the two phases. RJL2997

Figure 575. Pale yellow fibrous sklodowskite on matrix from Musonoi, Congo. RJL4517

uranophane) and a different relationship between the optic axes and the direction of elongation of the crystals. When sklodowskite occurs together with kasolite and soddyite, it can be distinguished by its distinctly lighter yellow color.

Sklodowskite occurs in the oxidation zone of uranium deposits, often replacing uraninite, and generally associated with other secondary uranium species. In the US, it occurs at the New Haven mine, Crook Co., Wyoming; in Utah at the Oyler Tunnel claim, near Fruita, and at Honeycomb Hills, Juab Co. It is found at the Midnight mine, Stevens Co., Washington, and in the Grants area, New Mexico. In Mexico, it is found as bright yellow spherical tufts on transparent gypsum at Charcas, San Luis Potosi; it is also reported from Naica, Chihuahua. Fine specimens have been reported from the Pedra Preta magnesite mine, Brumado, Bahia, Brazil, in association with nováčekite and zeunerite.

At the type locality in the Congo, it forms bright yellow spherical nodules of minute fibers, radiating clusters of pale yellow acicular crystals, and massive yellowish white replacements of uraninite cubes. Associates include hydrous uranium oxides such as becquerelite, billietite, schoepite, and curite, along with other uranyl silicates and rutherfordine. At Kalongwe, it occurs with secondary minerals of uranium and copper (cuprosklodowskite and vandenbrandeite). Sklodowskite is also found at Luiswishi, Musonoi, and Swambo.

Sklodowskite is monoclinic with a=17.28, b=7.30, c=6.56 Å, and β=105°53'. In that study, Gorman (1957) pointed out that although sklodowskite is the Mg analog of uranophane, and there might be some Ca↔Mg substitution, structural differences would seem to preclude a complete isomorphous series between the two minerals.

Figure 576. A sample from the TL at Swambo, Congo, showing cavities containing swamboite on soddyite (amber) along with massive to microcrystalline curite (orange). RJL904

Swamboite $UH_6(UO_2)_6(SiO_4)_6{\cdot}30H_2O$

Swamboite was named for the type locality at Swambo, a prospect site about 36 km west of Shinkolobwe, Congo (Deliens and Piret 1981). It typically forms minute pale yellow acicular crystals, elongated on [010], which are transparent to translucent and often have a beveled termination. The crystals are generally less than about 1 mm long and 50 µm thick and occur as isolated needles, radiating groups and rosettes, and felted masses resembling uranophane. Swamboite is not fluorescent in UV.

Figure 577. Detail of specimen in the previous photo, showing yellow needles of swamboite on blocky soddyite. Minute red-orange curite needles are visible near the bottom. FOV is ~1 cm wide. RJL904

At the TL, swamboite is relatively common in samples from the oxidation zone. It is associated with crested aggregates of soddyite crystals and orange needles of curite, both of which are often stained black by hydrous cobalt oxides.

Deliens and Piret (1981) noted that the structure is monoclinic, and ignoring weak XRD reflections gives a "small cell" with a'=8.82, b'=7.00, c'=6.67Å, and β=102.0°. The smaller cell shows the relationship to the uranophane group. They also note that its relationship to the rest of the group is somewhat analogous to that of vanmeersscheite in the phosphuranylite group, in that uranium is the only metal ion, and other cations are absent. Note that swamboite is currently getting a detailed re-examination, and the accepted formula may need revision (P. C. Burns, personal communication).

Figure 578. Acicular uranophane forming a divergent spray 3 mm tall in matrix, from the Faraday mine, Bancroft, Ontario. RJL4547

Uranophane $Ca[(UO_2)(SiO_3OH)]_2{\cdot}5H_2O$

Uranophane was described from Miedzianca (Kupferberg), Silesia, Poland (Websky 1853), where it occurred both as dense pseudomorphs after uraninite and as tufted aggregates in the oxidized zone of the copper veins. Material from Wölsendorf, Bavaria, was later described under the name *uranotile*; subsequent analyses showed that the two species are identical. It forms acicular crystals elongated on [010], often terminated by an oblique face that forms angles of 95° to (100) and 93° to (001).

The crystals may form radiating to spherical rosettes, felted masses and coatings, or comprise a massive yellow layer typically associated with orange-brown "gummite."

Localities where uranophane forms alteration rims with gummite include: Grafton, New Hampshire; Avondale, Pennsylvania; the Spruce Pine district, North Carolina; and in the Black Hills, South Dakota. Many localities in the Grants district, New Mexico, have produced excellent uranophane specimens; the occurrences in the Todilto limestone at Haystack Butte are noteworthy.

At Wölsendorf, Bavaria, uranophane occurs abundantly as crusts of acicular crystals and as fibrous nodules in cavities in massive purple fluorite. Elsewhere in Germany, it is found at Schneeberg, Saxony, and at Oberpfalz.

In Canada, at the Faraday mine, Bancroft, Ontario, the silky crystals form radial clusters to several cm. In Chihuahua, Mexico, uranophane occurs at La Virgen mine, Puerto del Aire. In Australia it is found at Mt. Painter in the Flinders Range, South Australia.

Uranophane is widespread in the deposits of southern Shaba; it has been collected from Kalongwe, Kambove, Kolwezi, Luiswishi, Musonoi, Swambo, and Shinkolobwe. The felted masses can be several centimeters and are often the base on which other secondary uranium species crystallize.

Figure 579. Bright yellow, lath-like uranophane on pink, Co-rich dolomite from the Musonoi mine, Congo. RJL1133

Figure 580. Bright yellow velvety uranophane thickly covering black uraninite from Shinkolobwe, Congo. RJL800

Figure 581. Uranophane needles to ~10 mm long, on matrix from Milan, New Mexico. RJL740

Figure 582. Isolated sprays of uranophane on massive calcite from Poison Springs, New Mexico. SEM examination revealed minute rosettes of tyuyamunite growing on the sides of some of the uranophane crystals. RJL766

It is preferentially associated with hydrous uranium oxides, kasolite, and rutherfordine adjacent to primary uraninite. At Swambo and Kolwezi it is often associated with soddyite. At Musonoi, colorful specimens of bright yellow uranophane on pink cobaltoan dolomite are found.

Uranophane is monoclinic with a=15.858, b=6.985, c=6.641 Å, and β=97°33'. The crystal structure was reported by various groups with increasing precision as better samples and techniques became available (Gorman and Nuffield 1955; Smith, Gruner, and Lipscomb 1957; Stohl and Smith 1981). Some authors refer to uranophane as uranophane-α, to emphasize the dimorphous relationship to uranophane-β. The two structures differ in the topology of the SiO_4 tetrahedra, discussed in the structure section above.

Figure 583. Yellow, acicular uranophane forming 1-mm spherical masses on matrix, from the Margaritas mine, Peña Blanca, Chihuahua, Mexico. RJL2727

Figure 584. Spherical uranophane aggregate ~2 mm across, from the Mammoth mine, Presidio Co., Texas. RJL2804

Figure 585. Crust of massive uranophane ("gummite") on pegmatite from the Chalk Mountain mine, Spruce Pine district, North Carolina. RJL2892

Uranophane-β $Ca[(UO_2)(SiO_3OH)]_2 \cdot 5H_2O$
Uranophane-β was described under the name β-uranotile by Nováček (1935) from Jachymov, Bohemia, Czech Republic (Steinocher and Nováček 1939; Frondel 1958), where it forms crusts of tiny needles on altered uraninite in association with liebigite and calcite. It has been found as velvety or bristly coatings, or radial to fanlike aggregates of acicular crystals. More compact or feltlike fiber aggregates are often seen as a replacement of uraninite. Individual needles are often somewhat thicker than those of uranophane or sklodowskite, elongated on the z-axis with a square or rectangular cross section, sometimes terminated with a relatively large {001} face. The three minerals cannot be reliably distinguished by appearance alone; either measurement of the optical constants or X-ray crystallography will provide a reliable identification.

Uranophane-β has the same occurrences and associations as uranophane, but is less common. Uranophane appears to be the more stable phase at ordinary temperature and pressure, based on the observation that when uranophane-β is finely crushed it may revert to uranophane.

The Faraday mine, near Bancroft, Ontario, has produced many excellent specimens of radiating, canary-yellow needles in a matrix of brecciated and hematitized granite, associated with clear gypsum. It has also been reported from the Evans-Lou mine, St-Pierre-de-Wakefield, Papineau Co., Quebec.

Noteworthy American localities include: the Dunton mine, Newry, Maine; the Ruggles mine, Grafton, New Hampshire; Haddam Neck, East Hampton, Connecticut; Jim Thorpe (Mauch Chunk), Pennsylvania; Crown Point Center, New York; the Grants District, New Mexico; the Silver Star District, Mineral Co., Nevada; and the Cameron area, Coconino Co., Arizona.

In Germany it has been reported from: the Menzenschwand uranium deposit in the Black Forest; the Wölsendorf District in Bavaria; the Ellwieler, Katharina I, and Katharina II mines and the Schweisweiler-Winnweiler uranium deposit in Rhineland-Palatinate; and the Mansfeld Basin in Saxony-Anhalt. French localities include: the Bigay mine, Auvergne; Le Bauzot mine, Burgundy; Mas d'Alary and Rabejac in Lodeve; Margnac and Margnac II mines in Compreignac; and Les Bois-Noirs mine in Loire.

In Africa, excellent examples are found at Musonoi and Shinkolobwe, Congo, and at the Rössing mine and the nearby Goanikontes claim, Namibia.

Crystal structure refinement (Viswanathan and Harneit 1986) confirmed the basic structural features, with a=13.966, b=15.443, c=6.632 Å, and β=91.38°. However, "bond-strength sums, calculated using the refined interatomic distances, reveal that the apical oxygens of all SiO_4 tetrahedra must be $(OH)^-$ ions and that the H bonds play as important a role as Ca in linking sheets that have the formula $[(UO_2)_2(SiO_3OH)_2]^{2-}$. Hence the structural formula should be $Ca(UO_2)_2(SiO_3OH)_2 \cdot 5H_2O$ instead of $Ca(UO_2)(UOOH)(SiO_4)(SiO_3OH) \cdot 4H_2O$ as suggested earlier."

Figure 587. Relatively large, dark amber uranophane-β crystals in divergent sprays, standing in a 3 × 3 cm vug, from Rössing, Namibia. RJL2568

Species with U:Si = 2:5 - Weeksite Group

Minerals of the weeksite group have the general formula: $\mathbf{M}(UO_2)_2(Si_5O_{13}) \cdot nH_2O$, where **M** = Ca, Mg, Ba, Th, or K+Na, and n = 4-6. The group includes coutinhoite, haiweeite, metahaiweeite, and weeksite.

Structure and morphology

In the original description of weeksite, Outerbridge et al. (1960) proposed the formula $K_2(UO_2)_2Si_6O_{15} \cdot 4H_2O$. Yeremenko, Il'menov, and Azimi (1977) studied two generations of weeksite from Afghanistan. For the first, the chemical analysis

Figure 588. Detail of the specimen in the previous figure. RJL2568

Figure 586. A superb cabinet specimen, with gemmy yellow uranophane-β needles, completely covering a 7 × 10 cm surface, from Rössing, Namibia. RJL905

Figure 589. Golden-yellow uranophane-β crystals to several mm long, lying flat on the rock surface, from Rössing, Namibia. RJL3170

Figure 590. The weeksite structure, viewed along [010] to show the chains of UO_7 polyhedra (blue-gray), cross-linked by SiO_4 groups to form a sheet.

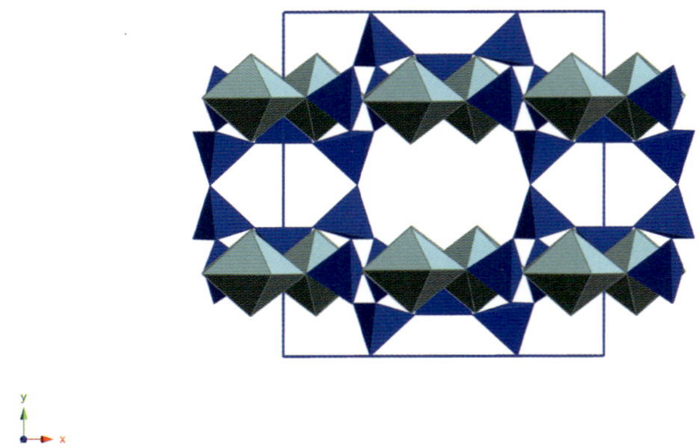

Figure 591. The structure viewed along [001], showing how the sheets are, in turn, cross-linked to form an open framework.

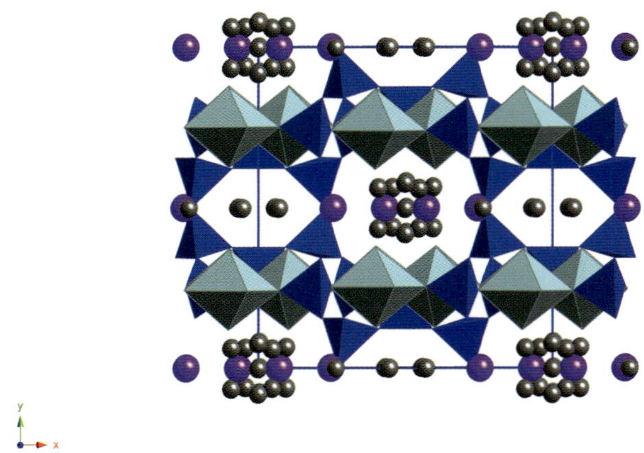

Figure 592. The full structure showing the locations of K (purple) and H_2O (gray) within the channels of the framework.

gave the empirical formula $(K_{1.04}Na_{0.63}Ca_{0.34}Ba_{0.12}Al_{0.05}Mg_{0.04}Sr_{0.02})_{\Sigma 2.24}(UO_2)_{1.67}(Si_5O_{13.10})\cdot 3.35H_2O$, and for the second $(K_{1.13}Na_{0.73}Mg_{0.05}Al_{0.04}Ca_{0.01}Sr_{0.01}Ba_{0.01})_{\Sigma 1.98}(UO_2)_{1.86}(Si_5O_{12.93})\cdot 3.45H_2O$. The authors reported optical data and lattice parameters for a monoclinic structure with *a*=9.63, *b*=7.12, *c*=7.15 Å, and β=111°56'. However, their optical and X-ray data are both in disagreement with the conventional data for weeksite. Baturin and Sidorenko (1985) reported the crystal structure of a weeksite sample from Afghanistan and proposed the formula $K_2(UO_2)_2Si_5O_{13}\cdot 3H_2O$. They found four sites occupied by $[K_{0.31}Na_{0.18}(H_2O)_{0.50}]$ and one by H_2O.

Jackson and Burns (2001) refined the structure of weeksite from the Anderson mine, Yavapai County, Arizona, and suggested the correct formula should be $K_{1.26}Ba_{0.25}Ca_{0.12}(UO_2)_2Si_5O_{13}\cdot H_2O$, or ideally, $K_2(UO_2)_2Si_5O_{13}\cdot H_2O$. For one formula unit, they found two sites partly occupied by K, one site occupied mainly by Ba and H_2O, and one site fully occupied by H_2O. However, they noticed that some cations were obviously displaced from their corresponding special positions, indicating that the results represent an average structure.

The structural formula was further revised by Fejfarova et al (2012), also using material from the Anderson mine. They demonstrated that weeksite is monoclinic, space group *C2/m*, with *a*=14.196, *b*=14.229, *c*=9.631 Å, and β=111.578°. The previously reported orthorhombic unit cell is attributed to twinning of the true monoclinic cell.

The general topology of the structure involves chains of edge-sharing uranyl polyhedra, which in turn share edges with SiO_4 tetrahedra. The uranyl silicate chains are linked to crankshaft-like chains of SiO_4 tetrahedra, creating layers that are connected by the sharing of vertices of some tetrahedra to form an open framework. The remaining components K, Ba, Ca, and H_2O lie within the channels in the framework (Jackson and Burns 2001).

Formation and geochemistry

These minerals are relatively rare and generally form in arid environments, possibly from the evaporation of Si-rich waters of relatively high pH, in which dissolved polymeric silicate species could reach fairly high concentrations. In contrast to the uranyl sulfates and carbonates, the weeksite-group minerals have much less tendency to dissolve on re-exposure to fresh water. Like uranophane, upon exposure to carbonate-free water, the minerals might preferentially lose some Ca and Si, thereby altering to more U-rich species such as uranophane or soddyite. Conversely, in more alkaline, carbonate-rich waters, these minerals might preferentially lose U, altering to cryptocrystalline or amorphous silica (Finch and Murakami 1999).

The minerals

Coutinhoite $Th_xBa_{(1-2x)}(H_2O)_y(UO_2)_2Si_5O_{13}\cdot H_2O$

Coutinhoite, ideally $Th_xBa_{(1-2x)}(H_2O)_y(UO_2)_2Si_5O_{13}\cdot H_2O$, where $0 \leq x \leq 0.05$ and $0 \leq y \leq (2+x)$, was described from Galileia, Minas Gerais, Brazil, where it occurs as a secondary hydrothermal mineral in the Corrego do Urucum granitic pegmatite. It forms irregular aggregates of minute curved scales and flaky crystals up to ~10 µm long and 0.5 µm thick. Crystals are transparent to translucent yellow, with waxy to silky luster, and are not fluorescent. It is intimately associated with weeksite, phosphuranylite, uranocircite, and meta-uranocircite on muscovite and microcline (Atencio, Carvalho, and Matioli 2004). The mine is known as Lavra do Urucum or Lavra do Tim; a general description of the locality is given by Cassedanne (1986).

Crystal structure refinement using powder XRD data indicated that coutinhoite is orthorhombic with a=14.17, b=14.19, and c=35.75 Å, and the space group is "probably *Cmmb*." The authors caution that, "No single-crystal study was done, due to the crystal dimensions. Although the unit cell of coutinhoite is similar to that of weeksite, we are not completely convinced they are actually isostructural," (Atencio, Carvalho, and Matioli 2004).

They further note that, "Weeksite has an open uranyl silicate framework structure, and the most likely location for Th^{4+} is in the channels that run through the structure. However, we cannot see how the bonding requirements of Th can be met within the channels, as the apical O atoms of the Si tetrahedra are bridging, and the uranyl ion O atoms can only form weak bonds. The substitution of Th^{4+} at a site normally occupied by monovalent (K) or divalent (Ba) cations suggests some considerable restructuring of this site to avoid overbonding of O atoms, presumably involving extra H_2O molecules. Significant distortion of the framework would be expected induced by the incorporation of a strong Lewis base such as Th^{4+}."

Figure 593. A patch of bright yellow coutinhoite, ~ 1 mm wide, on muscovite from the TL at the Urucum mine, Galiléia, Minas Gerais, Brazil. RJL2650

Haiweeite $Ca(UO_2)_2(Si_5O_{12})(OH)_2 \cdot 6H_2O$
Metahaiweeite $Ca(UO_2)_2Si_6O_{15} \cdot nH_2O$
Haiweeite was described from the Haiwee Reservoir, Coso Mountains, Inyo Co., California (McBurney and Murdoch 1981). It forms very small spherulitic aggregates of minute acicular crystals or small flake-like grains. The material is typically pale yellow to greenish yellow and is weakly fluorescent dull green in UV. It occurs on fracture surfaces in granite and in voids in the nearby lake-bed sediments. The outward appearance of the material is typically earthy, although under the stereo microscope individual needles may often be seen.

Far better examples of the species occur in Brazil: at Perus, north of São Paulo, Brazil, it is found along fractures

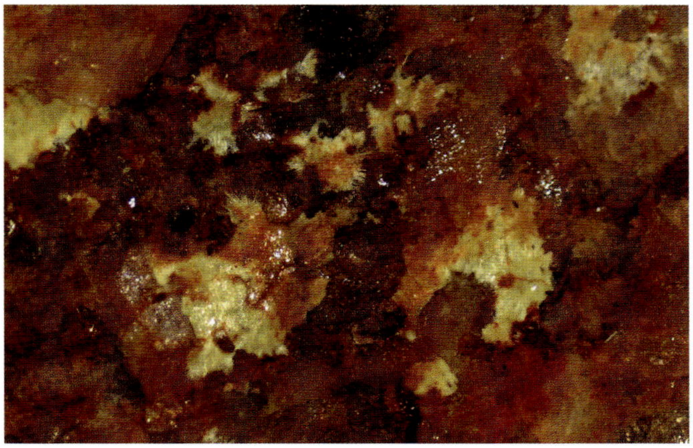

Figure 594. Powdery to acicular haiweeite in flattened aggregates on granite from the TL, Haiwee Reservoir, Coso Mountains, Inyo Co., California. This locale is also referred to as the Coso uranium mine. RJL695

Figure 595. Yellow, acicular haiweeite forming 1-2 mm balls on calcite, with uranophane and fluorite, from Teofilo Otoni, Minas Gerais, Brazil. RJL2719

in a tourmaline granite associated with autunite, meta-autunite, uranophane, uranophane-β, phosphuranylite, torbernite, meta-torbernite, and uranium opal (de Camargo and Svisero 1969; Atencio and Hypolito 1994). At Teofilo Otoni, Minas Gerais, it is associated with calcite, uranophane, and fluorite.

It also occurs in crevices in gneiss near Bad Gastein, Salzburg, Austria, and in the Ningyotoge deposit Tottori Pref., Japan. It is associated with uranophane at the Mammoth mine (Quinn Creek mine, Valentine mine), Presidio Co., Texas; and at the Margaritas mine, Peña Blanca, Mexico.

Metahaiweeite, $Ca(UO_2)_2Si_6O_{15}{\cdot}nH_2O$, where n < 5, is a species described mainly from studies of synthetic material in which haiweeite was artificially dehydrated in the laboratory by heating. Its properties are similar to those of haiweeite and the two phases are reported to be intimately associated at the type locality.

The structure refinement (Burns 2001) indicated that the topology is somewhat similar to that of weeksite, but in haiweeite there are no silicate linkages between adjacent uranyl silicate sheets to form a framework like that of weeksite. Instead, the sheets in haiweeite are connected only through Ca in the interlayer. Because certain features of the structure reported by Burns suggested that the model actually represented an average structure, Plášil et al. (2013) conducted a further refinement that included weak reflections. They concluded that the structure is orthorhombic, with a doubling of one of the previously reported cell parameters, giving a=18.300, b=14.233, and c=17.919 Å.

Figure 596. Pale yellow balls of fibrous haiweeite with bright yellow lath-like uranophane, from the Mammoth mine, Presidio Co., Texas. RJL2805

Figure 597. Off-white spherical blebs of haiweeite, with yellow, acicular uranophane on matrix, from the Margaritas mine, Peña Blanca, Chihuahua, Mexico. RJL2727

Weeksite $K_2(UO_2)Si_5O_{13}4H_2O$

Weeksite was described from the Autunite No. 8 claim, Thomas Mountains, Utah (Outerbridge et al. 1960). It typically forms small spherules of radiating lath-like crystals, which are yellow with a waxy to silky luster. At the TL, weeksite occurs in opal veinlets in rhyolite with carbonates and gypsum, and as replacements of pebbles in tuffaceous conglomerate.

Weeksite has been reported from a number of American localities, including: lining cavities in welded tuff at the Mammoth mine, near Presidio, Texas; in a sandstone-type deposit at the Jackpile mine, New Mexico; the Anderson mine, Mojave Co., Arizona; the Goodwill claim, Utah; and at Haiwee Reservoir, California. It also occurs at the Margaritas deposit, Sierra Peña Blanca, Chihuahua, Mexico; at Les Bois Noirs, France; Yinnietharra, Western Australia; and localities in Japan and Russia.

A small number of excellent specimens were found at the Goanikontes claim, an ore prospect near Rössing, Namibia, not far from the better-known uranium mine.

As previously noted, at a locality in Afghanistan, Yeremenko, Il'menov, and Azimi (1977) studied two generations of weeksite that formed in sandstone surrounding carbonatite near the old vent of a volcanic dome. The early generation forms felted masses and druses of acicular crystals to 0.5 mm long. The later generation forms compact spherulitic aggregates with a radial fibrous texture, as well as fine-grained masses and crusts. The two generations have similar compositions, but the earlier one is much richer in Ba and Sr.

Figure 598. Weeksite crystals forming golden-yellow radiating aggregates, less than ~1 mm wide, from the Anderson mine, Yavapai Co., Arizona. RJL3140

Figure 599. Bright yellow weeksite aggregates associated with minute orange carnotite crystals, less than ~1 mm wide, from the Anderson mine, Yavapai Co., Arizona. RJL280

Figure 600. Golden-yellow weeksite forming small, rounded aggregates with quartz and calcite, from near Rössing, Namibia. RJL2704

Species with U:Si = 1:7 - Uranosilite

Uranosilite $(UO_2)Si_7O_{15}$

Uranosilite is a very rare and incompletely characterized uranyl silicate. It was first mentioned as an "unknown siliciferous mineral" (Walenta 1974) accompanying the description of studtite. It is orthorhombic with needle-like crystals that are only visible in the SEM at high magnification. The crystals are acicular on [001], intimately intergrown with uranophane and studtite on quartz and hematite at Menzenschwand, Schwartzwald, Germany. It is yellowish white with vitreous luster. There is no discernable cleavage. Heating to 900°C causes no change in the powder diffraction pattern, but reduces the refractive indices to the range 1.54–1.55. This change is attributed to the loss of "zeolitic water" (Walenta 1983).

Uranosilite has also been reported from Mina Huemul, Pampa Amarilla, Malargüe Dept., Mendoza, Argentina (Gunnar Färber analysis).

Figure 601. Detail of the specimen in the previous figure, using multifocus montage (six layers). RJL2704

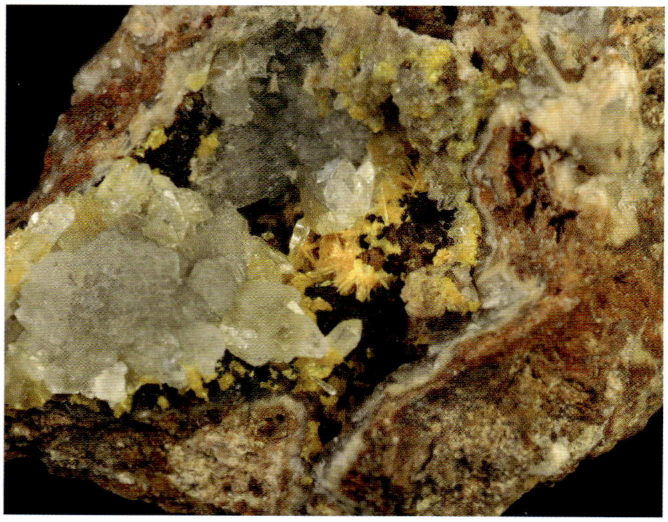

Figure 602. A typical matrix specimen from near Rössing, Namibia, showing tiny, golden aggregates of weeksite associated with longer, straw-yellow boltwoodite along with calcite and colorless quartz. RJL2864

Figure 603. A brilliantly colored association of bright yellow acicular weeksite and orange boltwoodite on dark calcite, from the Rössing area, Namibia. RJL3830

Figure 604. Minute yellow blebs of uranosilite, with coffinite and weeksite, from Mendoza, Argentina. RJL3809

OTHER MINERALS CONTAINING ESSENTIAL U OR TH

Minerals treated here are mostly U(IV) and/or Th(IV) compounds other than the primary oxide and silicate minerals treated earlier, along with REE-dominant minerals that contain essential U or Th. They include a number of groups and about fifty species.

Table 55. Other minerals containing essential U or Th
U(IV) and Th(IV) niobates, tantalates, and titanates
Brannerite series
Polycrase group
Aeschynite series
Samarskite group
Pyrochlore group
Crichtonite group
Petscheckite and liandratite
Aspedamite
U(IV) and Th(IV) phosphates
Rhabdophane group
Monazite group
Lermontovite
U(IV) and Th(IV) silicates
Steacyite group
Chevkinite group
Steentrupine
Thornasite
Karnasurtite
Umbozerite
Cipriianiite
Th(IV) carbonates
Thorbastnäsite
Tuliokite
Others
Vapnikite
Vorlanite

Brannerite series

Brannerite forms a series with thorutite, and both are monoclinic after heating; orthobrannerite is orthorhombic but it is *not* the orthorhombic polymorph of brannerite. These minerals are mixed oxides involving Th, U, REE, Ca, and Ti, with or without Fe. All are metamict in the natural state.

Brannerite $(U,Ca,Y,Ce)(Ti,Fe)_2O_6$

Brannerite was described from Kelley Gulch, Stanley Basin, Idaho, where it was found in a placer deposit being worked for gold. Kelley Gulch runs through granite cut by pegmatites, which are the likely source of the material. Individual grains show a prismatic habit to varying degrees, and have a paper-tin brownish yellow surface alteration. Unaltered material is black with a high luster and a conchoidal fracture. Based on detailed chemical analysis and comparison to other complex U oxides, the authors concluded, "Brannerite may be described as a complex titanate of uranium with smaller quantities of rare earths and other metals, in which the weight of uranium oxides exceeds that of titanium oxide" (Hess and Wells 1920). The surface alteration layer was later shown by XRD to be anatase (Pabst 1954).

In the San Bernardino Mountains, California, lenticular to nearly spherical brannerite nodules occur in granite gneiss. The smaller (1–20 g) nodules lie along the foliation and consist almost entirely of brannerite surrounded by a thin layer of biotite. Larger (up to 200g) nodules are surrounded by biotite and also contain small biotite plates, suggesting that the brannerite has replaced the biotite. It is theorized that the brannerite "formed at great depth under conditions of high pressure and temperature," (Hewett, Stone, and Levine 1957).

Figure 605. Quartz-pebble conglomerate containing disseminated brannerite from the Pronto uranium mine, Blind River district, Ontario, Canada. RJL4529

Brannerite is an important ore mineral in the quartz pebble conglomerate of the Elliot Lake U district, Ontario. It is a minor but still significant ore mineral in some hydrothermal deposits in the Beaverlodge district, Saskatchewan. At the Richardson gold mine, near Madoc, Ontario, discrete black grains of brannerite are found in crystalline calcite, associated with muscovite, tourmaline, pyrite, native gold, and minor uraninite (Steacy, Plant, and Boyle 1974). Brannerite masses to ~2 cm occur in tremolite rock associated with gneiss in South Sherbrooke Township, Ontario (Moddle 1957).

Whittie (1954) applied the name *absite* to a Th-rich brannerite (~13 wt. % ThO_2) that occurs in irregular grains in brecciated rock at Crockers Well, South Australia. The name is an acronym for airborne scintillometer, with which the deposit had been discovered. Povilaitis (1963) reviewed the physical and chemical properties of the brannerite series and suggested that *lodochnikite* and *absite* are identical with brannerite. Those names are therefore discredited as species names.

Efforts to understand the structure and crystal chemistry of brannerite typify the analytical difficulties that arise with minerals that are naturally metamict. Patchett and Nuffield (1960) noted that, "The x-ray powder diffraction patterns of ignited brannerite from Blind River, Ontario almost invariably show the presence of rutile or anatase and other minerals intimately associated with the brannerite. The appearance of the grains is affected by these impurities. The variable chemical constitution and physical properties of the grains have influenced some geologists to doubt the existence of the mineral in this region. They suggest that the act of heating, which is necessary to produce the x-ray pattern, causes the mineral to be synthesized from one or more mineral phases in the grains . . ." Acting on this idea, they successfully synthesized brannerite by arc-melting a pellet of UO_2 + $2TiO_2$ and determined that the structure is monoclinic. The cell dimensions and space group are compatible with the forms observed on natural crystals by Pabst (1954), giving some confidence that heating can restore the original structure. Patchett and Nuffield do, however, caution that fresh material is most suitable for such restoration, because as the metamict crystal undergoes weathering, disorder increases and the composition might change.

More precise structure refinement, also using synthetic UTi_2O_6, confirmed the monoclinic structure, space group *C2/m*, with *a*=9.812, *b*=3.770, *c*=6.925 Å, and β=118.96° (Szymanski and Scott 1982).

Orthobrannerite $U^{4+}U^{6+}Ti_4O_{12}(OH)_2$
Orthobrannerite was described from the weathering residue of a biotite pyroxene syenite in Yunan, China, where it forms black prismatic crystals to ~8 × 12 mm, bounded by {001}, {120}, {021}, {110}, and {140} forms. The prism faces are striated along [001]. Optical goniometry indicated that the mineral is orthorhombic. Natural material is metamict; after heating to 1000°C, the powder pattern is similar to that of synthetic $UTiO_5$. Indexing the powder pattern to an orthorhombic unit cell gives *a*=7.37, *b*=11.67, and *c*=6.33 Å, in good agreement with the axial ratios calculated from goniometric measurements (X-ray Laboratory 1978).

Note that in the abstract published in American Mineralogist, the reviewer comments, "The name orthobrannerite is a poor choice for the mineral, as the name implies an orthorhombic polymorph of brannerite."

Thorutite $(Th,U,Ca)Ti_2(O,OH)_6$
Thorutite was discovered in 1947 at Kutyur-Tyube (Kattar-Tyube) thorium deposit, near Urusai Peak, Sokh River basin, Alai Range, Southern Kyrgyzstan (Pekov 1998). It forms black, short prismatic crystals to 2 × 1 cm and grains in microcline-nepheline veins hosted by muscovitized syenite.

Figure 606. Massive Th-rich brannerite ("absite") ~1 cm wide, from Crockers Well, South Australia. Note the alteration crust (an earthy mixture of anatase and other phases) and the resinous appearance of the fracture surfaces. RJL4530

Figure 607. A small chip of thorutite, ~1 mm long, from the TL near Urusai Peak, Kyrgyzstan. Mounted in clear resin for microprobe work, it has a brown, somewhat resinous appearance. RJL4527

Kyrgyzstan, Zardalek

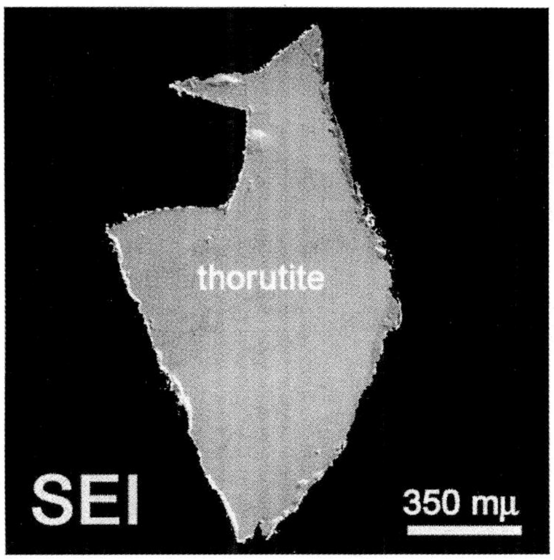

Figure 608. SEM image and EDS analysis of the thorutite sample in the previous figure (courtesy of A. J. Nikischer, Excalibur Mineral Co.)

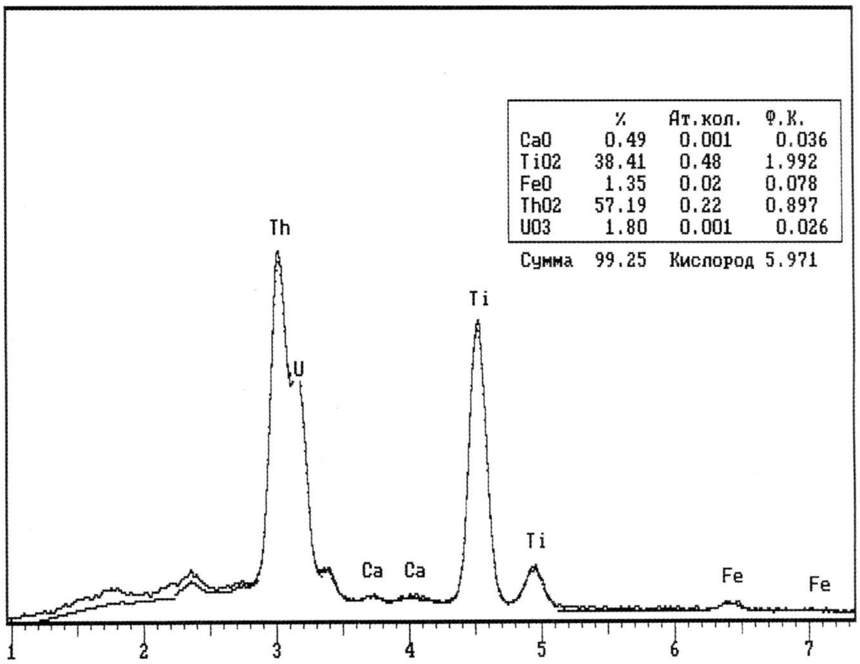

Associated species include thorite, zircon, calcite, barite, and galena (Gotman and Khapaev 1958).

Natural thorutite is metamict; synthetic $ThTi_2O_6$ is monoclinic, with *a*=9.822, *b*=3.824, *c*=7.036 Å, and β=118.84°.

Polycrase Group

The polycrase group consists of mixed oxides with the general formula $\mathbf{AB_2}O_6$, where **A** = Y, REE, U, Th, Ca; **B** = Ti (polycrase), and Nb, Ta (euxenite); and O is sometimes replaced by OH. The structure is similar to that of columbite, except that the **A** site is a distorted cube in polycrase or euxenite, and an octahedron in columbite (Weitzel and Schröcke 1980). In the euxenite members, which have (Nb,Ta) dominant on the **B** site, charge compensation is achieved by substitution of Ca at **A**: $Ca^{2+} + Nb^{5+} \rightarrow REE^{3+} + Ti^{4+}$ (Finch and Murakami 1999).

Of all the minerals in the group, only uranopolycrase is U-dominant; however, accepted formulas for polycrase-(Y), kobeite-(Y), and yttrocrasite-(Y) indicate essential, but not dominant, U or Th at **A**. Furthermore, most of the others are also metamict because of significant levels of U,Th present as impurities.

Members of the series are orthorhombic, and typically form stout prismatic crystals, sometimes flattened on {010}. The faces are often striated. Twinning on {201} is common; the twins are flattened on {010} and striated along [001] (Palache, Berman, and Frondel 1944).

These minerals typically occur as accessory minerals in granites and granite pegmatites. They are fairly widespread in small quantities, but are not easily distinguished from one another except by microprobe analysis. Localities that yield materials of most interest to collectors will be emphasized here.

Euxenite-(Y) $(Y,Ca,Ce)(Nb,Ta,Ti)_2O_6$
Euxenite was named by Scheerer (1840) based on a sample from Jölster, Norway. He described it as brownish black, translucent reddish-brown in thin splinters, with a metallic luster. It shows "no trace of cleavage or crystallinity." Brøgger (1906, pp. 82–97) reported goniometric measurements and chemical analyses for euxenite and polycrase and noted the crystallographic similarities to columbite.

In granite pegmatites, euxenite-(Y) is typically associated with biotite, muscovite, ilmenite, monazite, zircon, xenotime, magnetite, allanite, beryl, gadolinite, thorite, uraninite, "betafite," and columbite. Palache, Berman, and Frondel (1944) note that euxenite and polycrase have been noted at over 100 localities in Norway alone. In addition to its occurrence at the TL, it is found at Huntila, near Pitkäranta, Finland; and in Greenland at Karrakungak and at Nuk.

In Madagascar it is found in K-rich pegmatites, associated with monazite, beryl, columbite, magnetite, "betafite," scapolite, chrysoberyl, and native Bi. Localities include: at Ambatofotsikely; on Mount Vohambohitra; and at Samiresy. Detailed examination of an "ampangabeité" crystal from Ampangabé showed it to be euxenite (Ewing, Snetsinger, and Bunch 1977).

Good crystals occur in some Canadian pegmatites. Ellsworth (1927a) described the Th-rich variety *lyndochite* from a dike in Lyndoch Township, Renfrew Co., Ontario, where it is associated with microcline, quartz, black mica, smoky quartz, beryl, columbite, black tourmaline, and fluorite. Fleischer (1966) considered *lyndochite* to be a variety of aeschynite, rather than a variety of euxenite, based in part on analytical data on other materials (Butler 1957; Horne and Butler 1965). He noted, among other things, that thermal treatment of aeschynite can produce the euxenite structure. (This again serves as a warning that conclusions based on "repaired" metamict structures must always be approached with some caution.) However, Ercit (2002) re-examined material from the original locale at the Quadeville beryl mine and concluded that *lyndochite* is a Th-rich variety of euxenite-(Y) as originally reported

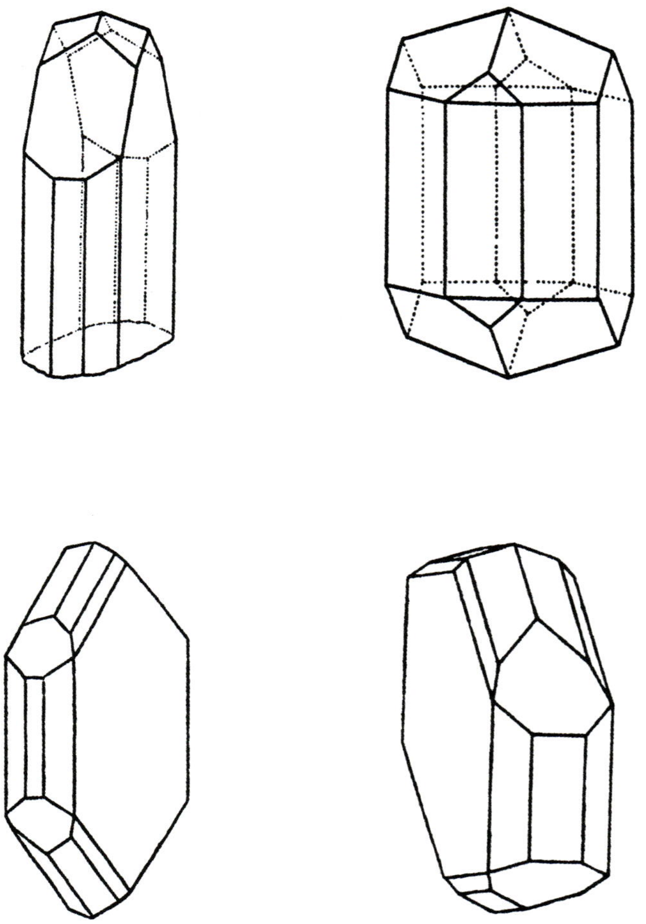

Figure 609. Morphology of some natural euxenite and polycrase crystals, modified from Goldschmidt (1916, 1920). Top: euxenite from Kragerø, Norway (left) and Hvaler, Norway (right). Bottom: polycrase from Hitterø, Norway (left) and Marietta, South Carolina (right).

Figure 610. A thumbnail-sized group of tabular euxenite-(Y) crystals from Betafo, Madagascar. RJL3077

Figure 611. Crude, terminated euxenite-(Y) crystal, ~35 mm tall, from Adi-Voandelake, Madagascar. Note the brown surface alteration and the small chipped area at lower left where the glassy, dark brown interior is exposed. RJL2899

by Ellsworth. Ercit also notes that the canonical discriminant model of Ewing (1976) supports this conclusion as well.

Euxenite has contributed to our understanding of radiometric dating. Metamict euxenite crystals from a pegmatite in Grenville-province rocks near St. Pierre de Wakefield, Quebec, show conflicting ages, depending on which isotope pairs are used in the calculation: $^{207}Pb/^{206}Pb$ gives 1 Ga; $^{207}Pb/^{235}U$ gives 710 Ma; $^{206}Pb/^{238}U$ gives 620 Ma; and $^{208}Pb/^{232}Th$ gives 550 Ma. The discrepancies are attributed to two factors: 1. differential loss of U relative to Th from the outer zones of the crystal; and 2. overall loss of Pb, with some evidence that more ^{208}Pb is lost compared to ^{207}Pb and ^{206}Pb (Robinson et al. 1963).

Tanteuxenite-(Y) $(Y,Ca,Ce)(Ta,Nb,Ti)_2(O,OH)_6$

Tanteuxenite was described from the Pilbara goldfield, Western Australia (Simpson 1928). At Cooglegong, masses up to 200 g were found, associated with gray quartz, in angular boulders shed from a pegmatite vein. The masses show some traces of a flattened prismatic structure and a conchoidal fracture. Fracture surfaces are brownish-black with a bright redinous luster, whereas external surfaces are coated with a thin gray alteration layer. Analysis indicated ~3.35% UO_3. At Woodstock Station, material with similar properties contains ~4% UO_3 and a smaller excess of Ta over Nb or Ti, with somewhat less Ca, as would be expected from charge balance considerations. Simpson further noted the probable occurrence of the mineral in alluvial tin rejects at Eleys and at Mount Francisco, but no analytical data were given.

Tanteuxenite-(Y) has also been noted at: Mattawa, Ontario; as crystals to 4 cm in pegmatite at Craveggia, Val Vigezzo, Italy; and with cassiterite, Mn-tantalates, simpsonite, and rankamaite in tin placers at Lina, Kivu province, Congo.

Polycrase-(Y) $(Y,Ca,Ce,U,Th)(Ti,Nb,Ta)_2O_6$

Polycrase was described by Scheerer (1844) from Rasvåg, Hitterö, Norway, where it forms tabular crystals in granite, associated with gadolinite, allanite, and other REE-rich accessory minerals. Crystals are opaque, black to brownish black, with resinous to bright submetallic luster, and are typically metamict. Analyses of polycrase from a half-dozen localities worldwide show ~5-25% U (as UO_2, UO_3, or U_3O_8) and ~0-5% ThO_2 (Palache, Berman, and Frondel 1944).

Notable localities include: near Marietta, South Carolina; on Baringer Hill, Llano Co., Texas; Idaho City, Idaho; in Calvin Township, Nipissing Co., Ontario; in pegmatites in Val Codera, Lombardy, Italy; and numerous localities in Norway and Madagascar.

Tomašić et al. (2004) studied aeschynite-(Y) and polycrase-(Y) by XRD and Raman spectroscopy after recrystallization at temperatures from 400 to 1000°C. They noted changes in Raman spectra, where increased heating led to the development of distinctive Raman bands, and conclude that, "Raman spectra of both minerals are generally similar, a reflection of their similar chemical and structural properties, but the differences observed probably arise because of the different way the coordination polyhedra are stacked. These differences are the most pronounced in the wavenumber region from 200 to 550 cm^{-1}, which is suitable for fingerprint recognition of the mineral phases with the aeschynite or euxenite structure."

Uranopolycrase $(U,Y)(Ti,Nb,Ta)_2O_6$

Uranopolycrase was described from the "Fonte del Prete" pegmatite vein, near San Piero in Campo, Elba, Italy (Aurisicchio et al. 1993). It forms opaque, brown-red crystals with adamantine luster. Crystals are tabular on {100} and elongated on [001]. They are found in cavities in the fine-grained intermediate zone of the pegmatite, near its contact with the surrounding granodiorite, and also as inclusions in beryl crystals associated with stilbite. Major associated minerals are quartz, orthoclase, albite, lepidolite, and elbaite. Other Nb-Ta minerals noted from this vein include euxenite-(Y), manganocolumbite, polycrase-(Y), titanowodginite, and uranmicrolite.

The empirical formula $(U_{.62}Y_{.29}Th_{.07}Mn_{.03}Ca_{.02}Nd_{.01})(Ti_{1.46}Nb_{.36}Ta_{.12})O_6$, based on microanalysis, suggests that charge balance is achieved via the coupled substitution $U^{4+} + Ti^{4+} \rightarrow Y^{3+} + Nb^{5+}$.

Natural crystals are metamict; after heating for 10 h at 900°C, the structure is orthorhombic, with a=14.51, b=5.558, and c=5.173Å.

Yttrocrasite-(Y) $(Y,Th,U)Ti_2(O,OH)_6$
Yttrocrasite was described from a pegmatite in Burnet Co., Texas (Hidden and Warren 1906). The original material consisted of a rough orthorhombic crystal with a dull alteration product on the surface. It has a conchoidal fracture, and fresh fracture surfaces are black, with a pitchy to resinous luster. Thin fragments are amber to light yellow. Chemical analysis showed 8.75 wt. % ThO_2 and ~2 wt. % UO_2, in reasonable agreement with values calculated from the observed radioactivity.

Yttrocrasite remains an inadequately described mineral. Analytical work reported by W. W. Crook in the 1970s was thoroughly discredited by Peacor et al. (1982), who state that, "This raises serious questions regarding the validity of all the data in the paper in question. Whether or not yttrocrasite is a valid member of the euxenite group therefore still remains to be shown."

Kobeite-(Y) $(Y,U)(Ti,Nb)_2(O,OH)_6$ (?)
Kobeite-(Y) was described from a pegmatite at Shirashi, Kobe-Mura, Kyoto Pref., Japan, where it forms metamict irregular brown-black grains and prismatic crystals, associated with parisite, zircon, monazite, chevkinite, xenotime, and biotite. It may be identical with polycrase.

Hutton (1957a) described similar material that forms slender prismatic crystals in a cobble of graphic granite found in gravels of the Paringa River, South Westland, New Zealand. Although no chemical analysis was reported, Hutton argues for the validity of kobeite based on the fact that the New Zealand material gave nearly identical XRD patterns to the Japanese material, both as-received and after various heat treatments, whereas the XRD patterns are very different from other euxenite-polycrase or aeschynite-priorite minerals.

Aeschynite Series
Minerals of the aeschynite group are orthorhombic mixed oxides with the general formula $\mathbf{AB}_2O_6$, where **A** = REE, Ca, Th, U; **B** = Ti, Nb, Ta; and OH may substitute for some of the O. In the various compositions, charge balance is generally maintained by substitutions of the type Ca^{2+} + $(Nb,Ta)^{5+} \rightarrow (Y,REE)^{3+} + Ti^{4+}$. Most are metamict because of either essential U,Th or at least significant impurity levels of those elements at the **A** site.

Aeschynite-(Ce) $(Ce,Nd,Ca,Th)(Ti,Nb)_2(O,OH)_6$
Aeschynite was named by Berzelius (1830, pp. 195-96) based on samples sent by Menge from Miask, Ural Mountains, Russia. Pekov (1998) identifies the exact locality as the group of pits now called Pit Nos. 75–76 (or "pits behind Dolgiye Mosty"). There, it forms prismatic crystals up to 10 cm long, associated with feldspar, mica, nepheline, magnetite, zircon, and pyrochlore, in an alkaline pegmatite. The crystals were originally mistaken for gadolinite. Pekov notes that aeschynite-(Ce) is found at many localities in the Ilmeny Mountains, but those from Pit Nos. 75-76 remain the best examples.

Berzelius provided a rough chemical analysis and commented that blowpipe analysis can't reliably distinguish Ti and Zr. He therefore chose the name based on the Greek άίσχυνω (= I am ashamed).

Natural material contains as much as 17 wt. % ThO_2 and is normally metamict. After heating to 1000-1100°C, the structure is orthorhombic, with *a*=5.37, *b*=11.08, and *c*=7.56 Å (Aleksandrov 1962).

Aeschynite-(Nd) $(Nd,Ce,Ca,Th)(Ti,Nb)_2(O,OH)_6$
Aeschynite-(Nd) was described from the Bayan Obo district, Inner Mongolia, China, where it forms tabular to prismatic crystals, to several mm, and equigranular to radiating clusters. It occurs in veins in slate and metamorphosed dolomite, associated with aegirine, riebeckite, barite, fluorite, albite, and phlogopite. Crystals are metamict, in various shades of brown to black, with adamantine luster and conchoidal fracture. After heating to 800°C, the material gives a powder XRD pattern typical of the orthorhombic structure of aeschynite minerals (Zhang and Tao 1982).

Aeschynite-(Y) $(Y,Ca,Fe,Th)(Ti,Nb)_2(O,OH)_6$
Aeschynite-(Y) is the currently accepted name for the member with Y dominant at **A** and Ti dominant at **B**. Material of this composition was described by Brøgger (1906, pp. 98–116) under the names *blomstrandine* and *priorite* based on samples found at Urstad, Hitterö (Island), Norway, where it forms large, black crystals in pegmatite. Crystals are metamict, with typical conchoidal fracture and a vitreous to greasy luster on fracture surfaces. The type material contains 7.69 wt. % ThO_2 and ~4 wt. % UO_2.

Similar compositions were reported by Prior (1899) from alluvial tin deposits in the Embabaan district, Swaziland. That material contains ~2.6 wt. % UO_2 and ~0.6 wt. % ThO_2.

Niobo-aeschynite-(Ce) $(Ce,Ca,Th)(Nb,Ti)_2(O,OH)_6$
Niobo-aeschynite-(Ce) was described from the Vishevnye Gory (Cherry Mountains), Southern Urals, Russia (Zhabin, Mukhitdinov, and Kazakova 1960). It forms black, prismatic crystals in quartz-arfvedsonite veinlets cutting finites in the contact aureole of miaskitic intrusive rocks of the Vishnevogorskii alkaline massif. Associated minerals include calcite and titanite. The empirical formula is Nb-dominant on the **B** site: $(REE_{.58}Ca_{.29}Th_{.03}OH_{.15})(Nb_{1.05}Ti_{.79}Fe_{.12}Si_{.02}Al_{.02})_2O_6$. The major REE on **A** are Ce:Nd:La = 1.68:1:0.56, so the type material is niobo-aeschynite-(Ce). The reported actinide content is ~2.5 wt. % ThO_2.

Nonmetamict niobo-aeschynite-(Ce) has been reported as sand-sized grains in alluvial concentrates from the Tofty tin belt, central Alaska. The grains are generally tabular, slender to stubby prisms. The reported actinide content is 1.33 wt. % ThO_2 and 0.01 wt. % U_3O_8 (Rosenblum and Mosier 1975).

Niobo-aeschynite-(Y) [(Y,REE),Ca,Th](Nb,Ta,Ti,Fe)$_2$(O,OH)$_6$
Niobo-aeschynite-(Y) was described from the Bearlake Diggings, Lot 9, concession X, Monmouth Township, Haliburton Co., near Gooderham, Ontario (Bermanec et al. 2008). Associated minerals include apatite, amphibole, feldspar, biotite, calcite, quartz, monazite, pyrite, and U-rich thorite. It forms translucent, deep brownish-red, subhedral crystals up to about 1 cm. Crystals have a conchoidal fracture and are naturally metamict, with 12.1 wt. % ThO_2 and 0.6 wt. % UO_2.

Analysis of the type material shows that Y is dominant over any single REE, as expressed by the empirical formula $[(Y_{.19}Ce_{.11}Nd_{.11}Sm_{.03}Gd_{.02})Ca_{.31}Th_{.18}U_{.009}Mn_{.006}]$ at **A**. Nb is barely dominant, with $(Nb_{.94}Ti_{.92}Ta_{.07}Fe_{.11})$ at **B**.

After heating to 1000°C, it is orthorhombic, with *a*=5.279, *b*=10.996, and *c*=7.443 Å.

Tantal-aeschynite-(Y) (Y,Ce)(Ta,Ti,Nb)$_2$O$_6$
Tantal-aeschynite-(Y) was described from the Raposa pegmatite, São Jose do Sabugi, Paraíba, Brazil, where it forms idiomorphic prismatic crystals to several cm long, associated with red microcline, biotite, columbite, and beryl. Crystals are brownish-black to black, and are metamict, with a resinous luster and conchoidal fracture. The empirical formula is $[Ca_{.28}(Y,REE)_{.425}(Th,U)_{.11}](Ta_{.75}Ti_{.62}Nb_{.595}Fe_{.05})O_6$ with (Y,REE) ~$Y_{.172}Ce_{.095}Nd_{.048}La_{.034}$.

After heating to 1300°C, the structure is orthorhombic with *a*=5.34, *b*=10.97, and *c*=7.38 Å (Adusimilli, Kieft, and Burke 1974).

Note that the "tantal-aeschynite" of Kornetova, Aleksandrov, and Kazakova (1963) is *not* Ta-dominant at **B** and should therefore be considered Ta-rich aeschynite.

Figure 612. Niobo-aeschynite-(Y), ~1 cm wide, from Quadeville, Lyndoch Township, Renfrew Co., Ontario. Note the resinous appearance and glassy fracture surfaces typical of metamict crystals (Gunnar Färber analysis). RJL3732

Samarskite Group
Samarskite-group minerals are ordered, mixed-oxide compounds having the general formula $A^{+3}B^{+5}O_4$, and a structure that is related to that of α-PbO_2 (Warner and Ewing 1993). In the formula, **A** = Ca, U, Th, Y, REE, and Fe; and **B** = Nb, Ta, and Ti. The compositional variations within this group were recently reviewed (Hanson et al., 1999) and based on that analysis, it was proposed that the group includes three possible species based on the occupancies of the **A** site. If REE+Y is dominant, the mineral should be named samarskite-(REE+Y), where the actual suffix used will be the dominant rare earth cation, in accordance with Levinson notation. If U+Th is dominant, the species is named ishikawaite. If Ca is dominant, the mineral is named calciosamarskite. In all cases, Nb > Ta and Ti on the **B** site.

Based on studies of synthetic material, Sugitani, Suzuki, and Nagashima (1985) suggested that low- and high-temperature polymorphs exist, with a transition at 950°C. In this model, the low-temperature polymorph is orthorhombic and the high-temperature form, apparently because of an ordering phenomenon, has a doubled *b* cell dimension and distortion to a monoclinic structure. One caveat is that these minerals occur in pegmatites that normally crystallize at temperatures far below those used by Sugitani, Suzuki, and Nagashima in their synthesis runs (Hanson et al. 1999).

Analysis of these minerals can be difficult because they are usually metamict and may be altered on the surface. The crystal structure can be restored by annealing (Sugitani, Suzuki, and Nagashima 1984), but again it is not always certain that the restored sample is identical in all respects to the mineral that crystallized originally. None can be identified by visual inspection or through simple field tests. Short of actual microprobe analysis of the sample in question, the collector must often rely on whatever published information exists on the specific locality and make an educated guess.

Calciosamarskite (Ca,Y,Fe,U,Th,)(Nb,Ta,Ti)O$_4$
Calciosamarskite was named by Ellsworth (1928a,b) based on examples from the Woodcox mine, Hybla, Ontario, and from Parry Sound, Ontario. Both sites represent rare-element pegmatites containing significant radioactivity. On the Woodcox mine samples, Ellsworth (1928a) states: "The mineral seems to be most nearly related to samarskite though containing very much more lime than has been previously

found in samarskite. All the radioactive minerals found in the calcite-bearing pegmatites of the Haliburton-Bancroft area tend to be high in lime, doubtless because of the universal presence of dissolved calcium carbonate in the magmas of that area. Thus this mineral perhaps represents the type of samarskite produced by a magma rich in lime. Considering the term samarskite as embracing a group of minerals in which considerable percentage variations of the constituents may conceivably occur, due to isomorphous replacement, the mineral here described seems to represent the case where iron is partly replaced by lime, and the name calciosamarskite is suggested as a suitable designation for this variety." At both localities, the mineral is jet black with submetallic luster and subconchoidal to irregular fracture. At Parry Sound it occasionally forms rough square prisms to ~ 1 × 2 cm.

It was also reported at Wiseman's mica mine in Mitchell Co., North Carolina, associated with columbite, uranmicrolite, and muscovite; Spruce Pine, and other localities in North Carolina; and the Hidden Valley claims, Custer Co., Colorado. Some other occurrences include: the Trigg Hill pegmatite, East Pilbara, Western Australia; the Ilmen Natural Reserve, Southern Urals, Russia; and several localities in the Czech Republic.

The status of this mineral remained in some doubt until the systematic study of the group by Hanson et al. (1999) established the validity of calciosamarskite using material from Mitchell Co., North Carolina, which would presumably be regarded as the TL for the species.

Calciosamarskite is monoclinic with a=5.632, b=9.912, c=5.221 Å, and β=93.87° (Hanson et al. 1999).

Ishikawaite $(U,Fe,Y,Ca)(Nb,Ta,Ti)O_4$

Ishikawaite was described from the Ishikawa District, Fukushima, Iwaki Prefecture, Japan, where it occurs in pegmatite. The mineral forms black, glassy, anhedral masses with a vitreous luster. The mineral is partially or completely metamict and has a conchoidal fracture (Shibata and Kimura 1922).

Recently, samples resembling the classic description of "samarskite" were collected from various localities in the Topsham Pegmatite District, Maine; these sites included the Yedlin Location, the Russell Brothers quarry, Stand Pipe Hill, and the Swamp No. 1 Quarry. In all cases, analysis showed U+Th greater than REE+Y so these materials are all properly regarded as ishikawaite (Hanson, Simmons, and Falster 1998).

Ishikawaite has been reported from several locales in Egypt. In the Abu Rushied gneissose granite, dark grains ~0.5 mm wide, assaying ~50% Nb_2O_5 and 26% UO_2, occur with columbite, Hf-rich zircon, and dark Li-mica (zinnwaldite). Based on the structural formula, the U content of individual grains ranged from 0.12 to 0.61 apfu (Raslan 2008). In fine-grained cataclastic rocks (mylonites) in Wadi Sikait, ishikawaite contains ~47% U, 10% Fe, 4% Ca, and 26% Nb (Saleh et al. 2012).

Figure 613. Tabular brown ishikawaite crystal, ~3 × 5 mm, in pegmatite from the Yedlin quarry, Maine. RJL1792

The Honeycomb Hills rhyolite, Utah, represents the product of differentiation in a highly evolved magma, and contains accessory minerals normally associated with rare-element pegmatites. There, ishikawaite occurs as euhedral prismatic microphenocrysts up to 50 m in diameter, associated with fergusonite, columbite, fluocerite, thorite, monazite, and zircon (Congdon and Nash 1991).

Ishikawaite is monoclinic, with a=5.652, b=9.934, c=5.243 Å, and β=93.95° (Hanson et al. 1999).

Samarskite-(Y) $(Y,Ce,U,Th,Ca,Fe)(Nb,Ta,Ti)O_4$

Samarskite-(Y) was originally described (under the name "uranotantal") by Gustav Rose in 1840, from the Blyumovskaya Pit (now Pit no. 50), Ilmeny Mountains, Southern Urals, Russia. Some confusion existed as to its true composition, with one analyst claiming the existence of a new metal ("ilmenium") in the mineral. This was eventually shown to be a mixture of Nb, Ta, and W [see Pekov (1998) for a concise historical review]. Rose's brother, Heinrich, published a rigorous analysis of material supplied by V. E. Samarskii-Bykhovets and gave it the name samarskite, rejecting the earlier names *uranotantal* and *yttroilmenite* (Rose 1847).

At the TL, samarskite-(Y) forms crystals to 4 mm long and grains embedded in feldspar. Parallel growths with ferrocolumbite are common. Associated species include quartz, biotite, muscovite, spessartine-almandine, magnetite, and ilmenorutile (Pekov 1998).

Figure 614. Samarskite-(Y) forming a lustrous black mass with a few crystal faces, from the McKerney mine, Mitchell Co., North Carolina. RJL2639

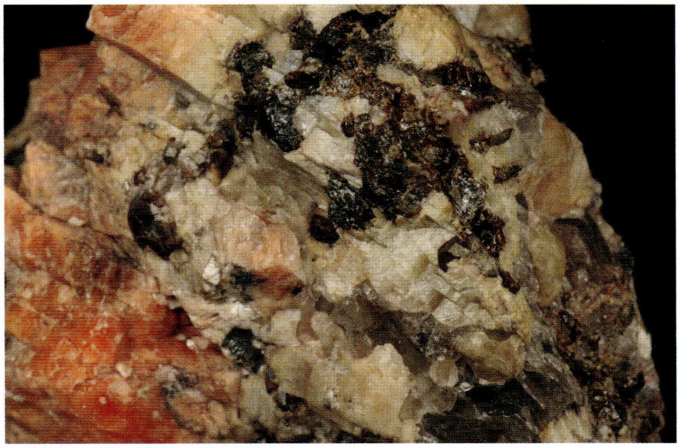

Figure 615. A hand specimen of pegmatite with resinous brown masses of samarskite-(Y) in pink feldspar from Mitchell Co., North Carolina. An old label from the Cleveland Museum of Natural History indicates 12.55% U, and the specimen displays significant radioactivity. RJL3673

Figure 616. A samarskite-(Y) crystal from Catipo-Tete, Mozambique; most of the surface is covered by a brown alteration layer, but glassy black fracture surfaces are exposed in several places. RJL2640

Reported localities include: the McKerney mine, Mitchell Co., North Carolina; Catipo-Tete, Mozambique; with almandine and magnetite at Antanamalaza, Madagascar; and Divina de Uba, Minas Gerais, Brazil. A "uranium-rich samarskite" from Kunar, Afghanistan, is very similar to the material from Ishikawa, but its composition places it within the range of samarskite-(Y) as presently defined.

Samarskite-(Y) is monoclinic with a=5.714, b=9.931, c=5.217 Å, and β=93.47° (Hanson et al. 1999).

Samarskite-(Yb) $(Yb,Ce,U,Th,Ca,Fe)(Nb,Ta,Ti)O_4$

Samarskite-(Yb) was described from the Little Patsy NYF-type granitic pegmatite, South Platte district, Jefferson Co., Colorado. It forms black anhedral masses up to tens of centimeters across, intimately intergrown with ferrocolumbite and zircon. It occurs in later-stage replacement units at the core-margin contact within the pegmatite, associated with ishikawaite, samarskite-(Y), zircon, columbite, monazite, and Y-rich fluorite.

The mineral as found is metamict; after heating to 1100°C for 12 h in a reducing atmosphere, the recrystallized material is monoclinic, with a=5.689, b=9.915, c=5.120 Å, and β=93.16° (Simmons, Hanson, and Falster 2006).

Pyrochlore Group

The pyrochlore group is a series of cubic oxides that contain essential amounts of niobium, tantalum, or titanium. Pyrochlore was first described in 1826, microlite in 1835, and "hatchettolite" in 1877. Over the next 100 years, many analyses were reported and because of the large number of compositional variations, the number of named species and varieties began to proliferate, along with several proposed schemes for classifying the group. Hogarth (1961) analyzed pyrochlore and betafite from several Canadian occurrences and noted, "The relationship of pyrochlore and betafite has long been disputed. X-ray powder patterns of pyrochlore and ignited betafite are very similar but chemical compositions are usually quite different. ... All betafite and some pyrochlore are metamict, a state which has hindered X-ray investigation. Crystallization is known to take place with evolution of heat when the minerals are ignited." The IMA Subcommittee on Nomenclature of the Pyrochlore Group proposed a classification scheme for the group (Hogarth 1977), which defined several species based on the essential role of U. More recently, a new scheme was approved (Atencio et al. 2010; Christy and Atencio 2013) in which pyrochlore became a supergroup. In this reorganization, which is followed here, there are 28 recognized species, of which only two have essential U, viz., oxynatropyrochlore and oxynatrobetafite.

Because Hogarth (1977) defined betafite as U-rich, Ti-dominant pyrochlore, and all published analyses show Ca>U, by the current rules betafite becomes the Ca-rich, Ti-dominant pyrochlore oxycalciobetafite. U-dominant material found thus far only on the moon would be defined as oxyuranobetafite (Atencio et al. 2010).

Analyses of uranmicrolite (or "djalmaite") show that they don't contain enough U to accord them species status. Uranmicrolite was first described as "djalmaite," based on material from Fazenda Posse, Brejauba district, Minas Gerais, Brazil (Guimaraes 1941). Crystals of uranmicrolite usually form octahedra modified by {113} and {001}, and are typically metamict. The color may be yellowish brown, greenish brown, or brownish black with a greasy luster. At the TL, "uranmicrolite" occurs in a granite pegmatite associated with columbite, magnetite, samarskite, garnet, beryl, tourmaline, and various bismuth minerals. It has been reported from Jooste's pegmatite, Okangava Ost, near Karibib, Namibia. Both uranmicrolite and uranpyrochlore have been found as small anhedral crystals in a pegmatite deposit at Parelhas, Rio Grande do Norte, Brazil (Robinson and Wegner 1998). The name *tantalohatchettolite* is a synonym for *uranmicrolite* and should also not be used.

Most "uranpyrochlores" are not U-dominant, except for a sample reported from the Khibiny massif, Kola Peninsula, Russia. Uranpyrochlore was first described as "hatchettolite" (Smith 1877; Allen 1877) based on material from Mitchell County, North Carolina. It may occur in massive form or as metamict crystals, which are generally octahedral modified by {113} and {001}. The color may be black, yellowish brown, red, or red-brown with an adamantine to resinous luster. At the TL, "uranpyrochlore" occurs with samarskite in mica mines. It has been reported from a pegmatite at the Woodcox mine, Hybla, Ontario, Canada, associated with columbite, cyrtolite, titanite, and amazonite. Near Kasenda, Fort Portal area, Uganda, it is found in association with zircon, pyrochlore, monazite, rutile, anatase, and calcite in a calcite-bearing tuff associated with a carbonatite intrusion. Other reported occurrences include Madagascar and Bavaria, as well as with *uranmicrolite* in Brazil (Robinson and Wegner 1998). The currently ac-

Figure 617. A small mass of uranpyrochlore ("ellsworthite") in pink feldspar, from an unspecified location in Canada (likely Hastings Co., Ontario). Note the resinous luster and somewhat glassy fracture typical of metamict phases. RJL259

Figure 618. A 15-mm uranpyrochlore crystal from Monteagle, Ontario, Canada. The accompanying label identified it as "ellsworthite.". RJL4538

cepted definition of uranpyrochlore encompasses compositions previously described as *ellsworthite* and *hatchettolite* and these names should not be used. *Samiresite* as originally described by Lacroix is a metamict material whose composition corresponds to plumbian uranpyrochlore, and the name should therefore be dropped.

"Bismutopyrochlore" is discredited, but some reported compositions might qualify as oxynatropyrochlore. It was reported from the Mika pegmatite vein, which lies in the central part of the Rangkul Highlands, Eastern Pamirs, Tajikistan (Chukanov et al., 1999). The mineral is strongly radioactive and metamict; after heating to 700°C a powder diffraction pattern typical of the pyrochlore structure was obtained. Crystal fragments showed {111}, {110}, and {211} with the octahedron dominant. The mineral is black, and greenish-brown in thin edges. It forms on lepidolite in miarolytic cavities in the pegmatite, associated with microcline, albite, quartz, schorl, elbaite, topaz, muscovite, apatite, fluorite, calcite, barite, beryl, cassiterite, and hambergite.

Many collectors will be familiar with "betafite" crystals from Madagascar or, more often, from Bancroft, Ontario. This material is described here, with two caveats: 1. betafite *per se* is no longer a recognized species: and 2. the original betafite of Lacroix apparently represents heavily altered material.

Betafite was described (Lacroix 1912) from Betafo, Madagascar. It forms sharp octahedral crystals, often modified by {011} or {113}, which are usually metamict. At the TL, crystals as large as 15 cm across have been found. Crystals are opaque, black to brown, and generally a few centimeters across. Fresh crystals have a waxy or greasy luster; altered material often has a dull, yellowish surface layer that is somewhat hydrated and depleted in uranium (Frondel 1958).

Betafite is found as a primary mineral in pegmatites, associated with fergusonite, thorite, rutile, euxenite, cyrtolite, allanite, beryl, titanite, and magnetite. Excellent, large crystals are found with zircon at the Silver Crater Mine, Bancroft, Ontario. In the United States it occurs at the Brown Derby No. 1 mine, Gunnison, Colorado, with gahnite, monazite, and columbite-tantalite species. Small crystals are found with uranoan zircon (*cyrtolite*) in the Cady Mountains near Hector, California; also in the Pidlite pegmatite, Mora County, New Mexico. Exceptional specimens have also been found at Mina El Muerto, Telixtlahuaca, Mexico.

Several previously named varieties or poorly defined minerals lie within the compositional ranges associated with betafite (per Hogarth 1977). Some of these varieties include: *aluminobetafite*, *mendeleyevite*, *tantalbetafite*, and *titanbetafite*. Those names are no longer accepted as proper species and should not be used.

The crystal structure of non-metamict material from the Vico volcanic complex, Italy, was reported by Cámara et al. (2004). Caprilli et al. (2006) studied the crystal chemistry of materials from this locale and several other sites within the Roman Comagmatic Region, and showed the ranges of compositions and examples of compositional zoning within small crystals.

Crichtonite Group

Minerals of the crichtonite group are trigonal oxides with the general formula $\mathbf{ABC}_{18}\mathbf{T}_2\varphi_{38}$, where **A** = Ba, K, Pb, Sr, La, Ce, Na, Ca; **B** = Mn, Y, U, Fe, Zr; **C** = Ti, Fe, Cr, Nb, V, Mn (dominantly Ti); T = Fe, Mg, Zn; and φ = O, OH, F

Figure 619. Simple octahedral crystal of "betafite" with pale tan surface alteration from the type locale, Betafo, Madagascar. This specimen has an interesting history: In 1907 LaCroix collected a large lot of crystals, on which he based the definition of the species. The samples were taken to Denmark and later sent to England during World War II. Fifty years later the lot was auctioned and these historic samples became available to collectors. RJL2564

Figure 620. Brown intergrown "betafite" crystals, ~2 cm wide, from Bancroft, Ontario. RJL213

Figure 621. "Betafite" crystal group from the Silver Crater mine, Bancroft, Ontario. RJL337

(dominantly O). One can see the possibility for wide variability in compositions, and only Ti and Fe are systematically present. Because small amounts of water might be present as OH, and small amounts of F are often reported, it isn't possible to reliably calculate Fe^{2+}/Fe^{3+} ratios from microprobe analysis. Defining the correct formula thus requires independent measurement of Fe^{2+}/Fe^{3+} and/or water content (Wülser, Brugger, and Meisser 2004).

Thirteen species are currently recognized, of which five contain essential U, viz, cleusonite, davidite-(Ce), davidite-(La), dessauite-(Y), and mapiquiroite.

Cleusonite $(Pb,Sr)(U^{4+},U^{6+})(Fe^{2+},Zn)_2(Ti, Fe^{2+},Fe^{3+})_{18}(O,OH)_{38}$
Cleusonite was described from two occurrences in Valais, Switzerland, (Cleuson and Bella Tolla summit), where it occurs in greenschist-facies metamorphosed gneiss of the Mont Fort and Siviez-Mischabel Nappes (Wülser et al. 2005). It forms black, opaque, cm-sized tabular crystals with bright submetallic luster. Crystals typically consist of multiple hexagonal prisms and rhombohedra, and are generally twinned. They tend to be partially metamict and fractured as a result of the swelling associated with metamictization.

At Cleuson, associated minerals include quartz, chlorite, calcite, albite, microcline, tourmaline, fluorapatite, zircon, ilmenite, hematite, pyrite, titanite, rutile, chrichtonite, monazite-(Ce), and native Au. Cleusonite sometimes contains small (30 μm) inclusions of clausthalite, chalcopyrite, and uraninite.

At Bella Tolla, it forms flattened masses parallel to the schistosity of the gneiss, and in cross-cutting Alpine quartz veins. Associated species include quartz, albite, chalcopyrite, uraninite, Hg-rich tennantite, pyrite, hematite, and magnetite.

Material previously described as "U-rich senaite" from Alinci, Macedonia, and "plumbodavidite" from Huanglongpu, China, are also cleusonite.

Cleusonite is trigonal. Natural material is partially metamict, with a=10.576 and c=21.235 Å. Heat treament causes the cell parameters to contract somewhat: a=10.4189 and c=20.942 Å (800°C treatment); a=10.385 and c=20.901 Å (1000°C treatment).

Davidite series
Davidites are members of the crichtonite group that Wülser, Brugger, and Meisser (2004) classify in the magmatic (±metasomatic) genetic group; they typically occur in granodiorite, granite, alkali-granite, and syenite pegmatites, or in high-temperature (granulite facies) metamorphic rocks. Both are strongly radioactive and both are normally found in a metamict state.

Zhabin et al. (1963) compared the analyses of so-called "ferutite" and "ufertite" with those of davidites from the Cherry Mountains, Russia; Radium Hill, Australia; Mozambique; and Arizona. They concluded that the materials are identical, and therefore *ferutite* and *ufertite* are discredited as synonyms of davidite.

Lumpkin, Blackford, and Colella (2013) studied the rate at which the davidite structure becomes metamict with increasing α decay dose, which has practical implications for its usefulness as a host material for nuclear waste. They noted that natural davidite is commonly altered to rutile, ilmenite, titanite, and other phases.

Pabst (1961) determined the structure of davidite from Pima Co., Arizona, and from Iveland, Norway, after heating at temperatures up to 1375°C to restore the crystallinity. He reported a hexagonal (rhombohedral) structure with a=10.37 and c=20.97 Å. Later work (Gatehouse, Grey, and Kelly 1979), using material from the Pandora Prospect, Pima Co., Arizona, heated in air at 920°C for 2 h, confirmed the results of Pabst, with a=10.376 and c=20.991 Å.

Davidite-(La) $(La,Ce,Ca)(Y,U)(Ti,Fe)_{20}O_{38}$
Davidite-(La) was described (as the original "davidite") from Radium Hill, near Olary, South Australia, where it was associated with rutile, ilmenite, and carnotite. It has a vitreous luster and may be massive or as anhedral grains and occasionally larger crystals. Mawson (1906) noted in his original report, "Pegmatite lodes, of the character of those at Radium Hill, often carry tin and wolfram, though so far these substances have not been reported from the locality, and the absence of even traces of them in the analyses suggests that, likely, the

ore body is a pegmatite of a basic rock, and that, in all probability, such minerals will be found entirely absent. It may be mentioned that this type of ore deposit does not usually develop pitchblende, but uraniferous titanates, niobates, and tantatates, and thorium minerals may be expected."

Radium Hill material displays a rough cuboidal habit. More distinct crystals occur at Mavusi, in the Tete District, Mozambique, either tabular on {0001} or pyramidal with ditrigonal pyramids dominant. The crystals from Mozambique occasionally form irregular clusters and can be large: one crystal fragment was reported to weigh about 10 kg (Frondel 1958).

Davidite-(La) is a fairly common accessory mineral in uranium-rich granite pegmatites as well as in syenitic rocks. It also occurs in pegmatite-like phases of calcite-apatite-diopside-titanate rocks of the Grenville type. Recently davidite-(La) has been obtained from Bektau-Ata, Kazakhstan, where it forms rough brown-black crystals that become gray with surface alteration.

Davidite-(Ce) (Ce,La)(Y,U)$(Ti,Fe)_{20}O_{38}$
Davidite-(Ce) was described from Tuftane, Iveland, Norway, where it occurs in a pegmatite assocated with thortveitite, gadolinite, and euxenite (Neumann and Sverdrup 1960). Davidite-(Ce) may be black or dark brown, with a vitreous to submetallic luster. Crystal habits are similar to those of davidite-(La).

Dessauite-(Y) (Sr,Pb)(Y,U)$(Ti,Fe^{3+})_{20}O_{38}$
Dessauite-(Y) was described from the Buca della Vena mine, Apuan Alps, Tuscany, Italy, where it occurs in cavities in calcite veins. It forms black, mm-sized rhombohedral crystals, tabular on {001}, associated with derbylite, hematite, rutile, karelianite, siderite, and calcite. Formation of the assemblage is attributed to hydrothermal fluids circulating through a hematite-barite orebody hosted in dolomite, in connection with an Alpine metamorphic event (Orlandi et al. 1997). The geology and mineralogy of the Buca della Vena mine are described in detail by Orlandi and Checchi (1986).

Dessauite-(Y) is trigonal, with a=10.385 and c=20.921 Å; an unusual feature of its structure is the presence of additional, partially occupied, octahedral sites. Based on these observations, the authors revised the general formula for the crichtonite group to $\mathbf{ABC}_{18}\mathbf{T}_2O_{38}$ rather than the former $\mathbf{AM}_{21}O_{38}$ (Orlandi et al. 1997).

Figure 622. Thick tabular davidite-(La) crystal, ~4 cm across, from Bektau-Ata, Karagandy Province, Kazakhstan. RJL2575

Mapiquiroite (Sr,Pb)(U,Y)$Fe_2(Ti,Fe^{3+})_{18}O_{38}$
Mapiquiroite was described from the Buca della Vena and Monte Arsiccio mines, Apuan Alps, Tuscany, Italy, where it occurs in barite-pyrite-iron oxide ores. At Buca della Vena, it forms complex rhombohedral or tabular pseudohexagonal crystals to ~1 mm with black submetallic luster. It is found in quartz-adularia-barite veins in schist, associated with allanite-(Ce), anatase, destinezite, gypsum, monazite-(Ce), pyrite, rutile, and tourmaline. At Monte Arsiccio, it forms tabular, pseudohexagonal crystals, up to 5 mm, in quartz-carbonate veins in magnetite-rich dolostones, associated with barite, boulangerite, derbylite, siderite, and sphalerite (Biagioni et al. 2014).

Mapiquiroite is trigonal, with a=10.372 and c=20.875 Å (Buca della Vena); and a=10.385 and c=20.894 Å (Monte Arsiccio). It is isostructural with other members of the crichtonite group.

Petscheckite and Liandratite
These are two metamict phases that were found and described together; liandratite appears to be an alteration product of petscheckite.

Petscheckite $U^{4+}Fe^{2+}(Nb,Ta)_2O_8$
Petscheckite was described from the Antsakoa I pegmatite near Berere, Madagascar. It forms black crystals up to 4 cm long × 2 cm thick, in the mineralized zone of the pegmatite, which was mined for beryl and Ta/Nb-bearing minerals from around 1950 to 1960. Associated species include rutile var. *strüverite*, monazite, ilmenite, garnet, tourmaline, quartz, and feldspar. The crystals have a pale alteration crust and are metamict.

After heating to 650°C, powder XRD data gave a hexagonal unit cell with a=6.42 and c=4.02 Å, which corresponds to the hexagonal compound $U^{6+}Ta_2O_8$ synthesized ca. 1960 by Gasparin. However, in petscheckite the uranium is present as U^{4+}, so the structure also contains Fe^{2+} to complete the charge balance. The authors discuss the further alteration of the material to yield a hydrous or "hydroxy-petscheckite" phase with the approximate formula $U^{4+}(Fe^{3+},\square)(Nb,Ta)_2O_7(OH)$ (Mücke and Strunz 1978).

Liandratite $U^{6+}(Nb,Ta)_2O_8$

Liandratite was described from the Antsakoa I pegmatite near Berere, Madagascar. It forms yellow to brown crusts to ~2 mm thick surrounding petscheckite (Mücke and Strunz 1978). Liandratite has also been reported from the Kempton beryl mine, Western Australia.

Liandratite is hexagonal, with a=6.36 and c=4.10 Å. Natural material is metamict; after rapid heating to ~650°C, the structure is identical to that of synthetic $U^{6+}Ta_2O_8$. There is an empty site in this structure, which is occupied by Fe in petscheckite. Mücke and Strunz suggest that, "$U(Nb,Ta)_2O_8$ arises through complete oxidation of uranium and iron, where nearly all Fe^{3+} ions have been removed. This process occurred *in situ* and only on the surface of petscheckite crystals, which are therefore epitaxially overgrown by liandratite."

Figure 623. Dark brown, resinous mass of liandratite, ~2 cm across, from the Kempton beryl mine (Arthur River South pegmatite), Western Australia (Gunnar Färber analysis). RJL3730

Aspedamite $\square_{12}(Fe^{3+},Fe^{2+})_3Nb_4[Th(Nb,Fe^{3+})_{12}O_{42}]\{(H_2O),(OH)\}_{12}$

Aspedamite was described from the Herrebøkasa quarry, Aspedammen, Østfold, southern Norway. It forms small (<50 µm) cubic or dodecahedral crystals on a partly altered monazite crystal, associated with columbite-(Fe) and muscovite. Crystals are brownish orange with adamantine luster, and are not fluorescent. The mineral was found in an NYF-family pegmatite, close to the allanite-monzonite type of the REL-REE subclass. The main minerals are microcline, muscovite, and quartz; other associated minerals include fluorite, monazite-(Ce), xenotime-(Y), rutile, rynersonite, samarskite, beryl, spessartine, topaz, uraninite, thorite, and various secondary U minerals (Cooper et al. 2012).

Aspedamite is cubic, with a=12.9079 Å. The structure is based on the heteropolyanion $[\mathbf{DA}_{12}O_{42}]$ (where $\mathbf{D}$ = Th, and sum $\mathbf{A} = Nb_9Fe^{3+}{}_2Ti$), which consists of twelve face- and corner-sharing $\mathbf{A}O_6$ octahedra that surround the [12]-coordinated **D** cation. There are eight heteropolyanions at the corners of the unit cell, with an additional heteropolyanion at the center, forming an *I*-centered arrangement. Each heteropolyhedral cluster is decorated by eight **B** octahedra, each of which bridges two adjacent clusters along the body diagonals of

Figure 624. SEM image of aspedamite crystals on a mat of fibrous Al-Nb-Fe-Ti-Ca-K-bearing silicate. The scale bar is 50 µm. Photo courtesy of Frank Hawthorne, University of Manitoba.

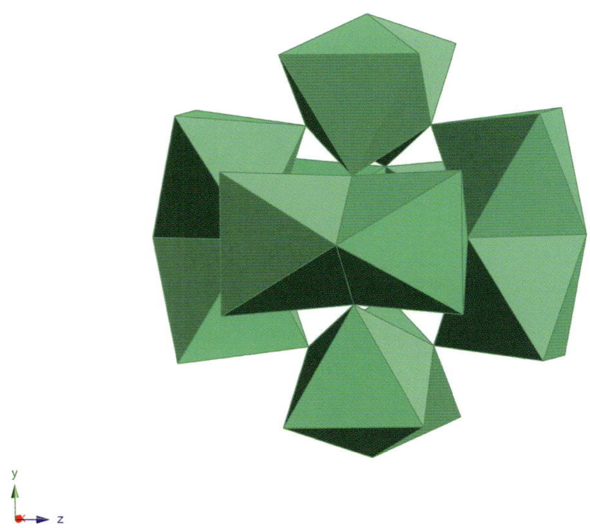

Figure 625. The $[DA_{12}O_{42}]$ heteropolyanion in aspedamite, showing edge-sharing AO_6 octahedra (A = Nb, Fe^{3+}, Ti, etc.) (green), which surround the D cation (Th^{4+}) hidden in the center.

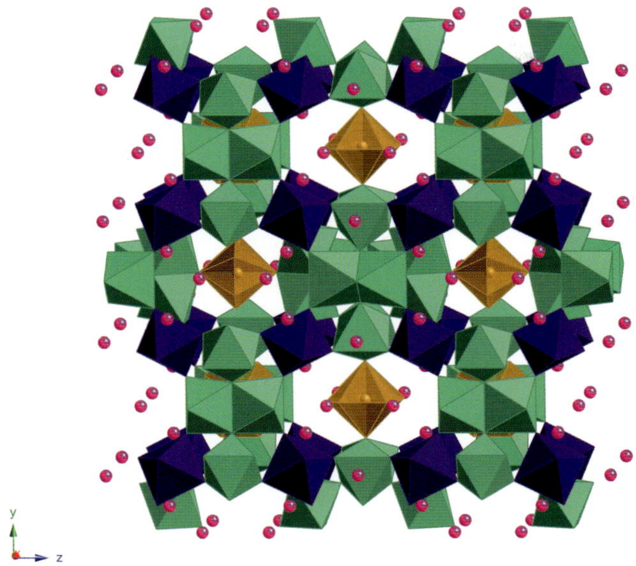

Figure 626. The structure of aspedamite viewed just off [100], showing how the polyanions are connected along [100] via CO_6 octahedra (C = Fe^{3+}, Fe^{2+}) (brown), and along [111] via BO_6 octahedra (B = Nb, etc.) (indigo). A' cations (□, Fe^{3+}) are not shown, and E cations (□, Mn^{2+}, etc.) are magenta.

the cell. Further intercluster linkage is provided by the *C* octahedra, which link pairs of adjacent clusters in the *a* direction. Aspedamite is isostructural with menezesite.

Rhabdophane Group

The rhabdophane group is a series of hexagonal or pseudo-hexagonal phosphates having the general formula $\mathbf{X}PO_4{\cdot}1\text{-}2H_2O$, where **X** = Ca, Ce, Fe^{3+}, La, Pb, or Th. They are the hydrated equivalents of the monazite group. Members that contain essential (Th,U) are brockite, grayite, ningyoite, and tristramite.

Brockite $(Ca,Th,Ce)PO_4{\cdot}H_2O$

Brockite was described from the Wet Mountains, Colorado, where it was found in a prospect pit near the Bassick mine, Querida, Custer Co. The host rock is deeply weathered granite with overlying volcanics. The Th minerals occur in veins localized along shear zones, and are associated with abundant hematite, barite, and carbonates (mainly calcite). Brockite forms fine-grained, massive reddish-brown aggregates in nodules to 35 mm in diameter, containing pyrite inclusions. The nodules, along with bits of brecciated quartz and albite, are enclosed in pale yellow earthy material that is mostly brockite. In thin section it appears mostly as radial aggregates of fibroid grains (Fisher and Meyrowitz 1962).

It was also identified by the same authors in earthy yellow coatings from the Hardwick mine, and from the Nightingale shaft, Custer Co., Colorado. There, aggregates of red K-feldspar represent potassic alteration zones cutting granitic rocks. Thorite, barite, and hematite fill the spongy matrix. Finely crystallized brockite has formed at the expense of thorite next to small cavities.

Grayite $(Th,Pb,Ca)PO_4{\cdot}H_2O$

Grayite was named by Bowie (1957) based on material found at the Gooddays mine, Mtoko, Zimbabwe. The occurrence and history are summarized as follows: "Geologists responsible for the running of the Atomic Energy Authority's Office in Salisbury, Southern Rhodesia, have completed their first year of service in the Federation of Rhodesia and Nyasaland, where they have done much to further the search for radioactive minerals. Pitchblende and kasolite have been recognized for the first time in Southern Rhodesia and a new thorium mineral has been discovered for which the name 'grayite' is proposed. The mineral is pale yellow with a powdery appearance and has a chemical composition approximating to thorium phosphate. Structurally it is related to rhabdophane and other hexagonal rare-earth phosphates. It occurs in a lithium-bearing pegmatite and is intimately associated with thorite."

Fleischer (1962) reviewed the status of several reported "rhabdophane" minerals and noted that in many cases the descriptions were inadequate to support the recognition of new species. He also noted that grayite gives an XRD pattern like that of rhabdophane, and when heated above 850°C, a monazite-like pattern.

Grayite has also been reported from Lavra de Itamotinga, Juaziero, Bahia, Brazil.

Ningyoite $(U,Ca)_2(PO_4)_2{\cdot}1\text{-}2H_2O$

Ningyoite was described from the unoxidized zone of the Ningyo-toge mine, Tottori Prefecture, Japan, where it forms coatings on pyrite and apatite or fills cracks and cavities in the ore. The ore is black or gray and very friable with abundant cavities, and because the ningyoite is very fine grained and intergrown with pyrite, clay, and other contaminants, precise chemical analysis was very difficult. Individual crystals are acicular to elongated lozenge-shaped, brownish green to brown, and slightly pleochroic. Powder XRD patterns of ningyoite, and of analogous synthetic material, resemble that of rhabdophane (Muto et al. 1959).

Ningyoite has also been reported in U deposits in southern British Columbia, where it represents the major U^{4+} phase present. "It occurs as accretionary stars, crystal mats and aggregate masses of spindle- or lozenge-shaped crystals that rarely exceed 2 mm in length. Occasionally it is present as single crystals in cleavage cracks of saléeite. The mineral forms in pyritiferous organic-rich fluvial sediments (Tyee deposit) or carbonaceous sandy mudstones (Blizzard deposit). . . . Optimum conditions for the formation of ningyoite are attained when weakly alkaline groundwaters leaching granitic basement complexes and containing high concenfations of U and PO_4 infiltrate pyritiferous, organic-matter-rich sediments" (Boyle et al. 1981).

Figure 627. Small black granules of ningyoite from the TL at Ningyo-toge, Tottori Prefecture, Japan. RJL523

Figure 628. Prismatic monazite crystal ~2 cm long, in pegmatite from Moss, Norway. RJL500

Tristramite $(Ca,U^{4+})PO_4 \cdot 2H_2O$
Tristramite was described from old dumps at Wheal Trewavas, Cornwall, England, where it forms fine-grained aggregates and coarser-grained infillings occupying fine fractures and cavities in pitchblende-quartz veins, associated with sulfides such as chalcopyrite, pyrite, sphalerite, galena, and marcasite. It has an acicular to fibrous habit, with crystals ranging from 5–80 μm long × 2–5 μm wide. Veinlets are generally less than 200 μm wide and crystals occur either as a cross-fiber infill or as radial aggregates, a typical habit of rhabdophane. The color is lustrous yellow to greenish-yellow but often shows color banding caused by microscopic goethite inclusions. The empirical formula contains some sulfate and carbonate, corresponding to ~ $(PO_4)_{.79}(SO_4)_{.12}(CO_3)_{.07}$. The original study noted that tristramite was also found on dumps at other Cu and Sn mines in Cornwall, including Trequeen, Wheal Alice, Wheal Butler, Wheal Damsel, and Wheal providence (Atkin, Basham, and Bowles 1983).

Monazite Group
Monazites are anhydrous phosphates with the general formula APO_4, where **A** = Ca, Th, or REE. Members of the group are cheralite, monazite-(Ce), monazite-(La), monazite-(Nd), and monazite-(Sm). Cheralite is the only member of the group that contains essential Th.

Monazite was named by Breithaupt (1829) from the Miask district, Russia. Crystals are normally tabular or prismatic, and occasionally large. As in other REE minerals, examples have been found with Ce, La, Nd, or Sm dominant, so there are presently four recognized species. Although none are U- or Th-dominant, most contain significant levels of these elements, and indeed monazite is the principal ore mineral of Th; this is no doubt partly a consequence of its occurrence in heavy sands, which greatly simplifies mining and concentration of the desired material. According to Frondel (1958): "Thorium is usually present in monazite in substitution for (Ce,La). The amount of Th present ordinarily ranges from a few weight percent ThO_2 up to 10 or 12 percent, and monazite with about 4 to 9 percent ThO_2 is common. The upper limit of substitution of thorium is not known, but may be near 30 weight percent ThO_2. Monazite free from thorium is very rare. As Th is quadrivalent any substitution for the trivalent rare earths may be compensated electrostatically in the crystal structure by a concomitant substitution. . . . Uranium has been frequently reported in monazite, but in amounts up to only about 1 weight percent U_3O_8."

The mechanisms of (U,Th) substitution in monazites were studied using samples of monazite-(Ce) from placer deposits in Western Australia. Regression analysis of microprobe results on many individual grains showed considerable variation, with ThO_2 ranging from 1.2 to 21.9 wt. % with a mean of 8.79 wt. %, and U ranging from <0.17 to 0.75 wt. % with a mean of 0.08 wt. %. There are two substitution mechanisms: $(Th,U)^{4+} + Ca^{2+} \rightarrow 2REE^{3+}$, and $(Th,U)^{4+} + Si^{4+} \rightarrow REE^{3+} + P^{5+}$, and both mechanisms occur together in both monazite and xenotime (Van Emden et al. 1997).

Cheralite $CaTh(PO_4)_2$
Cheralite was described from Kuttakuzhi, Travancore, India, where it occurs in a kaolinitized pegmatite dike associated with schorl, chrysoberyl, partly metamict U-rich zircon, and smoky quartz. It forms pale to dark green, brittle masses to ~ 5 cm across, with vitreous luster and uneven fracture. Chemical analysis gave the empirical formula $(REE_{.40}Ca_{.27}Th_{.28}U_{.04})(P_{.91}Si_{.08})O_4$. Bowie and Horne (1953) considered this to be an intermediate composition between

monazite and a hypothetical end member $CaTh(PO_4)_2$. They synthesized that compound and proposed a composition diagram with monazite-huttonite-$CaTh(PO_4)_2$ as the corners, with cheralite occupying a large range of compositions between monazite and $CaTh(PO_4)_2$.

Material falling within the $CaTh(PO_4)_2$ range was later found on the Brabant farm, Karibib district, Namibia, and named *brabantite* (Rose 1980). Brabantite was later discredited as the system was simplified to monazite-huttonite-cheralite (Linthout 2007).

Cheralite has been reported from dozens of localities worldwide. In the US, locales include: the Grizzly pegmatite, Marquette Co., Michigan; the Life Preserver mine, Nye Co., Nevada; the Wausau intrusive complex, Marathon Co., Wisconsin; and the Uranium King mine, Carbon Co., Wyoming. Some other occurrences include: the Pilanesberg alkaline complex, Northwest Province, South Africa; the Sahambano corundum deposit, Ilhosy district, Madagascar; the Ban Huay Xai sapphire mines, Laos; the Hagendorf South pegmatite, Gernamy; the Fluorite mine, Valzergues, France; in Finland in the Sokli carbonatite complex, Lappland, and the Viitaniemi pegmatite, Orivesi; the Lake Boga quarry, Australia; and a number of places in the Czech Republic.

Cheralite is monoclinic, with a=6.751, b=6.962, c=6.468 Å, and β=103°53' (Bowles, Jobbins, and Young 1980).

Lermontovite $U(PO_4)OH{\cdot}H_2O$

Lermontovite was discovered in 1948 in the Gremuchka ore zone, Beshtau uranium deposit, near Pyatigorsk, Northern Caucasus (Pekov 1998). It occurs as greenish-gray, earthy masses and botryoidal to spherulitic aggregates to 1.5 mm in cracks in a marcasite vein cutting granite porphyry. Associated species include nasturan, vrbaite, lorandite, opal, evansite, halloysite, pyrite, and a hydrous Mo sulfate.

Re-examination of a sample collected ca. 1957, using electron diffraction, indicated that the structure is orthorhombic, with a=9.74, b=19.0, and c=10.1 Å (Melkov et al. 1983).

Steacyite Group

The steacyite group includes five complex silicates with the general formula $\mathbf{AB_2C}[Si_8O_{20}](H_2O)_n$, where **A**=U,Th,REE; **B**=Ca,Na; and **C**=K,□. In arapovite, the silicate tetrahedra form units consisting of two four-membered rings, oriented one above the other in the z-direction and sharing apical oxygens. These units are separate from one another and are linked by sheets of $\mathbf{A}O_8$ and $\mathbf{B}O_8$ polyhedra parallel to (001). The topology of turkestanite and steacyite is similar to that of arapovite. In ekanite, the four-membered rings are interconnected to form puckered sheets lying parallel to (001). The silicate sheets are in turn linked by sheets of CaO_8 and ThO_8 polyhedra.

As with many other U, Th minerals, natural samples are often metamict, and structural studies typically require annealing to restore the crystallinity.

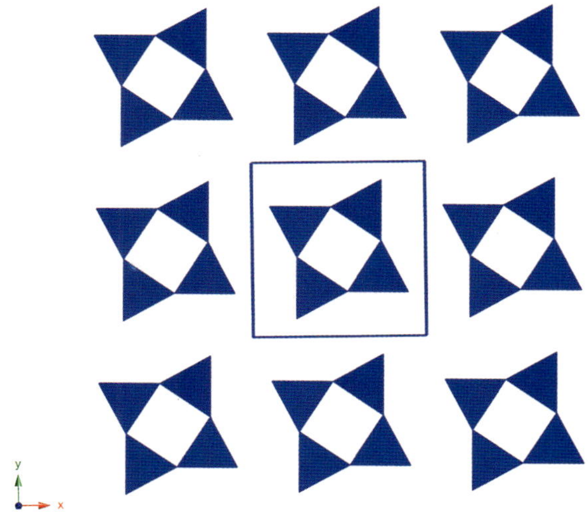

Figure 629. The silicate ring topology in arapovite, viewed along [001]. Each downwardly facing four-membered ring has another ring directly under it, facing upwardly and sharing apical oxygens. One unit cell is indicated at center.

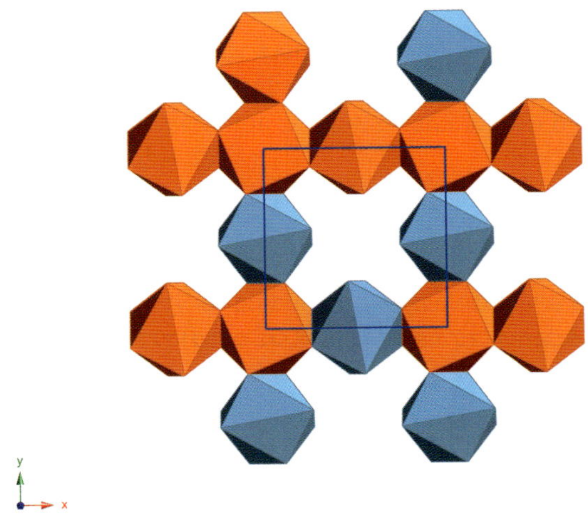

Figure 630. The sheet of 8-coordinated A and B polyhedra, viewed along [001].

Figure 631. The complete structure, viewed along [100] to show how the rings are bonded together by the A, B sheets.

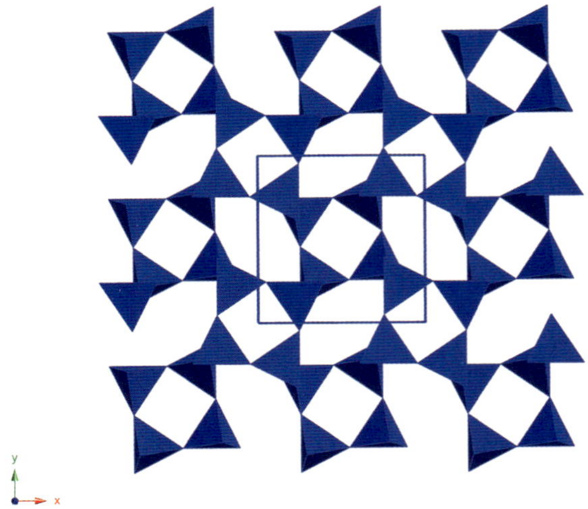

Figure 632. The puckered silicate sheet topology of ekanite, viewed along [001].

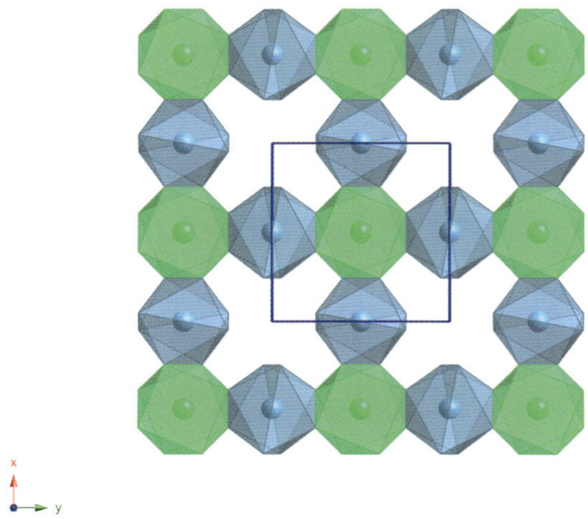

Figure 633. The sheet of 8-coordinated Ca (pale blue) and Th (green) polyhedra, viewed along [001].

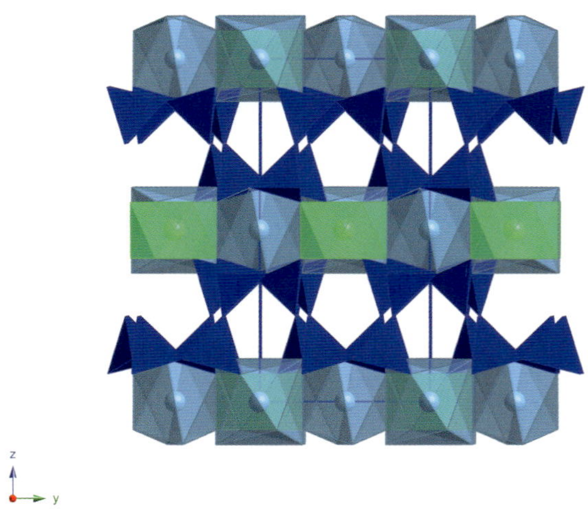

Figure 634. The complete structure, viewed along [100].

Arapovite $U^{4+}(Ca,Na)_2(K_{1-x}\square_x)[Si_8O_{20}]$

Arapovite, the U analog of turkestanite, was described from alkaline rocks of the Dara-i-Pioz moraine, Tien-Shan Mountains, Tajikistan, where it forms dark green zones (0.1–0.3 mm wide) in larger (to 1 cm) turkestanite crystals. The host is a polylithionite-aegirine-microcline rock in which associated minerals include stillwellite-(Ce), sogdianite, zektzerite, pyrochlore, hyalotekite, tadzhikite-group minerals, albite, and quartz. Arapovite is dark green, transparent in thin sections, with a vitreous to pitchy luster and conchoidal fracture (Agakhanov et al. 2004).

Natural arapovite is metamict; structure refinement was done after annealing 3.5 h at 1100°C in air. The structure is tetragonal, with a=7.551 and c=14.710 Å (Uvarova et al. 2004).

Figure 635. Small arapovite grain in probe mount; author's studied material from the TL at Dara-i-Pioz, Tajikistan. The dark green area is arapovite and the yellowish rim is turkestanite. RJL4557

Figure 636. Rectangular, dark green arapovite grain, ~6 mm long, in massive quartz, from Dara-i-Pioz, Tajikistan. RJL4624

Ekanite $ThCa_2[Si_8O_{20}]$
Ekanite was described from Eheliyagoda, Ratnapura, Sri Lanka, where it was found as somewhat prismatic pebbles in gem gravels. It is green, transparent to translucent, and typically metamict. Cabochons cut from this material occasionally show some asterism. The source rock from which the placer was derived hasn't been conclusively identified (Anderson et al 1961).

Ekanite has been noted at a number of places worldwide. Tiny but very gemmy green crystals occur in small cavities in volcanic rocks in Italy, at Pitigliano, Grosseto province, Tuscany, and at Vetralla, Vico volcanic province, Lazio; and in a similar environment at Bellerberg, Eifel Mountains, Germany. Massive ekanite occurs in the Murun alkaline massif, Siberia, Russia. Other occurrences include: straw-yellow to dark red crystals in syenite in the Tombstone Mountains, Yukon Territory, Canada (Szymanski et al. 1982); and as small yellow grains with sanbornite in metamorphic rocks at the Baumann prospect, Tulare Co., California.

Figure 637. A 6 × 8 mm oval faceted ekanite from Sri Lanka, weighing 1.5 ct., with a pleasing dark green color. The stone is very slightly radioactive, measuring ~160 cpm over background at contact. RJL4571

Ekanite is typically metamict. In the original report, Anderson et al. (1961) annealed the Sri Lankan material at 650-1000°C and obtained a tetragonal unit cell. Szymanski et al. (1982), using non-metamict ekanite from the Yukon Territory, further refined the structure on a tetragonal unit cell and reported a=7.483 and b=14.893 Å.

Iraqite-(La) $K_{0.6}Ca_2(La,Ce,Th)[Si_8O_{20}]$
Iraqite was described from Shakhi-Rash Mountain, Hero, Qala-Diza, northern Iraq, where it occurs in coarse granite in contact with dolomitic marble containing olivine and diopside. Unlike other members of the steacyite group, it contains significant REE; although it is somewhat radioactive from its content of Th and minor U, it isn't metamict. It forms pale greenish yellow grains, which have no clearly developed crystal faces but show three orthogonal cleavages.

The type specimen has La ≈ Ce >> Nd, Pr, Sm, Yt, and Gd, with La slightly greater than Ce ($\sim La_{.45}Ce_{.42}$) so under current nomenclatural rules, it is designated iraqite-(La). The REE content of iraqite is such that these elements not only fill more than half of the sites occupied by Th in ekanite, but also fill some of the Ca sites (Livingstone et al. 1976).

Iraqite-(La) is tetragonal, with a=7.61 and c=14.72 Å.

Steacyite $K_{0.6}(Na,Ca)_2Th[Si_8O_{20}]$
Steacyite was first noted in nepheline syenite veins of the Mont Saint-Hilaire pluton, Quebec, where it occurs in cavities in pegmatite veins, associated with nenadkevichite, analcime, and aegirine. It forms simple, dark brown, tetragonal crystals that rarely exceed 2 mm. Crystals show only the forms {100} and {001}, and are elongated on [001]. At the time, it was considered to be a form of ekanite (Perreault and Richard 1973).

Figure 638. A tiny green ekanite crystal in a small vug in volcanic rock from the Casa Collina quarry, near Pitigliano, Tuscany, Italy. RJL4559

Figure 639. Tiny steacyite crystal with characteristic square cross section, associated with golden, fibrous aegirine in a cavity in syenite from Mont St.-Hilaire, Quebec. RJL4564

The discovery of non-metamict ekanite in the Tombstone Mountains, Yukon Territory, led to a re-evaluation of the "ekanite" from Mont St.-Hilaire, as well as related compositions that had been named "kanaekanite." Based on a more thorough structural understanding, the material from Mont St.-Hilaire was redefined as the new species steacyite (Perrault and Szymanski 1982).

Other reported locales include: masses to >1 cm in Henderson Co., North Carolina; with serandite in nepheline syenite on Rouma Island, Los Archipelago, Guinea; with charoite and other phases on the Murun massif, Siberia; in alkaline rocks at Dara-i-Pioz, Tajikistan; and in basalt at Mendig, Eifel Mountains, Germany.

Steacyite is tetragonal, with *a*=7.58 and *c*=14.77 Å (Perrault and Sxymanski 1982).

Figure 640. Glassy yellow steacyite forming a pod ~8 mm wide, in charoite, with tinaksite and other minerals, from the Murun massif, Siberia, Russia. RJL4628

Figure 641. Rectangular green steacyite crystals to ~3 mm, in alkaline rock from Dara-i-Pioz, Tajikistan. RJL4629

Turkestanite $Th(Ca,Na)_2(K_{1-x}\square_x)Si_8O_{20}\cdot nH_2O$

Turkestanite was described from two localities, one in Tajikistan and one about 60 km ENE in Kyrgyzstan, and named for the area (Turkestan Ridge). It is the Ca analog of steacyite. In Kyrgyzstan, it occurs in albitized shale in the contact aureole of the Jelisu (Dzhelisu) massif, near the headwaters of the Kojaachkhan River. In Tajikistan, it was found in a boulder of microcline, pectolite, quartz, aegirine, and calcite at Dara-i-Pioz, Tien Shan (Pautov et al. 1997).

Turkestanite forms prismatic crystals, up to 5 mm at Dara-i-Pioz, and up to 50 mm at Jelisu. Color ranges from apple-green (Dara-i-Pioz) to shades of brown (Jelisu).

The REE enrichment patterns in turkestanite grains from evolved peralkaline granites in the Morro Redondo complex, Brazil, were studied by Vilalva and Vlach (2010), who reported enrichment of light REE over heavy REE, a strong negative Eu anomaly and positive Ce anomaly. They note that, "Observed compositional variations suggest the influence of coupled substitution mechanisms involving steacyite, a Na-dominant analogue of turkestanite, iraqite, a REE-bearing end-member in the ekanite-steacyite group, ekanite and some theoretical end-members. Turkestanite crystals were interpreted as having precipitated during post-magmatic stages in the presence of residual HFSE [high field-strength elements]-rich fluids carrying Ca, the circulation of which was enhanced by deformational events."

Turkestanite also occurs in the Ilimaussaq complex, Greenland, associated with albite, microcline, aegirine, narsarsukite, quartz, and eudialyte.

Turkestanite is tetragonal; structure refinement on a sample from the Jelisu massif gave *a*=7.579 and *c*=14.704 Å (Kabalov et al. 1998).

Chevkinite Group

This group includes a number of hard-to-distinguish titanosilicates, of which two contain essential thorium, viz., chevkinite-(Ce) and perrierite-(Ce). Chevkinite-(Ce) is related to perrierite-(Ce) and strontiochevkinite. Given the range of compositions and the fact that much natural material is metamict, the physical properties and crystal structure reported in the literature are frequently inconsistent. Studies on synthetic materials with carefully controlled compositions have provided some guidance on phase relationships and crystal chemistry, but care must be taken when applying their conclusions to naturally impure materials. Ito (1967) synthesized a large number of compositions and noted, "Chevkinite and perrierite apparently have very similar but not identical compositions, while having the same formula type, they are not polymorphs in the strict sense. They occupy regions of an extensive solid-solution series, separated by a two-phase field, in which the structural transition apparently is controlled by the average size of the ions occupying the several structural sites. A tentative phase boundary is given. X-ray spacing data, unit-cell contents, and thermal data are cited, as are observations on the recrystallization of metamict natural material."

The situation was further clarified by Segelstad and Larsen (1978), who studied chevkinite from three syenite pegmatites in the Oslo region, Norway, one of which also contains perrierite. They propose the general formula $\mathbf{A}_4\mathbf{BC}_4Si_4O_{22}$, where **A** = REE, Th, Ca, Sr, Na, K; **B** = Fe^{2+}, Mg, Mn, Ca; and **C** = Ti, Mg, Mn, Fe^{2+}, Fe^{3+}, Al. They present phase boundaries for the structural transitions as a function of the average size of the ions occupying the various structural sites, and note, "The results of the study tend to confirm that the phase change chevkinite-perrierite is controlled primarily by composition in that the size of the **B** and **C** cations exerts a strong control over the structure. In the chevkinite structure very large cations can therefore only be accomodated in the **A** positions either by increasing the average size of the **B** and **C** cations to open up the structure or by distorting it to give perrierite symmetry."

Bonatti (1959) noted the interesting crystallographic similarities between chevkinite, perrierite, and the epidotes, particularly allanite.

Chevkinite-(Ce)

$(Ce,La,Ca,Na,Th)_4(Fe,Mg)_2(Ti,Mg,Fe)_3Si_4O_{22}$

Chevkinite was described by Rose (1842) based on material brought by K. I. Lisenko from near Miask in the Ilmen Mountains, Russia, although the exact locale is not known. More material was found around 1916, as abundant, well-shaped crystals, at Pit No. 17, Ilmen Mountains (Pekov 1998).

Chevkinite-(Ce) forms anhedral to subhedral masses up to at least 10 cm, or short prismatic crystals with {100}, {001}, and {201} dominant, {110} minor, elongated on [010], or sometimes twinned on {001}. The crystals are black, dark brown, or dark reddish-brown with a vitreous or adamantine luster. Occasionally a very poor cleavage on {001} is seen; more often it shows conchoidal fracture typical of metamict materials.

The mineral is a widespread but minor accessory mineral in volcanic ash beds and rhyolites; it is more commonly an accessory in alkalic or peralkaline granites, granite pegmatites, nepheline syenites, and syenitic pegmatites.

It occurs in trace amounts in many silicic ash beds interbedded with terrestrial rocks of Eocene to Quaternary age (Young and Powers 1960). In a systematic study, Izett and Wilcox (1968) reported that, "Chevkinite was found in pumice fragments and airfall ash associated with large ash-flow sheets from several areas of Cenozoic volcanism in the western United States including Yellowstone Park, Wyoming, Jemez Mountains area, New Mexico, and southern and central Nevada. Several ash beds contain both chevkinite and allanite, but where they occur together chevkinite is generally much more abundant than the allanite. In the rocks and pumice fragments examined by the writers the chevkinite generally occurs as lustrous dark-reddish-brown, pencil-shaped euhedral crystals about 0.1 to 0.5 mm long, indistinguishable from perrierite crystals under the stereomicroscope and distinguishable only with difficulty under the petrographic microscope. Optical and X-ray powder data of chevkinite from the Pearlette Ash Member is given ... The chevkinite is geologically young 'nonmetamict' and gives excellent powder film data from unheated material."

Macdonald et al. (2012) reported analyses and site occupancies for samples of chevkinite-(Ce) from a number of different rock types in Russia and Mongolia. They noted that, "Despite the varied parageneses, the chevkinite-(Ce) compositions are similar to previously published data. Many crystals have strong internal compositional variations, partly produced during primary crystallization and partly during low-temperature hydrothermal alteration."

At the Clove Iron mine, Monroe, New York, chevkinite-(Ce) containing about 2% Th has been collected on the mine dump. The mineral forms rounded dark red-brown patches in a granitic pegmatite and is typically heavily altered and metamict.

Chevkinite occurs as a well-crystallized accessory mineral in fayalite-quartz syenite of the Devil's Slide ring dike near Stark, New Hampshire. An age determination of the mineral by the lead-alpha activity method gave 207 Ma compared with a maximum uncorrected age of 208 Ma from total Pb, U, and Th analyses. The calculated age of zircon from the same rock is 227 Ma. Jaffe, Evans, and Chapman (1956) suggest, "If the chevkinite contained a large amount of original lead of crystallization, its age would not be expected to show any agreement with the zircon age. The reasonably good agreement between both ages and the geology suggests that chevkinite may have potential value as an age mineral where it occurs as an accessory constituent of igneous rocks from other localities."

Recently, excellent specimens have been recovered from Chilas, Northern Areas, Pakistan, where lustrous black crystals

Figure 642. Lustrous black crystals of chevkinite-(Ce) to several cm long, on feldspar from Chilas, Northern Areas, Pakistan. This sample, ~6 × 10 cm across, is slightly radioactive, reading about twice background at a distance of ~1 cm. RJL2576

occur in veins in the pegmatite. Many crystals form rough intergrown clusters, often with contact damage where the vein is narrow, but from wider pockets undamaged, terminated crystals were collected. Associates include schorl, titanite, feldspar, and calcite.

Chevkinite-(Ce) is monoclinic; Sokolova et al. (2004), using annealed material from Mongolia, give a=13.400, b=5.723, c=11.057 Å, and β=100.537°. They note that IR spectra of unannealed crystals suggest significant radiation-induced damage. Calvo and Faggiani (1974) compared the structures of chevkinite and perrierite using synthetic materials.

Perrierite-(Ce) $(CeCa,Th)_4(Fe^{2+},Mg)Fe^{3+}_2(Ti,Fe^{3+})_2(Si_2O_7)_2O_8$
Perrierite-(Ce) was described from tuffaceous beach sands at Nettuno, Roma, Italy, where it occurs as grains and crystals up to 0.2 mm, which somewhat resemble epidote. It is fairly rare, ~0.1 %, but can easily be recognized by its marked pleochroism. The monoclinic crystals are prismatic, with a:b:c=2.047:1:2.380 and β=113°28' (Bonatti and Gottardi 1950).

Segelstad and Larsen (1978) reported the structure of perrierite from Norway as monoclinic with a=13.70, b=5.66, c=11.83 Å, and β=113.78°. These values are in good agreement with those of Bonatti (1959) for the type material: a=13.61, b=5.62, c=11.63 Å, and β=113.47°. One can also see the generally good agreement with the observed axial ratios derived from morphological studies reported initially by Bonatti and Galli (1950) (with a and c reversed).

Steenstrupine Series
The steenstrupines are complex REE-Th silicophosphates. Pekov, Yekimenkova, and Kononkova (1997) analyzed thorosteenstrupine from the Karnasurt mine, Lovozero Massif, and suggested that a continuous series extends between steenstrupine-(Ce) and thorosteenstrupine. They found compositions ranging continuously from $(REE_{5.40}Th_{.89})$ to $(Th_{3.81}REE_{2.59})$.

Steenstrupine-(Ce) $Na_{14}Ce_6Mn^{2+}Mn^{3+}Fe^{2+}_2(Zr,Th)(Si_6O_{18})_2(PO_4)_7{\cdot}3H_2O$
Steenstrupine was described from the Julianehaab district, Greenland, where it occurs in sodalite-syenite at Kangerdluarsuk, Ilimaussaq complex. It forms brown crystals and masses, associated with lepidolite and aegirine. In his original report, Lorenzen (1882) noted, "As the separation of thoria from the oxides of the cerium-group which is effected by sodium sulfates is not exact, and as moreover the composition of the mineral is very intricate, I shall not attempt to deduce any formula, until more analyses have been instituted."

Steenstrupine-(Ce) is typically found as masses and grains; crystals tend to be rhombohedral. Lorenzen noted some similarity to eudialyte crystals. This mineral has been noted at several localities in the Ilimaussaq complex, including at the TL and at Tunulliarfik. Other reported locales include: in sodalite xenoliths at Mont St.-Hilaire, Quebec; and as crystals to 4 cm in ussingite at the Umbozero mine, Mount Alluaiv, and at Mount Karnasurt, both in the Lovozero massif, Kola Peninsula, Russia.

Thorosteenstrupine $Na_{0\text{-}5}Ca_{1\text{-}3}(Th,REE)_6(Mn,Fe,Al,Ti)_{4\text{-}5}(Si_6O_{18})_2[(Si,P)O_4]_6(OH,F,O)_x{\cdot}H_2O$
Thorosteenstrupine was found at the Chergilen REE deposit on the southeastern slope of the Turana Range, left bank of the Verkhnii Mel'gin River, 60 km northwest of the town of Chekunda, Khabarovsk Territory, Eastern Siberia (Pekov 1998). It was described by Kuprianova, Stolyarova, and Sidorenko (1962) as forming dark brown to nearly black, platy crystals, typically 2–5 mm but occasionally up to 1 cm long, with rough faces. It has a greasy to vitreous luster and conchoidal fracture. It occurs in metasomatic veins, associated with albite, microcline, quartz, aegirine-augite, fluorite, thorite, and REE-rich miserite.

Natural material is X-ray amorphous and isotropic; after heating to 900°C for thirty minutes, the powder XRD pattern is different from those of steenstrupine-(Ce) and thorite. Further work by Pekov, Yekimenkova, and Kononkova (1997) supported the validity of thorosteenstrupine as a distinct mineral with the general formula given above.

Thornasite $Na_{12}Th_3[Si_8O_{19}]_4(H_2O)_{18}$
Thornasite was described from the DeMix quarry, Mont Saint-Hillaire, Quebec, where it forms anhedral, colorless to very pale green grains, embedded in brockite and associated with yofortierite, eudialyte, albite, and serandite. It displays a bright green fluorescence in SW and LWUV (Ansell and Chao 1987).

The structure is hexagonal (rhombohedral) with a=29.124 and c=17.260 Å. The structure consists of an interrupted framework with composition $[Si_8O_{19}]^{6-}$. The framework density is 15.4 atoms per 1000 Å^3, which is typical for

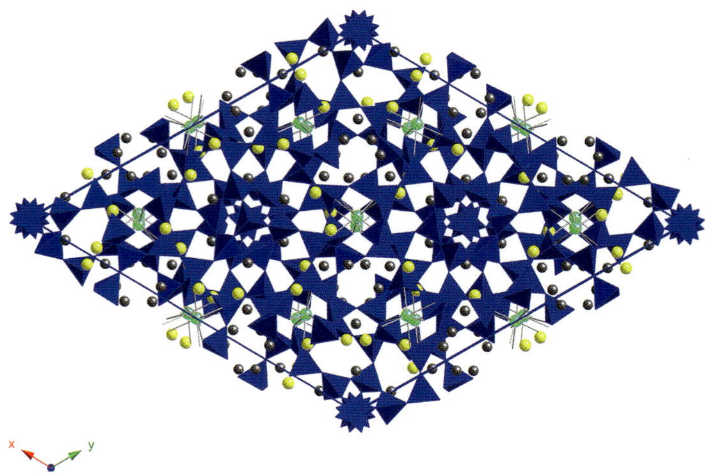

Figure 643. The structure of thornasite, viewed along [0001], showing the interrupted zeolite-like silicate framework, containing channels and cavities occupied by Na (yellow atoms) and water (gray). Th (green atoms with stick bonds) are coordinated to the apical oxygens on multiple tetrahedra in the interrupted framework.

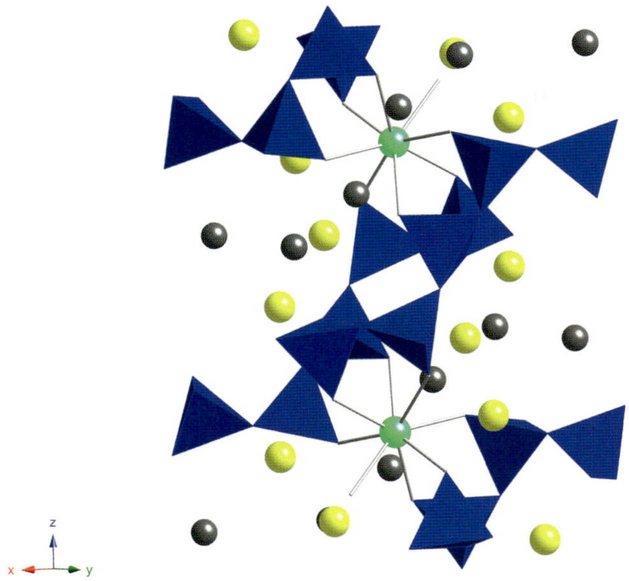

Figure 644. A detail to show how the free corners of tetrahedra in the interrupted framework are taken up by bonding to the Th ions. It isn't known with certainty whether or not the extra-framework Na ions are exchangeable with surrounding solutions, as in typical zeolites; however, at least some of the water appears to be zeolitic based on dehydration behavior reported by Ansell and Chao (1987).

microporous aluminosilicate materials. The framework density, together with the presence of a three-dimensional network of channels in the structure, permit thornasite to be designated as a new member of the zeolite structure family. The framework consists of cubic cages and distorted six-membered rings linked via common edges (Li, Krivovichev, and Burns 2000).

Karnasurtite-(Ce) (Ce,La,Th)(Ti,Nb)(Al,Fe)(Si_2O_7)$(OH)_4 \cdot 3H_2O$

Karnasurtite-(Ce) was discovered in 1947 at the Hackmanite Stock (Pegmatite No. 62), Mount Karnasurt, Lovozero massif, Russia (Pekov 1998). It forms individual plates or grains to ~1 cm and aggregates to 10 × 6 × 2 cm. The color is honey yellow when fresh, becoming pale yellow when altered; it has a greasy luster and is not fluorescent. Associated minerals include microcline, schizolite, natrolite, and epididymite; karnasurtite is sometimes replaced by poltlithionite. It also occurs nearby on Mount Punkuruaiv (Kuz'menko and Kozhanov 1959).

This mineral was called "kozhanovite" by Shilin (1956) in a paper on an unrelated mineral. No description was given, and the name "kozhanovite" was later officially discredited (International Mineralogical Association 1962).

Natural material is amorphous; after heating to 900°C, the powder XRD pattern is similar to that of huttonite (Kuz'menko and Kozhanov 1959).

Umbozerite $Na_3Sr_4ThSi_8(O,OH)_{24}$

Umbozerite was described from the alkaline complex at Mt. Karnasurt, Lovozero massif, Kola Peninsula, Russia, where it forms irregular masses and crude tetragonal crystals to ~1 × 3 mm in ussingite veinlets. Associated species include sphalerite, belovite, schizolite, and Nb minerals of the lomonosovite group. Individual needles are bottle green when fresh, turning greenish brown to brown on exposure. It is normally metamict and optically isotropic; after heating to 900°C the material gives a powder XRD pattern with many similarities to that of thorianite, but with additional diffraction peaks (Es'kova et al. 1974). It is named for Umbozero Lake, which lies between the Khibiny and Lovozero massifs (Pekov 1998).

Although umbozerite was originally described as a mineralogical rarity at Mt. Karnasurt, it was later shown to be a relatively widespread accessory mineral in hyperagpaitic rocks of the Lovozero massif. According to Khomyakov (1995): "This mineral has been observed in many ussingite and natrolite veins in the same area and at Mt. Kedykverpakh. This enabled us to gain a fuller picture of the modes of occurrence of this mineral and to supplement the data on its composition. . . . Most commonly, the mineral is found in the axial region of ussingite-natrolite veinlets or close to their selvages as composite aggregates made up of droplike grains 1–5 mm in diameter, surrounded by radiating acicular crystals. . . . Optically, it is isotropic, with n=1.640 . . . umbozerite from natural outcrops and opencast workings has n≈1.60 and, presumably, differs in composition from the fresh material."

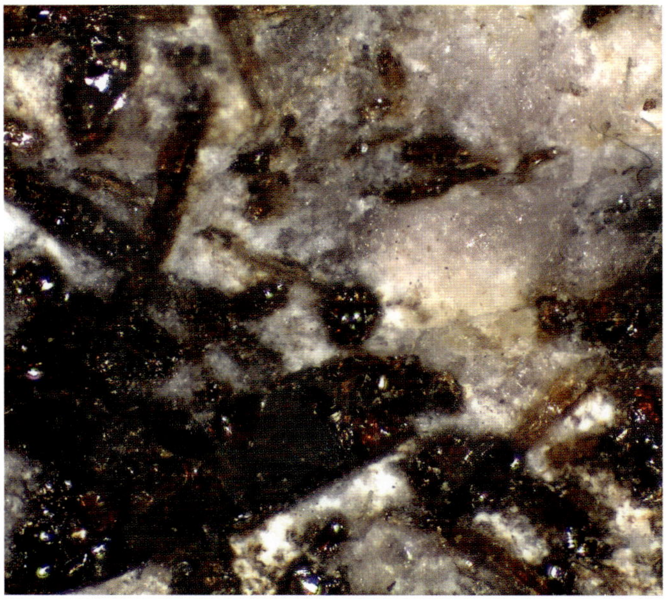

Figure 645. Umbozerite from Mt. Panquruaiv, Kola Peninsula, Russia, forming glassy brown grains and elongated crystals. FOV is ~4 mm wide. RJL1495

Ciprianiite $Ca_4[(Th,U)(REE)]Al\square_2(Si_4B_4O_{22})(OH,F)_2$
Cipriianiite is a mineral of the hellandite group, described from pyroclastic formations of the Roman Comagmatic Province, Latium, Italy (Della Ventura et al., 2002), where it is found within alkali-syenitic ejecta. The mineral forms tiny (under 0.5 mm) transparent brown euhedral crystals in miarolitic cavities and voids of the ejecta, which mainly consist of sanidine and plagioclase. Accessory minerals include britholite-(Ce), baddeleyite, thorite, fluorite, danburite, vonsenite, and other phosphate and silico-phosphate phases. It has a vitreous luster and is not fluorescent.

Its formation mineral is related to late-stage post magmatic hydrothermal fluids enriched in Zr, Ti, REE, and actinides.

Ciprianiite is monoclinic with a=19.059, b=4.729, c=10.291 Å, and β=111.32°. It is structurally similar to mottanaite-(Ce), another new member of the hellandite group discovered at the same time; because both were non-metamict, the occurrence afforded an opportunity to study the crystal chemistry of the hellandite group. Along with a critical re-evaluation of earlier data, this work clarified the role of vacancies and OH groups in the structures, the relation of hellandite to tadzhikite, and the general crystal chemistry of the group. Oberti et al. (2002) presented the re-defined nomenclature of the group and ideal formulas for the five currently-accepted members, of which only ciprianiite contains essential (Th,U).

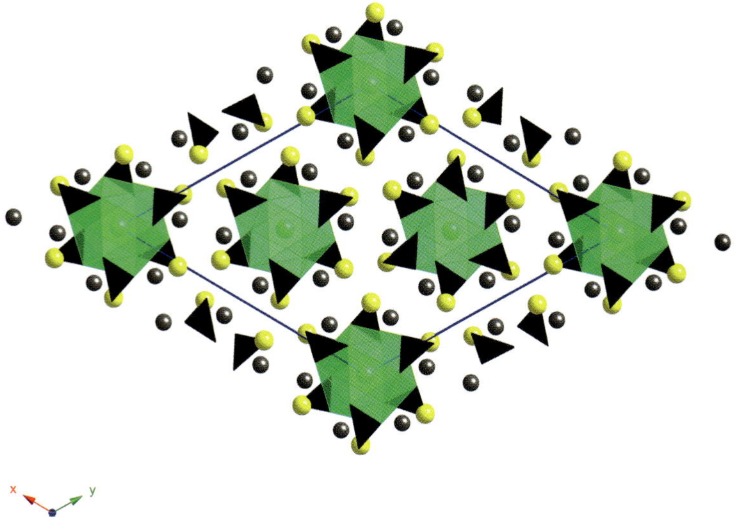

Figure 646. The tuliokite structure, viewed along [0001], showing the icosahedral coordination around the Th atoms (green), with each CO_3 group sharing two oxygens with the Th. Na (yellow atoms) are coordinated to water (gray) and to the remaining oxygens on the CO_3 groups.

Thorbastnäsite $Th(Ca,Ce)(CO_3)_2F_2{\cdot}3H_2O$
Thorbastnäsite was discovered in 1956 at the Pichikhol' alkaline massif, Balygtyg-Khem River, Sangilen Upland, Tuva, Siberia (Pekov 1998). It occurs in exocontact alkaline metasomatic rocks of a nepheline syenite intrusive; associated minerals include rinkite, zircon, pyrochlore, euxenite-(Y), and thorite. It forms brown, cryptocrystalline masses, pseudomorphous after Fe-rich thorite; its formation is attributed to the reaction of thorite with fluocarbonate solutions (Pavlenko et al. 1965).

It has also been reported from Point of Rocks, Colfax Co., New Mexico; and from Mont St.-Hilaire, Quebec. Thorbastnäsite is hexagonal, with a=6.99 and c=9.71 Å.

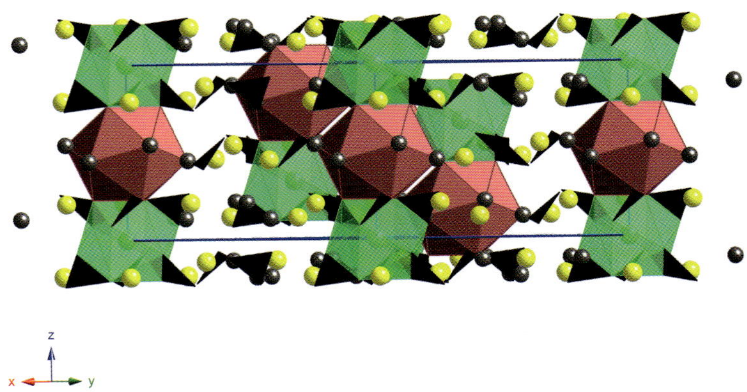

Figure 647. The structure viewed normal to the z-axis, showing the Ba (magenta) sharing three oxygens each with adjoining Th polyhedra and also coordinated to six water groups.

Tuliokite $Na_6BaTh(CO_3)_6{\cdot}6H_2O$
Tuliokite was described from the Kirovskii apatite mine, Mount Kukisvumchorr, Khibiny massif, Russia. It forms gray, prismatic crystals to ~4 mm, which have a vitreous luster and are not fluorescent. At the TL, it occurs in two somewhat different parageneses: In the first, it is associated with nepheline, cancrinite, aegirine, sidorenkoite, vinogradovite, villiaumite, and microcline, in a pegmatite vein 2–10 cm thick. In the second, it is associated with pirssonite, shortite, trona, thermonatrite, natron, and villiaumite in natrolite-aegirine-microcline veinlets 5–7 cm thick (Yakovenchuk et al. 1990).

Tuliokite is trigonal, with a=14.176 and c=8.605 Å. Each Th polyhedron is an icosahedron in which six CO_3 groups each share two oxygens with the Th. Each Ba icosahedron shares three oxygens each with the Th polyhedra directly above and below in the z-direction, with the other six apices coordinated to water groups. Na atoms are coordinated to water groups and to the remaining oxygens of the CO_3 groups (Yamnova, Puscharovskii, and Voloshin 1990).

Others

Vapnikite Ca_3UO_6
Vapnikite was described from Jabel Harmun in the Judean desert, West Bank, Palestinian Autonomy, Israel. It forms microscopic (20–30 μm) yellow-brown xenomorphic grains,

wedged between larnite, brownmillerite, and ye'elimite. The grains are transparent, with strong vitreous luster, and are generally not fluorescent, although a weak yellowish fluorescence is sometimes observed in both LW and SWUV (Galuskin et al. 2014).

The host rock is a larnite-rich pyrometamorphic pseudoconglomerate of the Hatrurim formation. The larnite rocks were formed by pyrogenic metamorphism caused by caustobiolith combustion at T > 800°C; a description of the geology of this locality and the hypotheses for how it was formed are given by Gauskina et al. (2014).

Vapkinite is the first naturally occurring example of an ordered double perovskite of the $\mathbf{A_2BB'}O_6$ type, in which **A**,**B** = Ca and **B'** = U^{6+}. A number of other ordered, U-bearing double perovskites have been synthesized. Vapnikite is monoclinic, with a=5.739, b=5.951, c=8.312 Å, and β=90.4°. It is one of the few U^{6+} minerals in which the coordination of the U doesn't follow the typical pattern in which the bonds to the "uranyl oxygens" are shorter than those to the other surrounding oxygens as shown in Figures 114–115. Instead, the U^{6+} occupies a small octahedral site.

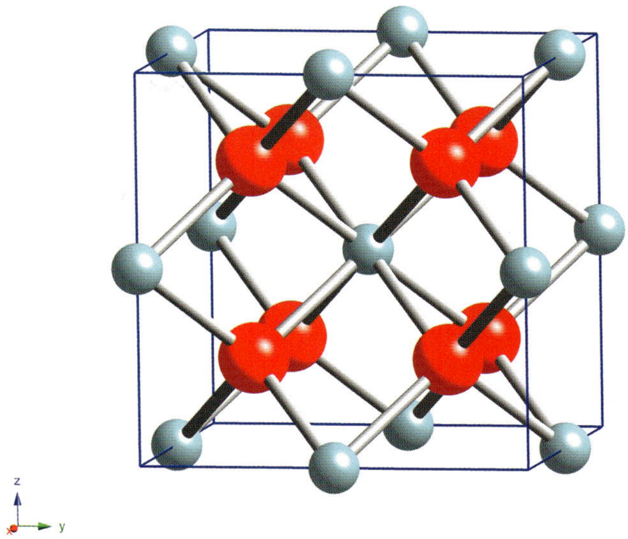

Figure 648. Ball and stick model of the structure of vorlanite, with red atoms representing oxygen and blue atoms representing the M site, occupied randomly by Ca and U atoms.

Vorlanite $CaU^{6+}O_4$

Vorlanite was described from Mount Vorlan, Upper Chegem caldera, in the Northern Caucasus, Kabardina-Balkaria, Russia, where it occurs in a calcareous skarn xenolith in ignimbrite. It forms aggregates of black, platy crystals up to 0.3 mm long, as an accessory mineral in a skarn composed of mm-sized lenses of larnite with rondorfite and wadalite inclusions. The larnite is partially replaced by chegemite, reinhardbraunsite, and minerals of the fluorellestadite-hydroxylellestadite series. Vorlanite is locally associated with elbrusite-(Zr), U-bearing lakargiite, perovskite, and srebodolskite. The assemblage is interpreted to have formed at 800-1000°C and shallow depth (i.e., low pressure) (Galuskin et al. 2011).

Vorlanite also occurs in larnite pyrometamorphic rocks of the Hatrurim formation, Jebel Harmun, Israel. There, vorlanite crystals are dark gray with a greenish hue in transmitted light. The unusual color of this material, compared to that of the type material, is assumed to be related to the presence of small amounts of U^{4+} (Galuskin et al. 2013).

Vorlanite has also been noted at the Nopal I uranium deposit (Sierra Peña Blanca, Mexico). There, it exhibits a cubic morphology, in agreement with its crystal structure. Vorlanite occurs as nanoscale crystals embedded in U-bearing opal, with a Ca/U ratio of ~1. Othmane et al. (2013) note, "Association with opal suggests that vorlanite formed at Nopal during late-stage U-mobilization under oxidizing conditions and low (<50°C) temperature. The presence of nanoscale uranate crystals in an environment largely dominated by uranyl silicates indicates that uranates may play a role in uranium scavenging at low temperature. In addition, the occurrence of vorlanite in the crystal shape consistent with its structure provides unique information on its conditions of formation."

Vorlanite is isometric, with a=5.3813 Å. The mineral is interesting in that it is isostructural with fluorite and uraninite, and is black (dark red in thin section), which is unusual for minerals containing hexavalent uranium The structure can be derived from that of uraninite, UO_2, via the replacement $Ca^{2+} + U^{6+} \rightarrow 2U^{4+}$, with Ca,U distributed randomly, i.e., Ca and U are not ordered on the **M** sites and the average **M**-O bond length is 2.33 Å.

Synthetic $CaUO_4$ has a rhombohedral structure, with an ordered arrangement of Ca,U on the M site. Furthermore, vorlanite crystals in the type specimens show trigonal morphology, suggesting that they are actually pseudomorphs (or more specifically paramorphs) after rhombohedral $CaUO_4$, with the phase transition possibly driven by radiation damage (Galuskin et al. 2011). This argument is supported by annealing experiments in which vorlanite has heated to temperatures above 750°C and transformed irreversibly to rhombohedral $CaUO_4$, indicating that the rhombohedral phase crystallized first in the skarn and became the precursor from which vorlanite formed. Galuskin et al. (2012) note, "Vorlanite most probably originated due to pseudomorphic transformation of rhombohedral $CaUO_4$, caused by disordering of cations and uranyl-bonds that resulted from α-decay events of uranium. The ease of the transition from rhombohedral $CaUO_4$ to vorlanite and the fast reversal transition during heating can be explained by the similarity of both structures. Formation of vorlanite prevents $CaUO_4$ from undergoing metamictization."

MINERALS FREQUENTLY CONTAINING U OR TH AS IMPURITIES

The likelihood that a mineral will contain radioactive elements as direct substitutional impurities will depend on several factors:

First, the structure of the mineral must be compatible with relatively large cations of high charge. Table 56 presents ionic radii of selected cations for comparison to those of U, Th, and Ra. One can see, for example, that Hf^{4+} and Zr^{4+} are somewhat smaller than U^{4+}, so this substitution would require a significant accommodation if Zr or Hf are octahedrally coordinated. On the other hand, zircon is somewhat unusual in having Zr^{4+} in a very large, 7-coordinated site that provides sufficient space for U or Th. The Ca^{2+} ion is virtually the same size as U^{4+}, but the charge difference requires that a mechanism for charge compensation be available, typically by a coupled substitution such as $U^{4+} + \square \rightarrow 2Ca^{2+}$ or $U^{4+} + 2Na^{+} \rightarrow 3Ca^{2+}$. The trivalent REE ions have a reasonably good size match with U^{4+} and Th^{4+}, and the charge difference is only 1+, so the frequent substitution of U,Th in REE minerals is not surprising.

Table 56. Ionic radii for selected species

Cation	Radius, Å
Hf^{4+}	0.78
Zr^{4+}	0.79
U^{6+}	**0.80**
Y^{3+}	0.89
Eu^{3+}	0.95
U^{4+}	**0.97**
Ca^{2+}	0.99
Th^{4+}	**1.02**
La^{3+}	1.02
Ce^{3+}	1.03
Sr^{2+}	1.12
Pb^{2+}	1.20
Ba^{2+}	1.34
Ra^{2+}	**1.43**

Second, there must be geochemical conditions that will bring U or Th together with the other components from which a particular mineral will crystallize. One example of favorable conditions for substitution is a rare-element pegmatite in which U^{4+} and/or Th^{4+} are present along with Zr, REE, etc. As an example of unfavorable conditions, most common minerals that crystallize under atmospheric levels of oxygen will be very unlikely to incorporate U because under these conditions, uranium will form the $(UO_2)^{2+}$ ion, which is far too large to substitute for divalent metal cations in typical crystal structures. This is why one can see relatively pure gypsum crystallizing side-by-side with zippeite, because the uranyl ion can't be accommodated on the Ca^{2+} site in gypsum.

The examples that follow will illustrate these principles at work in various ways.

Allanite group

Allanite was named for the Scottish mineralogist Thomas Allan (1777–1833) based on type material from Greenland. Under modern nomenclature, that material is allanite-(Ce). Later workers described additional compositions that are now denoted allanite-(La) and allanite-(Y). Allanites are a common accessory mineral in granite, granodiorite, monzonite, syenite, and granitic pegmatite. They are also found in diorite and gabbro, as phenocrysts in acid volcanic rocks, and in a number of metamorphic rock types.

Many allanites contain uranium and thorium in addition to rare earth elements. According to the literature reviewed by Gieré and Sorensen (2004) the maximum reported ThO_2 content is 4.9 wt. %, which would correspond to about 0.07 Th apfu. The maximum UO_2 content is 0.82 wt. % in a crystal that also contained 1.09 wt. % ThO_2, corresponding to 0.02 apfu of each. Thus, the actinide (ACT) elements Th and U are relatively minor constituents in these minerals, and for classification purposes the sum of ACT and REE is used. The recommended chemical criterion to assign epidote-group minerals to the allanite and dollaseite subgroups is that REE + ACT > 0.5 apfu. The justification for bundling U and Th with REE is that the coupled heterovalent substitution mechanism is presumably: $(Th^{4+},U^{4+}) + 2\,M^{2+} \rightarrow Ca + 2\,M^{3+}$.

Some localities for well-analyzed U- and Th-rich allanites include: in pegmatite in granite, Kleben, Ukraine; in microcline pegmatite, Wilmot Pass, New Zealand; in hornblende-biotite granodiorite, Boulder Creek batholith, Colorado; in migmatitic gneiss, Pietrosu valley, Romania; in biotite granite, Daimonji-yama, Kyoto Prefecture; with zircon, xenotime, and monazite in pegmatite at Uchino-mura, Fukuoka Prefecture, Japan; at the contact of calc-granulite with microcline pegmatite,

Garividi, India; in plagioclase-biotite-hornblende gneiss, Shnipka, Belarus; in charnockite at Kondapalli, India; and Mn-rich in pegmatite at Ohari, Miyagi Prefecture, Japan (Deer, Howie, and Zussman 1986, Table 7, pp. 154-63)

Through decay of the radioactive elements many REE-bearing epidote minerals are partly or completely metamict. As noted earlier, metamict minerals are typically annealed to "heal" the radiation damage and thereby restore the crystal structure for X-ray analysis. The problem is that a metamict mineral is fairly reactive in the environment and over the years it may adsorb soluble elements or undergo ion exchange. Thus, the "restored" crystal might have a composition that differs significantly from that of the original crystal and in effect becomes a synthetic material. As stated by Armbruster et al. (2006): "There is at least some suspicion that such 'mineral' compositions are influenced by the experimenter and are not an unaltered product of nature. These problems are not specific of epidote-group minerals but are much more prominent in other mineral groups with higher concentrations of radioisotopes. . . . We recommend exercising caution with compositions of 'partly' metamict epidote-group minerals in naming new species, even if the 'faulty' lattice has been mended by subsequent heat treatment."

Columbite-Tantalite Series

Minerals of this group are orthorhombic niobates and tantalates with the general formula $\mathbf{AB_2O_6}$ where **A** = Fe, Mn; and **B** = Nb, Ta. Solid solution series exist between the end-member compositions.

These species are widespread accessory minerals in granite pegmatites. Radioactive varieties are known from several localities. In the uraniferous pegmatites of Brazil (Robinson and Wegner, 1998), the garimpeiros reportedly use smoky quartz as a guide to locate columbite and tantalite crystals; it is presumed that these crystals were the source of radioactivity that darkened the quartz. Ferrocolumbite from Ambatofotsikely, Madagascar, may contain several percent uranium. Older reports of varieties with as much as 10% uranium, for example, the "toddite" described by Ellsworth (1926) may in fact represent admixtures or fine intergrowths of several species, e.g., euxenite.

Fergusonite-Formanite Series

These minerals comprise a group of tetragonal rare earth tantalates and niobates. The general formula is $\mathbf{ABO_4}$ where **A** = Y,REE; and **B** = Nb (fergusonite) and Ta (formanite). Small to significant amounts of U and Th can replace REE at **A**. Fergusonite-beta is a related series of monoclinic minerals, dimorphous with fergusonite.

Varieties of fergusonite containing 0.8 to 7.2 wt. % UO_2 and 0.7 to 6.0 wt. % ThO_2 have been reported from Hakatamura and Naegi in Japan; Baringer Hill, Texas; Helle, Norway ("*bragite*"); and the Gole quarry, near Madawaska, Ontario. A variety of formanite containing 1 wt. % UO_3 and 1 wt. % ThO_2 occurs at Cooglegong, Western Australia.

Zircon

Zircon is a common accessory mineral in granitic and syenitic igneous rocks. Because of its hardness and chemical stability it is also a common detrital mineral. In granitic rocks, U and Th are often concentrated in the zircon because its crystal structure contains a very large coordination polyhedron around the Zr ion, and indeed zircon is isostructural with thorite. It is the only common rock-forming silicate that has such a structure hospitable to U^{4+} and Th^{4+} ions, so it effectively scavenges these elements from a crystallizing melt. U- and Th-rich varieties of zircon have been given names such as *cyrtolite*, *malacon*, *orvillite*, etc. The radioactive varieties are often partly or completely metamict from internal radiation damage; metamict forms tend to have lower density and are more easily altered by corrosive solutions (Deer at al. 1982). Metamict zircons can generally be restored to the normal crystal structure by heating to 1000–1450°C.

Fine crystals of zircon (var. *cyrtolite*) found at several sites near Bancroft, Ontario, typically contain 1–2 wt. % UO_2 and 4-5 wt. % H_2O. They are often lighter colored than other zircons and typically have curved faces. Well-formed crystals are found with fergusonite at the John Gole quarry. Doubly terminated crystals to 1.5 cm are found along with betafite in black humus in the Silver Crater Mine. Fine, large crystal clusters have been collected from the Davis nepheline quarry in Monteagle Township (Kennedy, 1979).

Because the zircon structure so readily accommodates U and Th, radioactive zircons occasionally turn up in the gem trade, and might show radiation levels of 1,300 cpm over background and some degree of metamictization of the crystal structure. Highly metamict material ("low zircon") is typically dark green. Such stones can be quite attractive, but are not recommended for jewelry that is to be worn regularly!

Figure 649. Detail of the radioactive zircon var. cyrtolite shown in Figure 26. This cabinet-sized specimen, from the Davis mine, Bancroft, Ontario, weighs 660 g and shows ~1360 cpm over background, measured at contact with the handheld Geiger counter. RJL2660

Figure 650. A suite of faceted zircons with a pleasing range of colors, totaling ~30 cts., selected from a parcel of stones discovered to be slightly radioactive. Individual stones show ~300 to 500 cpm over background; as expected, the green stones show higher activity than the brown ones. Although these are not highly radioactive, a conscientious jeweler using the conservative threshold of 150 cpm would therefore avoid selling these to a retail customer.

Opal and Agate

Many opals, particularly hyalite, contain enough uranium to show bright UV fluorescence. The actual U content might range from 0.000048% to as high as 0.12%. Heinrich (1958) notes that, "The U is not in the form of crystalline inclusions, but, having been adsorbed early in the process of the accumulation of siliseous gels, is highly dispersed, probably commonly as colloidal uranyl sulfate, phosphate, or silicate."

One can assume that the same mechanism causes the green UV fluorescence seen in agates from many localities.

Figure 651. Globular, colorless hyalite opal from Square Top Mountain, near Dalby, Queensland, Australia, covering the face of a matrix ~5 cm wide. RJL2011

Figure 652. Specimen in the previous photo, viewed under SWUV light.

Figure 653. A Crater agate from Patagonia, Argentina, illustrating the usual combination of translucent dark gray chalcedony highlighted by intense red layers of hematite. RJL2233

Figure 654. The sample in the previous photo, viewed in SWUV light. RJL2233

Xenotime

Minerals of the xenotime series, (REE)PO_4, are tetragonal and isostructural with zircon. Although many show little or no radioactivity, compositions up to ~5% UO_2 and up to 3% ThO_2 have been reported (Heinrich 1958).

Rare earth carbonates

According to Heinrich (1958), "most of the rare-earth carbonate minerals . . . are known to contain very small amounts of Th or Th and U." These include: ancylite, 0.2% ThO_2; bastnäsite, <1% Th+U and up to 0.39% Th; cordylite, 0.3% ThO_2; doverite, 1.62% ThO_2; parisite, trace to 0.002% U and Th; synchisite, 0.3% ThO_2, and tengerite, 0.3% ThO_2.
Mckelveyite from the TL in the Green River formation, Wyoming, contains 4.6% UO_2 (Milton et al. 1965). Some references inlude U in the formula as an essential constituent of this mineral, whereas others do not.

Figure 655. A thumbnail-sized crust of radioactive barite ("hokutolite") from Tamagawa hot springs (Shibukuro hot springs), Shibukuro River, Tazawako, Akita Prefecture, Tohoku Region, Honshu Island, Japan. Hokutolite is getting hard to obtain in Japan, as the hot spring is now a "scientific monument" and off-limits to collecting. The sample was no doubt fairly radioactive when it was formed, but now shows virtually no detectable activity as measured by the hand-held Geiger counter. RJL4634

Barite

"Radiobarite," a Pb-rich variety of barite presumably originally containing radium, along with U and Th, occurs in hot spring deposits at Hokuto, Taiwan, and at Shibukuro, Japan (Ohashi 1920). It is given the varietal name *hokutolite* from the locality. Freshly deposited material containing traces of radium is quite radioactive, but owing to the short half-life of radium, samples from old collections show little or no activity. Spring-deposited barite from Teplitz and Karlsbad, Czech Republic, contain 2 to 150 ppm U. Other locales include: Tyuya Muyun, Kyrgyzstan; the Bitter Creek mine, Colorado; in a tuffaceous spring deposit in the Limestone Hills area, Kiowa Co., Oklahoma; and at La Baja, Colombia. In some of these deposits it is possible that some or perhaps all of the U, Th, or Ra is adsorbed onto the surfaces of the crystallites rather than being incorporated directly in the barite crystal structure (Heinrich 1958).

Titanite

Titanite (sphene) sometimes contains small amounts of REE (up to a few wt. %). It is occasionally metamict, and samples containing ThO_2 and/or UO_2 have been reported in beach sand at Wainui Inlet, N. W. Nelson, New Zealand; and in pegmatite near Bystrice nad Pernštejnem, Moravia, Czech Republic. Metamict titanite from Pierreville, New York, contains "significant" Th and U, in addition to radiogenic lead (Higgins and Ribbe 1976).

Apatite

Radioactive apatites fall mainly into two categories: REE-bearing types and marine carbonate-fluorapatite. Primary igneous apatite typically contains 0.00x% U, whereas late-stage igneous apatite might contain 0.0x% U (Heinrich 1958).

A Th-bearing variety, *abukumalite*, occurs with euxenite in pegmatite at Iisaka, Abukuma range, Japan. REE-rich fluorapatite containing 0.15% ThO_2 and 0.32% U occurs in magnetite ore at Mineville, New York.

Carbonate-fluorapatite deposits containing U are well known, e.g., as cement and fracture fillings in sandstones and conglomerates that host the ore at the Lucky Mc mine, Wyoming. Some other locales include the Phosphoria formation of Montana, Idaho, Wyoming, and Utah, and the Bone Valley formation, Florida.

It is important to note that in many of these deposits, some U might represent microscopic particles of other mineral phases intermixed with the apatite. Baumer et al. (1983) synthesized chlorapatite with controlled additions of Mn, Ce, U, and Th, and showed that a few hundred ppm U could substitute directly in the apatite crystal structure, whereas U or Th could also form minute inclusions of other phases such as $CaTh(PO_4)_2$, ThO_2, and $CaU(PO_4)_2$ within the apatite crystals.

Several studies, using synthetic compositions, have noted that the apatite structure contains two different Ca sites, which form significantly different coordination environments. These differences are reflected in the site preferences of the various impurities (Fleet and Pan 1995; Fleet, Liu, and Pan 2000). In U-doped fluorapatite, U substitutes almost exclusively onto the Ca2 site and Th dominantly substitutes on Ca2 as well. In chlorapatite, Th substitutes on both the Ca1 and Ca2 sites. The preference of U and Th for both Ca sites in chlorapatite is attributed to the large increase in size and distortion of the Ca2 polyhedron caused by the substitution of Cl for F (Luo et al. 2009).

Figure 656. Dark red apatite containing ~2,000 ppm ThO_2, intergrown with magnetite, from Mineville, New York. Sample is ~1 cm thick, top to bottom.

APPENDIX 1

OBSOLETE NAMES, VARIETAL NAMES, AND SYNONYMS

Absite = Th-rich **brannerite**
Aldanite = U- and Pb-rich **thorianite**
Aluminobetafite = Al-rich **betafite**(?)
Ampangabeite = **samarskite-(Y)**
Annerodite = **samarskite-(Y)** with parallel overgrowths of columbite
Anthraxolite = a nickeliferous and uraniferous hydrocarbon
Arsenuranocircite = **metaheinrichite**
Auerlite = PO_4-rich thorite or thorogummite(?)[a]
Barium-phosphuranylite = **bergenite**
Bariumphosphoruranit = **uranocircite**
Beta-uranopilite = **metauranopilite**
Blomstrandite = possibly **uranpyrochlore** or **betafite**
Bragite = **fergusonite**
Bröggerite = Th-rich **uraninite**
Calciocarnotite = **tyuyamunite**
Calciosamarskite = metamict U-rich **yttropyrochlore**(?)
Calciothorite = Ca-rich thorite or thorogummite(?)[a]
Calciumarsenuranit = **uranospinite**
Calciumphosphoruranit = **autunite**
Carbocer = a material similar to "thucholite", high in Ce
Caryocerite = Th-rich **melanocerite-(Ce)**(?)
Ceruranopyrochlore = Ce-rich **pyrochlore**
Chalcolamprite = pyrochlore
Chalcolite = **torbernite**
Chalco-uranite = **autunite**
Chinkolobwite = **sklodowskite**
Chlorothorite = thorogummite[a]
Cleveite = REE-rich **uraninite**
Copper uranite = **torbernite**
Coracite = **uraninite**
Cordobaite = **brannerite**
Cousinite = possibly a Mg-umohoite
Cupferphosphoruranite = **torbernite**
Cuprouranit = **torbernite**
Cryptolite = **monazite**
Dakeite = **scroekingerite**
Dauberite = possibly **zippeite**
Delorenzite = **tanteuxenite-(Y)**
Diderichite = **rutherfordine**
Djalmaite = **uranmicrolite**
Droogmansite = inadequately described, may be related to **sklodowskite**
Edwardsite = **monazite**
Eisenpecherz = **uraninite**
Eliasite = gummite
Ellsworthite = **uranpyrochlore**
Ellweilerite = Ca-free **natrouranospinite**(?)
Enalite = U-rich **thorite** or thorogummite(?)
Eremite = **monazite**
Epi-ianthinite = **schoepite**
Eschynite = **aeschynite**
Eucrasite = REE-rich **thorite** or thorogummite(?)
Eytlandite = **samarskite-(Y)**
Fenghuangite = Th-rich **britholite**
Ferghanite = possibly leached or weathered tyuymunite
Ferrothorite = altered variety of thorite; high in Fe
Ferutite = **davidite**
Flutherite = **liebigite**
Freyalite = REE-rich **thorite**
Fritzscheite = ill-defined mineral, some specimens were **metatorbernite**
Fynchenite = Th-rich **britholite** (Russian transliteration of fenghuangite)
Gastunite = **haiweeite**
Gilpinite = **johannite**
Grahamite = a uraniferous hydrocarbon similar to "thucholite"
Gruner Glimmer = **torbernite**
Guadarramite = intergrowth of radioactive monazite and some ilmenite
Gummierz = an obsolete German name for gummite.
Gummite = general term for yellow and orange secondary uranium oxides
Hatchettolite = **uranpyrochlore**
Hjelmite = possibly a mixture of **tapiolite** and **pyrochlore**
Hokutolite = "radiobarite" containing traces of Pb and Ra
Hyblite = thorogummite[a]
Hydroeuxenite = **samarskite-(Y)**
Hydrogen autunite = **chernikovite**
Hydrosamarskite = altered **samarskite-(Y)**
Hydrothorite = thorogummite[a]
Hydroxyl-thorite = thorogummite[a]
Ianthite = **ianthinite**
Iron uranite = **bassettite**(?)
Jachymovite[c] = **cuprosklodowskite**
Janthinite = **ianthinite**
Kalk-uran-carbonat = **liebigite**
Kalk Uranglimmer = **autunite**
Kalk-Uranit = **autunite**
Kieselkupfer-Uranoxyd = **cuprosklodowskite**
Kochelite = **fergusonite**
Koppite = **pyrochlore**(?)
Korarfveite = impure **monazite**

Kryptolith = **monazite**
Kupferarsenuranit = **zeunerite**
Kupfer-Uranglimmer = **torbernite**
Kupferuranit = **metazeunerite**
Kieselkupfer-uranoxyd = **cuprosklodowskite**
Lambertite = **uranophane**
Lime-Uranite = **autunite**
Lodochnikite = **brannerite**
Lyndochite = variety of **aeschynite** that is Ca- and Th-rich and U-poor
Mackintoshite = thorogummite[a]
Magnesium-phosphoruranit = **saleeite**
Maitlandite = thorogummite(?)[a]
Manganosteenstrupine = **steenstrupine-(Ce)**
Marignacite = **ceriopyrochlore**
Mavudzite = **davidite**(?)
Medjidite = poorly defined uranium sulfates, some mixtures of known species
Mendeleyevite = Ti-rich **betafite**
Mengite = **monazite**
Metachalcolite = **metatorbernite**
Metakalkuranite = **meta-autunite**
Metakupferuranit = **metatorbernite**
Metasandbergerite = **metaheinrichite**
Metasimpsonite = **microlite**
Monazitoid = **monazite**
Nasturan = pitchblende
Natroautunite = **sodium autunite**
Nenadkevite = mixture of relict uraninite, boltwoodite, and uranium hydroxides
Nicolayite = thorogummite(?)[a]
Niobozirconolite = **zirkelite**
Nivenite = REE-rich **uraninite**
Nuevite = **samarskite-(Y)**
Oliveiraite = alteration product of euxenite
Orangite = **thorite**
Para-autunite = **meta-autunite II**
Paraschoepite = **schoepite**
Peligotite = probably **johannite**
Phosphocerite = **monazite**
Pilbarite = mixture of thorogummite[a] and **kasolite**
Pitchblende = a colloform, fine-grained variety of **uraninite**
Pittinite = gummite
Plumboniobite = Pb-rich **samarskite-(Y)**(?)
Priorite = **aeschynite-(Y)**
Risorite = **fergusonite**
Randite = a mixture of uranophane-β, tyuyamunite, and calcite
Renardite = probably **dewindtite**
Samiresite = Pb-rich metamict **uranpyrochlore**(?)
Sandbergerite = **heinrichite**
Shentulite = PO_4-rich **thorite**
Shinkolobwite = **sklodowskite**
Sipylite = **fergusonite**
Smirnovite = **thorutite**
Soddite = **soddyite**
Sogrenite = an organo-uranium complex like "thucholite"
Stasite = **dewindite**
Tantalbetafite = **betafite**
Tantalohatchettolite = **uranmicrolite**
Thoro-aeschynite = Th-rich **aeschynite**
Thor-uranin = **uraninite**
Thucholite = a mixture of **uraninite** with carbonaceous matter
Titano-obruchevite = **yttrobetafite**
Titanothucholite = Ti-rich thucholite
Toddite = mixture of **samarskite-(Y)** and **columbite**
Torberite = **torbernite**
Torit = **thorite**
Turnerite = **monazite**
Tyrite = **fergusonite**
Ufertite = **davidite**
Ulrichite[b] = old term for **uraninite**
Uranblüte = translated as "uranium bloom"; old German miner's name for **zippeite**
Uraconite = obsolete name used for various uranium sulfates
Uranatemnite = **uraninite**
Uranite = term used for members of the **autunite** group
Uran-Kalk-Kupfer-Carbonat = **voglite**
Uran-mica = **torbernite**
Uranochalcite = ill-defined uranium sulfate of doubtful validity
Uranocher = generic term used chiefly for uranium sulfates and some oxides
Uranolepidite = **vandenbrandeite**
Uranoniobite = **uraninite**
Uranopissite = **uraninite**
Uranotantalite = **samarskite-(Y)**
Uranothallite = **liebigite**
Uranothorianite = U-rich **thorianite**
Uranothorite = U-rich **thorite**
Uranotile = **uranophane**
Uranotile-beta = **uranophane-β**
Uranphyllit = **torbernite**
Uranvitriol = **johannite**
Urdite = **monazite**
Ursilite = probably **haiweeite**
Vietinghofite = Fe-rich **samarskite-(Y)**(?)
Voglianite = hydrous calcium-uranium sulfate of doubtful validity
Wiikite = a mixture of **yttropyrochlore** and other minerals
Wisaksonite = U-rich **thorite**
Yanshainshynite = PO_4-rich **thorite**(?)
Yttrogummite = Y-rich gummite (alteration product of Y-rich uraninite)
Yttrohatchettolite = **yttropyrochlore**
Yttro-ilmenite = **samarskite-(Y)**
Zirconolite = **zirkelite**

[a]Thorogummite has itself been discredited as a valid species
[b]Ulrichite is now a valid name for a mineral whose composition is $CaCu^{+2}(UO_2)(PO_4)_2 \cdot 4H_2O$
[c]Jachymovite is now a valid name for a mineral whose composition is $(UO_2)_8(SO_4)(OH)_{14} \cdot 13H_2O$

APPENDIX 2

Checklist of radioactive minerals			
Mineral	**Formula**[d]	**Color**	**UV Fluorescence**[a]
Primary Oxides			
Thorianite	ThO_2	black	
Uraninite	UO_2	black	
Primary Silicates			
Thorite	$(Th,U)SiO_4$	black to brown	
Huttonite	$ThSiO_4$	colorless to yellow-brown	pinkish white (SW)
Coffinite	$USiO_4$	black	
Secondary Hydrated Oxides			
Agrinierite	$K_2(Ca_{0.65}Sr_{0.35})[(UO_2)_2O_3(OH)_2]_2{\cdot}5H_2O$	orange	
Bauranoite	$BaU_2O_7{\cdot}4\text{-}5H_2O$	red-brown	
Becquerelite	$Ca(UO_2)_6O_4(OH)_6{\cdot}8H_2O$	lemon yellow to orange-yellow	
Billietite	$Ba[(UO_2)_6O_4(OH)_6]{\cdot}8H_2O$	yellow to yellow-brown	
Calciouranoite	$(Ca,Ba,Pb)U_2O_7{\cdot}5H_2O$	brown to brown-orange	
Clarkeite	$(Na,K,Ca,Pb)[(UO_2)O(OH)]{\cdot}H_2O$	dark brown	
Compreignacite	$K_2[(UO_2)_3O_2(OH)_3]_2{\cdot}7H_2O$	yellow	
Curite	$Pb_2U^{6+}{}_5O_{17}{\cdot}4H_2O$	deep orange to red-orange	
Fourmarierite	$PbU^{6+}{}_4O_{13}{\cdot}4H_2O$	red-orange to deep red	
Heisenbergite	$UO_2(OH)_2{\cdot}H_2O$	yellow to orange-brown	
Ianthinite	$UO_2{\cdot}5UO_3{\cdot}10H_2O$	purple-brown to black	
Masuyite	$Pb(UO_2)_3O_3(OH)_2{\cdot}3H_2O$	reddish orange to carmine red	
Metacalciouranoite	$(Ca,Na,Ba)U_2O_7{\cdot}2H_2O$	orange	
Metaschoepite	$UO_3{\cdot}nH_2O$	yellow	weak yellow-green
Metastudtite	$UO_4{\cdot}2H_2O$	pale yellow	
Metavandendriesscheite	$PbU_7O_{22}{\cdot}nH_2O$ $(n < 12)$	orange	
Paraschoepite	$UO_3{\cdot}2H_2O$ (?)	yellow	yellow-green
Paulscherrerite	$UO_2(OH)_2$	canary yellow	
Protasite	$Ba[(UO_2)_3O_3(OH)_2]{\cdot}3H_2O$	orange	
Rameauite	$K_2CaU^{6+}{}_6O_{20}{\cdot}9H_2O$	orange	
Richetite	$(Fe,Mg)_xPb_{8.57}[(UO_2)_{18}O_{18}(OH)_{12}{\cdot}41H_2O$	black to deep brown	
Sayrite	$Pb_2(UO_2)_5O_6(OH)_2{\cdot}4H_2O$	yellow-orange to red-orange	
Schoepite	$[(UO_2)_3O_2(OH)_{12}]{\cdot}12H_2O$	sulfur yellow to amber	yellow-green
Spriggite	$Pb_3[(UO_2)_6O_8(OH)_2]{\cdot}3H_2O$	orange	
Studtite	$UO_4{\cdot}4H_2O$	pale yellow to cream	moderate yellow-green (SW)
Uranosphaerite	$Bi(UO_2)O_2OH$	yellow-orange to reddish orange	
Vandenbrandeite	$Cu(UO_2)(OH)_4$	dark green	
Vandendriesscheite	$Pb_{1.5}(UO_2)_{10}O_6(OH)_{11}{\cdot}11H_2O$	orange to amber	

Wölsendorfite	$Pb_7(UO_2)_{14}O_{19}(OH)_4{\cdot}12H_2O$	orange-red to carmine	
Secondary Carbonates			
Agricolaite	$K_4(UO_2)(CO_3)_3$	yellow	
Albrechtschraufite	$Ca_4Mg(UO_2)(CO_3)_6F_2{\cdot}17H_2O$	yellow-green	
Andersonite	$Na_2Ca(UO_2)(CO_3)_3{\cdot}6H_2O$	yellow-green	yellow-green (SW,LW)
Astrocyanite-(Ce)	$Cu_2(Ce, Nd, La)_2(UO_2)(CO_3)_5(OH)_2{\cdot}1.5\ H_2O$	blue	
Bayleyite	$Mg_2(UO_2)(CO_3)_3{\cdot}18H_2O$	yellow	weak yellowish green
Bijvoetite-(Y)	$(Y,Dy)_2(UO_2)_4(CO_3)_4(OH)_6{\cdot}11H_2O$	yellow	
Blatonite	$UO_2CO_3{\cdot}H_2O$	canary yellow	strong green-yellow (SW)
Čejkaite	$Na_4(UO_2)(CO_3)_3$	pale yellow to beige	yellow to yellow-green
Fontanite	$Ca(UO_2)_3(CO_3)_4{\cdot}3H_2O$	deep yellow	pale green (LW)
Grimselite	$K_3Na(UO_2)(CO_3)_3{\cdot}(H_2O)$	yellow	
Joliotite	$(UO_2)(CO_3){\cdot}nH_2O$	yellow	
Kamotoite-(Y)	$Y_2U_4(CO_3)_3O_{12}{\cdot}14.5H_2O$	bright yellow	green (SW,LW)
Lepersonnite-(Gd)	$Ca(Gd,Dy)_2(UO_2)_{24}O_{12}(CO_3)_8(SiO_4)_4{\cdot}60H_2O$	bright yellow	
Liebigite	$Ca_2(UO_2)(CO_3)_3{\cdot}11H_2O$	green to yellowish green	green (SW,LW)
Linekite	$K_2Ca_3[(UO_2)(CO_3)]_2{\cdot}7H_2O$	yellowish green	
Metazellerite	$Ca(UO_2)(CO_3)_2{\cdot}3H_2O$	chalky yellow	
Oswaldpeetersite	$(UO_2)_2CO_3(OH)_2{\cdot}4H_2O$	canary yellow	
Rabbittite	$Ca_3Mg_3(UO_2)_2(CO_3)_6(OH)_4{\cdot}18H_2O$	pale green	creamy yellow (SW)
Roubaultite	$Cu_2(UO_2)_3(CO_3)_2O_2(OH)_2{\cdot}4H_2O$	green	
Rutherfordine	$UO_2(CO_3)$	pale yellow, yellow-brown to orange	
Schroeckingerite	$NaCa_3(UO_2)(CO_3)(SO_4)F{\cdot}10H_2O$	greenish yellow	bright yellow-green (SW,LW)
Shabaite-(Nd)	$Ca(Nd,Sm,Y)_2(UO_2)(CO_3)_4(OH)_6.6H_2O$	pale yellow	
Sharpite	$Ca(UO_2)_6(CO_3)_5(OH)_4{\cdot}6H_2O$	greenish yellow to olive green	
Swartzite	$CaMg(UO_2)(CO_3)_3{\cdot}12H_2O$	bright green	bright yellow-green
Urancalcarite	$Ca(UO_2)_3(CO_3)(OH)_6{\cdot}3H_2O$	bright yellow	
Voglite	$Ca_2Cu(UO_2)(CO_3)_4{\cdot}6H_2O$	grass green to emerald green	
Widenmannite	$Pb_2(UO_2)(CO_3)_3$	yellow	
Wyartite	$CaU^{5+}(UO_2)_2(CO_3)O_4(OH){\cdot}7H_2O$	purplish black	
Zellerite	$Ca(UO_2)(CO_3)_2{\cdot}5H_2O$	light lemon yellow	weak green (SW,LW)
Znucalite	$CaZn_{11}(UO_2)(CO_3)_3(OH)_{20}{\cdot}4H_2O$	white to pale yellow	yellow-green
Secondary Sulfates			
Zippeite Group			
Cobaltzippeite	$Co(H_2O)_{3.5}[(UO_2)_2(SO_4)O_2]$	orange-yellow to rust	
Magnesiozippeite	$Mg(H_2O)_{3.5}[(UO_2)_2(SO_4)O_2]$	yellow	
Marecottite	$Mg_3(H_2O)_{18}[(UO_2)_8(SO_4)_4O_6]{\cdot}10H_2O$	yellow-orange	
Natrozippeite	$Na_5(H_2O)_{12}[(UO_2)_8(SO_4)_4O_5(OH)_3]$	yellow	green
Nickelzippeite	$Ni(H_2O)_{3.5}[(UO_2)_2(SO_4)O_2]$	yellow-orange to brown-yellow	
Pseudojohannite	$Cu_3(OH)_2[(UO_2)_2O_4(SO_4)_2]{\cdot}12H_2O$	olive green	
Sejkoraite-(Y)	$(Y,Dy)_{2.22}H^+_{.34}[(UO_2)_8O_{88}O_7OH(SO_4)_4(OH){\cdot}26\ H_2O$	yellow to orange	
Zinczippeite	$Zn(H_2O)_{3.5}[(UO_2)_2(SO_4)O_2]$	yellow, orange, red-brown	

Zippeite	$K_3(H_2O)_3[(UO_2)_4(SO_4)_2O_3(OH)]$	bright yellow to orange-yellow	green
Other Sulfates			
Adolfpateraite	$K(UO_2)(SO_4)(OH)\cdot H_2O$	sulfur yellow to greenish yellow	green
Běhounekite	$U(SO_4)_2\cdot 4H_2O$	green	
Belakovskiite	$Na_7(UO_2)(SO_4)_4(SO_3OH)(H_2O)_3$	pale yellowish green	
Beshtauite	$(NH_4)_2(UO_2)(SO_4)_2\cdot 2H_2O$	light green	strong yellow-green
Bluelizardite	$Na_7(UO_2)(SO_4)_4Cl(H_2O)_2$	pale yellow	bright yellow-green
Bobcookite	$NaAl(UO_2)_2(SO_4)_4\cdot 18H_2O$	lime green to greenish yellow	bright greenish
Deliensite	$Fe^{2+}(UO_2)_2(SO_4)_2(OH)_2\cdot 3H_2O$	pale yellow	
Fermiite	$Na_4(UO_2)(SO_4)_3\cdot 3H_2O$	pale greenish yellow	bright greenish white
Jachymovite	$(UO_2)_8(SO_4)(OH)_{14}\cdot 13H_2O$	yellow	yellow
Johannite	$Cu(UO_2)_2(SO_4)_2(OH)_2\cdot 8H_2O$	green	
Leydetite	$Fe(UO_2)(SO_4)_2\cdot 11H_2O$	pale yellow to greenish	
Mathesiusite	$K_5(UO_2)_4(SO_4)_4(VO_5)\cdot 4H_2O$	greenish yellow	yellowish green
Meisserite	$Na_5(UO_2)(SO_4)_3(SO_3OH)(H_2O)$	pale green to yellowish green	yellow-green
Metauranopilite	$(UO_2)_6(SO_4)(OH)_{10}\cdot 5H_2O$	yellow	yellowish green
Oppenheimerite	$Na_2(UO_2)(SO_4)_2\cdot 3H_2O$	pale greenish yellow	bright yellowish green
Plášilite	$Na(UO_2)(SO_4)(OH)\cdot 2H_2O$	greenish yellow	bluish white
Rabejacite	$Ca(UO_2)_4(SO_4)_2(OH)_6\cdot 6H_2O$	yellow	light yellow
Uranopilite	$(UO_2)_6(SO_4)(OH)_{10}\cdot 12H_2O$	yellow	strong yellow-green
Wetherillite	$Na_2(Mg,Zn)(UO_2)_2(SO_4)_4\cdot 18H_2O$	pale greenish yellow	
Secondary Selenites			
Demesmaekerite	$Pb_2Cu_5(UO_2)_2(SeO_3)_6(OH)_6\cdot 2H_2O$	bottle green	
Derriksite	$Cu_4(UO_2)(SeO_3)_2(OH)_6$	green to bottle green	
Guilleminite	$Ba(UO_2)_3(SeO_3)_2O_2(H_2O)_3$	canary yellow	
Haynesite	$(UO_2)_3(SeO_3)_2(OH)_2\cdot 5H_2O$	yellow	bright green (SW,LW)
Larisaite	$Na(H_3O)(UO_2)_3(SeO_3)_2O_2\cdot 4H_2O$	yellow	green (SW)
Marthozite	$Cu(UO_2)_3(SeO_3)_3(OH)_2\cdot 7H_2O$	yellowish green to greenish brown	
Piretite	$Ca(UO_2)_3(SeO_3)_2(OH)_4\cdot 4H_2O$	lemon yellow	
Secondary Tellurites and Tellurates			
Cliffordite	UTe_3O_9	sulfur yellow	
Markcooperite	$Pb(UO_2)Te^{6+}O_6$	orange	
Moctezumite	$Pb(UO_2)(TeO_3)_2$	bright orange to orange-brown	
Schmitterite	$(UO_2)TeO_3$	pale yellow	
Secondary Arsenites			
Chadwickite	$H(UO_2)(AsO_3)$	yellow	
Secondary Molybdates			
Calcurmolite	$Ca(UO_2)_3(MoO_4)_3(OH)_2\cdot 11H_2O$	honey yellow	strong yellowish green
Cousinite	$Mg(UO_2)_2(MoO_4)(OH)_2\cdot 5H_2O$	black	
Deloryite	$Cu_4(UO_2)(MoO_4)_2(OH)_6$	dark green to black	
Iriginite	$(UO_2)(Mo_2O_7)\cdot 3H_2O$	canary yellow	
Moluranite	$H_4U^{4+}(UO_2)_3(MoO_4)_7\cdot 18H_2O$	black	
Mourite	$U^{4+}Mo^{6+}_5O_{12}(OH)_{10}$	violet	

Sedovite	$U(MoO_4)_2$	reddish brown	
Tengchongite	$Ca(UO_2)_6(MoO_4)_2O_5{\cdot}12H_2O$	yellow	
Umohoite	$[(UO_2)MoO_4(H_2O)](H_2O)$	black to bluish black	
Ichnusaite	$Th(MoO_4)_2{\cdot}3H_2O$	colorless	
Nuragheite	$Th(MoO_4)_2{\cdot}2H_2O$	colorless	
Secondary Tungstates, Niobates, and Titanates			
Carlosbarbosaite	$(UO_2)_2Nb_2O_6(OH)_2{\cdot}2H_2O$	cream to yellow	
Holfertite	$U^{6+}_{1.75}TiO_{7.5}Ca_{0.25}(H_2O)_3$	yellow	
Uranotungstite	$(Fe^{2+},Ba,Pb)(UO_2)(WO_4)(OH)_4{\cdot}12H_2O$	yellow to brown	
Secondary Phosphates and Arsenates			
Autunite Group			
Abernathyite	$K_2(UO_2)_2(AsO_4)_2{\cdot}8H_2O$	yellow	moderate yellow-green (SW,LW)
Arsenuranospathite	$HAl(UO_2)_4(AsO_4){\cdot}40H_2O$	white to pale yellow	
Autunite	$Ca(UO_2)_2(PO_4)_2{\cdot}10\text{-}12H_2O$	yellow to green	bright yellowish green (SW,LW)
Bassetite	$Fe^{2+}(UO_2)_2(PO_4)_2{\cdot}8H_2O$	green to brown	
Chernikovite	$(H_3O)_2(UO_2)_2(PO_4)_2{\cdot}6H_2O$	yellow	bright yellowish green (SW,LW)
Chistyakovaite	$Al(UO_2)_2(AsO_4)_2(F,OH){\cdot}6.5H_2O$	yellow	
Heinrichite	$Ba(UO_2)_2(AsO_4)_2{\cdot}10\text{-}12H_2O$	yellow to yellow-green	bright yellowish green (SW,LW)
Kahlerite	$Fe^{2+}(UO_2)_2(AsO_4)_2{\cdot}10\text{-}12H_2O$	lemon yellow to yellow-green	
Lehnerite	$Mn^{2+}(UO_2)_2(PO_4)_2{\cdot}8H_2O$	bronze-yellow	
Meta-ankoleite	$K_2(UO_2)_2(PO_4)_2{\cdot}6H_2O$	yellow	yellowish green (SW,LW)
Meta-autunite	$Ca(UO_2)_2(PO_4)_2.2\text{-}6H_2O$	yellow	bright yellowish green (SW,LW)
Metaheinrichite	$Ba(UO_2)_2(AsO_4)_2{\cdot}8H_2O$	yellow to green	bright yellowish green (SW,LW)
Metakahlerite	$Fe^{2+}(UO_2)_2(AsO_4)_2{\cdot}8H_2O$	sulfur yellow	
Metakirchheimerite	$Co(UO_2)_2(AsO_4)_2.8H_2O$	pale rose	
Metalodevite	$Zn(UO_2)_2(AsO_4)_2{\cdot}10H_2O$	pale yellow to green	yellowish green (SW,LW)
Metanatroautunite[b]	$Na_2(UO_2)_2(PO_4)_2{\cdot}8H_2O$	lemon yellow	yellowish green (SW,LW)
Metanovačekite	$Mg(UO_2)_2(AsO_4)_2{\cdot}4\text{-}8H_2O$	pale yellow to yellow	dull green (SW,LW)
Metarauchite	$Ni(UO_2)_2(AsO_4)_2{\cdot}8H_2O$	yellowish green	
Metatorbernite	$Cu(UO_2)_2(PO_4)_2{\cdot}8H_2O$	green	
Metauranocircite I	$Ba(UO_2)_2(PO_4)_2{\cdot}6\text{-}8H_2O$	yellow-green	yellow-green (SW,LW)
Metauranocircite II	$Ba(UO_2)_2(PO_4)_2{\cdot}6H_2O$	yellow-green	yellow-green (SW,LW)
Metauranospinite	$Ca(UO_2)_2(AsO_4)_2{\cdot}8H_2O$	lemon yellow to green	bright lemon yellow
Metazeunerite	$Cu(UO_2)_2(AsO_4)_2{\cdot}8H_2O$	pale green to emerald green	
Natrouranospinite	$(Na_2,Ca)(UO_2)_2(AsO_4)_2{\cdot}5H_2O$	lemon to straw yellow	yellowish green (SW,LW)
Novačekite	$Mg(UO_2)_2(AsO_4)_2{\cdot}12H_2O$	pale yellow to yellow	yellowish green (SW,LW)
Przhevalskite	$Pb(UO_2)_2(PO_4)_2{\cdot}4H_2O$	pale yellow to pale green	
Rauchite	$Ni(UO_2)_2(AsO_4)_2{\cdot}10H_2O$	yellowish green	
Sabugalite	$H_{0.5}Al_{0.5}(UO_2)_2(PO_4)_2{\cdot}8H_2O$	yellow	bright lemon yellow (SW,LW)

Saléeite	$Mg(UO_2)_2(PO_4)_2\cdot 10H_2O$	yellow	bright yellowish (SW,LW)
Threadgoldite	$Al(UO_2)_2(PO_4)2(OH)\cdot 8H_2O$	yellow	green (SW,LW)
Torbernite	$Cu(UO_2)_2(PO_4)_2\cdot 10\text{-}12H_2O$	dark emerald green	
Trögerite	$(H_3O)_2(UO_2)_2(AsO_4)_2\cdot 8H_2O$	lemon yellow	
Uramarsite	$(NH_4,H_3O)_2(UO_2)_2(AsO_4,PO_4)_2\cdot 6H_2O$	pale green	green
Uramphite	$(NH_4)_2(UO_2)_2(PO_4)_2\cdot 6H_2O$	pale green to bottle green	yellow-green
Uranocircite	$Ba(UO_2)_2(PO_4)_2\cdot 12H_2O$	greenish yellow to yellow	green (SW,LW)
Uranospathite	$HAl(UO_2)_4(PO_4)\cdot 40H_2O$	yellow to pale yellow-green	yellow-green
Uranospinite	$Ca(UO_2)_2(AsO_4)_2\cdot 10H_2O$	lemon yellow to green	bright lemon yellow
Zeunerite	$Cu(UO_2)_2(AsO_4)_2\cdot 10\text{-}16H_2O$	yellowish green to emerald green	
Phosphuranylite Group			
Althupite	$AlTh(UO_2)[(UO_2)_3O(OH)(PO_4)_2]_2(OH)_3\cdot 15H_2O$	yellow	
Arsenovanmeersscheite	$U^{4+}(UO_2)_3(AsO_4)_2(OH)_2\cdot 4H_2O$	yellow	
Arsenuranylite	$Ca(UO_2)_4(AsO_4)_2(OH)_4\cdot 6H_2O$	orange	
Bergenite	$(Ba,Ca)_2(UO_2)_3(PO_4)_2(OH)_4\cdot 5H_2O$	yellow	weak green (SW)
Dewindtite[c]	$Pb[H(UO_2)_3O_2(PO_4)_2]_2\cdot 12H_2O$	yellow	light green
Dumontite	$Pb_2(UO_3)_3O_2(PO_4)_2\cdot 5H_2O$	yellow to golden yellow	
Françoisite-(Ce)	$Ce(UO_2)_2(PO_4)_2O(OH)\cdot 6H_2O$	yellow	
Françoisite-(Nd)	$Nd(UO_2)_2(PO_4)_2O(OH)\cdot 6H_2O$	yellow	
Hügelite	$Pb_2(UO_2)_3(AsO_4)_2(OH)_4\cdot 3H_2O$	brown to orange-yellow	
Metavanmeersscheite	$U^{6+}(UO_2)_3(PO_4)_2(OH)_6\cdot 2H_2O$	yellow	strong green (SW,LW)
Mundite	$Al(UO_2)_3(PO_4)_2(OH)_3\cdot 5.5H_2O$	pale yellow	
Nielsbohrite	$K(UO_2)_3(AsO_4)(OH)_4\cdot H_2O$	yellow	
Phosphuranylite	$KCa(H_3O)_3(UO_2)_7(PO_4)_4$	deep yellow	
Phuralumite	$Al_2(UO_2)_3(PO_4)_2(OH)_6\cdot 10H_2O$	lemon yellow	
Phurcalite	$Ca_2(UO_2)_3O_2(PO_4)_2\cdot 7H_2O$	yellow	yellowish green (SW,LW)
Upalite	$Al(UO_2)_3O(OH)(PO_4)_2\cdot 7H_2O$	yellow-brown	
Vanmeersscheite	$U^{6+}(UO_2)_3(PO_4)_2(OH)_6\cdot 4H_2O$	yellow	strong pale green
Yingjiangite	$(K_2,Ca)(UO_2)_7(PO_4)_4(OH)_6\cdot 6H_2O$	yellow to golden yellow	
Walpurgite Group			
Hallimondite	$Pb_2(UO_2)(AsO_4)_2$	yellow	
Lakebogaite	$CaNa_2Fe^{3+}{}_2H(UO_2)_2(PO_4)_4(OH)_2(H_2O)_8$	lemon yellow	
Orthowalpurgite	$(BiO)_4(UO_2)(AsO_4)_2\cdot 2H_2O$	pale yellow	
Parsonsite	$Pb_2UO_2(PO_4)_2$	amber to yellow or brownish yellow	
Phosphowalpurgite	$(UO_2)Bi_4O_4(PO_4)_2\cdot 2H_2O$	brownish gray	
Ulrichite	$CaCu(UO_2)(PO_4)_2\cdot 4H_2O$	bright apple green to lime green	
Walpurgite	$(BiO)_4(UO_2)(AsO_4)_2\cdot 2H_2O$	yellow	
Other Phosphates and Arsenates			
Asselbornite	$(Pb,Ba)(UO_2)_6(BiO)_4(AsO_4)_2(OH)_{12}\cdot 3H_2O$	brown to yellow	
Chistyakovaite	$Al(UO_2)_2[(F,OH)(AsO_4)_2]\cdot 6.5H_2O$	yellow	
Coconinoite	$Fe_2Al_2(UO_2)_2(PO_4)_4(SO_4)(OH)_2\cdot 20H_2O$	pale yellow	
Dymkovite	$Ni(UO_2)_2(As^{3+}O_3)_2\cdot 7H_2O$	yellow	

Eylettersite	$Th,(H_3O)]Al_3[(PO_4),(H_4O_4)](OH,H_2O)_6$	white to cream	
Furongite	$Al_2(UO_2)(PO_4)_2(OH)_2{\cdot}8H_2O$	lemon yellow	yellow-green
Kamitugaite	$PbAl(UO_2)_5[(P,As)O_4]_2(OH)_9{\cdot}9.5H_2O$	yellow	
Moreauite	$Al_3(UO_2)(PO_4)_3(OH)_2{\cdot}13H_2O$	yellow-green	
Ranunculite	$HAl(UO_2)PO_4(OH)_3{\cdot}4H_2O$	golden yellow	
Seelite	$Mg(UO_2)_2(AsO_4)_2{\cdot}7H_2O$	yellow	
Šreinite	$Pb(BiO)_3(UO_2)_4(PO_4)_2(OH)_7{\cdot}4H_2O$	orange to yellow	
Štěpite	$U^{4+}(AsO_3OH)_2{\cdot}4H_2O$	emerald-green	
Triangulite	$Al_3(UO_2)_4(PO_4)_4(OH)_5{\cdot}5H_2O$	bright yellow	
Vochtenite	$(Fe^{2+},Mg)Fe^{3+}(UO_2)_4(PO_4)_4(OH){\cdot}12\text{-}13H_2O$	brown	
Vyacheslavite	$U^{4+}(PO_4)(OH){\cdot}nH_2O$	green	
Vysokýite	$U^{4+}[AsO_2(OH)_2]_4{\cdot}4H_2O$	light green	
Xiangjiangite	$(Fe^{3+},Al)(UO_2)_4(PO_4)_2(SO_4)_2(OH){\cdot}22H_2O$	yellow	
Secondary Vanadates			
Carnotite Group			
Carnotite	$K_2(UO_2)_2(VO_4)_2{\cdot}3H_2O$	orange, yellow to greenish yellow	
Curienite	$Pb(UO_2)_2V_2O_8{\cdot}5H_2O$	canary yellow	
Francevillite	$Ba(UO_2)_2V_2O_8{\cdot}5H_2O$	orange-yellow to olive green	
Margaritasite	$(Cs,K,H_3O)_2(UO_2)_2(VO_4)_2{\cdot}H_2O$	yellow	
Metatyuyamunite	$Ca(UO_2)_2V_2O_8{\cdot}3H_2O$	yellow	
Sengierite	$Cu_2(UO_2)_2(VO_4)_2(OH)_2{\cdot}6H_2O$	olive green	
Strelkinite	$Na_2(UO_2)_2V_2O_8{\cdot}6H_2O$	golden yellow to yellow-green	dull green
Tyuyamunite	$Ca(UO_2)_2V_2O_8{\cdot}5\text{-}8H_2O$	bright yellow to yellow-green	
Vanuranylite	$(H_3O,Ba,Ca,K)_{1.6}(UO_2)_2V_2O_8{\cdot}4H_2O$	yellow	
Other Vanadates			
Metavanuralite	$Al(UO_2)_2(VO_4)_2(OH){\cdot}8H_2O$	lemon yellow	
Rauvite	$Ca(UO_2)_2V_{10}O_{28}{\cdot}16H_2O$	purple to grayish black	
Uvanite	$U^{6+}{}_2V^{5+}{}_6O_{21}{\cdot}15H_2O(?)$	brown-yellow	
Vanuralite	$Al(UO_2)_2(VO_4)_2(OH){\cdot}11H_2O$	lemon yellow	
Secondary Silicates			
Uranophane Group			
Boltwoodite	$HK(UO_2)SiO_4{\cdot}1.5H_2O$	yellow to orange	dull green
Cuprosklodowskite	$Cu(UO_2)_2Si_2O_7{\cdot}6H_2O$	grass green to emerald green	
Kasolite	$Pb(UO_2)(SiO_4){\cdot}H_2O$	yellow to orange	
Natroboltwoodite	$(H_3O)(Na,K)(UO_2)SiO_4{\cdot}H_2O$	pale yellow	
Oursinite	$Co(UO_2)_2Si_2O_7{\cdot}6H_2O$	pale yellow	
Sklodowskite	$Mg(UO_2)_2Si_2O_7{\cdot}6H_2O$	yellow	
Swamboite	$U^{6+}H_6(UO_2)_6(SiO_4)_6{\cdot}30H_2O$	pale yellow	
Uranophane	$Ca(UO_2)_2[SiO_3(OH)]_2{\cdot}5H_2O$	yellow	very weak yellow-green
Uranophane-β	$Ca(UO_2)_2[SiO_3(OH)]_2{\cdot}5H_2O$	yellow to amber	
Weeksite Group			
Coutinhoite	$(Th,Ba)_{0.5}(UO_2)_2(Si_5O_{13}){\cdot}1\text{-}3.5H_2O$	yellow	
Haiweeite	$Ca(UO_2)_2Si_5O_{12}(OH)_2{\cdot}4.5H_2O$	pale yellow	dull green

Metahaiweeite	$Ca(UO_2)_2(Si_6O_{15})\cdot H_2O(?)$	pale yellow to green-yellow	weak green
Weeksite	$(K,Na)_2(UO_2)_2(Si_5O_{13})\cdot 3H_2O$	yellow	
Other Silicates			
Soddyite	$(UO_2)_2(SiO_4)\cdot 2H_2O$	canary yellow to amber	
Uranosilite	$(UO_2)Si_7O_{15}$	pale yellow	
Other Minerals Containing Essential U or Th			
Brannerite Series			
Brannerite	$(U,Ca,Y,Ce)(Ti,Fe)_2O_6$	black	
Orthobrannerite	$U^{4+}U^{6+}Ti_4O_{12}(OH)_2$	black	
Thorutite	$(Th,U,Ca)Ti_2(O,OH)_6$	black	
Polycrase Group			
Euxenite-(Y)	$(Y,Ca,Ce,U,Th)(Nb,Ta,Ti)_2O_6$	black	
Kobeite-(Y)	$(Y,U)(Ti,Nb)_2(O,OH)_6$	brown-black	
Polycrase-(Y)	$(Y,Ca,Ce,U,Th)(Ti,Nb,Ta)_2O_6$	black	
Tanteuxenite-(Y)	$(Y,Ce,Ca,U)(Ta,Nb,Ti)_2O_6$	brown to black	
Uranopolycrase	$(U,Y)(Ti,Nb,Ta)_2O_6$	brown-red	
Yttrocrasite-(Y)	$(Y,Th,U)(Ti)_2(O,OH)_6$	black	
Aeschynite Series			
Aeschynite-(Ce)	$(Ce,Nd,Ca,Th)(Ti,Nb)_2(O,OH)_6$	black	
Aeschynite-(Nd)	$(Nd,Ce,Ca,Th)(Ti,Nb)_2(O,OH)_6$	black	
Aeschynite-(Y)	$(Y,Ca,Fe,Th)(Ti,Nb)_2(O,OH)_6$	black	
Niobo-aeschynite-(Ce)	$(Ce,Ca,Th)(Nb,Ti)_2(O,OH)_6$	black	
Niobo-aeschynite-(Y)	$[(Y,REE),Ca,Th](Nb,Ta,Ti,Fe)_2(O,OH)_6$	black	
Tantal-aeschynite-(Y)	$(Y,Ce)(Ta,Ti,Nb)_2O_6$	black	
Samarskite Group			
Calciosamarskite	$(Ca,Y,Fe,U,Th,)(Nb,Ta,Ti)O_4$	black	
Ishikawaite	$(U,Fe,Y,Ca)(Nb,Ta,Ti)O_4$	black	
Samarskite-(Y)	$(Y,Ce,U,Th,Ca,Fe)(Nb,Ta,Ti)O_4$	black	
Samarskite-(Yb)	$(Yb,Ce,U,Th,Ca,Fe)(Nb,Ta,Ti)O_4$	black	
Pyrochlore Group[e]			
Uranmicrolite	$(U,Ca,Ce)_2(Ta,Nb)_2(O,OH,F)_7$	yellow-brown, green-brown to black	
Uranpyrochlore	$(U,Ca,Ce)_2(Nb,Ta)_2(O,OH,F)_7$	amber, brown to black	
Crichtonite Group			
Cleusonite	$(Pb,Sr)(U^{4+},U^{6+})(Fe^{2+},Zn)_2(Ti,Fe^{2+},Fe^{3+})_{18}(O,OH)_{38}$	black	
Davidite-(Ce)	$(Ce,La)(Y,U)(Ti,Fe)_{20}O_{38}$	black	
Davidite-(La)	$(La,Ce,Ca)(Y,U)(Ti,Fe)_{20}O_{38}$	black	
Dessauite-(Y)	$(Sr,Pb)(Y,U)(Ti,Fe^{3+})_{20}O_{38}$		
Mapiquiroite	$(Sr,Pb)(U,Y)Fe_2(Ti,Fe^{3+})_{18}O_{38}$	black	
Rhabdophane Group			
Brockite	$(Ca,Th,Ce)(PO_4)\cdot H_2O$	yellow to red-brown	
Grayite	$(Th,Pb,Ca)PO_4\cdot H_2O$	pale yellow to red-brown	
Ningyoite	$(U,Ca)_2(PO_4)_2\cdot 2H_2O$	brown to brownish green	
Tristramite	$(Ca,U^{4+})(PO_4)\cdot 2H_2O$	pale yellow to yellow-green	

Monazite Group			
Cheralite	$CaTh(PO_4)_2$	green to brown	
Steacyite Group			
Arapovite	$U^{4+}(Ca,Na)_2(K_{1-x}\square_x)[Si_8O_{20}]$	dark green	
Ekanite	$ThCa_2[Si_8O_{20}]$	green	
Iraqite-(La)	$K_{0.6}Ca_2(La,Ce,Th)[Si_8O_{20}]$	yellow-green	
Steacyite	$K_{0.6}(Na,Ca)_2Th[Si_8O_{20}]$	gray to dark brown	
Turkestanite	$Th(Ca,Na)_2(K_{1-x}\square_x)Si_8O_{20}\cdot nH_2O$	brown or apple green	
Chevkinite Group			
Chevkinite-(Ce)	$(Ce,La,Ca,Na,Th)_4(Fe^{2+},Mg)_2(Ti,Mg,Fe^{3+})_3Si_4O_{22}$	black	
Perrierite-(Ce)	$(CeCa,Th)_4(Fe^{2+},Mg)Fe^{3+}_2(Ti,Fe^{3+})_2(Si_2O_7)_2O_8$	green	
Steenstrupine Series			
Steenstrupine-(Ce)	$Na_{14}Ce_6Mn^{2+}Mn^{3+}Fe^{2+}_2(Zr,Th)(Si_6O_{18})_2(PO_4)_7\cdot 3H_2O$	red-brown to black	
Thorosteenstrupine	$Na_{0-5}Ca_{1-3}(Th,REE)_6(Mn,Fe,Al,Ti)_{4-5}(Si_6O_{18})_2[(Si,P)O_4]_6(OH,F,O)_x\cdot nH_2O$	dark brown to black	
Other Minerals			
Aspedamite	$\square_{12}(Fe^{3+},Fe^{2+})_3Nb_4[Th(Nb,Fe^{3+})_{12}O_{42}]\{(H_2O),(OH)\}_{12}$	brownish orange	
Ciprianiite	$Ca_4[(Th,U)(REE)]Al\square_2(Si_4B_4O_{22})(OH,F)_2$	black	
Karnasurtite-(Ce)	$(Ce,La,Th)(Ti,Nb)(Al,Fe)(Si_2O_7)(OH)_4\cdot 3H_2O$	honey yellow	
Lermontovite	$U(PO_4)OH\cdot H_2O$	greenish gray	
Liandratite	$U^{6+}(Nb,Ta)_2O_8$	yellow to brown	
Petscheckite	$U^{4+}Fe^{2+}(Nb,Ta)_2O_8$	black	
Thorbastnaesite	$Th(Ca,Ce)(CO_3)_2F_2\cdot 3H_2O$	brown	
Thornasite	$Na_{12}Th_3[Si_8O_{19}]_4(H_2O)_{18}$	pale green to beige	green(SW,LW)
Tuliokite	$Na_6BaTh(CO_3)_6\cdot 6H_2O$	gray	
Umbozerite	$Na_3Sr_4(Mn,Fe)ThSi_8O_{24}$	bottle-green to brown	
Vapnikite	Ca_3UO_6	yellow-brown	weak yellowish
Vorlanite	$CaU^{6+}O_4$	black	

[a]Reported fluorescence can be subjective and to some degree anecdotal; it is rarely studied or reported as precisely as other optical properties. However, UV fluorescence is an important characteristic of quite a few uranium minerals, and many collectors have a general interest in the phenomenon. In the table, SW and LW denote short-wave (254 nm) and long-wave (360 nm) UV respectively, when reported data include this information. In general, many fluorescent uranium minerals will display similar colors in SW and LW but the fluorescence will be more intense under SW illumination than under LW (Frondel 1958). Many of the minerals fluoresce green because this is an intrinsic property of the uranyl ion, which can even be seen when uranyl ions are dissolved in water (Robbins 1994). Copper-containing species such as torbernite, zeunerite, and others are generally not fluorescent because the copper ions quench the fluorescence. Occasionally a specimen will show anomalous fluorescence, either positive or negative. In the case of fluorescent torbernite, the crystals might be zoned, with a thin layer of autunite or other fluorescent member of the group on the surface. Many zippeites fluoresce more weakly than their synthetic counterparts, usually because of admixed soil or clay.

[b]Older references refer to this species as sodium autunite.

[c]Renardite is now regarded as synonymous with dewindtite.

[d]In structural formula "□" denotes a vacancy.

[e]The pyrochlore group includes the "betafites", which have been extensively reclassified and renamed; see the text for a more complete discussion of this group.

BIBLIOGRAPHY

SUGGESTED CORE LIBRARY FOR FURTHER READING

Beginning around 1948 the US Atomic Energy Commission supported a rigorous study of uranium and thorium minerals as part of a broader effort to understand the nature and geology of the raw materials for nuclear power programs. Among several noteworthy reports published by the USGS (Weeks and Thompson 1954; Frondel 1958; Frondel Fleischer, and Jones 1967), the seminal work of Clifford Frondel, *Systematic Mineralogy of Uranium and Thorium*, has rightly been described by Dr. Carl Francis of Harvard Museum as "the basis of our modern understanding of radioactive minerals." Another excellent volume from this period is *Mineralogy and Geology of Radioactive Raw Materials* (Heinrich 1958), which discusses the geology of several thousand deposits, organized by the type of formation.

The importance of uranium resources in colonial Africa led to another series of books, *The Secondary Uranium Minerals of Zaire*, which are highly recommended for any serious collector or student of U minerals. Although published in French, the mineralogical data and photomicrographs are easily comprehensible to English-speaking readers. The Congo (Zaire) has been the source of many superb uranium minerals and is well known in the collecting community, making these volumes an indispensable resource. The original volume (Guillemin, Destas, and Vaes 1958) contains detailed descriptions of the important minerals, along with twenty-seven color plates, and is occasionally available from rare book dealers. A completely updated version, along with two supplements, (Deliens, Piret, and Comblain 1981; Deliens, Piret, and Comblain 1984; and Deliens, Piret, and Van Der Meersche 1990), describes sixty-six minerals and their associations.

A recent review of the scientific literature can be found in *Uranium: Mineralogy, Geochemistry, and the Environment*, which was recently published by the Mineralogical Society of America as *Reviews in Mineralogy* Vol. 38 (Burns and Finch, 1999). This volume reviews such topics as crystal chemistry, systematics, paragenesis of deposits, analytical methods, and implications for technological problems such as contamination and the disposal of spent reactor fuel. The individual review papers provide further references to literally thousands of additional research reports, including many original descriptions of new U minerals.

Several excellent reviews have been published by the Mineralogical Association of Canada: *Uranium Deposits: Their Mineralogy and Origin*, Short Course Series Vol. 3 (Kimberley 1978); *Recent and Not-So-Recent Developments in Uranium Deposits and Implications for Exploration*, Short Course Series Vol. 39 (Cuney and Kyser 2008); and *Uranium - Cradle to Grave*, Short Course Series Vol. 43 (Burns and Sigmon 2013).

Many of the individual papers cited below are available online through several excellent databases. It is the author's opinion that listing the URL addresses for specific web sites is of limited value as these often change and web sites can sometimes disappear entirely. However, a few important resources are mentioned here, primarily because they are maintained by long-standing organizations and are likely to be available for the foreseeable future. Additional resources can be found using familiar internet search engines, but the reader is cautioned that many internet sites are personal endeavors by individual researchers or hobbyists and are not subject to the rigorous peer review expected of scientific publications.

Mineralogical Magazine and *American Mineralogist* maintain online archives of all issues published through 1999. Individual papers are available for free download as *pdf* files.

The RRUFF™ Project at the University of Arizona maintains two important data resources:

1. The American Mineralogist Crystal Structure Database is searchable by mineral species. For each mineral, all available structure determinations are presented, and the structure data can be downloaded in various formats including the *cif* files that served as input to the CrystalMaker® software used in this book to generate structure diagrams. The bibliographic citation for the paper in which a structure was published is listed in this database and many of those papers are available for download as *pdf* files.

2. The Database of Raman Spectra is searchable by mineral species and presents microprobe analyses, Raman spectral data, and X-ray powder diffraction data for thoroughly characterized specimens. Using the Search References tab, for each mineral, numerous references are listed and many are available directly for download. A particularly notable feature of this database is that some very old references (dating to the 1700s) have been scanned and are available in *pdf* format.

REFERENCES

Adusumilli, M. S., C. Kieft, and E. A. J. Burke 1974. Tantal-aeschynite, a new mineral of the aeschynite group from the Borborema region, north-eastern Brazil. *Mineralogical Magazine* 39:571–76.

Agakhanov, A. A., L. A. Pautov, Y. A. Uvarova, E. V. Sokolova, F. C. Hawthorne, V. Yu. Karpenko, Y. D. Dusmatov, and E. I. Semenov 2004. Arapovite, (U,Th)(Ca,Na)$_2$(K$_{1-x}$□$_x$) Si$_8$O$_{20}$·H$_2$O. New mineral. *New Data on Minerals. M.* 39:14–19.

Agrinier, H., F. Chantret, J. Geffroy, B. Hery, B. Bachet, and H. Vachey 1972. Une nouvelle espece minerale: la metalodevite (arseniate hydrate d'uranium et de zinc), *Bulletin de la Société Française de Minéralogie et de Cristallographie* 95:360–4 [see abstract (1974) *American Mineralogist* 59:210–11].

Aleksandrov, V. B. 1962. The crystal structure of aeschynite. *Doklady Akademii Nauk SSSR* 142:181–84.

Alekseeva, M. A., A. A. Chernikov, D. P. Shashkin, E. A. Kon´kova, and I. N. Gavrilova 1974. Strelkinite, a new uranyl vanadate. *Zapiski Vsesoyuznogo Mineralogicheskogo Obshchestva* 103 (5):576–80 [see abstract (1975) *American Mineralogist* 60:488–89].

Allen, O. 1877. Chemical constitution of hatchettolite and samarskite from Mitchell County, North Carolina. *American Journal of Science 3rd ser.*, 14:128–31.

Anderson, B. W., G. F. Claringbull, R. J. Davis, and D. K. Hill 1961. Ekanite, a new metamict mineral from Ceylon. *Nature* 190:997 [see abstract (1961) *American Mineralogist* 46:1516].

Ansell, V. E., and G. Y. Chao 1987. Thornasite, a new hydrous sodium thorium silicate from Mont St.–Hilaire, Quebec. *Canadian Mineralogist* 25:181–83.

Aniel, B., and J. Leroy 1985. The reduced uraniferous mineralizations associated with the volcanic rocks of the Sierra Peña Blanca (Chihuahua, Mexico). *American Mineralogist* 70:1290–97.

Appleman, D. E., and H. T. Evans 1965. The crystal structures of synthetic anhydrous carnotite, K$_2$(UO$_2$)$_2$V$_2$O$_8$, and its cesium analogue, Cs$_2$(UO$_2$)$_2$V$_2$O$_8$. *American Mineralogist* 50:825–42.

Armbruster, T., P. Bonazzi, M. Akasaka, V. Bermanec, C. Chopin, R. Giere, S. Heuss-Assbichler, A. Liebscher, S. Menchetti, Y. Pan, and M. Pasero 2006. Recommended nomenclature of epidote-group minerals, *European Journal of Mineralogy* 18: 551–67.

Armstrong, C. R., K. L. Nash, P. R. Griffiths, and S. B. Clark 2011. Synthesis and characterization of françoisite-(Nd): Nd[(UO$_2$)$_3$O(OH)(PO$_4$)$_2$]·6H$_2$O. *American Mineralogist* 96:417–22.

Arnott, R. J. 1950. X-ray diffraction data on some radioactive oxide minerals. *American Mineralogist* 35:386–400.

Atencio, D. 1988. Chernikovite, a new mineral name for (H$_3$O)$_2$(UO$_2$)$_2$(PO$_4$)$_2$·6H$_2$O superceding hydrogen autunite, *Mineralogical Record* 19:249–52.

Atencio, D., M. B. Andrade, A. G. Christy, R. Gieré, and P. M. Kartashov 2010. The pyrochlore supergroup of minerals: nomenclature. *Canadian Mineralogist* 48:673–98.

Atencio, D., and R. Hypolito 1994. Fosfatos e silicatos secundarios de uranio de Perus, São Paulo. *Revista Brasiliera de Geosciencias* 24:43–51.

Atencio, D., F. M. S. Carvalho, and P. A. Matioli 2004. Coutinhoite, a new thorium uranyl silicate hydrate from Urucum mine, Galileia, Minas Gerais, Brazil. *American Mineralogist* 89:721–24.

Atencio, D., R. Neumann, A. J. G. C. Silva, and Y. P. Mascarenhas 1991. Phurcalite from Perus, Sao Paulo, Brazil, and redetermination of its crystal structure. *Canadian Mineralogist* 29:95–105.

Atencio, D., A. C. Roberts, M. A. Cooper, L. A. D. Menezes Filho, J. M. V. Coutinho, J. A. R. Stirling, K. E. Venance, N. A. Ball, E. Moffatt, M. L. S. C. Chaves, P. R. G. Brandao, and A. W. Romano 2012, Carlosbarbosaite, ideally (UO$_2$)$_2$Nb$_2$O$_6$(OH)$_2$·2H$_2$O, a new hydrated uranyl niobate mineral with tunnels from Jaguaracu, Minas Gerais, Brazil: description and crystal structure. *Mineralogical Magazine* 76:75–90.

Atkin, D., I. R. Basham, J. F. W. and Bowles 1983. Tristramite, a new calcium uranyl phosphate of the rhabdophane group. *Mineralogical Magazine* 47:393–6; [see abstract (1984) *American Mineralogist* 69:813.]

Aurisicchio, C., P. Orlandi, M. Pasero, and N. Perchiazzi 1993. Uranopolycrase, the uranium-dominant analogue of polycrase-(Y), a new mineral from Elba Island, Italy, and its crystal structure. *European Journal of Mineralogy* 5:1161–65.

Axelrod, J. M., F. S. Grimaldi, C. Milton, and K. J. Murata 1951. The uranium minerals from the Hillside mine, Yavapai County, Arizona. *American Mineralogist* 36:1–22.

Bachet, B., C. Brassy, and A. Cousson 1991. Structure de Mg[(UO$_2$)(AsO$_4$)]$_2$·4H$_2$O. *Acta Crystallographica* C47:2013–15.

Back, M., R. Ramik and E. Fodi 2007. New occurrences of kamotoite-(Y) and phenakite, *Rocks & Minerals* 82:232–3.

Barczak, V. J. 1966. Schroeckingerite from Ambrosia Lake uranium district. *American Mineralogist* 51:929–30.

Bariand, P., B. Bachet, C. Brassy, O. Medenbach, M. Deliens, and P. Piret 1993. Seelite, a new uranium mineral from the Talmessi mine, Iran, and Rabejac, France. *Mineralogical Record* 24 (6):463–7.

Barton, P. B. Jr. 1958. Synthesis and properties of carnotite and its alkali analogues. *American Mineralogist* 43:799–817.

Baturin, S. V. and G. A. Sidorenko 1985. Crystal-structure of weeksite. *Doklady Akademii Nauk SSSR* 282:1132–36.

Baumer, A., R. Caruba, H. Bizouard, and A. Peckett 1983. Chlorapatite de synthese: substitution et inclusions de Mn, Ce, U et Th en traces. *Canadian Mineralogist* 21:567–73.

Beddoe-Stephens, B., and K. Secher 1982. Barian wolsendorfite from East Greenland. *Mineralogical Magazine* 46:130–32.

Belakovskiy, D., L. Pautov, E. Sokolova, F. C. Hawthorne, and A. Mokhov 2006. Holfertite a new hydroxyl-hydrated uranium titanate from Starvation Canyon, Thomas Range, Utah. *Mineralogical Record* 37 (4):311–17.

Belova, L. N. 1958. Arsenuranylite, the arsenic analogue of phosphuranylite. *Zapiski Vsesoyuznogo Mineralogicheskogo Obshchestva* 87:589–602; [see abstract (1959) *American Mineralogist* 44:208].

Belova, L. N. 1958a Arsenuranocircite, $Ba(UO_2)_2(AsO_4)_2 8H_2O$. *Proc. 2nd Geneva Conf. on Peaceful Uses of Atomic Energy 1958*, Vol. 2, 294; [see abstract (1959) *American Mineralogist* 44: 466].

Belova, L. N., A. I. Gorshkov, O. A. Doinikova, A. V. Mokhov, N. V. Trubkin, and A. V. Sivtsov 1993. New data on coconinoite. *Doklady Akademii Nauk SSSR* 329:772–75.

Belova, L. N., A. I. Gorshkov, O. A. Ivanova, A. V. Sivtsov, L. I. Lizorkina, and V. A. Voronikhin 1984. Vyacheslavite $U^{4+}(PO_4)(OH)\cdot nH_2O$ – a new uranium phosphate. *Zapiski Vsesoyuznogo Mineralogicheskogo Obshchestva* 113:360–65 [see abstract (1985) *American Mineralogist* 70:878].

Belova, L. N., B. I. Ryzhov, O. V Fedorov, and G. V. Lyubomilova 1985. Typomorphism and isomorphism of the wölsendorfite group of uranium hydroxides. *Izvestiya Akademii Nauk SSSR, Seriya Geologicheskaya* 2:65–72.

Berman, E. R. 1955. Unit cell dimensions of uraninite. *American Mineralogist* 40:925–27.

Berman, R. 1957. Studies of uranium minerals (XXIII): torbernite, zeunerite, and uranospherite. *American Mineralogist* 42:905–8.

Bermanec, V., N. Tomašić, G. Kniewald, M. E. Back, and G. Zagler 2008. Nioboaeschynite-(Y), a new member of the aeschynite group from the Bear Lake Diggings, Haliburton County, Ontario, Canada. *Canadian Mineralogist* 46:395–402.

Berning, J., R. Cooke, S. Hiemstra, and U. Hoffman 1976. The Rössing uranium deposit, South West Africa. *Mineralogical Magazine* 71:351–68.

Berzelius, J. 1830. Mineralogie. *Jahres-Bericht über die Fortschritte der Physischen Wissenschaften* 9:182–209.

Biagioni, C., P. Orlandi, M. Pasero, F. Nestola, and L. Bindi 2014. Mapiquiroite, $(Sr,Pb)(U,Y)Fe_2(Ti,Fe^{3+})_{18}O_{38}$, a new member of the crichtonite group from the Apuan Alps, Tuscany, Italy. *European Journal of Mineralogy* 26: 427–37.

Birch, W. D., W. G. Mumme, and E. R. Segnit 1988. Ulrichite – a new copper calcium uranium phosphate from Lake Boga, Victoria, Australia. *Australian Mineralogist* 3:125–34; [see abstract (1990) *American Mineralogist* 75:243].

Boisson, J. M., and J. C. Leydet 1998. Les mineraux uraniferes Français. *le Règne Minéral* Hors Serie IV:37–60.

Bonatti, S. 1959. Chevkinite, perrierite and epidotes. *American Mineralogist* 44:115–137.

Bonatti, S., and G. Gottardi 1950. Perrierite, nuovo minerale ritrovato nella sabbia di Nettuno (Roma). *Atti della Reale Accademia dei Lincei, Rendiconti della Classe di Scienze Fisiche, Matematiche e Naturali* 9:361–68; [see abstract (1951) *American Mineralogist* 36:926].

Borène, J., and F. Cesbron 1971. Structure cristalline de la curiénite $Pb(UO_2)_2(VO_4)_2\cdot 5H_2O$. *Bulletin de la Société Française de Minéralogie et de Cristallographie* 94:8–14.

Bowie, S. H. U. 1957. Summary of progress of the Geological Survey of Great Britian and the Museum of Practical Geology for the year 1956. Atomic Energy Division. Geological Survey of Great Britain 1957:66–67.

Bowie, S. H. U., and J. E. T. Horne 1953. Cheralite, a new mineral of the monazite group. *Mineralogical Magazine* 30:93–99.

Bowles, J. F. W., E. A. Jobbins, and B. R. Young 1980. A re-examination of cheralite. *Mineralogical Magazine* 43:885–88.

Boyle, D. R., A. L. Littlejohn, A. C. Roberts, and D. M. Watson 1981. Ningyoite in uranium deposits of south-central British Columbia: first North American occurrence. *Canadian Mineralogist* 19:325–31.

Braithwaite, R. S. W., and J. R. Knight .1990. Rare minerals, including several new to Britain, in supergene alteration of U-Cu-As-Bi-Co mineralisation near Dalbeattie, South Scotland. *Mineralogical Magazine* 54:129–131.

Branche, G., M. E. Ropert, F. Chantret, B. Morignat, and R. Pouget 1957. La francevillite, nouveau minéral uranifère. *Comptes Rendus Hebdomadaires des Séances de l'Académie des Sciences* 245:89–91 [see abstract (1958) American Mineralogist 43:180].

Branche, G., P. Bariand, F. Chantret, R. Pouget, and A. Rimsky 1963. La vanuralite, nouveau mineral. uranifere. *Comptes Rendus Hebdomadaires des Séances de l'Académie des Sciences* 256:5374–76; [see abstract (1963) *American Mineralogist* 48:1415–16].

Brasseur, H. 1948. Properties and chemical formula of fourmarierite. *American Mineralogist* 33:619–21.

Breithaupt, J. F. A. 1829 Neue krystallographische Bestimmung und mineralogische Charakteristik verschiedener Mineralspecien, II. Ueber den Monazit, eine neue Specie des Mineral-Reichs. *Journal für Chemie und Physik* 55:296–306.

Breithaupt, A. 1865. Mineralogische studien. 2. Fritzscheite und uranite überhaupt. *Berg- und Huttenmannische Zeit.* 24:302–3.

Brichard, H., and H. Brasseur 1958. Sur les autunites naturelles et synthétiques. *Bulletin de la Société Française de Minéralogie et de Cristallographie* 81:4–10.

Brøgger, W. C. 1906. Die mineralien der Südnorwegischen granitpegmatitgänge. *Skrifter udgivne af Videnskabs-Selskabet i Christiania* 1906, issue 6:1–162.

Brooker, E. J., and E. W. Nuffield 1952. Studies of radioactive compounds: IV - Pitchblende from Lake Athabasca, Canada. *American Mineralogist* 37:363–85.

Brophy, G. P., and P. F. Kerr 1953. Hydrous uranium molybdate in Marysvale ore, in Annual Report for June 30, 1952 to April 1, 1953, RME-3046, US Atomic Energy Commission, pp. 45–51.

Brugger, J., S. Ansermet, and A. Pring 2003. Uranium minerals from Mt. Painter, Northern Flinders Ranges, South Australia. *Australian Journal of Mineralogy* 9 (1):15–31.

Brugger, J., P. Burns, and N. Meisser 2003. Contribution to the mineralogy of acid drainage of Uranium minerals: Marecottite and the zippeite-group. *American Mineralogist* 88:676–85.

Brugger, J., S. V. Krivovichev, P. Berlepsch, N. Meisser, S. Ansermet, and T. Armbruster 2004. Spriggite, $Pb_3[(UO_2)_6O_8(OH)_2](H_2O)_3$, a new mineral with β-U_3O_8-type sheets: description and crystal structure. *American Mineralogist* 89:339–47.

Brugger, J., Wallwork, K., Meissner, N. Pring, A., Ondrus, P., and Cejka, J. (2006) Pseudojohannite from Jachymov, Musonoi, and La Creusaz: A new member of the zippeite group. *American Mineralogist* 91:929–36.

Brugger, J., N. Meisser, B. Etschmann, S. Ansermet, and A. Pring 2011. Paulscherrerite from the Number 2 Workings, Mount Painter Inlier, Northern Flinders Ranges, South Australia: "Dehydrated schoepite" is a mineral after all. *American Mineralogist* 96:229–40.

Bulakh, A. G., and A. A. Mazalov 1974. Accessory thorianite from the massifs of alkaline rocks and carbonatites of Turja Peninsula (Kola Peninsula). *American Mineralogist* 59:378–80.

Bültemann, H. W., and G. H. Moh 1959. Bergenit, ein neues Mineral der Phosphuranylit-Gruppe. *Neues Jahrbuch für Mineralogie, Monatshefte* 1959 (10):232–3; [see abstract (1960) *American Mineralogist* 45:909].

Burakov, B. E., E. E. Strykanova and E. B. Anderson 1996. Secondary uranium minerals on the surface of Chernobyl "lava". *MRS Proceedings* 465:1309.

Burns, P. C. 1997. A new uranyl oxide hydrate sheet in vandendriesscheite: Implications for mineral paragenesis and the corrosion of spent nuclear fuel. *American Mineralogist* 82:1176–86.

Burns, P. C. 1998a. The structure of richetite, a rare lead uranyl oxide hydrate. *Canadian Mineralogist* 36:187–99.

Burns, P. C. 1998b. The structure of compreignacite, $K_2[(UO_2)_3O_2(OH)_3]_2(H_2O)_7$. *Canadian Mineralogist* 36:1061–67.

Burns, P. C. 1998c. The structure of boltwoodite and implications of solid solution toward sodium boltwoodite. *Canadian Mineralogist* 36:1069–75.

Burns, P. C. 1999. A new complex sheet of uranyl polyhedra in the structure of wölsendorfite. *American Mineralogist* 84:1661–73.

Burns, P. C. 1999a. The crystal chemistry of uranium. *Reviews in Mineralogy* 38:23–90.

Burns, P. C. 2000. A new uranyl phosphate chain in the structure of parsonsite. *American Mineralogist* 85:801–5.

Burns, P. C. 2001. A new uranyl silicate sheet in the structure of haiweeite and comparison to other uranyl silicates. *Canadian Mineralogist* 39:1153–60.

Burns, P. C. 2001a. A new uranyl sulfate chain in the structure of uranopilite. *Canadian Mineralogist* 39:1139–46.

Burns, P. C. 2004. Advances in understanding of the crystal chemistry of hexavalent uranium. *Materials Research Society Symposium Proceedings* 802:DD3.2.1–12.

Burns, P. C. and Finch, R. 1999 *Uranium: Mineralogy, Geochemistry, and the Environment, Reviews in Mineralogy* 38, Mineralogical Society of America, Washington, DC.

Burns, P.C. and R. Finch 1999a. Wyartite: crystallographic evidence for the first pentavalent-uranium mineral. *American Mineralogist* 84:1456–60.

Burns, P. C. and J. M. Hanchar 1999. The structure of masuyite, $Pb[(UO_2)_3O_3(OH)_2](H_2O)_3$, and its relationship to protasite. *Canadian Mineralogist* 37:1483–91.

Burns, P. C. and K. A. Hughes 2003a. A new uranyl carbonate sheet in the crystal structure of fontanite, $Ca[(UO_2)_3(CO_3)_2O_2](H_2O)_6$. *American Mineralogist* 88:962–66.

Burns, P. C. and K. A. Hughes 2003b. Studtite, $[(UO_2)(O_2)(H_2O)_2](H_2O)_2$: The first structure of a peroxide mineral. *American Mineralogist* 88:1165–8.

Burns, P. C., and Y. Li 2002. The structures of becquerelite and Sr-exchanged becquerelite. *American Mineralogist* 87:550–57.

Burns, P. C., and G. E. Sigmon 2013. *Uranium - Cradle to Grave, Short Course Series* 43, Mineralogical Association of Canada, Quebec.

Burns, P. C., K. M. Deely, and L. A. Hayden 2003. The crystal chemistry of the zippeite group. *Canadian Mineralogist* 41:687–706.

Burns, P. C., M. L. Miller, and R. C. Ewing 1996. U^{6+} minerals and inorganic phases: A comparison and hierarchy of crystal structures. *Canadian Mineralogist* 34:845–880.

Burns, P. C., R. J. Finch, F. C. Hawthorne, M. L. Miller, and R. C. Ewing 1997. The crystal structure of ianthinite, $[U^{4+}_2(UO_2)_4O_6(OH)_4](H_2O)_5$: a possible phase for Pu^{4+} incorporation during the oxidation of spent nuclear fuel. *Journal of Nuclear Materials* 249:199–206.

Buryanova, E. Z., G. S. Strokova, and V. A. Shitov 1965. Vanuranylite, a new mineral. *Zapiski Vsesoyuznogo Mineralogicheskogo Obshchestva* 94:437–43 [see abstract (1966) *American Mineralogist* 51:1548].

Butler, J. R. 1957. Rare earths and thorium in lyndochite. *American Mineralogist* 42:671–76.

Butt, C. R. M., and J. Graham 1981. Sodian potassian hydroxonian meta-autunite: first natural occurrence of an intermediate member of a predicted solid solution series. *American Mineralogist* 66:1068–72.

Buttgenbach, H. 1924. Minéralogie du Congo Belge. La fourmariérite nouvell espèce minérale. *Annales de la Société Géologique de Belgique* 47:C41–C43.

Cairncross, B., and U. Bahmann 2006. Famous mineral localities: The Erongo Mountains, Namibia. *Mineralogical Record* 37 (5):361–470.

Calvo, C., and R. Faggiani 1974. A re-investigation of the crystal structures of chevkinite and perrierite. *American Mineralogist* 59:1277–85.

Cámara, F., C. T. Williams, G. Della Ventura, R. Oberti, and E. Caprilli 2004. Non-metamict betafite from Le Carcarelle (Vico volcanic complex, Italy): occurrence and crystal structure. *Mineralogical Magazine* 68:939–50.

Caprilli, E., G. Della Ventura, T. C. Williams, G. C. Parodi, and P. Tuccimei 2006. The crystal chemistry of non-metamict pyrochlore-group minerals from Latium, Italy. *Canadian Mineralogist* 44:1367–78.

Cassedanne, J. P. 1986. The Urucum pegmatite, Minas Gerais, Brazil. *Mineralogical Record* 17 (5):307–14.

Čejka, J., Z. Mrazek, and Z. Urbanec 1984. New data on sharpite, a calcium uranyl carbonate. *Neues Jahrbuch für Mineralogie, Monatsh.* 109–17 [see abstract (1985) *American Mineralogist* 70:220].

Čejka, J., J. Sejkora, Z. Mrázek, Z. Urbanec, and T. Jarchovský 1996. Jáchymovite, $(UO_2)_8(SO_4)(OH)_{14}\cdot 13H_2O$, a new uranyl mineral from Jáchymov, the Krusné hory Mts., Czech Republic, and its comparison with uranopilite. *Neues Jahrbuch für Mineralogie, Abhandlungen* 170:155–170 [see abstract (1997) *American Mineralogist* 82:207–8].

Cesbron, F. 1970. Nouvelles donnees sur la vanuralite. Existence de la metavanuralite. *Bulletin de la Société Française de Minéralogie et de Cristallographie* 93:242–48; [see abstract (1971) *American Mineralogist* 56:637].

Cesbron, F, and P. Bariand 1975. The uranium-vanadium deposit of Mounana, Gabon. *Mineralogical Record* 6 [5]:237–49.

Cesbron, F., and N. Morin 1968. Une nouvelle espece minerale: la curienite. Etude de la serie francevillite-curienite. *Bulletin de la Société Française de Minéralogie et de Cristallographie* 91:453–59 [see abstract (1969) *American Mineralogist* 54:1220].

Cesbron, F., W. L. Brown, P. Bariand, and J. Geffroy 1972. Rameauite and agrinierite, two new hydrated complex uranyl oxides from Margnac, France. *Mineralogical Magazine* 38:781–9 [see abstract (1973) *American Mineralogist* 58:805–7].

Cesbron, F., B. Bachet, and R. Oosterbosch 1965. La demesmaekerite, selenite hydrate d'uranium, cuiver, et plomb. *Bulletin de la Société Française de Minéralogie et de Cristallographie* 88: 422–5 [see abstract (1966) *American Mineralogist* 51:1815–16].

Cesbron, F., R. Oosterbosch, and R. Pierrot 1969. Une nouvelle espece minerale: la marthozite, uranylselenite de cuivrehydrate. *Bulletin de la Société Française de Minéralogie et de Cristallographie* 92:278–83 [see abstract (1970) *American Mineralogist* 55:533].

Cesbron, F., R. Pierrot, and T. Verbeek 1970. La roubaultite, $Cu_2(UO_2)_3(OH)_{10}.5H_2O$, une nouvelle espece minerale. *Bulletin de la Société Française de Minéralogie et de Cristallographie* 93:550–54 [see abstract (1972) *American Mineralogist* 57:1912].

Cesbron, F., R. Pierrot, and T. Verbeek 1971. La derriksite, $Cu_4(UO_2)(SeO_3)2(OH)_6{\cdot}H_2O$, une nouvelle espece mineral. *Bulletin de la Société Française de Minéralogie et de Cristallographie* 94:534–7 [see abstract (1972) *American Mineralogist* 57:1912–3].

Chen, Z., K. Luo, F. Tan, Y. Zhang, and X. Gu 1986. Tengchongite, a new mineral of hydrated calcium uranyl molybdate. *Kexue Tongbao* 31:396–401 [see abstract (1988) *American Mineralogist* 73:195–6

Chen, Z., Y. Huang, and X. Gu 1990. A new uranium mineral - yingjiangite. *Acta Mineral. Sinica* 10:102–5; [see abstract (1991) *American Mineralogist* 76:1731–2].

Chernikov, A. A., and N. I. Organova 1994. Sodium autunite and sodium meta-autunite. *Doklady Akademii Nauk SSSR* 338:368–71 [see abstract (1995) *American Mineralogist* 80:1329–30].

Chernikov, A. A., O. V. Krutetskaya, and N. I. Organova 1957. Natroautunite. *Soviet Journal of Atomic Energy* 3:901–5; [see abstract (1958) *American Mineralogist* 43:383].

Chernikov, A., O. Krutetskaya, and V. Sidel'nikova 1957. Ursilite - a new silicate of uranium. *Voprosy geologii urana* 1957:73–7; [see abstract (1959) *American Mineralogist* 44:464].

Chernikov, A. A., D. P. Shaskin, and I. N. Gavrilova 1975. Sodium boltwoodite. *Doklady Akad. Nauk SSSR* 221:195–7 [see abstract (1976) *American Mineralogist* 61:1054–5].

Chiappero, P. J., and H. Sarp 1993. New data and the second occurrence of znucalite. *Archs Sci. Genève* 46:291–301 [see abstract (1994) *American Mineralogist* 79:1214].

Chirvinsky, P. N. 1925. Tyuyamunite from the Tyuya-Muyun radium mine in Fergana. *Mineralogical Magazine* 20:287–95.

Christ, C. L. 1965. Phase transformations and crystal chemistry of schoepite. *American Mineralogist* 50:235–9.

Christ, C. L., and J. R. Clark 1960. Crystal chemical studies of some uranyl oxide hydrates. *American Mineralogist* 45:1026–61.

Christy, A. G., and D. Atencio 2013. Clarification of status of species in the pyrochlore supergroup. *Mineralogical Magazine* 77:13–20.

Chukanov, N. V., S. Möckel, G. A. Sidorenko, and V. A. Zaitsev 2009. Arsenuranospathite, $Al(UO_2)_2(AsO_4)_2F \cdot 20H_2O$: Formula revision and relationships with allied uranyl arsenates and phosphates. Neues Jahrbuch für Mineralogie, Abhandlungen: Journal of Mineralogy and Geochemistry 185:305–12.

Chukanov, N. V., D. Yu. Pushcharavosky, M. Pasero, S. Merlino, A. V. Barinova, S. Möckel, I. V. Pekov, A. E. Zadov, and V. T. Dubinchuk 2004. Larisaite, $Na(H_3O)(UO_2)_2(SeO_3)_2 \cdot 4H_2O$, a new uranyl selenite mineral from Repete mine, San Juan County, Utah, USA, *European Journal of Mineralogy* 16:367–74.

Chukanov, N. V., G. A. Sidorenko, I. S. Naumova, A. E. Zadov, and V. I. Kuz′min 2006. Chistyakovaite, a new mineral $Al(UO_2)_2(AsO_4)_2(F,OH) \cdot 6.5H_2O$. (Translated from *Doklady Akademii Nauk* 406:816–19). *Doklady Earth Sciences* 407:290–93.

Chukanov, N., A. Skrigitil', O. Kuz'mina, and A. Zadov 1999. Bismutopyrochlore $(Bi,U,Ca,Pb)_{1+x}(Nb,Ta)_2O_6(OH) \cdot nH_2O$ - a new mineral from the Mika pegmatite vein (Eastern Pamirs), *Zapiski Vserossiskogo Mineralogichesk-ogo Obshchestva* 128 [4]:36–41; [see abstract (2000) *American Mineralogist* 85:1561].

Clark, J. 1960. X-ray study of alteration in the uranium mineral wyartite. *American Mineralogist* 45:200–8.

Clark, J. and C. Christ 1956. Some observations on rutherfordine. *American Mineralogist* 41:844–50.

Coda, A., A. Della Giusta, and V. Tazzoli 1981. The structure of synthetic andersonite, $Na_2Ca[UO_2(CO_3)_3] \cdot xH_2O$ ($x \approx 5.6$). *Acta Crystallographica* B37:1496–1500.

Coleman, R. G., and D. E. Appleman 1957. Umohoite from the Lucky Mc mine, Wyoming. *American Mineralogist* 42:657–60.

Coleman, R. G., D. R. Ross, and R. Meyrowitz 1966. Zellerite and metazellerite, new uranyl carbonates. *American Mineralogist* 51:1567–78.

Congdon, R. D., and W. P. Nash 1991. Eruptive pegmatite magma: Rhyolite of the Honeycomb Hills, Utah. *American Mineralogist* 76:1261–78.

Cooper, M. A., and F. C. Hawthorne 1995. The crystal structure of guilleminite, a hydrated Ba-U-Se sheet structure. *Canadian Mineralogist* 33:1103–09.

Cooper, M. A., and F. C. Hawthorne 2001. Structure topology and hydrogen bonding in marthozite, $Cu^{2+}[(UO_2)_3(SeO_3)_2O_2](H_2O)_8$, a comparison with guilleminite, $Ba[(UO_2)_3(SeO_3)_2O_2](H_2O)_3$. *Canadian Mineralogist* 39:797–807.

Cooper, M. A., Y. A. Abdu, N. A. Ball, P. Černý, and F. C. Hawthorne 2012. Aspedamite, ideally $\square_{12}(Fe^{3+},Fe^{2+})_3Nb_4[Th(Nb,Fe^{3+})_{12}O_{42}]\{(H_2O),(OH)\}_{12}$, a new heteropolyniobate mineral species from the Herrebøkasa quarry, Aspedammen, Østfold, southern Norway: Description and crystal structure. *Canadian Mineralogist* 50:793–804.

Costin, D. A., A. Mesbah, N. Clavier, N. Dacheux, C. Poinssot, S. Szenknect, and J. Ravaux 2011. How to explain the experimental difficulties in the coffinite synthesis: hints from the study of $Th_{1-x}U_xSiO_4$ uranothroite solid solutions. *Inorganic Chemistry* 50:11117–26.

Cottle, J. M. 2014. In-situ U-Th/Pb geochronology of (urano) thorite. *American Mineralogist* 99:1985–95.

Crook, T., and G. S. Blake 1910. On carnotite and an associated mineral complex from South Australia. *Mineralogical Magazine* 15:271–84.

Cuney, M., and K. Kyser 2008. *Recent and not-so-recent developments in uranium deposits and implications for exploration*, *Short Course Series* 39, Mineralogical Association of Canada, Quebec.

Dahlkamp, F. J. 1993. *Uranium ore deposits.* Berlin: Springer-Verlag. 460 pp.

Dana, E. S. 1879. Examination of the North Carolina uranium minerals; by F. A. Genth. *American Journal of Science and Arts* 118:153–54.

De Abeledo, M. E. J., E. E. Galloni, and M. A. R. Benyacar 1968. Electron diffraction data for some members of the metatorbernite group. *American Mineralogist* 53:1028–33.

De Camargo, W. G. R, and D. P. Svisero 1969. Haiweeite, a new occurrence in Brazil. *American Mineralogist* 54:966–9.

Deditius, A. P., S. Utsonomiya, and R. C. Ewing 2008. The chemical stability of coffinite, $USiO_4 \cdot nH_2O$, $0<n<2$, associated with organic matter: A case study from Grants uranium region, New Mexico, USA. *Chemical Geology* 251:33–49.

Deditius, A. P., S. Utsonomiya, M. W. Wall, V. Pointeau, and R. C. Ewing 2009. Crystal chemistry and radiation-induced amorphization of P-coffinite from the natural fission reactor at Bangombé, Gabon. *American Mineralogist* 94:827–36.

Deer, W. A., R. A. Howie, and J. Zussman 1982. *Rock-Forming Minerals,* Volume 1A, *Orthosilicates*, Second Edition. New York: Longman, 919 pp.

Deer, W. A., R. A. Howie, and J. Zussman 1986. *Rock-forming minerals,* Volume 1B, *Disilicates and ring silicates*, Second Edition. Harlow: Longman Group UK Ltd.

Deliens, M. 1977. Review of the hydrated oxides of U and Pb, with new X-ray powder data. *Mineralogical Magazine* 41:51–7.

Deliens, M. 1992. Etude comparative des calcurmolites de Rabejac (Lodeve, Herault, France) et de l'Union Sovietique. *Allales de la Societe Geologique de Belgique* 115:91–91.

Deliens, M., and P. Piret 1978. La phurcalite, $Ca_2(UO_2)_3PO_2(OH)_4 \cdot 4H_2O$, nouveau mineral. *Bulletin de Minéralogie* 101:356-8 [see abstract (1979) *American Mineralogist* 64:243].

Deliens, M., and P. Piret 1979a. Les phosphates d'uranyl et d'aluminium de Kobokobo. IV. La threadgoldite, nouveau mineral. *Bulletin de Minéralogie* 102:338–41 [see abstract (1980) *American Mineralogist* 65:209].

Deliens, M., and P. Piret 1979b. Les phosphates d'uranyl et d'aluminium de Kobokobo. II. La phuralumite $Al_2(UO_2)_3(PO_4)_2(OH)_6 \cdot 10H_2O$ et l'upalite $Al(UO_2)_3(PO_4)_2(OH)_3$, nouveaux minereaux. *Bulletin de Minéralogie* 102:333–7 [see abstract (1980) *American Mineralogist* 65:208].

Deliens, M., and P. Piret 1979c. Ranunculite, $AlH(UO_2)(PO_4)(OH)_3 \cdot 4H_2O$, a new mineral. *Mineralogical Magazine* 43:321–3.

Deliens, M., and P. Piret 1981. La swamboite, nouveau silicate d'uranium hydrate du Shaba, Zaire. *Canadian Mineralogist* 19:553–7.

Deliens, M., and P. Piret 1981a. Les phosphates d'uranyl et d'aluminium de Kobokobo. V. La mundite, nouveau mineral. *Bulletin de Minéralogie* 104:669–71 [see abstract (1982) *American Mineralogist* 67:624].

Deliens, M., and P. Piret 1982. Les phosphates d'uranyl et d'aluminium de Kobokobo. VI. La triangulite $Al_3(UO_2PO_4)_4(OH)_5 \cdot 5H_2O$, nouveau mineral. *Bulletin de Minéralogie* 105:611–14 [see abstract *American Mineralogist* 69:212].

Deliens, M., and P. Piret 1982a. Bijvoetite et lepersonnite, carbonates hydrates d'uranyl et de terres rares de Shinkolobwe, Zaire. *Canadian Mineralogist* 20:231–38 [see abstract (1983) *American Mineralogist* 68:1248].

Deliens, M., and P. Piret 1983. L'oursinite $(Co_{0.86}Mg_{0.10}Ni_{0.04})O \cdot 2UO_3 \cdot 2SiO_2 \cdot 6H_2O$, nouveau mineral de Shinkolobwe, Zaire. *Bulletin de Minéralogie* 106:305–8 [see abstract (1984) *American Mineralogist* 69:567].

Deliens, M., and P. Piret 1984 La kamitugaite, $PbAl(UO_2)_5[(P,As)O_4]_2(OH) \cdot 9H_2O$, nouveau mineral de Kobokobo, Kivu, Zaire, *Bulletin de Minéralogie* 107:15–19 [see abstract (1985) *American Mineralogist* 70:437].

Deliens, M. and P. Piret 1984a. Urancalcarite, $Ca(UO_2)_3CO_3(OH)_6 \cdot 3H_2O$, a new mineral from Shinkolobwe, Shaba, Zaire. *Bulletin de Minéralogie* 107:21–4 [see abstract (1985) *American Mineralogist* 70:438–9].

Deliens, M., and P. Piret 1985 Les phosphates d'uranyle et d'aluminium de Kobokobo. VII. La moreauite, $Al_3(UO_2)(PO_4)_3(OH)_2 \cdot 13H_2O$, nouveau mineral. *Bulletin de Minéralogie* 108:9–13 [see abstract (1985) *American Mineralogist* 70:1330–1].

Deliens, M., and P. Piret 1985a Les phosphates d'uranyle et d'aluminium de Kobokobo. VIII. La furongite. *Annales de Societe Geologique de Belgique* 108365–68:9–13 [see abstract of related paper (1988) *American Mineralogist* 73:198].

Deliens, M., and P. Piret 1986. La kamotoite-(Y), un nouveau carbonate d'uranyle et terres rares de Kamoto, Shaba, Zaire. *Bulletin de Minéralogie* 109:643–7; [see abstract (1988) *American Mineralogist* 73:91].

Deliens, M. and P. Piret 1989. Shababite-(Nd), $Ca(REE)_2(UO_2)(CO_3)_4(OH)_6 \cdot 6H_2O$, a new mineral species from Kamoto, Shaba, Zaire. *European Journal of Mineralogy* 1:85–8; [see abstract (1990) *American Mineralogist* 75:433].

Deliens, M, and P. Piret 1990. L'astrocyanite-(Ce), $Cu_2(REE)_2(UO_2)(CO_3)_5(OH)_2 \cdot 1.5H_2O$, nouvelle espece mineral de Kamoto, Shaba, Zaire. *European Journal of Mineralogy* 2:407–11; [see abstract (1991) *American Mineralogist* 76:665].

Deliens, M. and P. Piret 1991. La haynesite, selenite hydrate d'uranyl, nouvelle espece minerale de la mine Repete, Comte de San Juan, Utah. *Canadian Mineralogist* 29:561–64; [see abstract (1997) *New Minerals 1990–1994*, p. 79, J. Mandarino, ed., The Mineralogical Record, Tucson].

Deliens, M., and P. Piret 1992. La fontanite, carbonate hydrate d'uranyl et de calcium, nouvelle espece minerale de Rabajac, Herault, France. *European Journal of Mineralogy* 4:1271–4; [see abstract (1997) *New Minerals 1990–1994*, p. 65, J. Mandarino, ed., The Mineralogical Record, Tucson].

Deliens, M., and P. Piret 1993. La rabejacite, $Ca(UO_2)_4(SO_4)_2(OH)_6 \cdot 6H_2O$, nouveau sulfate d'uranyle et de calcium des gîtes du Lodévois, Hérault, France. *European Journal of Mineralogy* 5:873–77 [see abstract (1994) *American Mineralogist* 79:572].

Deliens, M., Piret, P., and Comblain, G. 1981. *Les Mineraux Secondaires D'Uranium Du Zaire*, Tervuren, Belgium.

Deliens, M., Piret, P., and Comblain, G. 1984. *Les Mineraux Secondaires D'Uranium Du Zaire (Supplement)*, Tervuren Belgium.

Deliens, M., Piret, P., and Van Der Meersche, E. 1990. *Les Mineraux Secondaires D'Uranium Du Zaire (2nd Supplement)*, Tervuren, Belgium.

Della Ventura, G., P. Bonazzi, R. Oberti and L. Ottolini 2002. Ciprianiite and mottanaite-(Ce), two new minerals of the hellandite group from Latium (Italy). *American Mineralogist* 87:739–44.

Desautels, P. 1967. The morphology of mckelvyite. *American Mineralogist* 52:860–64.

DeVoto, R. 1978. Uranium in Phanerozoic sandstone and volcanic rocks. *MAC Short Course Series* 3:293–306.

Donnay, G., and J. D. H. Donnay 1955. Contribution to the crystallography of uranium minerals. *USGS Trace Elements Investigations Report 507*.

Doynikova, O. A., and G. A. Sidorenko 2006. Coconinoite and related alumouranyl phosphates. *New Data on Minerals* 41:108–17.

Dunn, P. and J. Mandarino 1988. Formal definitions of type mineral specimens. *Mineralogical Record* 19 (4):227–28.

Dunstan, W. 1904. The occurrence of thorium in Ceylon. *Nature* 69:510–11.

Dürrfeld, V. 1913. Über krystalle eines wasserhaltigen blei-zink-vanadinats von Reichenbach bei Lahr (Schwarzwald). *Zeitschrift für Krystallographie und Mineralogie* 51:278–9.

Edwards, R. and K. Atkinson 1986. *Ore Deposit Geology and its Influence on Mineral Exploration,* 466 pp., Chapman and Hall, London.

Ellsworth, H. V. 1927. Toddite - a new uranium mineral from the Sudbury district, Ontario. *American Mineralogist* 11:332–34.

Ellsworth, H. V. 1927a. Lyndochite - a new mineral of the euxenite-polycrase group from Lyndoch Township, Renfrew County, Ontario. *American Mineralogist* 12:212–18.

Ellsworth, H. V. 1928a. A mineral related to samarskite from the Woodcox mine, Hybla, Ontario. *American Mineralogist* 13:63–5.

Ellsworth, H. V. 1928b. A mineral related to samarskite from Parry Sound, Ontario. *American Mineralogist* 1366–68.

Elton, N. J., and J. J. Hooper 1995. Widenmannite from Cornwall, England: the second world occurrence. *Mineralogical Magazine* 58:339–41

Elton, N. J., J. J. Hooper, and A. E. Jeal 1994, Nováčekite and metanováčekite from Cornwall. *Mineralogical Magazine* 58:513–14.

Elton, N. J., J. J. Hooper, and G. Ryback 1994. Compreignacite: a second occurrence, from St. Just, Cornwall. *Mineralogical Magazine* 58:339–41

Embrey, P. and R. Symes 1987. *Minerals of Cornwall and Devon*, British Museum (Natural History), London.

Epshtein, G. Yu. 1959. On the molybdates of uranium - moluranite and iriginite, *Zapiski Vsesoyuznogo Mineralicheskogo Obshchestva* 88:564–70; [see abstract (1960) *American Mineralogist* 45:257–8].

Ercit, T. S. 2002. "Lyndochite" revisited: A cautionary note on discreditations. *Canadian Mineralogist* 40:1211–13.

Es'kova, E. M., E. I. Semenov, A. P. Khomyakov, A. N. Mer'kov, S. I. Lebedeva, and L. S. Dubakina 1974. Umbozerite - a new mineral. *Doklady Akademii Nauk SSSR* 216:169–71 [see abstract (1975) *American Mineralogist* 60:341].

Evans, H. T. 1950. Studies of uranium minerals (VI): walpurgite. *American Mineralogist* 1021–27.

Ewing, R. C. 1976. A numerical approach toward the classification of complex, orthorhombic, rare-earth, AB_2O_6-type Nb-Ta-Ti oxides. *Canadian Mineralogist* 14:111–19.

Ewing, R. C., K. G. Snetsinger, and T. E. Bunch 1977. Euxenite from Ampangabé, Madagascar. *Canadian Mineralogist* 15:92–6.

Federov, O. V. 1963. Second find of calcium uranyl molybdate in the USSR. *Zapiski Vsesoyuznogo Mineralicheskogo Obshchestva* 92:464–5; [see abstract (1964) *American Mineralogist* 49:1152–3].

Fedorov, O. V., B. I. Ryzhov, and G. V. Lyubomilova 1982. K-Sr variety of calciouranoite. *Doklady Akademii Nauk SSSR* 262:209–13.

Fejfarová, K., J. Plášil, H. Yang, J. Čejka, M. Dušek, R. T. Downs, M. C. Barkley, and R. Škoda 2012. Revision of the crystal structure and chemical formula of weeksite, $K_2(UO_2)_2(Si_5O_{13})\cdot 4H_2O$. *American Mineralogist* 97:750–54.

Finch, R.J., and R. C. Ewing 1997. Clarkeite: new chemical and structural data. *American Mineralogist* 82:607–19.

Finch, R. J., and F. C. Hawthorne 1998. Structural relations among schoepite, metaschoepite, and "dehydrated schoepite". *Canadian Mineralogist* 36:831–45.

Finch, R. J. and T. Murakami 1999. Systematics and paragenesis of uranium minerals. *Reviews in Mineralogy* 38:91–179.

Finch, R. J., P. C. Burns, F. C. Hawthorne, and R. C. Ewing 2006. Refinement of the crystal structure of billietite, $Ba[(UO_2)_6O_4(OH)_6(H_2O)_8$. *Canadian Mineralogist* 44:1197–1205.

Finch, R. J., M. A. Cooper, and F. C. Hawthorne 1996. The crystal structure of schoepite, $[(UO_2)_8O_2(OH)_{12}](H_2O)_{12}$. *Canadian Mineralogist* 34:1071–88.

Finch, R. J., M. A. Cooper, F. C. Hawthorne, and R. C. Ewing 1999. Refinement of the crystal structure of rutherfordine. *Canadian Mineralogist* 37:929–38.

Finch, R. J., F. C. Hawthorne, and R. C. Ewing 1998. Structural relations among schoepite, metaschoepite, and "dehydrated schoepite". *Canadian Mineralogist* 36:831–45.

Finch, W., P. Molina, S. Naumov, V. Ruzica, F. Barthel, V. Thoste, E., Muller-Kahle, M. Pecnik, and M. Thauchid 1995. World Distribution of Uranium Deposits, First Edition; International Atomic Energy Agency, scale 1:30,000,000.

Finney, J., and N. Rao 1967. The crystal structure of cheralite. *American Mineralogist* 52:13–19.

Fischer, E. 1946. Kristallform und Kristalelement des Walpurgins. *Zeitschrift Kristallographie.* 106:25–33.

Fisher, F., and R. Meyrowitz 1962. Brockite, a new calcium thorium phosphate from the Wet Mountains, Colorado. *American Mineralogist* 47:1346–55.

Fleet, M. and Y. Pan 1995. Site preference of rare earth elements in fluorapatite. *American Mineralogist* 80:329–35.

Fleet, M., X. Liu, and Y. Pan 2000. Rare-earth elements in chlorapatite $[Ca_{10}(PO_4)_6Cl_2]$: Uptake, site preference, and degradation of monoclinic structure. *American Mineralogist* 85:1437–46.

Fleischer, M. 1962. New data on rhabdophane group. *American Mineralogist* 47:419–20.

Fleischer, M. 1966. Rare earths in the aeschynite-priorite series. The status of lyndochite. *Mineralogical Magazine* 35:801–9.

Foord, E., W. Chirnside, A. Davis, F. Lichte, and K Esposito 1995. A new U-Ti-Ca-HREE hydrated oxide and associated niobian rutile from Topaz Valley, Utah. *Mineralogical Record* 26 (2):123–28.

Ford, M. A. 1993. Uranium in South Africa. *Journal of the South African Institute of Mining and Metallurgy* 93:37–58.

Förster, H.-J., and D. E. Harlov 1999. Monazite–(Ce) – huttonite solid solutions in granulite–facies matabasites from the Ivrea–Verbano Zone, Italy. *Mineralogical Magazine* 63:587–94.

Förster, H.-J., D. E. Harlov, and R. Milke 2000. Composition and Th - U - total Pb ages of huttonite and thorite from Gillespie's Beach, South Island, New Zealand. *Canadian Mineralogist* 38:675–84.

Friedel, C. and E. Cumenge 1899. Sur un nouveau minerai d'urane. *Bulletin de la Société Française de Minéralogie* 22:26–9.

Frondel, C. 1950. Studies of uranium minerals (I): parsonsite and randite. *American Mineralogist* 35:245–50.

Frondel, C. 1950a. Studies of uranium minerals (V): phosphuranylite. *American Mineralogist* 35:756–63.

Frondel, C. 1951. Studies of uranium minerals (VIII): Sabugalite, an aluminum-autunite. *American Mineralogist* 36:671–9.

Frondel, C. 1951a. Studies of uranium minerals (IX): saleeite and novacekite. *American Mineralogist* 36:680–6.

Frondel, C. 1952. Studies of uranium minerals (X): uranopilite. *American Mineralogist* 37:950–9.

Frondel, C. 1953. Hydroxyl substitution in thorite and zircon. *American Mineralogist* 38:1007–18.

Frondel, C. 1954. Bassetite and uranospathite. *Mineralogical Magazine* 30:343–53.

Frondel, C. 1958. *Systematic Mineralogy of Uranium and Thorium.* Geological Survey Bulletin 1064, U. S. Government Printing Office.

Frondel, C., and F. Cuttita 1954. Studies of uranium minerals (XIV): renardite. *American Mineralogist* 39:448–51.

Frondel, C. and J. Ito 1956. Boltwoodite, a new uranium silicate mineral. *Science* 124:931.

Frondel, C., and R. Meyrowitz 1956. Studies of uranium minerals (XIX): Rutherfordine, diderichite, and clarkeite. *American Mineralogist* 41:127–33.

Frondel, C., J. Ito, R. Honea, and A. Weeks 1976. Minerαlogy of the zippeite group. *Canadian Mineralogist* 14:429–36.

Frondel, J. W. 1951. Studies of uranium minerals (VII): zeunerite. *American Mineralogist* 36:249–55

Frondel, J. W., and F. Cuttitta 1953. Studies of uranium minerals (XII): The status of billietite and becquerelite. *American Mineralogist* 38:1019–24.

Frondel, J. W., and F. Cuttitta 1954. Studies of uranium minerals (XVI): An alteration product of ianthinite. *American Mineralogist* 39:1018–20.

Frondel, J. W., M. Fleischer, and R. Jones 1967. *Glossary of Uranium- and Thorium-Bearing Minerals, 4th Edition.* Geological Survey Bulletin 1250, US Government Printing Office.

Frost, R. L., J. Čejka, and M. J. Dickfos 2009. Raman spectroscopic study of the uranyl minerals vanmeersscheite $U(OH)_4[(UO_2)_3(PO_4)_2(OH)_2]·4H_2O$ and arsenouranylite $Ca(UO_2)[(UO_2)_3(AsO_4)_2(OH)_2]·(OH)_2.6H_2O$. *Spectrochimica Acta Part A: Molecular and Biomolecular Spectroscopy* 71 (5):1799–1803.

Frost, R. L., J. Čejka, M. L. Weier, W. Martens, and D. A. Henry 2005. Vibrational spectroscopy of selected natural uranyl vanadates. *Vibrational Spectroscopy* 39:131–38.

Frost, R. L., S. J. Palmer, and J. Čejka 2011. The application of Raman spectroscopy to the study of the uranyl mineral coconinoite, $Fe_2Al_2(UO_2)_2(PO_4)_4(SO_4)(OH)_2·20H_2O$. *Spectroscopy Letters* 44 (6):381–87.

Frost, R. L., M. L. Weier, G. A. Ayoko, W. Martens, and J. Čejka 2006. An XRD, SEM, and TG study of a uranopilite from Australia. Mineralogical Magazine 70:299–307.

Gaboreau, S., M. Cuney, D. Quirt, D. Beaufort, P. Patrier, and R. Mathieu 2007. Significance of aluminum phosphate-sulfate minerals associated with U unconformity-type deposits: The Athabasca basin, Canada. *American Mineralogist* 92:267–80.

Gahn, J. G., J. Berzelius, C. Wallman, and H. P. Eggertz 1817. Examination of some minerals found in the neighbourhood of Fahlun, and of their situation. *Annals of Philosophy* 9: 452–60.

Gaines, R. V. 1965. Moctezumite, a new lead uranyl tellurite. American Mineralogist 50:1158–63.

Gaines, R. V. 1969. Cliffordite, a new tellurite mineral from Moctezuma, Sonora, Mexico. *American Mineralogist* 54:697–701.

Gaines, R. V. 1971. Schmitterite - a new uranyl tellurite from Moctezuma, Sonora. *American Mineralogist* 56:411–15.

Gallagher, M. J., and D. Atkin 1966. Meta–ankoleite, hydrated potassium uranyl phosphate. *Bulletin Geological Society of Great Britain* 25:49–54 [see abstract (1967) *American Mineralogist* 52:560].

Galuskin, E. V., T. Armbruster, I. O. Galuskina, B. Lazic, A.Winiarski, V. M. Gazeev, P. Dzierzanowski , A. E. Zadov, N. N. Pertzev, R. Wrzalik, A. G. Gurbanov, and J. Janeczek 2011. Vorlanite $(Ca(U^{6+})O_4$ – A new mineral from the Upper Chegem caldera, Kabardino–Balkaria, Northern Caucasus, Russia. *American Mineralogist* 96:188–96.

Galuskin, E. V., I. O. Galuskina, L. S. Dubrovinsky, and J. Janeczek 2012. Thermally induced transformation of vorlanite to "protovorlanite": Restoration of cation ordering in self-irradiated $CaUO_4$. *American Mineralogist* 97:1002–4.

Galuskin, E. V., I. O. Galuskina, J. Kusz, T. Armbruster, K. M. Marzec, P. Dzierżanowski P, and M. Murashko 2014. Vapnikite Ca_3UO_6 —a new double-perovskite mineral from pyrometamorphic larnite rocks of the Jabel Harmun, Palestinian Autonomy, Israel. *Mineralogical Magazine* 78:571–81.

Galuskin, E. V., J. Kusz, T. Armbruster, I. O. Galuskina, K. Marzec, Y. Vapnik , and M. Murashko 2013. Vorlanite, $(CaU^{6+})O_4$, from Jabel Harmun, Palestinian Autonomy, Israel. *American Mineralogist* 98:1938–42.

Galuskina, I. O., Y. Vapnik, B. Lazic, T. Armbruster, M. Murashko, and E. V. Galuskin 2014. Harmunite $CaFe_2O_4$: A new mineral from the Jabel Harmun, West Bank, Palestinian Autonomy, Israel. *American Mineralogist* 99:965–75.

Galy, J., and G. Meunier 1971. A propos de la cliffordite UTe_3O_8. Le système UO_3–TeO_2 à 700°C. Structure cristalline de UTe_3O_9. *Acta Crystallographica* B27:608–16.

Gatehouse, B. M., I. E. Grey, and P. R. Kelly 1979. The crystal structure of davidite. *American Mineralogist* 64:1010–17.

Gaubert, M. P. 1904. Produits de déshydratation de quelques phosphates et orientation du chlorure de baryum sur les minéraux du groupe de l´autunite. *Bulletin de la Société Française de Minéralogie* 27:222–33.

Gauthier-Lafay, F., and F. Weber 1989. The Francevillian (Lower Proterozoic) uranium deposit of Gabon. *Economic Geology* 84:2267–85.

Gauthier-Lafay, F., F. Weber, and H. Ohmoto 1989. Natural fission reactors of Oklo. *Economic Geology* 84:2286–95.

Genth, F. A. 1879. Examination of the North Carolina uranium minerals. *American Chemical Journal* 1:87–93.

George, D. R. 1951. Thorite from California, a new occurrence and variety. *American Mineralogist* 36:129–32.

George-Aniel, B., J. Leroy, and B. Poty 1991. Volcanogenic uranium mineralizations in the Sierra Pena Blanca distric, Chihuahua, Mexico: three genetic models. *Economic Geology* 86:233–48.

Giere, R., and S. Sorensen 2004. Allanite and other REE-rich epidote-group minerals. *Reviews in Mineralogy and Geochemistry* 46:431–93.

Ginderow, D., and F. Cesbron 1983a. Structure de la demesmaekerite, $Pb_2Cu_5(SeO_3)_6(UO_2)_2(OH)_6 \cdot 2H_2O$. *Acta Crystallographica* C39:824–27.

Ginderow, D., and F. Cesbron 1983b. Structure da la derriksite, $Cu_4(UO_2)(SeO_3)_2(OH)_6$. *Acta Crystallographica* C39:1605–7.

Ginderow, D., and F. Cesbron 1985. Structure de la roubaultite, $Cu_2(UO_2)_3(CO_3)_2O_2(OH)_2 \cdot 4H_2O$. *Acta Crystallographica* C41:654–57 [see abstract (1986) *American Mineralogist* 72:1282].

Goodel, P. C. 1981. Geology of the Peña Blanca uranium deposits, Chihuahua, Mexico, in *Uranium in volcanic and volcanoclastic rocks*, P. Goodell and A. Waters, eds., AAPG Studies in Geology 13:275–91.

Gordillo, C. E., E. Linares, R. O. Toubes, and H. Winchell 1966. Huemulite, $Na_4MgV_{10}O_{28} \cdot 24H_2O$, a new hydrous sodium and magnesium vanadate from Huemul mine, Mendoza Province, Argentina. *American Mineralogist* 51:1–13.

Gorman, D. H. 1957. Studies of radioactive compounds, IX - sklodowskite. *Canadian Mineralogist* 6:52–60.

Gorman, D. H., and E. W. Nuffield 1955. Studies of radioactive compounds: VIII - uranophane and beta-uranophane. *American Mineralogist* 634–45.

Gotman, Y. D., and I. A. Khapaev 1958. Thorutite—a new mineral of the group of titanites of thorium. *Zapiski Vsesoyuznogo Mineralogicheskogo Obshchestva* 87:201–2; [see abstract (1958) *American Mineralogist* 43:1007].

Granger, M. M., and J. Protas 1965. Étude de la structure cristalline de la compreignacite. *Bulletin de la Société Française de Minéralogie et de Cristallographie* 88:211–13.

Gross, E. B., A. S. Corey, R. S. Mitchell, and K. Walenta 1958. Heinrichite and metaheinrichite, hydrated barium uranyl arsenate minerals. *American Mineralogist* 43:1134–43.

Gruner, J. W. 1954. The chemical formula of clarkeite. *American Mineralogist* 39:836–38.

Guillemin, C., A. Destas, and J. Vaes 1958. *Mineraux D'Uranium Du Haut Katanga*, Tervuren, Belgium.

Guillemin, C., and J. Protas 1959. Ianthinite et wyartite. *Bulletin de la Société Française de Minéralogie et de Cristallographie* 82:80–86 [see abstract (1959) *American Mineralogist* 44:908].

Guimaraes, C. 1941. Djalmaite, a new radioactive mineral. *American Mineralogist* 26:343–46.

Guo, X., S. Szenknect, A. Mesbah, S. Labs, N. Clavier, C. Poinssot, S. V. Ushakov, H. Curtius, D. Bosbach, R. C. Ewing, P. Burns, N. Dacheux, and A. Navrotsky 2015. Thermodynamics of Formation of Coffinite, $USiO_4$. *Proceedings of the National Academy of Sciences 112:6551–55.*

Haidinger, W. 1830. On johannite, a new mineral species. *The Edinburgh Journal of Science* 3, 306–10.

Haidinger, W. 1845. *Handbuch der Bestimmenden Mineralogie*. Vienna: Braumüller and Seidel.

Hallimond, A. F. 1915. On bassetite and uranospathite, new species hitherto classed as autunite. *Mineralogical Magazine* 17:221–36.

Hallimond, A. F. 1916. The crystallography and dehydration of torbernite. *Mineralogical Magazine* 17:326–39.

Hallimond, A. F. 1920. Meta-torbernite I, its physical properties and relation to torbernite. *Mineralogical Magazine* 19:43–7.

Hamilton, P.-K., and P. F. Kerr 1959. Umohoite from Cameron, Arizona. *American Mineralogist* 44:1248–60.

Hansley, P. L., and J. J. Fitzpatrick 1989. Compositional and crystallographic data on REE-bearing coffinite from the Grants uranium region, northwestern New Mexico. *American Mineralogist* 74:263–70.

Hanson, S. L., W. B. Simmons, and A. U. Falster 1998. Nb-Ta-Ti oxides in granitic pegmatites from the Topsham Pegmatite District, Southern Maine. *Canadian Mineralogist.* 36:601–8.

Hanson, S. L., W. B. Simmons, A. U. Falster, E. E. Foord, and F. E. Lichte 1999. Proposed nomenclature for samarskite-group minerals: new data on ishikawaite and calciosamarskite. *Mineralogical Magazine* 63:27–36.

Hawthorne, F. C., R. J. Finch, and R. C. Ewing 2006. The crystal structure of dehydrated wyartite, $Ca(CO_3)[U^{5+}(U^{6+}O_2)_2O_4(OH)](H_2O)_3$. *Canadian Mineralogist* 44:1379–85.

Haynes, P. E. 1993. The Repete mine, San Juan County, Utah. *Rocks & Minerals* 68 (6):398–400.

Hazen, R. M., R. C. Ewing, and D. A. Sverjensky 2009. Evolution of uranium and thorium minerals. *American Mineralogist* 94:1293–1311.

Heinrich, E. W. 1958. *Mineralogy and Geology of Radioactive Raw Materials*, McGraw-Hill, New York.

Henry, D. A., W. Birch, P. Stockman, and A. van der Heyden 1999. Saléeite and associated minerals from the Ranger No. 3 deposits, Jabiru, Northern Territory. *Australian Journal of Mineralogy* 5:19–24.

Henry, D. A., R. E. Pogson, and P. A. Williams 2005. Threadgoldite from the South Alligator Valley uranium field, Northern Territory, Australia: second world occurrence. *Australian Journal of Mineralogy* 11:7–11.

Hess, F. L. 1922. Uranium-bearing asphaltite sediments of Utah. *Engineering and Mining Journal-Press* 114:272–76.

Hess, F. L. 1924. Rauvite, a new uranium-vanadium mineral. *US Geological Survey Bulletin - Contributions to Economic Geology* 750D:68–70.

Hess, F. L. 1924. New and known minerals from the Utah-Colorado carnotite region. *US Geological Survey Bulletin* 750:63–78 [see abstract (1925) *American Mineralogist* 10:133].

Hess, F. L., and W. T. Schaller 1914. Pintadoite and uvanite, two new minerals from Utah: A preliminary note. *Journal of the Washington Academy of Sciences* 4:576–79.

Hess, F. L., and R. C. Wells 1920. Brannerite, a new uranium mineral. *Journal of the Franklin Institute* 189:225–37.

Hewett, D. F., J. Stone, and H. Levine 1957. Brannerite from San Bernardino County, California. *American Mineralogist* 42:30–38.

Hidden, W. E., and J. B. Mackintosh 1889. A description of several yttria and thoria minerals from Llano County, Texas. *American Journal of Science* 138:474–86.

Hidden, W. E., and C. H. Warren 1906. On yttrocrasite, a new yttrium-thorium-uranium titanate. *American Journal of Science* 172:515–19.

Hiemstra, S. A. 1955. Baddeleyite from Phalaborwa, eastern Transvaal. *American Mineralogist* 40:275–82.

Higgins, J. B., and P. H. Ribbe 1976. The crystal chemistry and space groups of natural and synthetic titanites. *American Mineralogist* 61:878–88.

Hlousek, J., and J. Tvrdy 2002. Interesting secondary minerals from Jachymov, *Lapis* 27 (7–8):44–66 (in German).

Hoeve, J. and T. Sibbald 1978. Mineralogy and geological settings of unconformity-type uranium deposits in northern Saskatchewan. *MAC Short Course* 3:457–74.

Hogarth, D. D. 1951 Studies of radioactive compounds: II—meta-zeunerite, uranophane, kasolite, and cuprosklodowskite in Canada. *American Mineralogist* 36:411–14.

Hogarth, D. D., 1961. A study of pyrochlore and betafite. *Canadian Mineralogist* 6:610–33.

Hogarth, D. D. 1977. Classification and nomenclature of the pyrochlore group. *American Mineralogist* 62:403–10.

Hogarth, D. D., and E. W. Nuffield 1954. Studies of radioactive compounds: VII - phosphuranylite and dewindtite. *American Mineralogist* 39:444–7.

Hollinger, P. 1992. Les nouvelles zones de reaction d'Oklo: Datation U-Pb et caracterisation in-situ des produits de fission a l'anayseur ionique. CEREM, Note technique DEM No. 01/92, Grenoble. 42 pp.

Honea, R. M. 1961. New data on boltwoodite, an alkali uranyl silicate. *American Mineralogist* 46:12–25.

Horne, J. E. T., and J. R. Butler 1965. A second occurrence of lyndochite. *Mineralogical Magazine* 34:237–48.

Hughes, K.-A., and P. C. Burns 2003. A new uranyl carbonate sheet in the crystal structure of fontanite, $Ca[(UO_2)_3(CO_3)_2O_2](H_2O)_6$. *American Mineralogist* 88:962–66.

Hughes, K.-A., P. C. Burns, and U. Kolitsch 2003. The crystal structure and crystal chemistry of uranosphaerite, $Bi(UO_2)O_2OH$, *Canadian Mineralogist* 41:677–85

Hughes Kubatko, K.-A., and P. C. Burns 2004. The crystal structure of a novel uranyl tricarbonate, $K_2Ca_3[(UO_2)(CO_3)_3]_2(H_2O)_6$. *Canadian Mineralogist* 42: 997–1003.

Hunan 230 Laboratory, Hunan 305 Geological Team and X-ray Laboratory, Wuhan Geological Institute 1976. Furongite - a new uranium mineral found in China. *Acta Geologica Sinica* 2:203–4 [see abstract (1978) *American Mineralogist* 63:425].

Hunan 230 Institute and X-ray Laboratory, Wuhan Geologic College 1978. Xiangjianite - a new uranium mineral discovered in China. *Scientia Geologica Sinica* 2:183–88 [see abstract (1979) *American Mineralogist* 64:466].

Hurlbut, C. S. Jr. 1954. Studies of uranium minerals (XV): Schroeckingerite from Argentina and Utah. *American Mineralogist* 39:901–7.

Hutton, C. O. 1957. Sengierite from Bisbee, Arizona. *American Mineralogist* 42:408–11.

Hutton, C. O. 1957a. Kobeite from Paringa River, South Westland, New Zealand. *American Mineralogist* 42:342–53.

IAEA 2009. World distribution of uranium deposits (UDEPO) with uranium deposit classification. IAEA-TECDOC-1629. Vienna: International Atomic Energy Agency, 115 pp.

International Mineralogical Association 1962. International Mineralogical Association: Commission on new minerals and mineral names. *Mineralogical Magazine* 33:260–63.

Ito, J. 1967. A study of chevkinite and perrierite. *American Mineralogist* 52:1094–1104.

Izett, G. A., and R. E. Wilcox 1968. Perrierite, chevkinite, and allanite in Upper Cenozoic ash beds in the western United States. *American Mineralogist* 53:1558–67.

Jackson, J. M., and P. C. Burns 2001. A re-evaluation of the structure of weeksite, a uranyl silicate framework mineral. *Canadian Mineralogist* 39:187–95.

Jaffe, H. W., A. M. Sherwood, and M. J. Peterson 1948. New data on schroeckingerite. *American Mineralogist* 33:152–7.

Jaffe, H. W., H. T. Evans, and R. W. Chapman 1956. Occurrence and age of chevkinite from the Devil's Slide fayalite-quartz syenite near Stark, New Hampshire. *American Mineralogist* 41:474–86.

Janeczek, J. 1999. Mineralogy and geochemistry of natural fission reactors in Gabon. *Reviews in Mineralogy* 38:321–92.

Janeczek, J., and R. Ewing 1996. Phosphatian coffinite with rare earth elements and Ce-rich françoisite-(Nd) from sandstone beneath a natural fission reactor at Bangombé, Gabon. *Mineralogical Magazine* 60:665–9.

Jensen, M., and J. Hermann 2012. Uranium mineralization at the Apex mine, Lander County, Nevada. *Rocks & Minerals* 87:270–76.

Kabalov, Yu., E. Sokolova, L. Pautov, and J. Schneider 1998. Crystal structure of a new mineral, turkestanite: a calcium analogue of steacyite. *Crystallography Reports* 43:584–88; [see abstract (2001) *New Minerals 1995–1999*, J. Mandarino, ed., *Canadian Mineralogist Special Publication* 4:249].

Kamhi, S. R. 1959. An X-ray study of umohoite. *American Mineralogist* 44:920–25.

Kampf, A. R., A. V. Kasatkin, J. Čejka, and J. Marty 2015a. Plášilite, $Na(UO_2)(SO_4)(OH)\cdot 2H_2O$, a new uranyl sulfate mineral from the Blue Lizard mine, San Juan County, Utah, USA. *Journal of Geosciences* 60:1–10.

Kampf, A. R., J. Plášil, A. V. Kasatkin, and J. Marty 2014. Belakovskiite, $Na_7(UO_2)(SO_4)_4(SO_3OH)(H_2O)_3$, a new uranyl sulfate mineral from the Blue Lizard mine, San Juan County, Utah, USA. *Mineralogical Magazine* 78:639–49.

Kampf, A. R., J. Plášil, A. V. Kasatkin, and J. Marty 2015b. Bobcookite, $NaAl(UO_2)_2(SO_4)_4(H_2O)_{18}$, and wetherillite, $Na_2Mg(UO_2)_2(SO_4)_4\cdot 18H_2O$, two new uranyl sulfate minerals from the Blue Lizard mine, San Juan County, Utah, USA. *Mineralogical Magazine* 79:in press

Kampf, A. R., J. Plášil, A. V. Kasatkin, and J. Marty, 2015c. Fermiite, $Na_4(UO_2)(SO_4)_3\cdot 3H_2O$, and oppenheimerite, $Na_2(UO_2)(SO_4)_2\cdot 3H_2O$, two new uranyl sulfate minerals from the Blue Lizard mine, San Juan County, Utah, USA. *Mineralogical Magazine* 79:in press

Kennedy, I., 1979. Some interesting radioactive minerals from the Bancroft Area, Ontario. *Mineralogical Record* 10 (3):153–58.

Khomyakov, A. P. 1995. *Mineralogy of hyperagpaitic alkaline igneous rocks*. Oxford:Clarendon Press. 248 pp.

Khosrawan-Sazedj, F. 1982. On the space group of threadgoldite. *Tschermaks Mineral. Petrol. Mitteil.* 30:111–15.

Kimberley, M. M. 1978. Uranium Deposits: Their Mineralogy and Origin, *Short Course Series* 3, Mineralogical Association of Canada, Quebec.

Klepper, M. and G. Wyant 1956. Uranium provinces. U. S. Geological Survey Professional Paper 300, 17–25.

Kolitsch, U., and G. Giester 2001. Revision of the crystal structure of ulrichite, $CaCu^{2+}(UO_2)(PO_4)_2\cdot 4H_2O$. *Mineralogical Magazine* 65:717–24.

Kopchenova, E. V., and K. V. Skvortsova 1957. Sodium uranospinite. *Doklady Akademii Nauk SSSR* 114:634–6; [see abstract (1958) *American Mineralogist* 43:383–4].

Kopchenova, E. V., K. V. Skvortsova, N. I. Silantieva, G. A. Sidorenko, and L. V. Mikhailova 1962. Mourite, a new supergene uranium-molybdenum mineral. *Zapiski Vsesoyuznogo Mineralogicheskogo Obshchestva* 91 (1):67–71; [see abstract (1962) *American Mineralogist* 47:1217].

Kornetova, V. A., V. B. Aleksandrov, and M. E. Kazakova 1963. (Tantal-aeschynite). *Trudy Mineralogicheskaya Muzeya Akademiya Nauk SSSR* 14:108–21; [see discussion by Fleischer (1974) *American Mineralogist* 59:1331–32].

Krause, W., H. Effenberger, and F. Brandstätter 1995. Orthowalpurgite, $(UO_2)Bi_4O_4(AsO_4)_2\cdot 2H_2O$, a new mineral from the Black Forest, Germany. *European Journal of Mineralogy* 7:1313–24; [see abstract (2001) *New Minerals 1995–1999*, J. Mandarino, ed., *Canadian Mineralogist Special Publication* 4:180].

Krivovichev, S. V., and P. C. Burns 2000. Crystal structure of uranyl molybdates. I. The structure and formula of umohoite. *Canadian Mineralogist* 38:717–26.

Krivovichev, S. V., and P. C. Burns 2000a. Crystal structure of uranyl molybdates. II. The crystal structure of iriginite. *Canadian Mineralogist* 38:847–51.

Krivovichev, S. V., and J. Plašil 2013. Mineralogy and crystallography of uranium. *MAC Short Course Series* 43:15–119.

Kubatko, K.-A., K. B. Helean, A. Navrotsky, and P. C. Burns 2005. Thermodynamics of uranyl minerals: Enthalpies of formation of rutherfordine, UO_2CO_3, andersonite, $Na_2CaUO_2(CO_3)_3(H_2O)_5$, and grimselite, $K_3NaUO_3(CO_3)_3H_2O$. *American Mineralogist* 90:1284–90.

Kupriyanova, I. I., T. I. Stolyarova, and G. A. Sidorenko 1962. A new thorium silicate - thorosteenstrupine. *Zapiski Vserossijskogo Mineralogicheskogo Obshchestva* 91:325–30; [see abstract (1963) American Mineralogist 48, 433–34].

Kuz'menko, M. V., and S. I. Kozhanov 1959. The new mineral karnasurtite. *Trudy Institut Mineralogii, Geokhimii, Kristallokhimii Redkikh Elementov, Akademia Nauk SSSR* 2:95–8; [see abstract (1960) *American Mineralogist* 45:1133–34].

Kyser, K., and M. Cuney 2008. Unconformity-related uranium deposits. *MAC Short Course Series* 39:161–291.

Lacroix, A., 1912. Quelques nouvelles observations sur les mineraux uraniferes de la province de l'Itasy, Madagascar. *Bulletin Societe Français de Mineralogie* 35:233–35.

Langford, F. 1978. Uranium deposits in Australia. *MAC Short Course Series* 3:205–16.

Langmuir, D. 1978. Uranium solution-mineral equilibria at low temperatures with applications to sedimentary ore deposits. *MAC Short Course* 3:17–55.

Larsen, E. S., and G. V. Brown 1917. Gilpinite, a new uranium mineral from Colorado. *American Mineralogist* 2:75–9.

Larsen, E. S., and H. Berman 1926. The identity of gilpinite and johannite. *American Mineralogist* 11:1–5.

Larsen, E. S., and F. A. Gonyer 1937. Dakeite, a new uranium mineral from Wyoming. *American Mineralogist* 22:561–3.

Lauf, R. J., T. B. Lindemer, and R. L. Pearson 1984. Out-of-reactor studies of fission product - silicon carbide interactions in HTGR fuel particles. *Journal of Nuclear Materials* 120:6–30.

Leo, G. W. 1960. Autunite from Mt. Spokane, Washington. *American Mineralogist* 45:99–128.

Leroy, J. 1978. The Margnac and Fanay uranium deposits of the La Crouzille district (western Massif Cental, France): Geologic and fluid inclusion studies. *Economic Geology* 73:1611–1634

Li, Y., and P. C. Burns 2000a. Investigations of crystal-chemical variability in lead uranyl oxide hydrates. I. Curite. *Canadian Mineralogist* 38:727–35.

Li, Y., and P. C. Burns 2000b. Investigations of crystal-chemical variability in lead uranyl oxide hydrates. II. Fourmarierite. *Canadian Mineralogist* 38:737–49.

Li, Y., and P. C. Burns 2001. The crystal structure of synthetic grimselite, $K_3Na(UO_2)(CO_3)_3{\cdot}(H_2O)$. *Canadian Mineralogist* 39:1147–51.

Li, Y., P. C. Burns, and R. A. Gault 2000. A new rare-earth-element uranyl carbonate sheet in the structure of bijvoetite-(Y). *Canadian Mineralogist* 38:153–62.

Li, Y., S. V. Krivovichev, and P. C. Burns 2001. The crystal structure of $Na_4(UO_2)(CO_3)_3$ and its relationship to schröckingerite. *Mineralogical Magazine* 65:297–304.

Linthout, K. 2007 Tripartite division of the system $2REEPO_4$—$CaTh(PO_4)_2$—$2ThSiO_4$, discreditation of brabantite, and recognition of cheralite as the name for members dominated by $CaTh(PO_4)_2$. *Canadian Mineralogist* 45:503–8.

Livingstone, A., D. Atkin, D. Hutchison, and H. M. Al-Hermezi 1976. Iraqite, a new rare-earth mineral of the ekanite group. *Mineralogical Magazine* 40:441–45.

Locock, A. J., and P. C. Burns 2003. The structure of hugelite, an arsenate of the phosphuranylite group, and its relationship to dumontite. *Mineralogical Magazine* 67:1109–20.

Locock, A. J., and P. C. Burns 2003a. The crystal structure of synthetic autunite, $Ca[(UO_2)(PO_4)]_2(H_2O)_{11}$. *American Mineralogist* 88:240–4.

Locock, A. J., and P. C. Burns 2003b. The crystal structure of bergenite, a new geometrical isomer of the phosphuranylite group. *Canadian Mineralogist* 41:91–101.

Locock, A. J., and P. C. Burns 2003c. Crystal structures and synthesis of the copper-dominant members of the autunite and meta-autunite groups: torbernite, zeunerite, metatorbernite and metazeunerite. *Canadian Mineralogist* 41:489–502.

Locock, A. J., P. C. Burns, and T. M. Flynn 2004. Divalent transition metals and magnesium in structures that contain the autunite-type sheet. *Canadian Mineralogist* 42:1699–1718.

Locock, A. J., P. C. Burns, and T. M. Flynn 2005.Structures of strontium- and barium-dominant compounds that contain the autunite-type sheet. *Canadian Mineralogist* 43:721–33.

Locock, A. J., P. C. Burns, and T. M. Flynn 2005a. The role of water in the structures of synthetic hallimondite, $Pb_2[(UO_2)(AsO_4)_2](H_2O)_n$ and synthetic parsonsite, $Pb_2[(UO_2)(PO_4)_2](H2O)_n$, $0 \leq n \leq 0.5$. *American Mineralogist* 90:240–46.

Locock, A. J., P.C. Burns, M. J. M. Duke, and T. M. Flynn 2004. Monovalent cations in structures of the meta-autunite group. *Canadian Mineralogist* 42:973–66.

Locock, A. J., W. S. Kinman, and P. C. Burns 2005. The structure and composition of uranospathite, $Al_{1-x}\square_x[(UO_2)(PO_4)]_2(H_2O)_{20+3x}F_{1-3x}$, $0 < x < 0.33$, a non-centrosymmetric fluorine-bearing mineral of the autunite group, and of a related synthetic lower hydrate, $Al_{0.67}\square_{0.33}[(UO_2)(PO_4)]_2(H_2O)_{15.5}$. *Canadian Mineralogist* 43:989–1003.

Lorenzen, J. 1882. On some minerals from the sodalite-syenite in the Juliane-haab district, south Greenland. *Mineralogical Magazine* 5:49–70.

Lumpkin, G. R., M. G. Blackford, and M. Colella 2013. Chemistry and radiation effects of davidite. *American Mineralogist* 98:275–78.

Luo, Y., J. M. Hughes, J. Rakovan, and Y. Pan 2009. Site preference of U and Th in Cl, F, and Sr apatites. *American Mineralogist* 94:345–51.

Lyon, R. J. P. 1956. Anomalous fluorescence in torbernite from Rum Jungle, N. T., Australia. *American Mineralogist* 41:789–92.

Macdonald, R., B. Bagiński, P. Kartashov, D. Zozulya, and P. Dzierżanowski 2012. Chevkinite-group minerals from Russia and Mongolia: new compositional data from metasomatites and ore deposits. *Mineralogical Magazine* 76:535–49.

Maras, A., G. Parodi, G. Della Ventura, and D. Ohnenstetter 1995. Vicanite-(Ce): a new Ca-Th-REE borosilicate from the Vico volcanic district (Latium, Italy). *European Journal of Mineralogy* 7:439–46; [see abstract (2001) *New Minerals 1995–1999*, J. Mandarino, ed., *Canadian Mineralogist Special Publication* 4:256].

Marckwald, W. 1906. Ueber Uranerze aus Deutsch-Ostafrika. *Zentralblatt für Mineralogie, Geologie und Paläontologie.* 1906:761–63.

Markl, G. and S. Wolfsried 2011. *Das Uran von Menzenschwand.* Munich: Christian Weise Verlag. 143 pp.

Mawson, D. 1906. On certain new mineral species associated with carnotite in the radio-active ore body near Olary. *Transactions of the Royal Society of South Australia* 30:188–93.

Mayer, H., and K. Mereiter 1986. Synthetic bayleyite, $Mg_2[UO_2(CO_3)_3]·18H_2O$: Thermochemistry, crystallography and crystal structure. *Tschermaks Mineralogische und Petrographische Mitteilungen* 35:133–46.

McBurney, T. C., and J. Murdoch 1959. Haiweeite, a new uranium mineral from California. *American Mineralogist* 44:839–43.

McColl, D. 2008. Thorian brannerite crystals from Crocker Well, South Australia. *Australian Journal of Mineralogy* 14:69–71.

McMillan, R. 1978. Genetic aspects and classification of important Canadian uranium deposits. Mineralogical Association of Canada, *MAC Short Course* 3:187–204.

Meisser, N., J. Brugger, S. Ansermet, P. Thelin, and F. Bussy 2010. Françoisite-(Ce), a new mineral species from La Creusaz uranium deposit (Valais, Switzerland) and from Radium Ridge (Flinders Ranges, South Australia): description and genesis. *American Mineralogist* 95:1527–32.

Meixner, H. 1953. Kahlerit, ein neues Mineral der Uranglimmergruppe, aus der Huttenberger Lagerstätte. *Der Karinthin* 23:277–80; [see abstract (1954) *American Mineralogist* 39:1038].

Melkov, V. G., L. N. Belova, A. I. Gorshkov, O. A. Ivanova, V. A. Sivtsov, and V. A. Boronnikhin 1983. New data on lermontovite. *Mineralogiceskij Zhurnal* 5:82–7; [see abstract (1984) *American Mineralogist* 69:214–15].

Mélon, M. J. 1938. La sharpite, nouveau carbonate d'uranyle du Congo belge. *Bulletin des séances de l'institut Royal Colonial Belge* 9, 333–36 [see abstract (1939) *American Mineralogist* 24:658].

Mereiter, K. 1982. The crystal structure of walpurgite, $(UO_2)Bi_4O_4(AsO_4)_2·2H_2O$. *Tschermaks Mineralogische und Petrographische Mitteilungen* 30:129–39.

Mereiter, K. 1982a. Die Kristallstruktur des Johannits, $Cu(UO_2)_2(OH)_2(SO_4)_2·8H_2O$. *Tschermaks Mineralogische und Petrographische Mitteilungen* 30:47–57.

Mereiter, K. 1982b. The crystal structure of liebigite, $Ca_2UO_2(CO_3)_3·{\sim}11H_2O$. *Tschermaks Mineralogische und Petrographische Mitteilungen* 30:277–88.

Mereiter, K. 1984. The crystal structure of albrechtschraufite, $MgCa_4F_2[(UO_2)(CO_3)_3]_2·17H_2O$. *Acta Crystallographica* C40:247.

Mereiter, K. 1986a. Synthetic swartzite, $CaMg[UO_2(CO_3)_3]·12H_2O$, and its strontium analogue, $SrMg[UO_2(CO_3)_3]·12H_2O$:c rystallography and crystal structure. *Neues Jahrbuch für Mineralogie, Monatshefte* 1986:481–92.

Mereiter, K. 1986b. Crystal structure and crystallographic properties of a schröckingerite from Joachimsthal. *Tschermaks Mineralogische und Petrographische Mitteilungen* 35:1–18.

Mereiter, K. 2013. Description and crystal structure of albrechtschaufite, $MgCa_4F_2[UO_2(CO_3)_3]_2·17\text{-}18H_2O$. *Mineralogy and Petrology* 107:179–88.

Meunier, G., J. Galy 1973. Structure cristalline de la schmitterite synthétique $UTeO_5$. *Acta Crystallographica* B29:1251–55.

Miller, S. A., and J. C. Taylor 1986. The crystal structure of saléeite, $Mg[UO_2PO_4]_2·10H_2O$. *Zeitschrift für Kristallographie* 177:247–53.

Miller, M. L., R. J. Finch, P. C. Burns, and R. C. Ewing 1996. Description and classification of uranium oxide hydrate sheet anion topologies. *Journal of Materials Research* 11:3048–56.

Mills, S. J. 2004. Metanatroautunite ('sodium autunite') from the Lake Boga Granite, Victoria. *Australian Journal of Mineralogy* 10:29–39.

Mills, S. J., W. D. Birch, U. Kolitsch, W. G. Mumme, and I. R. Grey 2008. Lakebogaite, $CaNaFe^{3+}_2H(UO_2)_2(PO_4)_4(OH)_2(H_2O)_8$, a new uranyl phosphate with a unique crystal structure from Victoria, Australia. *American Mineralogist* 93:691–7.

Mills, S. J., A. R. Kampf, and W. D. Birch 2012. The crystal structure of metanatroautunite, $Na[(UO_2)(PO_4)](H_2O)_3$, from the Lake Boga Granite, Victoria, Australia. *American Mineralogist* 97:735–38.

Milne, I. H., and E. W. Nuffield 1951. Studies of radioactive compounds: I - vandenbrandeite. *American Mineralogist* 36:394–410.

Milton, C., B. Ingram, J. Clark, and E. Dwornik 1965. Mckelveyite, a new hydrous sodium barium rare-earth uranium carbonate mineral from the Green River Formation, Wyoming. *American Mineralogist* 50:593–612.

Moddle, D. A. 1957. Brannerite from Eastern Ontario. *Canadian Mineralogist* 6:155–57.

Mrose, M. E. 1950. Studies of uranium minerals (III): saléeite from Schneeberg, Saxony. *American Mineralogist* 35:525–30.

Mrose, M. E. 1951. Synthetic sodium- ammonium- and hydrogen-uranospinite. *American Mineralogist* 36:322.

Mrose, M. E. 1953. Studies of uranium minerals (XIII): synthetic uranospinites. *American Mineralogist* 38:1159–68.

Mücke, A. 1988. Lehnerit $Mn[(UO_2)(PO_4)]_2.8H_2O$ ein neues Mineral aus dem Pegmatit von Hagendorf/Oberpfalz. *Aufschluss* 39:209–17 [see abstract (1990) *American Mineralogist* 75:1433].

Mücke, A., and H. Strunz 1978. Petscheckite and liandratite, two new pegmatite minerals from Madagascar. *American Mineralogist* 63:941–46.

Muto, T., R. Meyrowitz, and A. Pommer 1959. Ningyoite, a new hydrous uranous phosphate mineral from Japan. *American Mineralogist* 44:633–50.

Nash, J., H. Granger, and S. Adams 1981. Geology and concepts of genesis of important types of uranium deposits. *Economic Geology* 75th Anniversary Volume, 63–116.

Naudet, R. 1991. *Oklo: Des Reacteurs Nucliaires Fossiles. Etude Physique.* Eyrolles, Paris, 685 pp.

Navrotsky, A., T. Shvareva, and X. Guo 2013. Thermodynamics of uranium minerals and related minerals. *MAC Short Course Series* 43:147–64.

Nekrasova, Z. A. 1958. A hydrous uranyl ammonium phosphate (uramphite), $NH_4(UO_2)(PO_4){\cdot}3H_2O$. *Voprosy geologii urana, Atomizdat* 1957:67–72; [see abstract (1959) *American Mineralogist* 44:464].

Nenadkevich, K. A. 1912 Tyuyamunite - a new mineral. *Bulletin de l'Académie Impériale des Sciences de St.-Pétersbourg* 6:945–46.

Neumann, H.,and T. L. Sverdrup 1960. Davidite from Tuftan, Iveland. *Norsk Geologisk Tidsskrift* 40:277–88.

Nininger, R. 1954. *Minerals for Atomic Energy*, Van Nostrand, New York, 367 pp.

Novácek, R. 1935. Study on some secondary uranium minerals. *Vestniku Královské České Společnosti Nauk* 7:1–36.

Nováček, R. 1939. The identity of dakeite and schroeckingerite. *American Mineralogist* 24:317–23.

Nuffield, E. W., and I. H. Milne 1953. Studies of radioactive compounds: VI - meta-uranocircite, *American Mineralogist* 38:476–88.

Oberti, R., G. Della Ventura, L. Ottolini, F. C. Hawthorne, and P. Bonazzi 2002. Re-definition, nomenclature, and crystal-chemistry of the hellandite group. *American Mineralogist* 87:745–52.

OECD 2008. *Uranium 2007: Resources, Production and Demand.* Organisation for Economic Co-operation and Development, Paris.

OECD 2012. *Uranium 2011: Resources, Production and Demand.* Organisation for Economic Co-operation and Development, Paris.

Ohashi, R. 1920. Note on the plumbiferous barytes from Shibukuro, prefecture of Akita, Japan. *Mineralogical Magazine* 19:73–6.

Ondruš, P., F. Veselovsky, and R. Rybka 1990. Znucalite, $Zn_{12}(UO_2)Ca(CO_3)_3(OH)_{22}{\cdot}4H_2O$, a new mineral from Pribram, Czechoslovakia. *Neues Jahrbuch für Mineralogie, Monats.* 1990:393–400 [see abstract (1997) *New Minerals 1990–1994*, p. 229, J. Mandarino, ed., The Mineralogical Record, Tucson].

Ondruš, P., R. Skala, F. Veselovsky, J. Sejkora, and C. Vitti 2003. Čejkaite, the triclinic polymorph of $Na_4(UO_2)(CO_3)$ - a new mineral from Jachymov, Czech Republic. *American Mineralogist* 88:696–93.

Ondruš, P., F. Veselovsky, A. Gabasova, J. Hloušek, and V. Srein 2003a. Supplement to secondary and rock-forming minerals of the Jachymov ore district, *Journal of the Czech Geological Society* 48 [3–4]:149–55.

Orlandi, P., and F. Checchi 1986. The Buca della Vena mine, Tuscany, Italy. *Mineralogical Record* 17 (4):261–68.

Orlandi, P., C. Biagioni, L. Bindi, and F. Nesola 2014. Ichnusaite, $Th(MoO_4)_2{\cdot}3H_2O$, the first natural thorium molybdate: Occurrence, description, and crystal structure. *American Mineralogist* 99:2089–94.

Orlandi, P., C. Biagioni, L. Bindi, and S. Merlino 2015. Nuragheite, $Th(MoO_4)_2{\cdot}H_2O$, the second natural thorium molybdate and its relationships with ichnusaite and synthetic $Th(MoO_4)_2$. *American Mineralogist* 100:267–73.

Orlandi, P., M. Pasero, G. Duchi, and F. Olmi 1997. Dessauite, $(Sr,Pb)(Y,U)(Ti,Fe^{3+})_{20}O_{38}$, a new mineral of the crichtonite group from Bucca della Vena mine, Tuscany, Italy. *American Mineralogist* 82:802–11.

Othmane, G., T. Allard, N. Menguy, G. Morin, I. Esteve, M. Fayek, and G. Calas 2013. Evidence for nanocrystals of vorlanite, a rare uranate mineral, in the Nopal I low-temperature uranium deposit (Sierra Peña Blanca, Mexico). *American Mineralogist* 98:518–21.

Outerbridge, W. F., M. H. Staatz, R. Meyrowitz, and A. M. Pommer 1960. Weeksite, a new uranium silicate from the Thomas Range, Juab County, Utah. *American Mineralogist* 45:39–52.

Pabst, A. 1952. The metamict state. *American Mineralogist* 37:137–57.

Pabst, A. 1954. Brannerite from California. *American Mineralogist* 39:109–17.

Pabst, A. 1961. X-ray crystallography of davidite. *American Mineralogist* 46:700–18.

Pabst, A., and C. O. Hutton 1951. Huttonite, a new monoclinic thorium silicate, with an account of its occurrence, analysis, and properties. *American Mineralogist* 36:60–69.

Pagoaga, M. K., D. E. Appleman, and J. M. Stewart 1986. A new barium uranyl oxide hydrate mineral, protasite. *Mineralogical Magazine* 50:125–8 [see abstract (1987) *American Mineralogist* 72:224].

Pagoaga, M. K., D. E. Appleman, and J. M. Stewart, 1987. Crystal structures and crystal chemistry of the uranyl oxide hydrates becquerelite, billietite, and protasite. *American Mineralogist* 72:1230–38.

Palache, C., H. Berman, and C. Frondel 1944. The system of mineralogy, 7th edition, Vol. I. New York: Wiley, 834 pp.

Paschis, J. A. 2009. *Uranium ore microscopy*. Boulder: Morning Sun-Ferry Press, 68pp.

Patchett, J. E., and E. W. Nuffield 1960. Studies of radioactive compounds: X - The synthesis of crystallography of brannerite. *Canadian Mineralogist* 6:483–90.

Pauliš, P., S. Kopecký, and P. Černý 2007. *Uranium Minerals of the Czech Republic and Their Occurrences*, (in Czech with extended abstract in German). Two volumes; 252 pp.

Pautov, L. A., A. A. Agakhanov, E. V. Sokolova, and Y. K. Kabalov 1997. Turkestanite $Th(Ca,Na)_2(K_{1-x}\square_x)Si_8O_{20}\cdot nH_2O$ - a new mineral. *Zapiski Vserossijskogo Mineralogicheskogo Obshchestva* 126:45–55; [see abstract (1998) *American Mineralogist* 83:1348–49].

Pavlenko, A. S., L. P. Orlova, M. V. Akhmanova, and K. I. Tobelko 1965. A thorium fluocarbonate, thorbastnaesite. *Zapiski Vsesoyuznogo Mineralogicheskogo Obshchestva* 94:105–13; [see abstract (1965) *American Mineralogist* 50:1505].

Peacor, D. R., W. B. Simmons, E. J. Essene, and E. W. Heinrich 1982. New data on and discreditation of "texasite," "albrittonite," "cuproartinite," "cuprohydromagnesite," and "yttromicrolite," with corrected data on nickelbischofite, rowlandite, and yttrocrasite. *American Mineralogist* 67:156–69.

Pekov, I. 1998. *Minerals First Discovered on the Territory of the Former Soviet Union*, Ocean Pictures Ltd., Moscow, 369 pp.

Pekov, I. 2007. New minerals from former Soviet Union countries, *Mineralogical Almanac,* 11.

Pekov, I., S. V. Krivovichev, V. O. Yapaskurt, N. V. Chukanov, and D. I. Belakovskiy 2014. Beshtauite, $(NH_4)_2(UO_2)(SO_4)_2\cdot 2H_2O$, a new mineral from Mount Beshtau, Northern Caucasus, Russia. *American Mineralogist* 99:1783–87.

Pekov, I. V., V. V. Levitskiy, S. V. Krivovichev, A. A. Zolotarev, I. A. Bryzgalov, A. E. Zadov, and N. V. Chukanov 2012a. New nickel-uranium-arsenic mineral species from the oxidation zone of the Belorechenskoye deposit, Northern Caucasus, Russia: I. Rauchite, $Ni(UO_2)_2(AsO_4)_2\cdot 10H_2O$, a member of the autunite group. *European Journal of Mineralogy* 24, 913–22.

Pekov, I. V., V. V. Levitskiy, S. V. Krivovichev, A. A. Zolotarev, N. V. Chukanov, I. A. Bryzgalov, and A. E. Zadov 2012b. New nickel-uranium-arsenic mineral species from the oxidation zone of the Belorechenskoye deposit, Northern Caucasus, Russia: II. Dymkovite, $Ni(UO_2)_2(As^{3+}O_3)_2\cdot 7H_2O$, a seelite-related arsenite. *European Journal of Mineralogy* 24:923–30.

Pekov, I. V., I. A. Yekimenkova, and N. N. Kononkova 1997. Thorosteenstrupine from Lovozero massif and the steenstrupine-(Ce) - thorosteenstrupine isomorphous series. *Zapiski Vserossijskogo Mineralogicheskogo Obshchestva* 126:35–44; [see abstract (1998) *American Mineralogist* 83, 1350].

Perrault, G., and P. Richard 1973. L'ekanite de Saint-Hilaire, P.Q.. *Canadian Mineralogist* 11:913–29.

Perrault, G., and J. T. Szymanski 1982. Steacyite, a new name, and a re-evaluation of the nomenclature of "ekanite"-group minerals. *Canadian Mineralogist* 20:59–63.

Phillips, W., H. J. Brooke, and W. H. Miller 1852. Autunite. *An Elementary Introduction to Mineralogy*, Longman, Brown, Green, and Longmans (London) p. 519.

Pierrot, R., J. Toussaint, and T. Verbeek 1965. La guilleminite, une nouvelle espece minerale. *Bulletin de la Société Française de Minéralogie et de Cristallographie* 88:132–35 [see abstract (1965) *American Mineralogist* 50:2103].

Piilonen, P. C., R. Rowe, G. Poirier, J. D. Grice, and A. M. McDonald 2013. Crystal structure determination of a unique Nb-Ti-rich thorite from Mont Saint-Hilaire, Quebec, with comments on 'thorogummite'. *Canadian Mineralogist* 51:597–612.

Piret, P. 1985. Structure cristalline de la fourmarierite, $Pb(UO_2)_4O_3(OH)_4\cdot 4H_2O$. *Bulletin de Mineralogie* 108:659–65 [see abstract (1987) *American Mineralogist* 72:229].

Piret, P., and J. P. Declerq 1983. Crystal structure of upalite $Al[(UO_2)_3O(OH)(PO_4)_2]\cdot 7H_2O$. An example of a mimetic twin. *Bulletin de Mineralogie* 106:383–89 [see abstract (1984) *American Mineralogist* 69:1196].

Piret, P., and M. Deliens 1981. Nouvelles données sur la bergenite holotype. *Bulletin de Minéralogie* 104:16–18.

Piret, P., and M. Deliens 1982. La vanmeersscheite $U(UO_2)_3(PO_4)_2(OH)_6\cdot 4H_2O$, et la metavanmeersscheite, $U(UO_2)_3(PO_4)_2(OH)_6\cdot 2H_2O$, nouveaux minereaux. *Bulletin de Mineralogie* 105:125–8 [see abstract (1982) *American Mineralogist* 67:1077].

Piret, P., and M. Deliens 1984. New data for richetite $PbO\cdot 4UO_3\cdot 4H_2O$. *Bulletin de Mineralogie* 107:581–5 [see abstract (1985) *American Mineralogist* 70:1335].

Piret, P, and M. Deliens 1987. Les phosphates d'uranyle et d'aluminium de Kobokobo. IX. L'althupite, $AlTh(UO_2)[(UO_2)_3O(OH)(PO_4)_2](OH)_3\cdot 15H_2O$, nouveau mineral; proprietes et structure cristalline. *Bulletin de Mineralogie* 110:65–72 [see abstract (1988) *American Mineralogist* 73:189–90].

Piret, P., and J. Piret-Meunier 1988. Nouvelle determination de la structure cristalline de la dumontite $Pb_2[(UO_2)_3O_2(PO_4)_2]\cdot 5H_2O$. *Bulletin de Mineralogie* 111:439–42 [see abstract (1989) *American Mineralogist* 74:1403].

Piret, P., and J. Piret-Meunier 1991. Composition chimique et structure cristalline de la phosphuranylite $Ca(UO_2)[(UO_2)_3(OH)_2(PO_4)_2]_2\cdot 12H_2O$. *European Journal of Mineralogy* 3:69–77 [see abstract (1989) *American Mineralogist* 76:1734–5].

Piret, P., and J. Piret-Meunier 1994. Structure de la seelite de Rabejac (France). *European Journal of Mineralogy* 6:673–77.

Piret, P., J.-P. Declerq, and D. Wauters-Stoop 1979. Structure of threadgoldite. *Acta Crystallographica* B35:3017–20.

Piret, P., J.-P. Declerq, and D. Wauters-Stoop 1980. Crystal structure of sengierite. *Bulletin de Mineralogie* 103:176–8; [see abstract (1981) *American Mineralogist* 66:220].

Piret, P., M. Deliens, J. Piret-Meunier, and G. Germain 1983. La sayrite, $Pb_2[(UO_2)_5O_6(OH)_2]\cdot 4H_2O$, nouveau mineral; proprietes et structure cristalline. *Bulletin de Mineralogie* 106:299–304 [see abstract (1984) *American Mineralogist* 69:568].

Piret, P., M. Deliens, M., and J. Piret-Meunier 1988. La françoisite-(Nd), nouveau phosphate d'uranyl et de terres rares; proprietes et structure crystalline. *Bulletin de Mineralogie.* 111:443–9 [see abstract (1990) *American Mineralogist* 75:241].

Piret, P., J. Piret-Meunier, and J.-P. Declerq 1979. Structure of phurcalite. *Acta Cryst.* B35:1880–82.

Piret, P., J. Piret-Meunier, and M. Deliens 1990. Chemical composition and crystal structure of dewindtite $Pb_3[H(UO_2)_3O_2(PO_4)_2 \cdot 12H_2O$. *European Journal of Mineralogy* 2:399–405 [see abstract (1991) *American Mineralogist* 1734–5].

Piret-Meunier, J., and P. Piret 1982. Nouvelle determination de la structure cristalline de la becquerelite. *Bulletin de Mineralogie* 105:606–10 [see abstract (1984) *American Mineralogist* 69:214].

Plant, J. A., P. R. Simpson, B. Smith, and B. F. Windley 1999. Uranium ore deposits - products of the radioactive earth. *Reviews in Mineralogy* 38:255–319.

Plášil, J., J. Cejka, J. Sejkora, P. Škácha, V. Goliáš, P. Jarka, F. Laufek, J. Jehlicka, I. Nemec, and L. Strnad 2010. Widenmannite, a rare uranyl lead carbonate: occurrence, formation and characterization. *Mineralogical Magazine* 74:97–110.

Plášil, J., M. Dušek, M. Novák, J. Čejka, I. Císařová, and R. Škoda 2011a. Sejkoraite-(Y), a new member of the zippeite group containing trivalent cations from Jáchymov (St. Joachimsthal), Czech Republic: Description and crystal structure refinement. *American Mineralogist* 96:983–91.

Plášil, J., K. Fejfarová, M. Novák, M. Dušek, R. Škoda, J. Hloušek, J. Čejka, J. Majzlan, J. Sejkora, V. Machovič, and D. Talla 2011b, Běhounekite, $U(SO_4)_2(H_2O)_4$, from Jáchymov (St. Joachimsthal), Czech Republic: the first natural U^{4+} sulphate. *Mineralogical Magazine* 75, 2739–53.

Plášil, J., K. Fejfarová, J. Čejka, M. Dušek, R. Škoda, and J. Sejkora 2013. Revision of the crystal structure and chemical formula of haiweeite, $Ca(UO_2)_2(Si_5O_{12})(OH)_2 \cdot 6H_2O$. *American Mineralogist* 98:718–23.

Plášil, J., K. Fejfarová, M. Dušek, R. Škoda, and J. Rohlíček 2013b. Revision of the symmetry and the crystal structure of čejkaite, $Na_4(UO_2)(CO_3)_3$. *American Mineralogist* 98:549–53.

Plášil, J., K. Fejfarová, J. Hloušek, R. Škoda, M. Novák, J. Sejkora, J. Čejka, M. Dušek, F. Veselovský, P. Ondruš, J. Majzlan, and Z. Mrázek 2013c. Štěpite, $U(AsO_3OH)_2 \cdot 4H_2O$, from Jáchymov, Czech Republic: the first natural arsenate of tetravalent uranium. *Mineralogical Magazine* 77, 137–152.

Plášil, J., K. Fejfarova, J. Sejkora, J. Čejka, M. Novák, R. Škoda, J. Hloušek, M. Dušek, and I. Cısarová 2013f. Lınekite, IMA 2012–066. CNMNC Newsletter No. 15, February 2013, page 7; *Mineralogical Magazine* 77:7.

Plášil, J., K. Fejfarová, K. S. Wallwork, M. Dušek, R. Škoda, J. Sejkora, J. Čejka, F. Veselovský, J. Hloušek, N. Meisser, and J. Brugger 2012. Crystal structure of pseudojohannite, with a revised formula, $Cu_3(OH)_2[(UO_2)_4O_4(SO_4)_2](H_2O)_{12}$. *American Mineralogist* 97:1796–1803.

Plášil, J., K. Fejfarová, R. Skála, R. Škoda, N. Meisser, J. Hloušek, I. Císařová, M. Dušek, F. Veselovský, J. Čejka, J. Sejkora, and P. Ondruš 2012c. The crystal chemistry of the uranyl carbonate mineral grimselite, $(K,Na)_3Na[(UO_2)(CO_3)_3](H_2O)$, from Jáchymov, Czech Republic. *Mineralogical Magazine* 76:443–53.

Plášil, J., J. Hauser, V. Petříček, N. Meisser, S. J. Mills, R. Škoda, K. Fejfarová, J. Čejka, J. Sejkora, J. Hloušek, J.-M. Johannet, V. Machovič, and L. Lapčák 2012b. Crystal structure and formula revision of deliensite, $Fe[(UO_2)_2(SO_4)_2(OH)_2](H_2O)_7$. *Mineralogical Magazine* 76:2837–60.

Plášil, J., J. Hloušek, R. Škoda, M. Novák, J. Sejkora, J. Čejka, F. Veselovský, and J. Majzlan 2013d. Vysokýite, $U^{4+}[AsO_2(OH)_2]_4 \cdot 4H_2O$, a new mineral from Jáchymov, Czech Republic. *Mineralogical Magazine* 77:3055–66.

Plášil, J., J. Hloušek, F. Veselovský, K. Fejfarová, M. Dušek, R. Škoda, M. Novák, J. Čejka, J. Sejkora, P. Ondruš 2012a. Adolfpateraite, $K(UO_2)(SO_4)(OH)(H_2O)$, a new uranyl sulphate mineral from Jáchymov, Czech Republic. *American Mineralogist* 97:447–54.

Plášil, J., A. R. Kampf, A. V. Kasatkin, and J. Marty 2014c. Bluelizardite, $Na_7(UO_2)(SO_4)_4Cl(H_2O)_2$, a new uranyl sulfate mineral from the Blue Lizard mine, San Juan County, Utah, USA. *Journal of Geosciences* 59:145–58.

Plášil, J., A. R. Kampf, A. V. Kasatkin, J. Marty, R. Škoda, S. Silva, and J. Čejka 2013e. Meisserite, $Na_5(UO_2)(SO_4)_3(SO_3OH)(H_2O)$, a new uranyl sulfate mineral from the Blue Lizard mine, San Juan County, Utah, USA. *Mineralogical Magazine* 77:2975–88.

Plášil, J., J. Sejkora, J. Cejka, M. Novák, J. Viñals, P. Ondruš, F. Veselovský, P. Škácha, J. Jehlicka, V. Goliáš, and J. Hloušek 2010a. Metarauchite, $Ni(UO_2)_2(AsO_4)_2 \cdot 8H_2O$, from Jáchymov, Czech Republic, and Schneeberg, Germany: A new member of the autunite group. *Canadian Mineralogist* 48:335–50.

Plášil, J., A. V. Kasatkin, R. Škoda, M. Novák, A. Kallistová, M. Dušek, R. Skála, K. Fejfarová, J. Čejka, N. Meisser, H. Goethals, V. Machovič, and L. Lapčák 2013a. Leydetite, $Fe(UO_2)(SO_4)_2(H_2O)_{11}$, a new uranyl sulfate mineral from Mas d'Alary, Lodève, France. *Mineralogical Magazine* 77:429–41.

Plášil, J., L. Palatinus, J. Rohlícek, L. Houdková, M. Klementova, V. Goliáš, and P. Škácha 2014a. Crystal structure of lead uranyl carbonate mineral widenmannite: Precession electron-diffraction and synchrotron powder-diffraction study. *American Mineralogist* 99:276–82.

Plášil, J., F. Veselovský, J. Hloušek, M. Šák, J. Sejkora, J. Čejka, P. Škácha, and A. V. Kasatkin 2014. Mathesiusite, $K_5(UO_2)_4(SO_4)_4(VO_5)(H_2O)_4$, a new uranyl vanadate-sulfate from Jáchymov, Czech Republic. *American Mineralogist* 99:625–32.

Pointeau, V., A. P. Deditius, F. Miserqe, D. Renock, U. Becker, J. Zhang, N. Clavier, N. Dacheaux, C. Poinssot, and R. C. Ewing. 2009. Synthesis and characterization of coffinite. *Journal of Nuclear Materials* 393:449–58.

Povilaitis, M. M. 1963. On the new mineral lodochnikite, absite, and thorutite. *Zapiski Vsesoyuznogo Minerologicheskogo Obskchestva* 92:113–23; [see abstract (1963) *American Mineralogist* 48:1419–20].

Prior, G. T. 1899. Minerals from Swaziland: Niobates and titanates of the rare earths, chemicallt allied to euxenite and fergusonite; cassiterite, monazite, etc. The "aeschynite" from Hitterö. *Mineralogical Magazine* 12:96–101.

Protas, J. 1957. La wölsendorfite, nouvelle espece uranifere. *Comptes Rendus Hebdomadaires des Séances de l'Académie des Sciences* 244:2942–44; [see abstract (1957) *American Mineralogist* 42: 919].

Protas, J. 1957a. Proprietes et synthese d'un oxyde hydrate d'uranium et du calcium de Shinkolobwe, Katanga. *Comptes Rendus Hebdomadaires des Séances de l'Académie des Sciences* 244:91–3; [see abstract (1957) *American Mineralogist* 42:920].

Protas, J. 1964. Une nouvelle espece minerale: la compreignacite, $K_2O{\cdot}6UO_3{\cdot}11H_2O$. *Bulletin de la Société Française de Minéralogie et de Cristallographie* 87:365–71; [see abstract (1965) *American Mineralogist* 50:807–8].

Pulou, R. 1957. Méta-uranocircite d'Entraygues (Aveyron). *Bulletin de la Société Française de Minéralogie et de Cristallographie* 80:32–8.

Pushcharovsky, D. Y, R. K. Rastsvetaeva, and H. Sarp 1996. Crystal structure of deloryite, $Cu_4(UO_2)[Mo_2O_8](OH)_6$. *Journal of Alloys and Compounds* 239:23–26.

Raslan, M. F. 2008. Occurrence of ishikawaite (uranium-rich samarskite) in the mineralized Abu Rushied gneiss, Southeastern Desert, Egypt. *International Geology Review* 50:1132–40.

Rey, A., S. Utsunomiya, J. Gimenez, I. Casas, J. de Pablo, and R. C. Ewing 2009. Stability of uranium (VI) peroxide hydrates under ionizing radiation. *American Mineralogist* 94:229–35.

Reyes-Cores, M., M. E. Montero-Cabrera, M. Renteria-Villalobos, L. Fuentes-Montero, L. Fuentes-Cobas, E. F. Herrera-Peraza, H. Esparza Ponce, and A. Rodreguez-Pineda 2007. Radioactive mineral samples from the northwest of Chihuahua City, Mexico. *Revista Mexicana de Fisica* S53:23–28.

Rich, R. A., H. D. Holland, and U. Petersen 1977. *Hydrothermal uranium deposits*. New York: Elsevier, 264 pp.

Robinson, G. and R. Wegner 1998. Mineralogy of the Boqueirazinho pegmatite, Parelhas, Rio Grande do Norte, Brazil. *Mineralogical Record* 29 (3):193–97.

Robinson, S. C., and A. P. Sabina 1955. Uraninite and thorianite from Ontario and Quebec. *American Mineralogist* 40:624–33.

Robinson, S. C., W. D. Loveridge, J. Rimsaite, and J. Van Peteghem 1963. Factors involved in discordant ages of euxenite from a Grenville pegmatite. *Canadian Mineralogist* 7:533–46.

Rogova, V. P., L. N. Belova, G. N. Kiziyarov, and N. N. Kuznetsova 1973. Bauranoite and metacalciouranoite - new minerals of the group of hydrous uranium oxides. *Zapiski Vsesoyuznogo Mineralogicheskogo Obshchestva*102:75–81 [see abstract (1973) *American Mineralogist* 58:1111].

Rogova, V. P., L. N. Belova, G. N. Kiziyarov, and N. N. Kuznetsova 1974. Calciouranoite - a new hydroxide of uranium. *Zapiski Vsesoyuznogo Mineralogicheskogo Obshchestva* 103:108–9 [see abstract (1975) *American Mineralogist* 60:161].

Rose, D. 1980. Brabantite, $CaTh[PO_4]_2$, a new mineral of the monazite group. *Neues Jahrbuch für Mineralogie - Monatschefte* 1980:247–57 [see abstract (1981) *American Mineralogist* 66:878].

Rose, G. 1842. Tschewkinit. *Mineralogisch-Geognostische Reise nach dem Ural, dem Altai und dem Kaspischen Meere* 2:513–14.

Rose, H. 1847. Ueber die Zusammensetzung des Uranotantals und des Columbits vom Ilmengebirge in Sibirien. *Annalen der Physik und Chemie* 71:157–69.

Rosenblum, S., and E. L. Mosier 1975. Nonmetamict nioboeschynite-(Ce) from Alaska. *American Mineralogist* 60:309–15.

Rosenzweig, A., and R. R. Ryan 1975. Refinement of the crystal structure of cuprosklodowskite, $Cu[(UO_2)_2(SiO_3OH)_2]{\cdot}6H_2O$. *American Mineralogist* 60:448–53.

Rosenzweig, A., and R. R. Ryan 1977. Vandenbrandeite $Cu(UO_2)(OH)_4$. *Crystal Structure Communications* 6:53–6.

Ross, C. S., E. P. Henderson, and E. Posnjak 1931. Clarkeite; a new uranium mineral. *American Mineralogist* 16:213–20.

Ross, M. 1963. The crystallography of meta-autunite-(I). *American Mineralogist* 48:1389–93.

Ross, M., and H. T. Evans 1964. Studies of the torbernite minerals (I): the crystal structure of abernathyite and the structurally related compounds $NH_4(UO_2AsO_4){\cdot}3H_2O$ and $K(H_3O)(UO_2AsO_4)_2{\cdot}6H_2O$. *American Mineralogist* 49:1578–1602.

Ross, M., and H. T. Evans 1965. Studies of the torbernite minerals (III): role of the interlayer oxonium, potassium, and ammonia ions, and water molecules. *American Mineralogist* 50:1–12.

Ross, M., H. Evans, and D. Appleman 1964. Studies of the torbernite minerals (II): the crystal structure of meta-torbernite. *American Mineralogist* 49:1603–21.

Ross, V. 1955. Studies of uranium minerals (XXI): synthetic hydrogen-autunite. *American Mineralogist* 40:917–19.

Ross, V. 1956. Studies of uranium minerals (XXII): synthetic calcium and lead uranyl phosphate minerals. *American Mineralogist* 41:915–26.

Rudnitskaya, L. S. 1959. Calcium molybdate of uranium. *Yadernoe Goryuchee i Reaktornye Metally* 3:160–61.

Ryback, G., and P. C. Tandy 1992. Eighth supplementary list of British Isles minerals (English). *Mineralogical Magazine* 56:261–75.

Saleh, G. M., S. A. Abdallah, A. A. Abbas, N. A. Dawood, and M. A. Rashad 2012. Uranium mineralizations of Wadi Sikait mylonites, Southeastern Desert, Egypt. *Journal of Geology and Mining Research* 3:86–104.

Sarp, H., J. Bertrand, and J. Deferne 1983. Asselbornite, $(Pb,Ba)(UO_2)_6(BiO)_4[(As,P)O_4]_2(OH)_{12}\cdot 3H_2O$, a new uranium, lead, and barium hydrous arsenate. *Neues Jahrbuch für Mineralogie, Monatschrift* 1983:117–23 [see abstract (1984) *American Mineralogist* 69:565].

Sarp, H., and P. J. Chiappero 1992. Deloryite, $Cu_4(UO_2)(MoO_4)_2(OH)_6$, a new mineral from the Cap Garonne mine near le Pradet, Var, France. *Neues Jahrbuch für Mineralogie, Monatschrift* 1992:58–64; [see abstract (1992) *American Mineralogist* 77:1305].

Scheerer, T. 1840. Ueber den euxenite, eine neue mineralspecies. *Annalen der Physik und Chemie* 50:149–53.

Scheerer, T. 1844. Polykras und malakon, zwei neue Mineral-species. *Annalen der Physik und Chemie* 62:429–43.

Schindler, M., P. Mandaliev, F. C. Hawthorne, and A. Putnis 2006. Dissolution of uranyl-oxide-hydroxy-hydrate minerals. I. Curite. *Canadian Mineralogist* 44:415–31.

Schoep, A. 1921 La curite, nouveau minéral radioactif. *Comptes Rendus Hebdomadaires des Séances de l'Académie des Sciences* 173:1186–87 [see abstract (1922) *American Mineralogist* 7:128].

Schoep, A. 1921a. Sur la kasolite, nouveau mineral radioactif. *Comptes Rendus Hebdomadaires des Séances de l'Académie des Sciences* 173:1476–77 [see abstract (1922) *American Mineralogist* 7:128–9].

Schoep, A. 1922. La soddite, nouveau mineral radioactif. *Comptes Rendus Hebdomadaires des Séances de l'Académie des Sciences* 174:1066–7 [see abstract (1922) *American Mineralogist* 7:179–80].

Schoep, A. 1922a. Sur la dewindtite, nouveau mineral radioactif. *Comptes Rendus Hebdomadaires des Séances de l'Académie des Sciences* 174:623–5 [see abstract (1922) *American Mineralogist* 7:162].

Schoep, A. 1922b. Sur la becquerélite, nouveau minéral radioactif. *Comptes Rendus Hebdomadaires des Séances de l'Académie des Sciences* 174:1240–42 [see abstract (1922) *American Mineralogist* 7, 179].

Schoep, A. 1923. Sur la parsonsite, nouveau mineral radioactif. *Comptes Rendus Hebdomadaires des Séances de l'Académie des Sciences* 176:171–3; [see abstract (1923) *American Mineralogist* 8:150].

Schoep, A. 1924. La dumontite, nouveau mineral radioactif. *Comptes Rendus Hebdomadaires des Séances de l'Académie des Sciences* 179:693; [see abstract (1925) American Mineralogist 10:131].

Schoep, A. 1924a. Sur la sklodowskite, nouveau minéral uranifère; ses analogies avec l'uranotile. *Bulletin de la Société Française de Minéralogie* 47:162–72;

Schoep, A. 1924b La sklodowskite, nouveau Minéral radioactif. *Comptes Rendus Hebdomadaires des Séances de l'Académie des Sciences* 179:413–5 [see abstract (1925) *American Mineralogist* 10:132].

Schoep, A. 1926. Over janthiniet, een nieuw uranium mineraal uit Katanga. *Natuurwetenschappelijk Tijdschrift* 7:97–9 [see abstract (1927) *American Mineralogist* 12:355–6].

Schoep, A. 1928. La renardite, nouveau mineral uranifere. *Bulletin de la Société Française de Minéralogie* 51:1–6; [see abstract (1929) *American Mineralogist* 14:244].

Schoep, A. 1932. La vandenbrandeite, un nouveau minéral uranifére. *Annales du Musée du Congo Belge* 1:24–31 [see abstract (1933) *American Mineralogist* 18:179–80].

Schoep, A., and S. Stradiot 1947. Paraschoepite and epiianthinite, two new uranium minerals from Shinkolobwe (Belgian Congo). *American Mineralogist* 32:344–50.

Schoep, A., and S. Stradiot 1948. Additional data on the properties of becquerelite and billietite. *American Mineralogist* 33:503–7.

Schrauf, A. 1873. Schröckingerit, ein neues Mineral von Joachimsthal. *Tschermaks Mineralogische und Petrographische Mitteilungen* 1:137–38.

Schwab, P.-N. 1998. French torbernite. *Mineralogical Record* 29 (5):493–4.

Segalstad, T. V., and A. O. Larsen 1978. Chevkinite and perrierite from the Oslo region, Norway. *American Mineralogist* 63:499–505.

Sejkora, J., and J. Čejka 2007. Šreinite from Horní Halže, the Krušné hory Mountains, Czech Republic, a new mineral species, its comparison with asselbornite from Schneeberg, and new data for asselbornite. *Neues Jahrbuch für Mineralogie, Abhandlungen* 184:197–206.

Sejkora, J., J. Čejka, J. Hloušek, M. Novák, and V. Šrein 2004. Phosphowalpurgite, the (PO_4)-dominant analogue of walpurgite, from Smrkovec, Slakovský Les Mountains, Czech Republic. *Canadian Mineralogist* 42:963–72.

Shibata, Y., and K. Kimura 1922. Ishikawaite, a new mineral from Ishikawa, Iwaki province. *Journal of the Chemical Society of Japan* 43:648–49 [in Japanese; see abstracts (1922, 1923) *American Mineralogist* 7:197 and 8:230–31].

Shilin, L. L. 1956. Karpinskyite - a new mineral. *Doklady Akademii Nauk S.S.S.R.* 107,737–39; [see abstract (1957) *American Mineralogist* 42:119–20].

Sidorenko, G. A., N. I. Chistyakova, N. V. Chukanov, I. S. Naumova, and V. A. Rassulov 2005. Calcurmolite: new data on chemical composition and constitution the mineral. *New Data on Minerals* 40:29–36.

Sidorenko, G. A., N. V. Chukanov, and I. S. Naumova 2004. New data on sodium uranospinite. *Zapiski Vserossijskogo Mineralogicheskogo Obshchestva* 133:88–94.

Sidorenko, G. A., N. V. Chukanov, N. I. Chistyakova, G. I. Bebeshko, A. E. Zadov, and I. S. Naumova 2007. Uramarsite, $(NH_4,H_3O)_2(UO_2)_2(AsO_4,PO_4)_2 \cdot 6H_2O$: a new mineral of the meta-autunite group. *Doklady Earth Sciences* 415A:965–69 [see abstract (2009) *American Mineralogist* 94:399–408].

Sidorenko, G. A., R. K. Rastsvetaeva, N. V. Chukanov, N. I. Chistyakova, and A. V. Barinova 2003. Umohoite: new data on its mineralogy and crystal-chemistry. *Zapiski Vserossijskogo Mineralogicheskogo Obshchestva* 132:73–80.

Simmons, W. B., S. H. Hanson, and A. U. Falster 2006. Samarskite-(Yb): a new species of the samarskite group from the Little Patsy pegmatite, Jefferson County, Colorado. *Canadian Mineralogist* 44:1119–25.

Simpson, E. S. 1928 Tanteuxenite (Sp. nov.) Pilbara Goldfield, N.W. Div. *Journal of the Royal Society of Western Australia* 14:45–48; [see abstract (1929) *American Mineralogist* 14:340].

Singh, Y., S. Bagora, R. Viswanathan, P. V. Ramesh Babu, and P. S. Parihar 2014. A new occurrence of thorianite from syenitic pegmatite near Bhaluchuan, Odisha. *Journal Geological Society of India* 83:252–58.

Skála, R., P. Ondruš, F. Veselovský, I. Císařová, and J. Hloušek 2011. Agricolaite, a new mineral of uranium from Jáchymov, Czech Republic. *Mineralogy and Petrology* 103:169–75.

Skvortsova, K. V., and G. A. Sidorenko 1965. Sedovite, a new supergene mineral of uranium and molybdenum. *Zapiski Vsesoyuznogo Mineralogicheskogo Obshchestva* 94:548–54 [see abstract (1966) *American Mineralogist* 51:530].

Smith, D. 1959. An x-ray crystallographic study of schroeckingerite and its dehydration product. *American Mineralogist* 44:1020–5.

Smith, D. K., J. W. Gruner, and W. L. Lipscomb 1957. The crystal structure of uranophane $[Ca(H_3O)_2](UO_2)_2(SiO_4)_2 \cdot 3H_2O$. *American Mineralogist* 42:594–618

Smith, J. 1877. Examination of American minerals; No. 6. *American Journal of Science* 3rd ser., v. 13:365.

Smith, J. L. 1848. Two new minerals, - medjidite (sulphate of uranium and lime) - liebigite (carbonate of uranium and lime). *American Journal of Science and Arts* 55:336–8.

Smith, M. L., J. Marinenko, and A. D. Weeks 1971. Comparison of mourite from Karnes County, Texas with mourite from the USS.R. *American Mineralogist* 56:163–73.

Smits, G. 1989. (U,Th)-bearing silicates in reefs of the Witwatersrand, South Africa. *Canadian Mineralogist* 27:643–55.

Soboleva, M. V., and I. A. Pudovkina 1957. *Uranium Minerals Handbook*, pp. 249–51, Moscow [see abstract (1958) *American Mineralogist* 43:378–84].

Sokolova, E., F. C. Hawthorne, D. I. Belakovskiy, and L. A. Pautov 2005. The OD (order-disorder) structure of holfertite, a hydrated uranyl titanate mineral from Searle Canyon, Thomas Range, Utah, USA. *Canadian Mineralogist* 43:1545–52.

Sokolova, E., F. C. Hawthorne, G. Della Ventura, and P. M. Kartashov 2004. Chevkinite-(Ce): crystal structure and the effect of moderate radiation-induced damage on site-occupancy refinement. *Canadian Mineralogist* 42:1013–25.

Steacy, H. R., A. G. Plant, and R. W. Boyle 1974. Brannerite associated with native gold at the Richardson Mine, Ontario. *Canadian Mineralogist* 12:360–63.

Steinocher, V., and R. Nováček 1939. On β-uranotile. *American Mineralogist* 24:324–38.

Stephenson, D. A. 1964. Iriginite from South Dakota. *American Mineralogist* 49:408–14.

Stern, T. W., and C. S. Annell 1954. A second locality of novacekite. *American Mineralogist* 39:675–6.

Stern, T. W., L. R. Stieff. M. N. Girhard, and R. Meyrowitz 1956. The occurrence and properties of metatyuyamunite, $Ca(UO_2)_2(VO_4)_2 \cdot 3\text{-}5H_2O$. *American Mineralogist* 41:187–201.

Stieff, L. R., T. W. Stern, and A. M. Sherwood 1956. Coffinite, a uranous silicate with hydroxyl substitution: A new mineral. *American Mineralogist* 41:675–88.

Stohl, F.V., and D. K. Smith 1981. The crystal chemistry of the uranyl silicate minerals. *American Mineralogist* 66:610–25.

Sugitani, Y., Y. Suzuki, and K. Nagashima 1984. Recovery of the original samarskite structure by heating in a reducing atmosphere. *American Mineralogist* 69:377–79.

Sugitani, Y., Y. Suzuki, and K. Nagashima 1985. Polymorphism of samarskite and its relationship to other structurally related Nb-Ta oxides with the α-PbO_2 structure. *American Mineralogist* 70:856–66.

Suzuki, Y., and J. Banfield 1999. Geomicrobiology of uranium. *Reviews in Mineralogy* 38:393–432.

Suzuki, Y., T. Sato, H. Isobe, T. Kogure, and T. Murakami 2005. Dehydration processes in the meta-autunite group minerals meta-autunite, metasaleeite, and metatorbernite. *American Mineralogist* 90:1308–14.

Swihart, G. H., P. K. Sen Gupta, E. O. Schlemper, M. E. Back, and R. V. Gaines 1993. The crystal structure of moctezumite $[PbUO_2](TeO_3)_2$. *American Mineralogist* 78:835–9.

Szenknect, S., D. T. Kostin, N. Clavier, A. Mesbah, C. Poinssot, P. Vitorge, and N. Dacheux 2013. Fron uranothorites to coffinite: A solid solution route to the thermodynamic properties of $USiO_4$. *Inorganic Chemistry* 52:6957–68.

Szymanski, J. T., and J. D. Scott 1982. A crystal-structure refinement of synthetic brannerite, UTi_2O_6, and its bearing on rate of alkaline-carbonate leaching of brannerite in ore. *Canadian Mineralogist* 20:271–9.

Szymanski, J. T., D. R. Owens, A. C. Roberts, H. G. Ansell, and G. Y. Chao 1982. A mineralogical study and crystal-structure determination of nonmetamict ekanite, $ThCa_2Si_8O_{20}$. *Canadian Mineralogist* 20:65–75.

Takano, Y. 1961. X-ray study of autunite. *American Mineralogist* 46:812–22.

Takubo, J., Y. Ukai, and T. Minato 1950. Studies on the minerals containing rare elements (Part II) A new mineral found in Kobe-mura, Kyoto Prefecture, Japan. *Journal of the Geological Society of Japan* 56:509–13; [see abstract (1951) *American Mineralogist* 36:925–27].

Thompson, M. E., B. Ingram, and E. B. Gross 1956. Abernathyite, a new uranium mineral of the metatorbernite group. *American Mineralogist* 41:82–90.

Thompson, M. E., A. D. Weeks, and A. M. Sherwood 1955. Rabbittite, a new uranyl carbonate from Utah. *American Mineralogist* 40:201–6.

Thoreau, J., and J. F. Vaes 1932. La saléeite, nouveau mineral uranifere. *Bulletin de la Societé belge de Géologie* 42:96–9 [see abstract (1934) *American Mineralogist* 19:36].

Threadgold, I. M. 1960. The mineral composition of some uranium ores from the South Alligator River area, Northern Territory. *Mineragraphic Investigations Technical Paper 2*, 53 pp., Commonwealth Scientific and Industrial Research Organization, Australia.

Tomašić, N., A. Gajović, V. Bermanec, and M. Rajić 2004. Recrystallization of metamict Nb-Ta-Ti-REE complex oxides: a coupled X-ray-diffraction and Raman spectroscopy study of aeschynite-(Y) and polycrase-(Y). *Canadian Mineralogist* 42:1847–57.

Uvarova, Y. A., E. Sokolova, F. C. Hawthorne, A. A. Agakhanov, and L. A. Pautov L A 2004. The crystal structure of arapovite, $U^{4+}(Ca,Na)_2(K_{1-x}\square_x)[Si_8O_{20}]$, x~0.5, a new mineral species of the steacyite group from the Dara-I-Pioz Moraine, Tien-Shan Mountains, Tajikistan. *Canadian Mineralogist* 42:1005–11.

Vaes, J. F. 1933. Sur un mineral de Kalongwe (Katanga). *Annales de la Société Géologique de Belgique* 56:331–2 [see abstract (1934) *American Mineralogist* 19:235].

Vaes, J. F. 1947 Six nouveaux minereaux d'urane provenant de Shinkolobwe (Katanga). *Annales de la Société Géologique de Belgique* 70:B212–29 [see abstract (1948) *American Mineralogist* 33, 384–5].

Vaes, J. F. 1958 Cousiniet, een nieuw uraanmineral. *Geologie en Mijnbouw* 20:449; [see abstract (1959) *American Mineralogist* 44:910].

Vaes, J. and P. F. Kerr 1949. Sengierite: a preliminary description. *American Mineralogist* 34:109–20.

Van Emden, B., M. R. Thornber, J. Graham, and F. J. Lincoln 1997. The incorporation of actinides in monazite and xenotime from placer deposits in Western Australia. *Canadian Mineralogist* 35:95–104.

Van Wambeke, L. 1958. Contribution a l'etude de la mineralisation radioactive de la pegmatitre de Kobokobo et description d'une nouvelle espece minerale radioactive de la serie phosphuranylite-renardite: la kivuite. *Bull Soc. Belg. Geol., Paleontol., et Hydrol.* 67:383–403; [see abtract (1959) *American Mineralogist* 44:1326–7].

Van Wambeke. L. 1972. Eylettersite, un nouveau phosphate de thorium appartenant à la série de la crandallite. *Bulletin de la Société Française de Minéralogie et de Cristallographie* 95:98–105; [see abtract (1974) *American Mineralogist* 59:208].

Veselovsky, F., P. Ondruš, and J. Hloušek 1997. History of secondary minerals discovered in Jachymov (Joachimsthal). *J. Czech Geological Soc.* 42/4:115–22.

Vilalva, F. C. J., and S. R. F. Vlach 2010. Major- and trace-element composition of REE-rich turkestanite from peralkaline granites of the Morro Redondo Complex, Graciosa Province, south Brazil. *Mineralogical Magazine* 74:645–58.

Viswanathan, K., and O. Harneit 1986. Refined crystal structure of beta-uranophane, $Ca(UO_2)_2(Si_3OH)_2 \cdot 5H_2O$. *American Mineralogist* 71:1489–93.

Vochten, R., and M. Deliens 1998. Blatonite, $UO_2CO_3 \cdot H_2O$, a new uranyl carbonate monohydrate from San Juan County, Utah. *Canadian Mineralogist* 36:1077–81; [see abstract (2001) *New Minerals 1995–1999*, J. Mandarino, ed., *Canadian Mineralogist Special Publication* 4:28].

Vochten, R., and J. Pelsmaekers 1983. Synthesis, solubility, electrokinetic properties and refined crystallographic data of sabugalite. *Physics and Chemistry of Minerals* 9:23–29.

Vochten, R., and K. Van Springel 1996. A natural ferrous substituted saléeite from Arcu su Linnabaru, Capoterra, Cagliari, Sardinia. *Mineralogical Magazine* 60:647–51.

Vochten, R., N. Blaton, O. Peeters, and M. Deliens 1996. Piretite, $Ca(UO_2)_3(SeO_3)_2(OH)_2 \cdot 4H_2O$, a new calcium uranyl selenite from Shinkolobwe, Shaba, Zaire. *Canadian Mineralogist* 34:1317–22; [see abstract (2001) *New Minerals 1995–1999*, J. Mandarino, ed., *Canadian Mineralogist Special Publication* 4:194].

Vochten, R., N. Blaton, and O. Peeters 1997. Deliensite, $Fe(UO_2)_2(SO_4)_2(OH)_2 \cdot 3H_2O$, a new ferrous uranyl sulfate hydroxyl hydrate from Mas d'Alary, Lodeve, Herault (France). *Canadian Mineralogist* 35:1021–25; [see abstract (2001) *New Minerals 1995–1999*, J. Mandarino, ed., *Canadian Mineralogist Special Publication* 4:58].

Vochten. R., E. De Grave, and J. Pelsmaekers 1984. Mineralogical study of bassetite in relation to its oxidation. *American Mineralogist* 69:967–78.

Vochten, R., E. De Grave, and J. Pelsmaekers 1986. Synthesis, crystallographic and spectroscopic data, solubility, and electrokinetic properties of metakahlerite and its Mn analogue. *American Mineralogist* 71:1037–44.

Vochten, R., M. Deliens, and O. Medenbach 2001. Oswaldpeetersite, $(UO_2)_2CO_3(OH)_2 \cdot 4H_2O$, a new basic uranyl carbonate mineral from the Jomac uranium mine, San Juan County, Utah, U. S. A. *Canadian Mineralogist* 39:1685–89.

Vogl, J. F. 1853 Drei neue Mineral-Vorkommen von Joachimsthal. *Jahrbuch der Kaiserlich-Königlichen Geologischen Reichsanstalt* 4:220–23.

Walenta, K. 1958. Die secondaren Uranmineralien des Schwartzwaldes. Preliminary report. Techn. Hochschule, Stuttgart; [see abstract (1959) *American Mineralogist* 44:466].

Walenta, K. 1958a Die secondaren Uranmineralien des Schwartzwaldes. *Jahreschreft geol. Landesamt Baden-Wurttemberg* 3:17–51; [see abstract (1960) *American Mineralogist* 45:254].

Walenta, K. 1965. Hallimondite, a new uranium mineral from the Michael Mine near Reichenbach (Black Forest, Germany). *American Mineralogist* 50:1143–57.

Walenta, K. 1972. Grimselit, ein neues Kalium-Natrium-Uranylkarbonat aus dem Grimselgebiet (Oberhasli, Kt. Bern, Schweiz). *Schweiz. Mineral. Petrogr. Mitt.* 52:93–108 [see abstract (1973) *American Mineralogist* 58:139].

Walenta, K. 1974. On studtite and its composition. *American Mineralogist* 59:166–71.

Walenta, K. 1976. Widenmannit und joliotite, zwei neue Uranylkarbonat-mineralien aus dem Schwarzwald. *Schweiz. Mineral. Petrog. Mitt.* 56:167–85 [see abstract (1977) *Mineral. Abstr.* 28:208].

Walenta, K. 1978. Uranospathite and arsenuranospathite. *Mineralogical Magazine* 42**:**117–28.

Walenta, K. 1983. Uranosilite, a new mineral from the uranium deposit at Menzenschwand (Southern Black Forest). *Neues Jahrbuch fur Mineralogie, Monatschrift* 1983:259–69; [see abstract (1984) *American Mineralogist* 69:408–9].

Walenta, K. 1985. Uranotungstite, a new secondary uranium mineral form the Black Forest, West Germany. *Tschermaks Mineralogische and Petrographische Mitteilungen* 28:25–30 [see abstract (1986) *American Mineralogist* 71:1547].

Walenta, K. 1998 Chadwickit, ein neues Uranylarsenit von Wittichen im Schwartzwald. *Aufschluss* 49:253–7 [see abstract (1999) *American Mineralogist* 84:1195].

Walenta, K., and W. Wimmenauer 1961. Die Mineralbestand des Michaelganges im Weiler bei Lahr (Schwartzwald). *Jahrshefte geol. Landesamtes Baden-Wurttemberg*, 4:7–37; [see abstracts (1962) *American Mineralogist* 47:415 (widenmannite) and 418–19 (hügelite)].

Walenta, K. and T. Theye 2007. Arsenovanmeerscheit, ein neues Uranmineral von der Uranlagerstätte Menzenschwand im südlichen Schwartzwald. *Aufschluss* 58:159–64.

Walenta, K., and T. Theye 2012. Heisenbergite, a new uranium mineral from the uranium deposit of Menzenschwand in the Southern Black Forest, Germany. *Neues Jahrbuch fur Mineralogie - Abhandlungen: Journal of Mineralogy and Geochemistry* 189:117–23.

Walenta, K., F. Hatert, T. Theye, F. Lissner, and K. Röller 2009. Nielsbohrite, a new potassium uranyl arsenate from the uranium deposit of Menzenschwand, southern Black Forest, Germany. *European Journal of Mineralogy* 21:515–520

Walker, T. L. 1923. Schoepite, a new mineral from Kasolo, Belgian Congo. *American Mineralogist* 8:67–9.

Warner, J. K., and R. C. Ewing 1993. Crystal chemistry of samarskite. *American Mineralogist* 78:419–24.

Websky, M. 1853. Uber die geognostischen Verhaltnisse der Erzlagerstatten von Kupferberg. *Deutsche geol. Gesell. Zeitschr.*, Band 11: 384.

Weeks, A. and M. Thompson 1954. Identification and Occurrence of Uranium and Vanadium Minerals from the Colorado Plateaus. US Geological Survey Bulletin 1009-B.

Weisbach, A. 1871. Vorläufige Mittheilung. *Neues Jahrbuch für Mineralogie, Geologie und Palaontologie* 1871:869–70.

Weisbach, A. 1873. Neue Uranerze von Neustädtel bei Schneeberg (Freiberger Jarbuch). *Neues Jahrbuch für Mineralogie, Geologie und Palaontologie* 1873:314–17.

Weisbach, A. 1877. Mineralogische Mittheilungen. *Jahrbuch für das Berg- und Hüttenwesen im Königreiche Sachsen* 1877:42–53.

Weitzel, H., and H. Schröcke 1980. Kristallstrukturverfeinerungen von euxenit, $Y(Nb_{0.5}Ti_{0.5})_2O_6$, und M-fergusonit, $YNbO_4$. *Zeitschrift für Kristallographie* 152:69–82.

Wenrich, K., P. Modreski, R. Zielinski, and J. Seeley 1982. Margaritasite: a new mineral of hydrothermal origin from the Pena Blanca Uranium District, Mexico. *American Mineralogist* 67:1273–89.

Whittie, A. W. G. 1954. Absite - a new mineral related to brannerite. *South Australia Department of Mines, Mining Review No. 97 for 1952*, pp. 99–106; [see abstract (1956) *American Mineralogist* 41:166].

Wülser, P.-A., J. Brugger, and N. Meisser 2004. The crichtonite group of minerals: a review of the classification. *Bulletin de liaison de la Société Francaise de Minéralogie et Cristallographie* 16:76–77.

Wülser, P. A., N. Meisser, J. Brugger, K. Schenk, S. Ansermet, M. Bonin, and F. Bussy 2005. Cleusonite, (Pb,Sr) $(U^{4+},U^{6+})(Fe^{2+},Zn)_2(Ti,Fe^{2+},Fe^{3+})_{18}(O,OH)_{38}$, a new mineral species of the crichtonite group from the western Swiss Alps. *European Journal of Mineralogy* 17:933–42.

X-ray Laboratory, Peking Institute of Uranium Geology, Wuhan Geological College (1978) Orthobrannerite - a new mineral of the brannerite group. *Acta Geologica Sinica* 52:241–51, Chinese with English abstract; [see abstract (1979) *American Mineralogist* 64:656].

Yakovenchuk, V. N., Y. A. Pakhomovskiy, A. V. Voloshin, A. N. Bogdanova, N. A. Yamnova, and D. Y. Pushcharovskiy 1990. Tuliokite $Na_6BaTh(CO_3)_6 \cdot H_2O$ - a new hydrous carbonate of sodium, barium, and thorium, from alkalic pegmatites of the Khibiny massif (Kola Peninsula). *Mineralogiceskij Zhurnal* 12:74–8; [see abstract (1992) American Mineralogist 77:209].

Yamnova, N. A., D. Y. Pushcharovskii, and A. V. Voloshin 1990. Crystal structure of tuliokite a new natural Na, Ba, Th carbonate. *Doklady Akademii Nauk SSSR* 310:99–102; [see abstract (1991) American Mineralogist 76:668].

Yeremenko, G. K., Ye. S. Il'menov, and N. A. Azimi 1977. Find of weeksite-group minerals in Afghanistan. *Doklady Akademii Nauk SSSR Earth Science Sections* 237:226–28 [translated from *Doklady Akademii Nauk* SSSR 237:1191–93].

Young, E. J., and H. A. Powers 1960. Chevkinite in volcanic ash. *American Mineralogist* 45:875–81.

Young, E., A. Weeks, and R. Meyrowitz 1966. Coconinoite, a new uranium mineral from Utah and Arizona. *American Mineralogist* 51:651–63.

Zhabin, A. G., G. N. Mukhitdinov, and M. E. Kazakova 1960. Paragenetic associations of accessory rare-element minerals in exocontact fenitized rocks of the miaskite intrusive of Vishnevye Gor. *Trudy Institut Mineralogii, Geokhimii, Kristallokhim. Redkikh Elementov, Akademiia Nauk SSSR* 4:51–73.

Zhabin, A. G., A. A. Voronkov, E. B. Khalezova, and M. E. Kazahova 1963. New data on accessory davidite from Vishnevye Gory in the Urals and the so-called uferite and ferutite. *Trudy Institut Mineralogii, Geokhimii, Kristallokhimii Redkikh Elementov, Akademiia Nauk SSSR* 15:110–20; [see abstract (1964) *American Mineralogist* 49:447].

Zhang, P., and K. Tao 1982. Aeschynite-(Nd). *Scientia Geologica Sinica* 1982, issue 4:424–28; [see abstract (1984) *American Mineralogist* 69:565].

Zwaan, P. C., C. E. S. Arps, and E. de Grave 1989. Vochtenite $(Fe^{2+},Mg)Fe^{3+}(UO_2)_4(PO_4)_4(OH) \cdot 12\text{-}13H_2O$, a new uranyl phosphate mineral from Wheal Basset, Redruth, Cornwall, England. *Mineralogical Magazine* 53:473–78; [see abstract (1990) *American Mineralogist* 75:1212].

INDEX OF MINERAL NAMES

Page numbers in **bold** indicate photographs.
Names of varieties or questionable species are *italicized.*

INDEX OF MINERAL LOCALITIES

INDEX OF GENERAL TOPICS

CrystalViewer™

Bring crystal structures to life with this free software for Mac and Windows computers.

Through a special arrangement with CrystalMaker Software Ltd., the CrystalViewer™ program has been made available for free download to purchasers of this book. This also includes access to a library of crystal files for all structures shown in the book. CrystalViewer™ permits interactive rotation, scaling, and measurement of crystal and molecular structures, providing a richer appreciation of their details and improved three-dimensional understanding.

Visit our website at:
www.schifferbooks.com/crystalviewer/index.php
and follow the instructions to download the files.

About the software developers. CrystalMaker Software Ltd. develops a range of award-winning professional programs for visualizing crystal amd molecular structures and their diffraction properties.

The flagship program, CrystalMaker®, lets users design new structures, create animations, record video, and export 3D graphics. Originally conceived with teaching in mind, the software is also a powerful research tool and includes a comprehensive mineral structures library. To find out more, including free demonstration software, please visit www.crystalmaker.com.

CrystalMaker®
SOFTWARE

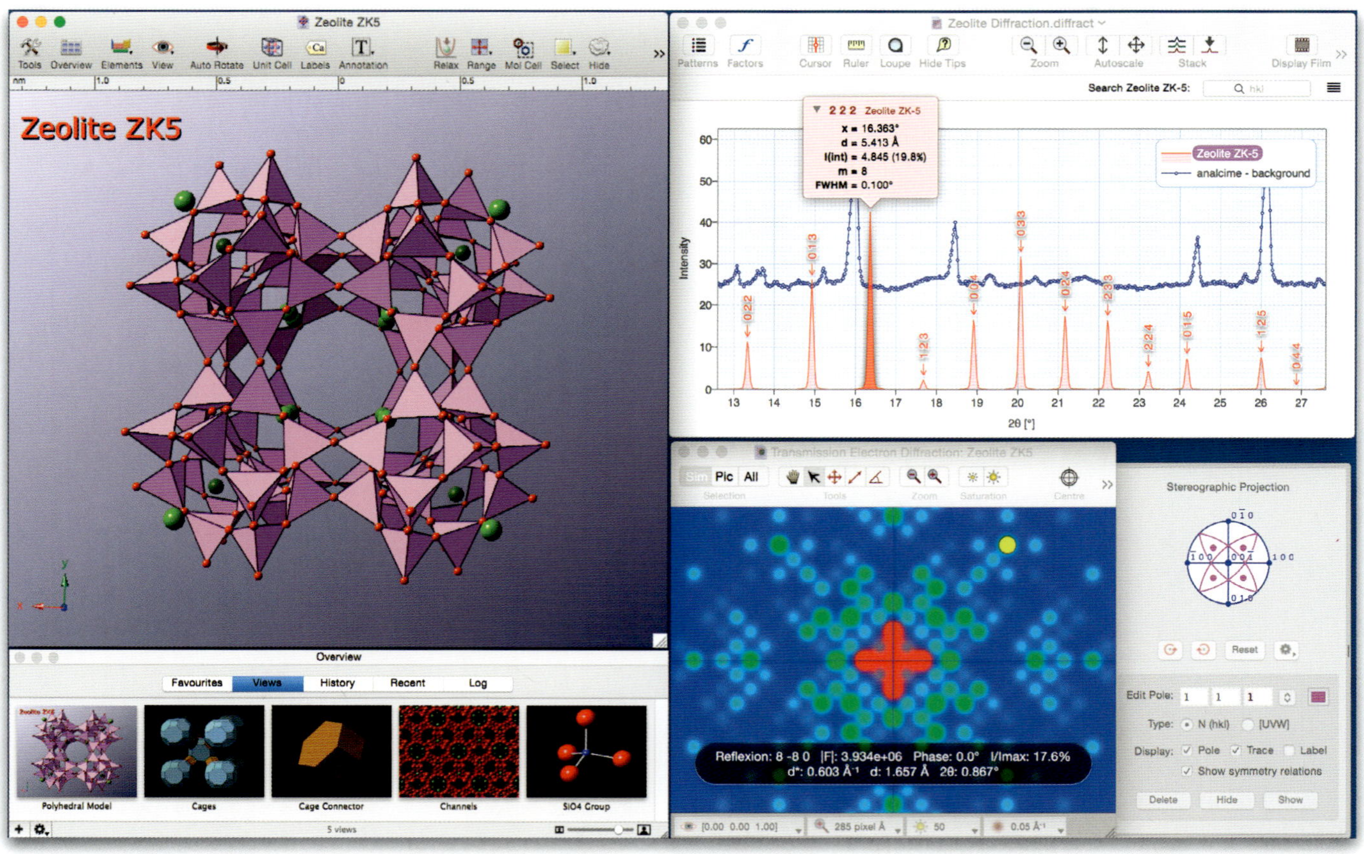

The CrystalMaker software suite offers programs for crystal/molecular structures modeling and diffraction.